Elementary Mathematical Models

AMS/MAA | TEXTBOOKS

VOL **50**

Elementary Mathematical Models

Providence, Rhode Island

Committee on Books
Jennifer J. Quinn, Chair

MAA Textbooks Editorial Board
Stanley E. Seltzer, Editor

Bela Bajnok	Suzanne Lynne Larson	Jeffrey L. Stuart
Matthias Beck	John Lorch	Ron D. Taylor, Jr.
Heather Ann Dye	Michael J. McAsey	Elizabeth Thoren
William Robert Green	Virginia Noonburg	Ruth Vanderpool
Charles R. Hampton		

2010 *Mathematics Subject Classification.* Primary 00-01, 39-01.

About the Cover: The cover illustrates an example from Chapter 3, with a quadratic model for summer arctic ice cap size. The underlying data, shown as dots on the graph, represent the extent of arctic ice on September 1 each year from 1979 through 2012.

For additional information and updates on this book, visit
www.ams.org/bookpages/text-50

Library of Congress Cataloging-in-Publication Data
Names: Kalman, Dan, author. | Forgoston, Sacha, 1974– author. | Goetz, Albert, 1947– author.
Title: Elementary mathematical models: An accessible development without calculus / Dan Kalman, Sacha Forgoston, Albert Goetz.
Description: Second edition. | Providence, Rhode Island: MAA Press, an imprint of the American Mathematical Society, [2019] | Series: AMS/MAA textbooks; volume 50 | Subtitle varies from previous edition. | Includes bibliographical references and index.
Identifiers: LCCN 2018053676 | ISBN 9781470450014 (alk. paper)
Subjects: LCSH: Mathematical models. | Mathematical models–Textbooks. | Mathematical analysis. | Mathematical analysis–Textbooks. | AMS: General – Instructional exposition (textbooks, tutorial papers, etc.). msc | Difference and functional equations – Instructional exposition (textbooks, tutorial papers, etc.). msc | General – Instructional exposition (textbooks, tutorial papers, etc.). msc | Difference and functional equations – Instructional exposition (textbooks, tutorial papers, etc.). msc
Classification: LCC QA401 .K24 2019 | DDC 511/.8–dc23
LC record available at https://lccn.loc.gov/2018053676

Copying and reprinting. Individual readers of this publication, and nonprofit libraries acting for them, are permitted to make fair use of the material, such as to copy select pages for use in teaching or research. Permission is granted to quote brief passages from this publication in reviews, provided the customary acknowledgment of the source is given.

Republication, systematic copying, or multiple reproduction of any material in this publication is permitted only under license from the American Mathematical Society. Requests for permission to reuse portions of AMS publication content are handled by the Copyright Clearance Center. For more information, please visit www.ams.org/publications/pubpermissions.

Send requests for translation rights and licensed reprints to reprint-permission@ams.org.

© 2019 by the authors. All rights reserved
The American Mathematical Society retains all rights
except those granted to the United States Government.
Printed in the United States of America.

∞ The paper used in this book is acid-free and falls within the guidelines
established to ensure permanence and durability.
Visit the AMS home page at https://www.ams.org/

10 9 8 7 6 5 4 3 2 1   24 23 22 21 20 19

Contents

Preface to Second Edition	vii
Note for Students	xv
1 Sequences and Number Patterns	1
1.1 Number Patterns	4
1.1 Exercises	8
1.2 Position Numbers, Graphs, and Subscript Notation	14
1.2 Exercises	22
1.3 Difference and Functional Equations	27
1.3 Exercises	32
2 Arithmetic Growth Models	39
2.1 Properties of Arithmetic Growth	40
2.1 Exercises	51
2.2 Applications of Arithmetic Growth	57
2.2 Exercises	78
2.3 Linear Functions and Equations	82
2.3 Exercises	97
2.4 Applying Linear Functions and Equations	101
2.4 Exercises	115
3 Quadratic Growth	121
3.1 Properties of Quadratic Growth	121
3.1 Exercises	137
3.2 Applications of Quadratic Growth	142
3.2 Exercises	162
3.3 Quadratic Functions and Equations	168
3.3 Exercises	186
3.4 Quadratic Models for Revenue and Profit	190
3.4 Exercises	196
4 Geometric Growth	201
4.1 Properties of Geometric Growth Sequences	201
4.1 Exercises	214
4.2 Applications of Geometric Growth Sequences	219
4.2 Exercises	236
4.3 Exponential Functions	247

4.3 Exercises	264
4.4 Applications of Exponential Functions	270
4.4 Exercises	283
4.5 More About *e*	290
4.5 Exercises	294

5 Mixed Growth Models — 297

5.1 Properties of Mixed Growth Sequences	297
5.1 Exercises	312
5.2 Applications of Mixed Growth Sequences	319
5.2 Exercises	347

6 Logistic Growth — 359

6.1 Properties of Logistic Growth Sequences	360
6.1 Exercises	382
6.2 Chaos in Logistic Growth Sequences	390
6.2 Exercises	407
6.3 Refined Logistic Growth	408
6.3 Exercises	431

Selected Answers to Exercises — 441

Bibliography — 501

Index — 505

Preface to Second Edition

Elementary Mathematical Models (EMM) is for students at the level of college algebra or precalculus. But it does not fit the mold of these or the other courses one typically sees at this level, including liberal arts mathematics and quantitative literacy courses. Because EMM has instructional goals in common with all of the above mentioned courses, it may be considered a hybrid of sorts. And because it is distinct from the standard courses, potential readers and instructors are entitled to a detailed description of EMM's underlying philosophy and goals, as well as the topics covered. The first part of this preface is provided for that purpose. There will follow a few remarks about instructional methods, experiences from EMM teachers, and new features of the second edition.

Students Served. EMM is intended to serve the same students as traditional college algebra, liberal arts mathematics, and quantitative literacy courses. These are students who have studied at least a year of algebra in high school, but who are not headed for calculus. They may need to take a mathematics course for a general education requirement. They may also need familiarity with the main ideas of college algebra for quantitative general education courses in areas outside of mathematics, such as science, economics, and business. For most of these students this will be the only mathematics course completed in college.

EMM has also been used as a mathematics content course in teacher education programs. In one instance it was part of a master's program in Mathematics Education for middle and high school teachers. The instructor found that the material was accessible to participants with an elementary education background, while still novel and engaging enough to interest those with an undergraduate mathematics major. Similarly, at another institution, EMM was part of a summer program for in-service middle school teachers seeking to add a mathematics endorsement to their teaching credentials. As an example at the undergraduate level, one college offered EMM as one of six options fulfilling a general education requirement. There, majors in the Education program were advised to take EMM because it covers topics closely related to middle and secondary school curricula, while also challenging students to deepen and extend their understanding of those topics.

Development Philosophy. The development of EMM reflects two important concerns. First, we want the course to possess an internal integrity. The aim is to make the material intrinsically interesting and worthwhile, to present a coherent story throughout the semester, and to convey something of the utility, power, and method of mathematics. Among other things, this demands genuine and understandable applications.

In formulating these goals for the first edition, Kalman was strongly influenced by the innovative statistics text by Freedman, Pisani, and Purves.[1] The authors' account[2] of the development of this book is a model of clarity and insight. It describes a succession of attempts to refine an introductory statistics course. Ultimately, the authors were forced to focus on a single fundamental question: What are the main ideas the field of statistics has to offer the intelligent outsider? Their answer to this question became the foundation for a new course in statistics. In the same way, EMM tries to capture some of the main ideas that arise in the applications of mathematics, and to present these ideas in an interesting and intelligible context.

Course development was also guided by a second fundamental concern, to emphasize the topics that students are most likely to meet in mathematical applications in other disciplines. We are thinking here of the most elementary applications, formulated in terms of arithmetic and simple algebraic operations: linear, quadratic, polynomial, and rational functions; square roots; and exponential and logarithmic functions. These functions are the building blocks for the simple models that appear in first courses in the physical, life, and social sciences. Many students in our course have learned about some or all of these functions in prior classes. The challenge is to make the material fresh and interesting for these students and, at the same time, accessible as a first exposure. EMM tries to do this by allowing each mathematical topic to appear naturally in the context of an application or simple recursive growth model.

Throughout EMM applications are analyzed using a common methodology involving difference equations. This point will be discussed in greater detail presently. It suffices here to indicate that difference equation methods are applicable to a number of problem contexts that (we hope) have obvious significance and relevance. The mathematical topics we cover arise as a consequence of studying difference equation models. The algebraic emphasis is restricted to what is really required for working with these simple models.

This is one distinction between EMM and standard college algebra or precalculus courses. Such courses typically aim to train students in techniques of algebraic manipulation, covered abstractly and in encyclopedic breadth. Students are asked to master algebraic operations that they will seldom see again, if ever, even in more advanced mathematics courses: the simplification of complicated rational functions, compound radical expressions with several different radicals, and logarithmic equations with variable bases, to name a few. EMM de-emphasizes algebra facility as an end itself.

That is not to say that we have eliminated all algebra. Far from it. Algebra is an important and powerful tool, and students should appreciate and remember that fact. But we want the students to make that observation themselves, from seeing algebra in action. The algebra that does appear in the book is presented in a way that makes it clear *why* algebra is needed, and what it contributes to formulating and analyzing models.

EMM is a hybrid of liberal arts mathematics courses and traditional college algebra courses. Many of the goals of instruction are like those of a liberal arts mathematics course. We wish students to experience and appreciate the power of mathematical methods, the role of aesthetics. We have attempted to present the material in a way that allows students to fully understand the logic behind our methods and how we

---

[1] David Freedman, Robert Pisani, and Roger Purves. *Statistics.* W. W. Norton, New York, 1978.
[2] See the first 6 pages of the introduction to the instructors manual for the text.

know our conclusions are valid. We also hope they will have a positive experience studying this material, and recognize the applicability of mathematics to topics with evident relevance to students' lives. At the same time, by emphasizing modeling and elementary functions, EMM prepares students for the mathematical topics that will arise in their other courses.

EMM also shares goals with courses in quantitative literacy. We hope that students will develop and retain an understanding of the modeling methodology that underlies so many real applications of mathematics. However, whereas many quantitative literacy courses emphasize topics that students might apply in their own daily lives, EMM is more concerned with providing a basis for understanding how mathematical methods lead to conclusions about climate, public health, ecology, and other significant concerns. We do not necessarily hope students will be practitioners of applied mathematics. We do want them to be informed consumers of the conclusions of applied mathematics. As one aspect of this, we have a consistent emphasis on identifying and critiquing assumptions of models, and on the limitations these assumptions impose on results derived from models.

Based on the concerns discussed above, the development of EMM was guided by the following principles:

- Introduce new mathematical operations in the context of a believable problem or plausible growth assumption;

- Weave all of the topics into an integrated whole;

- Provide as a theme a methodology that can be used in many problem contexts;

- Emphasize conceptual understanding of how the mathematics contributes to solving problems over technical mastery of each mathematical topic for its own sake.

The second and third principles are reflected in the choice of topics for the course. A brief discussion of the organization of topics will be presented next.

Course Outline. In overview, EMM concerns discrete and continuous models of growth. Each application starts with a discrete model built around a sequence $\{a_n\}$. There is also a simple hypothesis describing the way successive terms depend on preceding terms. One example of such a hypothesis is that each new term a_{n+1} can be obtained by adding a fixed constant to the preceding term, a_n. The algebraic expression of this hypothesis,

$$a_{n+1} = a_n + \text{a constant},$$

is a *difference equation*. Throughout the course we consider a succession of simple hypotheses: the terms of the sequence increase by a constant amount; the terms increase by varying amounts which themselves increase by a constant amount; the terms increase by a constant percentage; and so on. Each of these hypotheses finds expression as a particular kind of difference equation. Each new class of difference equation is studied using a common methodology, as follows.

- Formulate a family of difference equations;

- Develop *solutions* to the difference equations;

- Study the behavior of the solutions.

While the models are initially formulated in terms of discrete variables, the solutions of the difference equations can generally be interpreted in the context of continuous variables.

The course starts with a discussion of sequences and number patterns, and introduces difference equations and appropriate algebraic formalisms for describing sequences. Next we introduce arithmetic growth, before proceeding to build up successively more complicated models: quadratic growth, geometric growth, mixtures of arithmetic and geometric growth, and finally logistic growth. In the penultimate section, EMM discusses the chaotic behavior that can occur in logistic models. Finally, we consider whether chaos might be an artifact of the assumptions of our logistic growth model. By modifying one assumption in a natural way, we derive a revised version of logistic growth that avoids chaos and leads to the standard family of continous logistic growth functions.

Interspersed among the discussions of the various growth models are units on the families of functions that appear as solutions to difference equations. Thus, the study of arithmetic growth leads into a unit on linear functions; quadratic growth models provide a setting for studying the properties of quadratic functions, and so on.

Unifying Themes. Throughout the course, several themes are touched on repeatedly. Obviously, the formulation of a model in terms of a difference equation, and the development of a solution to that difference equation has a constant presence. Another theme pertains to the distinct roles played by difference equations and their solutions in studying discrete models. It is repeatedly stressed that simple recursive patterns, expressed as difference equations, are a natural way to describe a sequence $\{a_n\}$, while expressing a_n as a function of n is a powerful tool for analyzing the sequence. The use of systematic investigation and patterns to discover the solutions of difference equations is a third common theme. In particular, students repeatedly see the formulation and solution of parametric families of difference equations by such methods. Numerical and graphical methods are used systematically in all the topics.

The progression of topics demonstrates another theme, the incremental nature of modeling. We repeatedly see earlier models refined or modified to obtain later, more realistic models. Students are encouraged to think critically about the assumptions incorporated in our models, and how inevitable simplifications can dilute the accuracy of conclusions derived from models. They see a dramatic evolution of simple growth models, from arithmetic, to geometric, to discrete logistic, and ultimately, to a refined version of discrete logistic growth.

Finally, the most fundamental theme of EMM is the applicability of the mathematical topics. Students recognize this from the start, because the entire course evolves out of the investigation of real problems. In this second edition we have increased the emphasis on modeling situations using real data, about topics that students recognize as important. And by the end of the course, they are studying sophisticated models with unexpected and nontrivial behaviors. The power of algebra to answer important questions about these models is always on display.

Syllabus. EMM was designed as a three-credit-hour course extending over a semester with 14 weeks of instruction. In several semesters of class testing, meeting 75 minutes twice per week, Kalman found it possible to cover almost the entire text, omitting or mentioning only briefly sections 3.4 and 4.5. An advantage of this approach is that

Preface to Second Edition

students see the full range of models, with the final two topics providing a nice capstone for the course.

Some instructors may prefer to cover less material in greater depth or at a slower pace. In particular, the inclusion of student projects would nicely complement the modeling emphasis of the course, but would be difficult to include if the entire text is covered. Omitting all or part of Chapter 6 would be one way to provide for additional instructional time for the remaining topics.

For use in a ten-week quarter, the first four chapters might constitute an appropriate syllabus. Or, in two quarters, the entire text might be covered with additional instructional time devoted to students projects.

The instructor's guide provides more detailed suggestions related to specific topics or sections of the text.

Teaching Methods. The methods used to teach EMM are independent of the course content. We believe the material could be successfully presented in a very traditional format. Our test classes have used a combination of teaching methods. The students listen to some lectures, and complete fairly traditional homework assignments. But they also work in groups and do reading and writing assignments. With the first edition, they also developed models based on data they found in magazines or newspapers, and investigated aspects of models in a computer laboratory. We have tried to develop the material for EMM in a way that lends itself to a variety of approaches.

Nevertheless, we feel that the material works best when there is a strong emphasis on reading and writing. There are reading comprehension questions throughout the text, and we work hard at the beginning of the course to get the students to actually *read* their math books. This is a new experience for many students, and it may require some coaching. As teachers, we are well aware that reading a math book with comprehension requires different strategies and skills than for other kinds of literature. Students generally do not know that this is the case.

Websites. Additonal resources are available at two websites. At a public site of ancillary resources [22] students and teachers can access a suite of Excel spreadsheets and a technology guide. Through the AMS, teachers can access an instructor's guide, a collection of *clicker questions*,[3] and a compilation of all exercises from the text with selected answers and solutions.

Technology. EMM lends itself to numerical exploration and experimentation. We highly recommend the use of calculators or computers for this purpose. This makes numerical and graphical analysis quicker and easier than paper and pencil methods, and produces results that are more accurate, both numerically and visually.

In our test classes, we have used a combination of graphing calculators and Excel spreadsheets. We have developed a suite of spreadsheets that allow students to enter data and parameters for various growth models, and instantly obtain graphs and tabular results. More details about spreadsheets, as well as calculators and other technology options, appear in the technology guide. The guide and the collection of spreadsheets are available online [22].

---

[3]These are multiple choice questions for in-class use as part of a peer-learning activity with a classroom response system (aka *clickers*). More detailed information is provided with the collection of questions.

Classroom Experience. Kalman has frequently taught EMM at American University for over 20 years. From 2015 to 2018 he used drafts of the new edition as they were produced. The first edition of EMM has also been used at a number of other institutions, including Columbus (Georgia) State University, Evergreen State College, Hollins University, Hood College, and Messiah College, among others. Anecdotal evidence from American University and the other institutions mentioned shows that the student reaction has been generally favorable. The students seem to appreciate the fact that the applications are so tightly integrated into the development of the ideas. They also have generally found the course to be demanding. But on the whole, they find the material within reach.

At American University the students have diverse backgrounds. Those whose preparation is weak find the text quite challenging. Better prepared students sometimes complain that the prose is repetitive or long-winded. In most cases, we have erred on the side of too little reliance on the power of algebra to summarize and justify conclusions, and students with a superior understanding of symbolic methods might sometimes perceive the exposition as inefficient. In spite of this, or maybe because of it, our students have generally liked the text.

For almost all of our students, EMM (or some alternative mathematics course) is a requirement. Some of these students are unmotivated or do not engage with the material. But even students who are engaged sometimes find the material quite difficult. In a class of thirty, it is not unusal to find one or two who really seem to make an honest effort but struggle to succeed in the course. Overall, though, we have found students to be interested in the material and to complete it successfully.

This conclusion is supported by student remarks on teaching evaluations. Each semester, there are a few students (perhaps three or four in a class of thirty) who make comments like these:

> *Best math course I ever took; I am usually awful at math but I really understood this course; first time I actually enjoyed a math course; I was dreading math but this turned out to be one of my favorite courses.*

There also are always a few students who object to the emphasis on writing and *thinking*. It is rare for a student to find the material so easy that it offers no intellectual challenge. On the other hand, the stronger students would find that a traditional college algebra course can be completed in a much more mechanical way and with much less in the way of conceptual demands. Perhaps these are the same students who complain: *Too much writing. I never had to write essays in a math class before.*

Changes in the Second Edition. This book is not so much a new edition of *Elementary Mathematical Models* as it is a *reboot*. We have retained the original aims, philosophy, and design principles as in the first edition. But almost the entire text has been rewritten, with updated examples and a greater emphasis on genuine applications using real data. Based on the experience of teachers over the past 20 years, we have reorganized the presentation, dividing chapters into separate sections, each with its own set of exercises. Some of the material from the first edition has been omitted, including a derivation of linear regression formulae, aspects of logarithms and logarithmic scales, and a unit on polynomials and rational functions. One new unit has been included, deriving the family of continuous logistic growth functions from a modified discrete logistic growth difference equation.

Preface to Second Edition

We have made an effort to include a richer collection of exercises in the second edition. Answers are provided for about half of the exercises (marked with an @ symbol), including in many cases a detailed solution. There are also separate sections of *Digging Deeper* exercises that are more challenging or sophisticated, or that go into greater depth.

With this edition we are also providing several ancillary resources for teachers and students. These include the previously mentioned teacher's guide, extended solutions manual, and a *clicker question* collection for teachers, as well as the technology guide and Excel spreadsheet collection for both teachers and students.

Acknowledgments. Both editions of this book were developed with the cooperation and support of the Department of Mathematics and Statistics at American University. We thank the Department for encouraging our efforts, providing opportunities to class-test the material, and continued support of our pedagogical vision.

Parts of the first edition were supported by American University Curriculum Development Grants, and the project also received support from American University's College of Arts and Sciences, and Academic Support Center. In particular, these organizations joined with the Department of Mathematics and Statistics in funding a 2008 project to develop an EMM Supplementary Problem Collection. Some of the problems from the Supplementary Problem Collection have been included in this second edition of EMM. We are grateful for all of the support the project has received from American University.

Several colleagues have contributed to this new edition. Jon Scott and Angela Hare were directly involved in parts of the project, reviewing and critiquing material, topics, and exercises. Adam Mills-Campisi, a former student of Dan Kalman, wrote exercises for the Supplementary Problem Collection mentioned above, including some that were incorporated in the new edition. Other colleagues contributed by sharing their experiences with EMM over the years, including Julie M. Clark, Ladnor Geissinger, Kitt Lumley, Elizabeth Mayfield, and Neal Nelson.

We are also grateful to those at the MAA and AMS who assisted with the editorial review and production of this volume. In particular, Stephen Kennedy and Stan Seltzer were generous with their time and advice. We appreciate the efforts of the AMS production staff, including Brian Bartling, John Brady Jr., Becky Rivard, Peter Sykes, and Christine Thivierge.

Note for Students

Welcome to Elementary Mathematical Models, EMM for short. If you are like most students you have been told for years how important mathematics is, and how widely applicable. But you have also probably seen little evidence in your own studies to support such claims. The topics you encounter in mathematics classes seem to have little direct application to your own life, and do nothing to illuminate how experts in science, engineering, and other fields apply mathematical methods. If this describes your past experiences with mathematics classes, then EMM is for you.

Our aim in designing this course is to provide a detailed look at how mathematics really gets applied. Over the course of the semester, you will learn about mathematical models: special realms within mathematics where problems and questions about the real world are investigated. Using familiar mathematical tools such as arithmetic, algebra, geometry, and functions, you will create, analyze, and critique the types of models used to predict population changes, study global warming, and much more.

This will require your attention, not just for the computations but also for the context of each model. Traditional mathematics textbooks include lots of worked out examples that can be copied to solve homework problems. In contrast, this book is contextually rich, with extended discussions of why and how the mathematical ideas arise and can be applied. The emphasis is on conceptual understanding more than on mastering a collection of standardized problem types.

Almost all of the homework sections include reading comprehension questions and application problems. The latter, sometimes called word problems, involve using mathematical ideas in a meaningful problem context. We refer to these as Problems in Context. Their solutions, and the answers to reading comprehension questions, are rarely as simple as a number or equation. Ideally, your answers will take the same form as the writing in the text: coherent and organized explanations of your ideas and lines of reasoning.

Below are a few friendly tips to help you be successful in the course.

(1) Read thoroughly. This text is written as a discussion in which ideas unfold. Skipping around or scanning for boxes will not help you get the most out of the text. Do not attempt the exercises until you have spent the time to really read the section.

(2) For repetitive calculations and for graphing, make consistent use of technology. This can be with a graphing calculator, spreadsheet software, or other applications. See the Technology Guide at http://emm2e.info for some possible technology tools.

(3) In the exercise sections, some problems are marked with the @ symbol. For each problem so marked, an answer or worked out solution is provided at the end of the

book in a Selected Answers to Exercises collection. We recommend you use these to check your work, or for guidance when you get stuck.

At the end of an EMM course, our students often say something like this: *This is the first math course where I really understood what was going on and why any of it matters.* We hope you have a similar experience, and that you find the material interesting and illuminating.

1

Sequences and Number Patterns

Overview. It is common knowledge that mathematics is supremely important in all sorts of applications. Indeed, a primary goal for this book is to demonstrate the close ties between applications and traditional methods of precalculus mathematics. It may therefore be surprising that we begin with the following observation: the methods and results of mathematics are not directly applicable to the real world.[1] Here the emphasis is on the word *directly*. Mathematics concerns idealizations that exist in our imaginations: perfectly round circles, perfectly straight lines, infinite sets. Therefore, to apply mathematics, we must create an imaginary world that approximates part of our real world. We use mathematics to answer questions about the imaginary world, and we hope that the questions about the real world have similar answers. The imaginary world is called a *mathematical model*.

Applications of mathematics are found everywhere, and invariably they involve mathematical models. To quote one authority,

> The world around us is filled with important, unanswered questions. What effect will rising sea levels have on the coastal regions of the United States? When will the world's human population surpass 10 billion? How much will it cost to go to college in 10 years? Who will win the next U.S. Presidential election? There are also other phenomena we wish to understand better. Is it possible to study crimes and identify a burglary pattern? What is the best way to move through the rain and not get soaked? How feasible is invisibility cloaking technology? Can we design a brownie pan so the edges do not burn but the center is cooked? Possible answers to these questions are being sought by researchers and students alike. Will they be able to find the answers? Maybe. The only thing one can say with certainty is that any attempt to find a solution

---

[1]This is closely related to the certainty of mathematical conclusions. Albert Einstein addressed the issue thus: "As far as the laws of mathematics refer to reality, they are not certain, and as far as they are certain, they do not refer to reality." [**17**, page 55]

requires the use of mathematics, most likely through the creation, application, and refinement of mathematical models.[2]

To understand how mathematics is really applied, you have to understand mathematical models. This book aims to provide an in-depth exposure to a particular kind of model involving traditional mathematical topics such as algebra, analysis, and functions. Although we will focus on a narrow subset of mathematical topics and techniques, readers will develop a general understanding of the methodology of mathematical modeling.

A Global Warming Example. Throughout we will be interested in phenomena or situations which produce streams of data. As an introductory example, we consider an important variable in the study of global climate change: the concentration of carbon dioxide (CO_2) in the atmosphere. We ask:

- Is the concentration of CO_2 increasing?

- If so, can we predict what it will be in the future?

Table 1.1. Carbon dioxide concentration in parts per million. A concentration of one part per million is the same as 0.0001%, so the entry of 381.90 for 2006 means 0.03819%.

Year	Carbon Dioxide Concentration
2006	381.90
2007	383.76
2008	385.59
2009	387.37
2010	389.85
2011	391.63
2012	393.82
2013	396.48

Because climate scientists have been interested in these questions for many years, they have developed methods for measuring CO_2 levels in the atmosphere, and for averaging these measurements over time and location. Annual average concentrations of carbon dioxide (CO_2) are given in Table 1.1. These represent averages over the entire atmosphere for each year from 2006 through 2013 [**10**].

The annual CO_2 concentrations illustrate what we mean by a stream of data. It can also be described as a numerical list or sequence. As long as we continue to make measurements, there will be one additional value each year. Our questions above therefore ask about future elements of the data stream. The basic assumption of our approach is that future values will be consistent with trends observed in the data we have already collected. And we recognize these trends as patterns in the data.

One very powerful kind of pattern is revealed visually by creating a graph, as in Figure 1.1. The data points form an orderly progression from lower left to upper right.

---

[2]From [**6**], a pamphlet produced as a guide for participants in a national competition in applied mathematics.

Chapter 1. Sequences and Number Patterns

In particular, each year's CO_2 concentration is higher than the preceding year's. This answers our first question: the levels of CO_2 concentration have been increasing, at least for the past eight years.

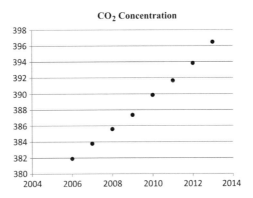

Figure 1.1. Data from Table 1.1 portrayed graphically.

For the second question we would like a better understanding of *how* the concentration has increased. Accordingly, we look for a numerical pattern in the table. As a simplification, let us momentarily disregard the fractional part of each value, ignoring the digits to the right of the decimal points.[3] A very simple pattern results: 381, 383, 385, etc. We recognize these as consecutive odd numbers, and it is easy to predict what will occur in the future. Assuming this pattern continues, the levels for 2014 and 2015 will be 397 and 399, respectively. Looking further ahead, an increase of 2 parts per million each year corresponds to 20 parts per million in a decade, or 200 in a century. This answers the second question, but subject to certain assumptions. The pattern we found only *approximates* the actual data, and even after rounding down the final data point doesn't fit the pattern. Our forecast of future CO_2 levels is correct as a projection of the future values of the *pattern*; it will only be valid if the pattern continues to be a reasonable approximation of the data in the future.

In this example, we see some of the key features of mathematical models. First, we narrowed the scope of a complex topic, climate change, to consider a single aspect, CO_2 concentration, represented by a stream of data values. Second, examining our data, we found and analyzed a pattern. But the pattern does not agree with the data exactly. We had to round the data values down to whole numbers, and even then, the last entry in the table breaks the pattern. Here, we consider the pattern to be part of a *model* for the actual data. Third, we use mathematical methods to explore the properties of the model. In the process we ignore the context of global warming, what was measured (CO_2 concentration), and the units of measurement (parts per million), seeing only a list of successive odd numbers. Finally, we interpret conclusions about *t* as predictions about future CO_2 levels, in effect assuming that future data remain close to the model. We can be confident about mathematical con rived within the model. How reliable these conclusions are in predictir concentrations is another matter.

---

[3]This converts our data values to *whole numbers*.

Indeed, as outlined above, mathematical models may seem a dubious means for understanding complex phenomena. And yet, they have proven to be remarkably effective in a wide range of contexts. Part of this success stems from our ability to refine and improve models. What we learn about the properties of the models extends our mathematical knowledge, and analyzing how closely models represent their real world counterparts points the way to improved models that are more accurate. The combination of these factors results in a kind of leverage, extending the power of modeling far beyond what might be expected.

You will witness aspects of these ideas as you progress through the book. We will find patterns that describe or approximate aspects of a real world problem, and then study these mathematical patterns in the abstract, outside of the context of any specific application. Among other things, we will see efficient methods for extending patterns far into the future. Then we will apply these methods in the context of real world problems. At the same time, as we proceed our mathematical knowledge will accumulate, allowing us to observe and analyze ever more sophisticated patterns.

As these remarks indicate, patterns will play a key role in all that follows. You will learn how to recognize patterns, how to analyze them, and how they contribute to our study of real world problems. In the rest of this first chapter we will introduce the basic concepts of number patterns, and the tools for describing and representing them.

1.1 Number Patterns

The patterns we consider here occur in ordered lists or progressions of numbers. Such a list is called a *sequence*, and the individual numbers are referred to as *terms* of the sequence. For example,

$$2, 4, 6, 8, \cdots$$

is a sequence; 2 is the first term, 4 is the second term, and so on. The three dots at the end of the line indicate that the sequence continues indefinitely.

In describing sequences, we sometimes refer to special kinds of numbers. For example, the *counting numbers* are $1, 2, 3, 4, \cdots$. Counting numbers together with zero make up the *whole numbers* and whole numbers together with their negatives constitute the *integers*. In symbols, the whole numbers are listed as $0, 1, 2, 3, \cdots$, while the list of integers extends indefinitely in both directions, $\cdots, -2, -1, 0, 1, 2, \cdots$.

In applications, we often think of a stream of data as producing a sequence. We begin with data we have already collected, and imagine that the sequence can be continued with future readings. The exact nature of the continuation may be unknown, as was the case with the CO_2 example. But conceptually at least we understand that the sequence will continue with no definite end.

At the same time, there is an evident pattern in the terms 2, 4, 6, 8, and, when asked for the next three terms, most observers would agree that they should be 10, 12, and 14. In the discussion of the CO_2 example, we saw both of these ideas, terms of a sequence obtained as part of a data stream, and terms obtained by extending an observed pattern. There is no certainty that an observed pattern will continue. But for now, the patterns and their continuations will be our main focus.

In many cases, we can extend a pattern by applying some rule. For the sequence $6, 8, \cdots$ each successive term can be found by adding two to the preceding term. So

1.1. Number Patterns

we can refer to either the pattern of successive even numbers, or the rule of repeatedly adding two.

Nonnumerical Patterns and Puzzles. This idea of a pattern need not be restricted to sequences of *numbers*. The terms might be letters, or words, or symbols of any sort, and finding patterns can take on a puzzle-like quality. Can you spot the pattern in the following sequence of letters?

$$o, t, t, f, f, s, s, e, n, t, \cdots$$

In this instance, it is more natural to understand the sequence in terms of a rule. For most people, the rule is difficult to find, if not impossible. And yet, when you discover it, or when it is pointed out, the rule is strikingly obvious. If you don't already know the rule, take a few minutes to try to find it. The solution is given at the bottom of the page.[4]

A similar puzzle was made famous in an episode of the Simpsons. For the sequence of symbols shown below, can you find a pattern or a rule?

Like the prior puzzle, this one has a pattern that is difficult to find for most people, but is strikingly obvious once it is recognized. Some readers will wish to solve this puzzle without aid, so no answer is given here.[5]

Some patterns, like the one in the prior figure, are pictorial or visual. Here is another example.

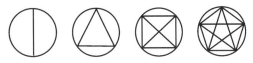

According to the pattern, what should the next figure be? Geometric patterns like this can also have numerical counterparts. For example, if you count the number of straight lines in each circle, left to right, you find 1, 3, 6, 10. Do you see a pattern here? If so, what is the next number? And would that agree with the number of lines in the next figure? Similarly, we can imagine that the lines divide each circle into a number of separate pieces or areas. Again we can count the number in each circle, obtaining the results 2, 4, 8, 16. And now we can ask the same questions: Do you see a pattern? Does the pattern you see correctly predict the number of areas in the next figure?

To further illustrate the large variety of possible patterns, we consider two more examples. In this one,

$$3, 7, -2, 3, 7, -2, 3, 7, -2, \cdots.$$

there is a pattern of repeated blocks of three terms. Similar repeating patterns arise in the decimal representations of fractions. For example, $5/37 = .135135135\cdots$.

---

[4]Write out the counting numbers as words, highlighting the first letter of each: **o**ne, **t**wo, **t**hree, **f**our, etc. The highlighted letters provide the pattern we are looking for. In the given sequence, the final *t* is the first letter of **t**en, so the next three letters should be *e* for **e**leven, *t* for **t**welve, and another *t*, for **t**hirteen.

[5]One place to find the answer is in [**49**, page 42].

Can you spot a simple pattern in the following terms?

$$1, 2, 2, 4, 3, 6, 4, 8, \cdots.$$

They can be understood as the result of shuffling together these two familiar sequences,

$$\begin{array}{ccccc} 1, & 2, & 3, & 4, & \cdots \\ 2, & 4, & 6, & 8, & \cdots. \end{array}$$

Patterns and Arithmetic Rules. The sequences discussed above serve to illustrate how broad the pattern concept is, and that patterns that are obvious when understood can be almost impossible to discover. But puzzle sequences are not typical of what we will work with in this book. Rather, we will usually be interested in number patterns that can be generated using arithmetic rules.

As an example, consider this sequence:

$$5, 8, 13, 20, 29, 40, \cdots.$$

These terms follow a simple pattern, though it may not be apparent to begin with. The key this time is to ask how each term must be modified to obtain the next term. Starting with 5, we can add 3 to get 8, and then add 5 to get 13, then 7 to get 20, and so on. This idea can be illustrated in a diagram as shown below.

$$5 \xrightarrow{+3} 8 \xrightarrow{+5} 13 \xrightarrow{+7} 20 \xrightarrow{+9} 29 \xrightarrow{+11} 40 \cdots.$$

Figure 1.2. The terms of this sequence increase by successive odd numbers.

Because the added amounts are successive odd numbers, it is easy to see how to extend the pattern. Following the 40, we have to add 13 to get 53, then add 15 to get 68.

This pattern can be formulated in terms of the rule, *add the next odd number*. Starting from the first term, we add 3, then we add the next odd number, 5, then the next odd number, 7, and so on. This type of pattern, which can be extended by applying an arithmetic rule, will be our focus throughout the book.

One Sequence, Two Rules. It is often the case that more than one rule can describe the same number sequence. For example, consider

$$2, 6, 12, 20, 30, 42, \cdots.$$

Using the same idea as in the preceding example, we might ask how much has to be added to each term to get the next. Using a diagram as before, we find

$$2 \xrightarrow{+4} 6 \xrightarrow{+6} 12 \xrightarrow{+8} 20 \xrightarrow{+10} 30 \xrightarrow{+12} 42 \cdots.$$

Here we observe a pattern in the added amounts, and formulate the rule *add the next even number*. Using this rule to extend the sequence, the next added amount will be 14 leading to a term of 56, then we will add 16 to reach a term of 72.

But this same sequence has another observable pattern, if you look at it the right way. Each term can be written as a multiplication of two smaller numbers, like so:

$$1 \cdot 2, 2 \cdot 3, 3 \cdot 4, 4 \cdot 5, 5 \cdot 6, 6 \cdot 7, \cdots.$$

1.1. Number Patterns

Looking at this formulation, we readily conclude that the next two terms should be $7 \cdot 8 = 56$ and $8 \cdot 9 = 72$, the same additional terms we found before. We can again formulate the pattern in the form of a rule, but while the examples make it pretty clear what the rule is, stating it in words is a bit of a challenge. We will return to this issue presently. For now, the point to understand is that a sequence may be governed by more than one rule.

It also sometimes happens that two different patterns are observed in one sequence, but extending the patterns produces different sets of additional terms. This is shown by the terms
$$5, 7, 11, 13, 17, 19, 23, \cdots.$$
Given these numbers, some students create the diagram
$$5 \xrightarrow{+2} 7 \xrightarrow{+4} 11 \xrightarrow{+2} 13 \xrightarrow{+4} 17 \xrightarrow{+2} 19 \xrightarrow{+4} 23 \cdots$$
and observe that the added amounts are alternating 2's and 4's. They expect the extended sequence to be
$$5, 7, 11, 13, 17, 19, 23, 25, 29, \cdots.$$
Others recognize the given numbers as consecutive primes, and therefore obtain this extended sequence:
$$5, 7, 11, 13, 17, 19, 23, 29, 31, \cdots.$$
How can we be sure which pattern is correct?

The answer is, there is no way to be sure. Given the first several terms of a sequence, there may be any number of valid patterns, and indicating the continuation of a pattern with three dots is therefore inherently ambiguous.[6] In real world problems, this ambiguity is unavoidable, and even beneficial. Sometimes, the key to understanding a problem is to recognize a new pattern that has not been observed before. But in analyzing and working with patterns, we need a way to describe the precise pattern we have in mind, without resorting to three dots. That is, our descriptions should not depend on the reader seeing the same pattern we see.

To illustrate this idea, let us again consider the sequence
$$5, 7, 11, 13, 17, 19, 23, \cdots.$$
and formulate descriptions of the two patterns recognized before. The first might be expressed this way:

> Starting with 5, each succeeding term is found by adding something to the term that came before. The first added amount is 2, the second is 4, and thereafter the added amounts alternate 2, 4, 2, 4, and so on.

Here is a description of the second pattern:

> The terms of the sequence are successive prime numbers, starting with 5.

Ideally, each description should communicate to the reader enough information to produce the sequence we have in mind, and only that sequence. Do these descriptions achieve the ideal?

One way to check would be to ask a random stranger to read the description and then write down the first six or seven terms of the sequence. Realistically, it is easy

---

[6] In fact, the Online Encyclopedia of Integer Sequences discusses 305 different sequences in which the given terms appear consecutively. See www.oeis.org.

to foresee problems with this approach, the first being that it is weird behavior to ask random strangers such questions. But aside from that, will the stranger know what is meant by a *sequence* or by *terms*?

In spite of these objections, it is useful as a thought experiment to ask whether a proposed description would be understandable and unambiguous to a random stranger. For future reference, we will call this the *random stranger test*.

To illustrate, consider again the following description from one of the earlier examples:

> Add the next even number.

This made sense in the context of the example, but by itself it fails the random stranger test: A random stranger could not use it to determine the first several terms of the sequence we started with. To improve the description we could include the first term of the sequence, but even then, our stranger would not know what the *next* even number should be. Taking this into account, we might formulate an improved description:

> A sequence begins with 2, and subsequent terms are found by adding consecutive even numbers. The first added even number is 4.

This is more complicated than the first description, but it includes a lot more information. Does it pass the random stranger test?

It is probably not practical to actually carry out a random stranger test. Rather, it is a mental exercise for judging the effectiveness of verbal pattern descriptions. On the other hand, you *can* subject your proposed pattern descriptions to a random *classmate* test. You will be asked to do this in the exercises.

People have grappled with the precise formulation of mathematical ideas for millenia. Unfortunately, as illustrated by patterns such as

$$1 \cdot 2, 2 \cdot 3, 3 \cdot 4, 4 \cdot 5, 5 \cdot 6, 6 \cdot 7, \cdots,$$

for even moderately complicated arithmetical procedures, everyday language is not very well suited to the task of description. That is one of the reasons why algebra was invented.[7] Accordingly, in the next section you will learn about some algebraic tools that will be useful in our study of number patterns.

1.1 Exercises

Reading Comprehension. Answer each of the following with a short essay. Give examples, where appropriate, but also answer the questions in complete sentences.

(1) Imagine that you are trying to describe our course to a distant relative who has never studied mathematical models. Write a short letter that you could send to this relative, explaining what is meant by a *mathematical model*.

(2) Think of a collection of data, different from the examples in the text, that you might find in a field you are interested in (perhaps your major) and discuss the types of questions you could answer by using the data to create a mathematical model.

---

[7]Our modern form of symbolic algebra evolved over a very long time. With roots dating to ancient civilizations in Greece, India, and China, among others, it appeared in Europe in the 1600s in something very like the form we use today. See [14].

1.1. Exercises

Please note, you are not being asked to collect the data or create the model but rather to discuss the possibilities. You should be as specific as possible about the type of data you are considering, the assumptions you would make about it, and the exact questions you would use a model to answer.

(3) @Explain what is meant by the expression *number sequence* and what benefits there are to determining a pattern for a number sequence.

(4) If a pattern is strikingly evident, is a verbal description necessary? Explain why or why not.

(5) @Can the terms of a sequence follow two different patterns? If so, give an example. Otherwise, explain why not.

(6) What is the goal of the *random stranger test*? When and why should it be used?

Math Skills.

(7) @Each item below lists several terms of a number sequence. For each one:
   i. Find a pattern in the sequence and use it to determine the next three terms. If the pattern allows, also use it to find the 100th term.
   ii. Write a verbal description of your pattern. Each description should permit someone to find the listed terms for the corresponding sequence, given at most one or two initial terms.
   iii. Subject your description to a random stranger test or a random classmate test (see page 8).

   a. @3, 6, 9, 12, 15, $\cdots$
   b. @1, 2, 4, 8, 16, $\cdots$
   c. @5, 7, 11, 17, 25, $\cdots$
   d. 64, 32, 16, 8, 4, $\cdots$
   e. @$\frac{1}{5}, \frac{2}{5}, \frac{3}{5}, \frac{4}{5}, 1, \cdots$
   f. 1, 4, 9, 16, 25, 36, $\cdots$
   g. 1, 3, 6, 10, 15, 21, $\cdots$
   h. @5, 8, 11, 14, 17, $\cdots$
   i. 1, 1, 2, 3, 5, 8, $\cdots$
   j. @5, 55, 555, 5555, 55555, $\cdots$
   k. $\frac{1}{2}, \frac{2}{3}, \frac{3}{4}, \frac{4}{5}, \cdots$

(8) Consider the visual pattern in Figure 1.3.
   a. Explain how the sequence 1, 3, 5, 7, $\cdots$ relates to the picture.
   b. Explain how the sequence 4, 10, 16, 22, $\cdots$ relates to the picture.

Figure 1.3. Figure for problem 8.

c. Identify another number sequence related to the picture, and explain what the relationship is.

(9) @Consider the sequence 3, 3, 5, 4, 4, 3, 5, 5, 4, 3, ⋯. Predict what comes next, and briefly explain why. Can you think of other ways to continue the sequence? Give as many examples as you can.

(10) Consider the sequence 3, 1, 4, 1, 5, ⋯. Determine the next three terms and give a verbal description of the pattern you used. Can you think of other ways to continue the sequence? Give as many examples as you can.

(11) @Consider the sequence 2, 1, 4, 4, 6, 9, 8, 16, ⋯. Determine the next three terms and give a verbal description of the pattern. If you find yourself stuck, look to the footnote for a hint.[8]

Problems in Context.

(12) Tuition. Over several years, tuition charges at a small private college increased consistently, as shown in this table:

Year	1	2	3	4	5
Tuition	15,750	16,775	17,770	18,925	20,470

Look for an approximate pattern in the table and use it to predict the tuition for years 6 and 12, similar to the way future levels of CO_2 were predicted in the global warming example at the start of the chapter (see page 2). How confident are you of the prediction for year 6? For year 12? Why? Write a short essay explaining your method and findings.

(13) @Coin triangles. Have you ever arranged coins together on a table? One arrangement that naturally presents itself is a triangle of the sort shown in Figure 1.4.

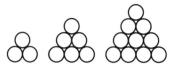

Figure 1.4. Figure for problem 13.

In the first triangle there are two coins on each side of the triangle, and three coins in all. Call that a triangle of side 2. Using similar terminology, the middle triangle

---

[8]Hint: This is a shuffled pattern. Try separating it as illustrated on page 6.

1.1. Exercises

has side 3 and the triangle on the right has side 4. We sometimes meet such arrangements in everyday life, such as in stacks of circular objects in a supermarket or hardware store, sets of bowling pins, or balls on a billiard table.

The numbers of coins in triangles of this sort are called *triangular numbers*. The simplest case is just one single coin, which can be considered a triangle of side 1. So the first triangular number is 1. Counting coins in the figure, we can see that the second, third, and fourth triangular numbers are 3, 6, and 10, respectively.

The triangular numbers form a sequence : 1, 3, 6, 10, $\cdots$.

- a. @Find the next three terms of the sequence by direct counting using actual coins or appropriate pictures.
- b. @Try to find a pattern in the form of a numerical rule. Use it to predict how many coins are in a triangle of side 10. Check your answer by direct counting.
- c. @Can your pattern be used to determine how many coins are in a triangle of side 100 without computing the number of coins in all the preceding triangles? If so, find the number. If not, explain why not.

(14) Ball pyramids. Similar to the arrangement of coins on a flat surface, it is possible to stack balls in triangular based pyramids of different sizes. Figure 1.5 shows such pyramids of side 2, 3, and 4. Notice that the layers of each pyramid are triangles.

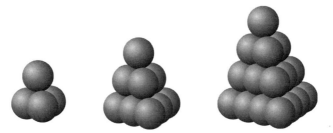

Figure 1.5. Figure for problem 14.

It is a little tricky to count the balls in each pyramid, because some cannot be seen in the figure, but by considering the layers separately, we can determine that the pyramid on the left has 4 balls, the one in the middle has 10 balls, and the one on the right has 20 balls. We can extend the visual pattern of the figure to the left by including a single ball, representing a pyramid of side 1.

The numbers 1, 4, 10, and 20 might be called pyramidal numbers because each counts the number of balls in a pyramid. But because these pyramids have triangular bases, they are more properly referred to as tetrahedra, and 1, 4, 10, and 20 are called *tetrahedral numbers*.

- a. Determine the next three tetrahedral numbers.
- b. Determine and describe a pattern in the sequence of tetrahedral numbers.

(15) @Spiderwebs. An imaginary spider spins a web in the following manner. The spider begins with some number of anchor points on a circular frame, and then strings a line from each anchor to all the other anchors. How many lines are needed? For

example, in Figure 1.6 there are six lines joining the four anchor points on the circle.

Figure 1.6. Figure for problem 15.

    a. @For a spiderweb with three points, how many lines would be needed? How about for five points? Two points?

    b. @Organize your results by entering them in a table like the one shown below.

Anchor Points	Lines of Web
1	
2	
3	
4	6
5	

    c. @Notice that the number 1 is included in the anchor point column in the table. Can you carry out the spiderweb construction process using 1 anchor point? If so, include the resulting number of lines in the table. If not, explain why.

    d. @Your results so far can be thought of as the start of a number sequence. Find a pattern in this sequence, and predict the next term.

    e. Mark 6 anchor points on a circle and draw lines from each point to all the other points. How many lines did you draw? Compare your result with your prediction from part *d*.

(16) Coin Tossing. If you toss a coin several times, recording the result each time as a head H or tail T, how many different outcomes can occur? For example, if you toss the coin just once, there are just two possibilities, H or T. But if you toss the coin twice, the possible outcomes are HH, HT, TH, and TT. Therefore there are four possibilities. Notice that we distinguish between HT and TH: throwing first a head and then a tail is not the same as throwing first a tail and then a head.[9]

    a. If a coin is tossed three times, how many different outcomes are possible? Four times?

---

[9]This distinction is important in analyzing certain probability questions, such as *If a coin is tossed six times, how likely is it to get three heads before getting two tails?*

1.1. Exercises

b. Organize your results by entering them in a table like the one shown below.

Tosses	Outcomes
1	2
2	
3	
4	16

c. Suppose you wish to include the number 0 in the *tosses* column in the table. Can you toss a coin zero times? If so, decide how many different outcomes are possible, and add a line to your table. If not, explain why.

d. Your results so far can be thought of as the start of a number sequence. Find a pattern in this sequence, and predict the next term.

e. Find a systematic way to list all the possible outcomes if a coin is tossed 5 times. For example, your list might begin $HHHHH, HHHHT, HHHTH, \cdots$. How many outcomes are included in your list? Does this agree with your prediction from part *d*?

(17) @Cutting a Pizza. Simon loves to cut up pizza pies. He has a special method. He begins with a number of anchor points on the circular edge of a pie, and draws a line from every point to each of the other points, just like the spiderweb diagram in problem 15. Then, he cuts along each line using a rolling pizza cutter.

Simon's pie cutting method is illustrated in Figure 1.7, with four anchor points. The diagram on the left shows the pizza after Simon has drawn all the possible lines. The diagram on the right shows the result after cutting along all the lines, producing 8 pieces of pizza.

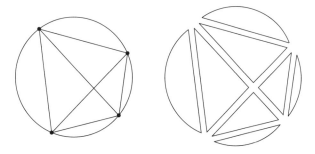

Figure 1.7. Figure for problem 17.

Of course, if Simon starts with 5 points, he will end up with a different number of pieces. In general, he would like to know how many pieces of pie are produced for any given number of anchor points. Help Simon analyze his pie cutting method by answering the following questions.

a. @In the figure with four anchor points on the circle, eight regions are formed. By changing where the points are located on the circle, can you get a different number of regions?

b. @Draw a circle with three anchor points on it, and draw all the lines between the anchors. How many pieces of pie will be created? By changing the locations of the points, can you get a different number of regions? If so, what is the maximum number of regions that can be formed?

c. @Repeat part *b* as often as necessary to complete the following table.

Anchor Points	Maximum Number of Pie Pieces
1	
2	
3	
4	8
5	

d. @Your results for the maximum number of regions can be thought of as the start of a number sequence. Find a pattern in this sequence, and predict the next term.

e. @Repeat part *b* using 6 anchor points. First draw a diagram with the anchors spaced equally around the circle, so that three lines all cross at the center of the circle. How many pieces of pizza does that produce? Does it match your prediction?

f. @Now modify your diagram from the preceding item by moving one anchor point slightly, so the lines will no longer meet at the center of the circle. Show that one additional piece of pie will be formed. With this arrangement, how many pieces of pie are produced? Does that match your prediction?

g. @By shifting other anchors, can you get a different number of pie pieces? What is the largest number of pieces you can obtain? Do any of these results match your prediction above?

1.2 Position Numbers, Graphs, and Subscript Notation

In describing a number pattern, we can work both with the numerical values of the terms, and also with the *positions* of the terms. For example, consider the pattern in the following sequence

$$2, 1, 4, 1, 6, 1, 8, 1, \cdots.$$

Here, we have two simple patterns, shuffled together. Looking only at the terms in the even positions, that is, the second term, fourth term, sixth term, and so on, we see all 1's. The remaining terms are the even numbers in increasing order, starting with 2. In describing this pattern, it is natural to refer to the positions of the terms.

But we have to be careful with terminology. It is tempting to refer to the even and odd terms of the sequence. But does that refer to the values of the terms, or the positions in which they appear? The third term is 4. Is that an even term (because 4 is even) or an odd term (because it occurs in an odd position)?

1.2. Position Numbers, Graphs, and Subscript Notation

Prompted by such questions, we will attach to each term of a sequence a *position number* indicating where the term is located in the sequence. Position numbers are integers, and it would seem most natural for the first term in the sequence to be in position 1, the second in position 2, and so on. But often equations work out more simply if we take the first term to have position 0, or some other number.

Whatever system we use can be easily indicated in a table, as shown here:

position	1	2	3	4	5	6
term	2	1	4	1	6	1

Referring to the table, we can identify the terms in even positions—they are the terms with position numbers equal to 2, 4, 6, and so on. At the same time half the terms are even numbers; they appear in positions 1, 3, 5, and so on. Sometimes, for emphasis, we refer to the *position* and the *value* of a term. Thus, the first even term in the table is in position 1 and has value 2. Tables of this sort, showing both position numbers and terms, will be referred to as *data tables*.

The Even Number Reciprocal Sequence. Position numbers can be useful in describing patterns. As an example, consider the sequence

$$\frac{1}{2}, \frac{1}{4}, \frac{1}{6}, \frac{1}{8}, \ldots .$$

We can describe the terms as reciprocals of successive even numbers. As we will refer to this example several times, let us agree to call it the *Even Number Reciprocal sequence*.

A more specfic description is possible using position numbers, assigned as in the following data table.

position	1	2	3	4	$\cdots$
term	$\frac{1}{2}$	$\frac{1}{4}$	$\frac{1}{6}$	$\frac{1}{8}$	$\cdots$

Now the pattern has a simple description:

Each term is the reciprocal of twice its position number.

This formulation has two benefits. First, it takes advantage of the visual arrangement of the data table, allowing us to verify at a glance that the description is correct. Second, it compactly specifies exactly what the pattern is. In fact, we can think of the description above as a rule for calculating terms. Conceptually, this passes the random stranger test. Without even looking at a table, the rule tells us what term will appear in the ninth position, or the ninety-ninth.

In fact, you probably recognize that this table has the same form as the ones used to discuss functions in other mathematics books. The position numbers are like x and the terms like $f(x)$. Thinking of the example in that way, we can express the rule using the equation $f(x) = \frac{1}{2x}$. However, since the position numbers are always whole numbers, it is the convention to represent them with n rather than x. Then the rule is exactly defined by this statement:

The term in position n is $\frac{1}{2n}$.

Notice how using position numbers and algebraic notation contribute to the clarity and simplicity of the description.

The Added Multiples of Three Sequence. A more complicated example is given by the sequence

$$5, 8, 14, 23, 35, 50, \cdots.$$

Try to spot a pattern in these numbers, before continuing.

As in an earlier example, a pattern can be found by looking at the differences between successive terms. Starting with 5, we add 3 to get the second term, then add 6 to get the third term, 9 to get the fourth term, and so on. Expressed in a diagram,

$$5 \xrightarrow{+3} 8 \xrightarrow{+6} 14 \xrightarrow{+9} 23 \xrightarrow{+12} 35 \xrightarrow{+15} 50 \cdots.$$

The pattern here is that the added amounts are multiples of 3, inspiring the name *Added Multiples of Three sequence.*

While this name will suffice for reference purposes, it does not indicate clearly *which* multiples of 3 are added to which terms. From the figure we can see that the first added amount is $1 \cdot 3$, the second is $2 \cdot 3$, the third is $3 \cdot 3$, and so on. How can we describe this compactly in a way that will pass the random stranger test? Here again it is useful to introduce position numbers, as in the following data table.

Table 1.2. Data table for the Added Multiples of Three sequence.

position	1	2	3	4	5	6	...
term	5	8	14	23	35	50	...

Now we can see that each added amount is three times the position number. That is, we add 3×1 to the term in position 1, 3×2 to the term in position 2, and so on. This gives rise to a verbal description of the rule:

Any term plus three times its position number produces the next term.

Though a bit complicated, this can be applied with no ambiguity, once we know the starting term. For example, since the term 5 appears in position 1, the added amount is $3 \cdot 1 = 3$, so the next term is $5 + 3 = 8$. Then, with 8 in position 2, we add $3 \cdot 2 = 6$ to get 14 as the next term. Continuing in this fashion, you can verify that the described rule produces all of the entries in the data table. Furthermore, using the term 50 with position number 6, we can compute $50 + 3 \times 6 = 68$ to extend the table to position 7.

There is an important distinction between this rule and the preceding rule for the Even Number Reciprocal sequence. We related the earlier rule to the function concept, writing

$$f(x) = \frac{1}{2x},$$

and thinking of the position number as x and the term as $f(x)$. That approach will not work with the rule we just found, which produces the term $f(x)$ using *both* the position number x and the *term preceding $f(x)$*. Although we can still think of Table 1.2 as representing a function $f(x)$, the *rule* we found does not translate into an equation for the function.

The verbal description above for the Added Multiples of Three sequence is correct and concise. Taken together with a starting value, it passes the random stranger test, provided the stranger understands about terms and position numbers and follows the directions carefully. But it is still awkward to work with. Shortly, we will develop an

1.2. Position Numbers, Graphs, and Subscript Notation 17

algebraic approach that is more convenient. But first we pause to consider another important tool for finding patterns—a visual representation using graphs.

Graphs of Number Sequences. A number sequence can be portrayed graphically in much the same way as functions and equations were in your earlier mathematical studies. As an example, let us graph the Added Multiples of Three sequence, using the data in Table 1.2. We group each position number with its corresponding term to create ordered pairs, $(1, 5), (2, 8), (3, 14)$, and so on. Then we graph these as points in the usual manner. The result is shown in Figure 1.8.

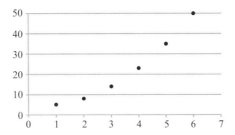

Figure 1.8. Graph for a sequence. Position numbers are on the horizontal axis and term values on the vertical axis, a convention we will follow throughout the text.

Graphs are important because some patterns have a clear visual aspect that can provide new insights about simple number patterns. Moreover, graphs can reveal visual patterns that are difficult to spot in a table of numerical values. To illustrate this point, the data table below shows several terms of a sequence. At a casual glance, almost no one would recognize an obvious pattern in the table.

position	1	2	3	4	5	6	7	8	9	10
term	4.00	6.89	6.04	6.86	6.08	6.84	6.10	6.82	6.13	6.80

But graphing the same number sequence reveals a striking visual pattern, one that *is* recognized at first glance. The pattern is enhanced by plotting points for more terms than appear in the table, and by connecting the plotted points with straight lines, as shown in Figure 1.9.

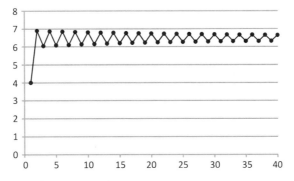

Figure 1.9. The graph of this number sequence has a striking visual pattern.

We will make frequent use of graphs and you will often be called upon to create them in assignments. For the greatest benefit, graphs should be plotted extremely accurately. Fortunately, technology makes the creation of accurate graphs almost effortless. The supplementary resources [22] for this text include an Excel spreadsheet called seqgraph.xlsx for graphing sequences, as well as a technology guide with information about alternative graphing tools.

Graphs allow us to recognize patterns visually. Earlier, we saw how position numbers helped us to describe patterns verbally. Next we will develop a means for describing patterns algebraically. This will prove to be the most convenient way to formulate descriptions that are succinct and unambiguous. As a first step, we introduce two useful notations.

Subscript Notation and Parenthesis Notation. When discussing number sequences, it is customary to use a shorthand notation consisting of a label (usually a single letter) and a number. For the Even Number Reciprocal sequence, we might choose the letter e (for *even*). Then e_1 denotes the term in position 1, e_2 denotes the term in position 2, and so forth. The 1 in e_1 is called a *subscript*, because it appears below the line of text.

Repeating our data table from page 15

n	1	2	3	4	$\cdots$
e_n	$\frac{1}{2}$	$\frac{1}{4}$	$\frac{1}{6}$	$\frac{1}{8}$	$\cdots$

we see that $e_1 = 1/2$, $e_3 = 1/6$, and $e_4 = 1/8$. Note that the rows of the data table are now labeled with n and e_n. Here, we think of n as a variable representing position number, and e_n as the corresponding term of the sequence.

Similarly, choosing t as a label for the Added Multiples of Three sequence, our data table becomes

n	1	2	3	4	5	6	$\cdots$
t_n	5	8	14	23	35	50	$\cdots$

and we can refer to specific terms such as $t_3 = 14$ and $t_5 = 35$.

In this notation, e and t are not variables in the usual sense. They refer not to a particular number, but rather to an entire sequence, the same way that a letter like f can be used to name a function. In that case f by itself does not indicate a numerical value, but $f(1)$ or $f(3)$ or $f(142)$ does. Similarly for a number sequence, t does not represent a numerical value until we combine it with a subscript, say 3, and write t_3.

We can use any letter to label a sequence. Sometimes we want to refer to a sequence repeatedly, and reserve a special label for that purpose. For the rest of this chapter, for example, e will always refer to the Even Number Reciprocal sequence and t to the Added Multiples of Three sequence. Other times a generic label such as a or b will stand for an unspecified sequence, in the same way that $f(x)$ can stand for an unspecified function. Generic labels can also be used for specific examples, sometimes representing several different examples in the same section, or even on the same page.

It is important to pay close attention to the position and size of symbols when subscripts are used. Compare t_{3+2} with t_3+2. For the first of these, because the symbols $3+2$ are all set below the line of type, and in the small type size, they are all part of the subscript attached to the t, and you are meant to recognize t_{3+2} as meaning the same as t_5, which is 35. In contrast, in $t_3 + 2$, only 3 appears in the subscript attached to t.

1.2. Position Numbers, Graphs, and Subscript Notation

This time, you are supposed to add 2 to the numerical value of t_3. Since $t_3 = 14$, $t_3 + 2$ means $14 + 2$ or 16. Continuing in the same vein, what does $t_6 - 4$ mean? Or t_{6-4}? The answers are given in a footnote, but don't peek until after you write down your own answers.[10]

You will be using subscripts throughout this course. The examples above illustrate the importance of reading subscript notation carefully. It is also important to write carefully. It can be very hard to tell in handwritten work what is part of a subscript, and what is not. As the examples illustrate, a mistake about what belongs within the subscript can completely alter the meaning of the symbols. To avoid errors, be careful in written work to clearly distinguish between symbols that are within a subscript and those that are not.

Subscript notation provides a convenient way to express relationships between terms of a sequence and corresponding position numbers. Continuing our discussion of the Added Multiples of Three sequence, recall that we found the following rule:

> Any term plus three times its position number produces the next term.

The initial term is $t_1 = 5$. Adding three times the position number means adding three times 1, and that produces the next term, t_2. So in symbols

$$t_1 + 3 \cdot 1 = t_2.$$

Applying the same rule again, t_2, plus three times its position number (2) produces t_3. Again we express this fact as an equation

$$t_2 + 3 \cdot 2 = t_3.$$

Continuing in the same way leads to an entire series of equations

$$t_3 + 3 \cdot 3 = t_4$$
$$t_4 + 3 \cdot 4 = t_5$$
$$t_5 + 3 \cdot 5 = t_6.$$

Observe how these equations all follow a clear pattern. And although the next equation in the pattern should begin with t_6, we can skip over that and go directly to another equation, say one that starts with t_{17}:

$$t_{17} + 3 \cdot 17 = t_{18}.$$

As this example shows, with the right notation a pattern in the terms of a sequence can lead to a pattern in a series of equations. We will see in the next section how this can be used to describe sequences efficiently and accurately.

As we have already seen, position numbers and terms can be expressed using parentheses, similar to the familiar $f(x)$ used with functions. In fact, the $f(x)$ notation and subscript notation are simply different ways to represent the same idea. In this book, terms of sequences will generally be represented with subscripts, such as a_1, a_2, a_3. But it would be equally valid to write $a(1), a(2), a(3)$. For discussion purposes, we will refer to these as *subscript notation* and *parenthesis notation*, respectively.

Revisiting the prior example, we can write all of the equations using parenthesis notation. Thus, for instance,

$$t_{17} + 3 \cdot 17 = t_{18}$$

---

[10]Answers: $t_6 - 4 = 50 - 4 = 46$ and $t_{6-4} = t_2 = 8$.

becomes
$$t(17) + 3 \cdot 17 = t(18).$$

Why should you have to learn two different notations for the same thing? One good reason is that both notations are in common usage. For example, there are calculators and computer applications with built-in functions for analyzing sequences, but some use subscripts and others use parenthesis notation. See the supplementary technology guide [23] for more details. Also, reference books and articles on sequences and related topics may use either notation. In the exercises you are asked to practice using both notations.

Regardless of the choice of subscripts or parentheses, the point of algebraic notation is to say in symbols exactly what our verbal pattern and rule descriptions say in words. The advantages of the algebraic formulation are compactness and exactness: algebraic equations are much shorter than verbal descriptions, and better suited for expressing a specific meaning. But there are also advantages to verbal descriptions. They can be more effective than algebra for conceptualizing key ideas, and they are accessible to more readers because they are expressed in everyday language. Both contribute to our ability to construct and analyze mathematical models, and you should develop a facility with both as you study this book.

Recursive and Direct Rules. Often, the patterns that we recognize in number sequences reveal how one term leads to the next. A good example of this is the Added Multiples of Three sequence, $5, 8, 14, 23, 35, 50, \cdots$.

We have already seen a rule for this sequence both in a diagram,
$$5 \xrightarrow{+3} 8 \xrightarrow{+6} 14 \xrightarrow{+9} 23 \xrightarrow{+12} 35 \xrightarrow{+15} 50 \cdots.$$
and in equations, such as
$$t_1 + 3 \cdot 1 = t_2$$
$$t_2 + 3 \cdot 2 = t_3$$
$$t_3 + 3 \cdot 3 = t_4$$
$$t_4 + 3 \cdot 4 = t_5.$$

We can use the first equation to find t_2, but only because we know what t_1 is. Thus, since $t_1 = 5$, then $t_1 + 3 = 8$ and that is t_2. In general, to find any term in this way we have to know the preceding term. Such a rule is called *recursive*, and using it repeatedly to generate the terms of the sequence is called *recursion*. In contrast, for the Even Number Reciprocal sequence, we found this rule: the term in position n is $\frac{1}{2n}$. Here we can compute any term in the sequence directly from the position number. For instance, we can see that the seventeenth term will be $\frac{1}{34}$, without having to first compute all the preceding terms. We will call this a *direct rule*.

One recurring idea in this book is that recursive rules are easier to recognize and direct rules are more convenient to use. For most of the cases we consider, we will begin with a sequence defined by a simple recursive rule, and learn how to construct a direct rule for that same sequence. In the exercises, you are asked to look for both kinds of rules.

When we represent a rule using position numbers and equations, the algebraic form of the equations shows whether the rule is recursive or direct. For the Added

1.2. Position Numbers, Graphs, and Subscript Notation

Multiples of Three sequence, a typical equation is

$$t_4 + 3 \cdot 4 = t_5.$$

Notice that two different terms of the sequence are involved; the t appears with two different subscripts. That is characteristic of the equation for a recursive rule. Usually, we write the equation in the reverse order,

$$t_5 = t_4 + 3 \cdot 4,$$

to emphasize the fact that we are finding t_5.

For the Even Number Reciprocal example we have a direct rule that leads to equations such as

$$e_5 = \frac{1}{2 \cdot 5}$$

and

$$e_{11} = \frac{1}{2 \cdot 11}.$$

This time only one term of the sequence appears in each equation, and again it is isolated on the left side of the equation. Equations for direct rules can always be expressed in this form. If we write these equations using parenthesis notation,

$$e(5) = \frac{1}{2 \cdot 5}$$

and

$$e(11) = \frac{1}{2 \cdot 11},$$

we recognize the function

$$e(n) = \frac{1}{2n}.$$

This is another characteristic of a direct rule: it can be expressed in the familiar form of a function. More specifically, we say that the preceding equation expresses the term $e(n)$ as a function of the position number n, because it permits us to compute $e(n)$ directly as soon as we know a specific value of n.

A visual distinction between recursive and direct rules can be observed in a modified form of data table. With a vertical arrangement, it is often convenient to create a diagram representing a computational rule, as shown in Figure 1.10.

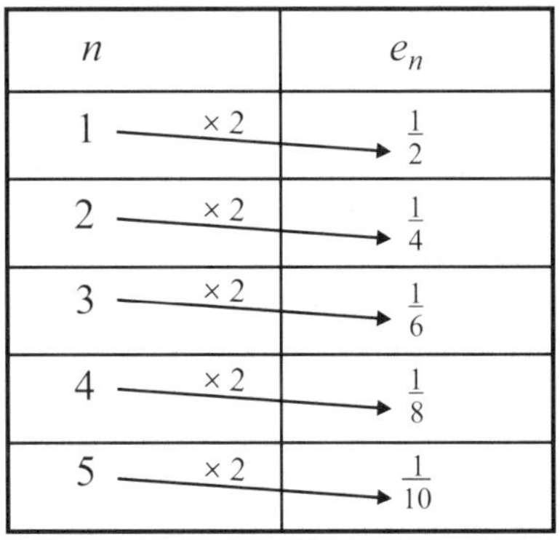

n		e_n
1	$\times 2$	$\frac{1}{2}$
2	$\times 2$	$\frac{1}{4}$
3	$\times 2$	$\frac{1}{6}$
4	$\times 2$	$\frac{1}{8}$
5	$\times 2$	$\frac{1}{10}$

Figure 1.10. Diagram for a direct rule.

The figure shows that each term is found by dividing 1 by twice the position number, and the arrows indicate how we actually use each position number in the calculations. Notice that each arrow extends horizontally, remaining within a single row of the data table. This shows that each term depends only on its own position number and reveals that the rule is direct.

Now look at Figure 1.11, which shows the computational diagram for the Added Multiples of Three sequence t_n. This is more complicated, and shows that each term depends on both a position number and a term from the preceding line. In particular, the arrows that extend vertically in the second column of the data table show that each term depends on the preceding term. This vertical dependence in the computational diagram is characteristic of a recursive rule. In contrast, for a direct rule, there are no arrows pointing vertically from one term to the next.

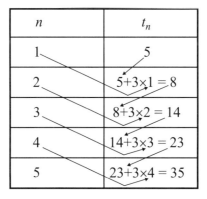

Figure 1.11. Diagram for a recursive rule.

1.2 Exercises

Reading Comprehension. Answer each of the following with a short essay. Give examples, where appropriate, but also answer the questions in complete sentences.

(1) @Define *position number* and explain how position numbers are used. Please include an example in your answer.

(2) What is a *data table* for a number sequence and how can it be used to create a *graph* for a number sequence? Please include an example in your answer.

(3) @What are the benefits of creating a graph for a number sequence?

(4) Discuss the uses of subscript and parenthesis notation. Give examples of each. Are the notations interchangeable? With which notation are you more familiar?

(5) Compare and contrast a *direct rule* with a *recursive rule*. Your answer should be in the form of an essay half a page or more in length. Include the definition for each kind of rule and an example of each.

1.2. Exercises

Math Skills. You will be using subscripts in these exercises. It is important to write carefully. As we saw in this section, a mistake about what belongs within the subscript can completely alter the meaning of the symbols.

(6) @For each part, use the described rule to create a data table and graph. The data table should include at least the first four terms of the sequence. Preferably, your graphs should be produced with technology (such as Excel or a graphing web page) and printed. Alternatively, your graphs may be created by hand using graph paper and plotting the points as accurately as possible.

   a. @Each term of the sequence is found by multiplying the preceding term by 10. The starting term is 8.
   b. Each term of the sequence is found by adding 5 to the preceding term. The starting term is 3.
   c. Each term of the sequence is found by adding 2 to its position number.
   d. Each term of the sequence is found by multiplying its position number by the preceding term. The starting term is 1.

(7) @For each part below, complete the following tasks:

   i. Create a data table showing position numbers as well as terms.
   ii. Create a graph. Preferably, your graphs should be produced with technology (such as Excel or a graphing web page) and printed. Alternatively, your graphs may be created by hand using graph paper and plotting the points as accurately as possible.
   iii. Find a rule for the sequence and describe it using the terminology of *position numbers* and *terms*, similar to the way the patterns are described in problem 6. Indicate whether the rule is *recursive* or *direct*.

   a. @1, 2, 4, 8, 16, $\cdots$
   b. 1, 4, 9, 16, 25, 36, $\cdots$
   c. -5, 2, 9, 16, 23, $\cdots$
   d. 17, 14, 11, 8, 5, $\cdots$
   e. 1000, 100, 10, 1, 0.1, $\cdots$
   f. 1, -2, 4, -8, 16, -32, 64, -128, $\cdots$
   g. 1, 1, 2, 3, 5, 8, $\cdots$
   h. $\frac{1}{2}, \frac{1}{3}, \frac{1}{4}, \frac{1}{5}, \cdots$

(8) @Consider the graphs below. Use each graph to create a data table for the sequence.

   a. @
   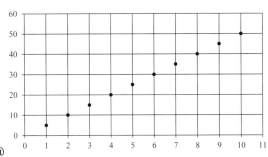

b.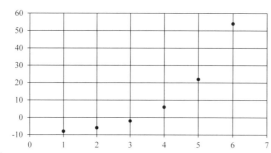

c.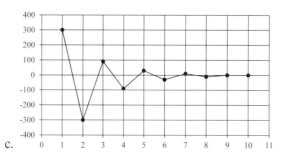

(9) For each part of problem 6, determine if the given rule is recursive or direct, and write down equations for the first 4 terms, as in the following example. Although the sample solution gives equations using both subscript and parenthesis notation, in your answers just use subscript notation for parts *a* and *b*, and parenthesis notation for parts *c* and *d*.

Rule: Each term is ten less than the previous term and the starting term is 300.

Solution: This rule is recursive. Equations for the first four terms follow.

$$a_1 = 300 \qquad\qquad a(1) = 300$$
$$a_2 = a_1 - 10 \qquad\qquad a(2) = a(1) - 10$$
$$a_3 = a_2 - 10 \qquad\qquad a(3) = a(2) - 10$$
$$a_4 = a_3 - 10 \qquad\qquad a(4) = a(3) - 10$$

(10) @For each of the tables below, the sequence values were generated using a computational rule. The arrows, circles and squares indicate how the rule works. For each table, work out the next three terms following the given rule.

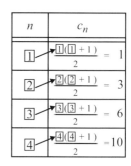

1.2. Exercises

(11) For each table in problem 10, the indicated computational rule can be expressed in the form of equations using subscript or parenthesis notation. For example, the lines of the first table are represented by the equations

$$a_1 = 10 \qquad a(1) = 10$$
$$a_2 = a_1 + 3 \qquad a(2) = a(1) + 3$$
$$a_3 = a_2 + 3 \qquad a(3) = a(2) + 3$$
$$a_4 = a_3 + 3 \qquad a(4) = a(3) + 3.$$

In a similar way find equations for the lines of the other two tables. Use subscript notation for sequence b and parenthesis notation for sequence c.

(12) @For each list of equations below:

i. Identify a pattern in the equations and use it to write the next two equations as well as the 10th equation.
ii. Create a table of values like the ones in problem 10, using arrows to indicate how the pattern you identified in part i can be used to fill in the table.
iii. Indicate whether the rule represented in part ii is recursive or direct.

@List a:
$a_0 = 17$
$a_1 = a_0 - 4$
$a_2 = a_1 - 4$
$a_3 = a_2 - 4$

List b:
$b_0 = 0 \cdot 1 + 5$
$b_1 = 1 \cdot 2 + 5$
$b_2 = 2 \cdot 3 + 5$
$b_3 = 3 \cdot 4 + 5$

List c:
$c_0 = 3$
$c_1 = c_0(c_0 - 1)$
$c_2 = c_1(c_1 - 1)$
$c_3 = c_2(c_2 - 1)$

(13) @Rewrite the equations in problem 12 using parenthesis notation rather than subscript notation.

Problems in Context.

(14) **School Enrollment.** A school board is trying to plan for the future. The board members have been studying enrollments in the school district and have found the following data:

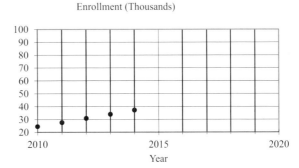

Year	Students
2010	24.4
2011	27.6
2012	30.8
2013	34.0
2014	37.2

In the table and on the graph, the number of students is expressed in units of thousands. That means for 2010 the enrollment was 24.4 thousands, or 24,400. Use the table and/or graph to predict the school enrollment for 2020 and explain your method. Include any assumptions you made.

(15) @For the parts of this problem, refer to the sequence shown in the following table.

n	0	1	2	3	4	5	6	7	8	9
a_n	15.6	0.7	14.0	1.5	12.4	2.3	10.8	3.1	9.2	3.9

   a. @Based on the table, predict the next three terms.

   b. @Create a graph for the data. (It is important that the graph be highly accurate, so either use a computer graphing application, or plot the points very carefully on graph paper.)

   c. @Based on the graph, predict the next three terms of the sequence.

   d. @In which prediction do you have more confidence, the one based on the data or the one based on the graph? Explain.

(16) Black Squirrels. Rosslyn has noticed the black squirrels in the District of Columbia and keeps track of how many she sees on her walk home from school each day.

Day	Jan. 1	Jan. 2	Jan. 3	Jan. 4	Jan. 5	Jan. 6
Number of Black Squirrels	3	5	3	4	4	7

Use the notation B_n to stand for the number of black squirrels Rosslyn saw n days after December 31, and answer the following questions.

   a. What is the numerical value of B_1?

   b. What is the numerical value of B_5?

   c. If you are told that $B_{33} = 2$, what date does that give information about?

   d. The number of black squirrels Rosslyn saw on January 25 was 17. Express this fact using B_n notation.

(17) @Central Park Temperatures. The National Weather Service recorded the following average temperatures for Central Park in New York City in May 2014.[11]

Day	May 1	May 2	May 3	May 4	May 5	May 6
Temperature (°F)	65	63	63	60	60	62

Use the notation T_n to stand for the temperature n days after April 30, and answer the following questions.

   a. @What is the meaning of the term T_0?

   b. @What is the numerical value of T_1?

   c. @What is the numerical value of T_3?

   d. @The average temperature for May 25 was 68 degrees Fahrenheit. Express this fact using the T_n notation.

   e. @Give an English statement interpreting the equation $T_{33} = 70$ in terms of a date and temperature.

---

[11] Data from the National Weather Service [42].

1.3. Difference and Functional Equations

(18) Renewable Energy. The U.S. Energy Information Administration [53] records and predicts United States renewable energy consumption. In the table below, energy consumption is expressed in units of quadrillions of BTUs (British Thermal Units). The figures for the last two years in the table are projections.

Year	2010	2011	2012	2013	2014	2015
Energy Consumption	7.37	8.33	8.09	8.61	8.80	9.09

In referring to these figures, let C_n represent the U.S. renewable energy consumption in year n. For example, C_{2010} represents consumption in 2010. Use this notation and the data in table to answer the following questions.

  a. What information does the value 8.33 in the table convey about energy consumption? Be as specific as possible.

  For the remaining parts of this problem, find the value of the given quantity and tell what it represents.

  b. C_{2010+3}
  c. $C_{2010} + 3$
  d. C_{2014-2}
  e. $C_{2014} - 2$

(19) @Revisiting Descriptions. Each of the following items provides a brief description of a sequence considered in a problem in context in Section 1.1. The original problems asked students to find patterns for the sequences. Now you are asked to write verbal descriptions of the patterns you found, and classify them as *recursive* or *direct*. Use the terminology of terms and position numbers, as appropriate.

Note: although a brief description of each problem context is provided here, it may be helpful to reread the detailed descriptions given with the original exercises, and to review your findings for each.

  a. @Coin Triangles: the number of coins needed to form triangles of increasing side length. The triangular numbers are 1, 3, 6, 10, $\cdots$.
  b. @Ball Pyramids: the number of balls needed to form triangular based pyramids of increasing side length. The tetrahedral numbers are 1, 4, 10, 20, $\cdots$.
  c. @Coin Tossing: the number of possible outcomes when a coin is tossed once, twice, three times, etc. The first several terms for this sequence are 2, 4, 8, 16, 32.

1.3 Difference and Functional Equations

In the preceding sections we have seen that patterns in number sequences can involve relationships between successive terms, or between terms and their position numbers, or both. Using subscript or parenthesis notation, these relationships were expressed in the form of equations. Now we will see how introducing a variable for position number leads to very clear and concise representations of patterns.

Let's think again about the Even Number Reciprocal sequence[12] $1/2, 1/4, 1/6, \cdots$, for which we found the direct computational rule

$$e(n) = \frac{1}{2n}. \tag{1.1}$$

We have our choice of parenthesis or subscript notation. It would be equally valid to write

$$e_n = \frac{1}{2n}. \tag{1.2}$$

In either form this is referred to as a *functional equation* because it expresses each term as a function of the position number. Indeed the first version of the equation is in the common algebraic form for representing functions. In general, functional equations represent direct rules.

Although the functional notation $e(n)$ will be familiar to most readers, we pause to emphasize one point. We might describe the terms of this sequence as follows:

$$e(1) = \frac{1}{2 \cdot 1}, \quad e(2) = \frac{1}{2 \cdot 2}, \quad e(3) = \frac{1}{2 \cdot 3}, \quad \text{and so on.}$$

Because the pattern here is pretty clear, most readers will agree on the meaning of *and so on:* replacing both 3's in the final equation with any other whole number results in another valid equation. We can express this algebraically by replacing the 3's with a variable, n, understood to stand for any whole number. Using one equation with n takes less space, and is more specific than writing three equations and leaving it to the reader to guess how the pattern is to be continued. This is the essential point in using algebra to describe a rule or pattern that extends to all the terms of a sequence.

Using the same approach, we can also represent a recursive rule with a single equation. To see how this works, consider again the Added Multiples of Three sequence,[13] $t_1 = 5, t_2 = 8, t_3 = 14, \cdots$. Recall that we described a rule for this sequence as follows:

Any term plus three times its position number produces the next term.

As we found earlier, this verbal description can be expressed using subscript notation and equations like these:

$$\begin{aligned} t_3 &= t_2 + 3 \cdot 2 \\ t_4 &= t_3 + 3 \cdot 3 \\ t_5 &= t_4 + 3 \cdot 4 \\ t_6 &= t_5 + 3 \cdot 5. \end{aligned} \tag{1.3}$$

These equations display a clear pattern that can be extended to other instances, such as

$$t_{17} = t_{16} + 3 \cdot 16. \tag{1.4}$$

To express this pattern most efficiently, we again use a variable for the position number. However, we have to be a little careful. Each equation has two different subscripts, and we cannot use the same letter for both of them. Looking at the right side of (1.4), observe that the position number 16 appears in two places. If we decide to use n for this position number, then the right side of the equation will be $t_n + 3n$. But now on

---

[12] Introduced on page 15.
[13] Introduced on page 16.

1.3. Difference and Functional Equations

the left the 17 is the next number after 16. We represent that by $n + 1$, and so replace t_{17} with t_{n+1}. In this way, we arrive at

$$t_{n+1} = t_n + 3n. \tag{1.5}$$

With the understanding that n can represent any position number, this single equation captures the pattern of Eqs. (1.3) exactly. The equation can also be written with parenthesis notation as

$$t(n+1) = t(n) + 3n.$$

This is an example of a *difference equation*. In general, one side of a difference equation is an isolated term of the sequence; on the other side will be a formula involving one or more preceding terms. The formula may also involve constants and/or the variable representing the position number (the n in (1.5)). Difference equations represent recursive rules in the same way that functional equations represent direct rules. We emphasize again that functional equations are much more convenient to use than difference equations. For example, with the functional equation $e_n = 1/(2n)$ we can find $e_{100} = 1/200$ immediately. In contrast, to find t_{100} using the difference equation $t_{n+1} = t_n + 3n$, we would have to compute all the preceding values $t_1, t_2, t_3, \cdots, t_{99}$. On the other hand, while an easily observed pattern revealed this difference equation, finding a functional equation for the t_n presents a greater challenge. This illustrates a theme we will see repeatedly: difference equations are easier to find than functional equations; but functional equations are easier to use. We will see a systematic method for finding functional equations for sequences like the added multiples of three sequence in Chapter 3.

In deriving a difference equation for the t_n sequence, we have used n for the position number on the right side of the equation. But it is equally valid to use it for the position number on the left. In the equation

$$t_{17} = t_{16} + 3 \cdot 16$$

if we think of 17 as being n then 16 must be $n-1$. That would lead to a different version of (1.5):

$$t_n = t_{n-1} + 3(n-1). \tag{1.6}$$

These different versions of the equation can be better understood as follows. Suppose you are looking at one specific term of the sequence, with position number n. If we refer to that as the *current* term, then

$\qquad t_n \quad$ is the current term,
$\qquad t_{n+1} \quad$ is the next term, and
$\qquad t_{n-1} \quad$ is the preceding term.

Thus, (1.5) has the verbal interpretation *the next term is the current term plus three times its position number,* whereas (1.6) means *the current term is the preceding term plus three times its position number.* These are both equally valid ways to describe the same rule. Most of the difference equations in the book are formulated in the same form as (1.5), which in many cases is algebraically a bit simpler.

Initial Terms. Because of their recursive nature, difference equations cannot be used to compute terms of a sequence without a starting value. This is an important distinction between difference equations and functional equations. A functional equation, by itself, completely determines a sequence. For example, consider a sequence $c_1, c_2, c_3, \cdots$ where we have the functional equation

$$c_n = 5 + 3n.$$

Substituting 1 for n shows

$$c_1 = 5 + 3 = 8.$$

In a similar fashion we can see that

$$c_2 = 11$$
$$c_3 = 14$$

and

$$c_4 = 17.$$

Using only the functional equation, we can produce every term of the sequence.

The same is not true of a difference equation. Consider this one:

$$a_{n+1} = a_n + 2.$$

Taking $n = 1$ leads to

$$a_2 = a_1 + 2,$$

but this cannot tell us the numerical value of a_2 unless we know what a_1 is. For this reason, difference equations are often combined with definitions for a particular term, referred to as an *initial term* or *initial value*. If we decide that the initial value should be $a_1 = 1$, then we can find

$$\begin{aligned}
a_2 &= a_1 + 2 &= 1 + 2 &= 3 \\
a_3 &= a_2 + 2 &= 3 + 2 &= 5 \\
a_4 &= a_3 + 2 &= 5 + 2 &= 7 \\
a_5 &= a_4 + 2 &= 7 + 2 &= 9,
\end{aligned}$$

and so on. This is the sequence of positive odd integers. On the other hand, if we define $a_1 = 2$, the difference equation will instead produce

$$\begin{aligned}
a_2 &= a_1 + 2 &= 2 + 2 &= 4 \\
a_3 &= a_2 + 2 &= 4 + 2 &= 6 \\
a_4 &= a_3 + 2 &= 6 + 2 &= 8 \\
a_5 &= a_4 + 2 &= 8 + 2 &= 10,
\end{aligned}$$

the sequence of positive even integers. Both sequences follow the rule expressed in the difference equation: each term is the preceding term plus two. This shows how one difference equation may produce two or more different sequences, depending on what the initial terms are.

Notationally, we sometimes write

$$a_{n+1} = a_n + 2; \quad a_1 = 1$$

or

$$a_{n+1} = a_n + 2; \quad a_1 = 2$$

to indicate the combination of a difference equation and an initial term. The first of these is a complete description of the sequence of positive odd integers, while the second describes the positive even integers.

1.3. Difference and Functional Equations

The initial term does not have to be a_1. For example, we might decide to begin our sequence of even integers with $a_0 = 2$. The combination

$$a_{n+1} = a_n + 2; \quad a_0 = 2$$

leads again to the sequence $2, 4, 6, \cdots$, but now with different position numbers. In some cases, difference or functional equations are simpler or more readily interpreted if we begin with an initial term other than $n = 1$. We will see this in the next chapter. For this chapter, it is enough to realize that you can assign position numbers starting with 0 or 1 or any other integer.

One Sequence, Many Descriptions. Subscript notation and difference and functional equations are widely applicable because they provide great flexibility in describing rules for the terms of a sequence. This very flexibility also increases the number of ways that a single sequence can be described. We saw this above in the example of even numbers when we assigned position numbers in two different ways. Here is another example of the same idea.

Consider the sequence $1, 2, 4, 8, 16, \cdots$. For most readers, the most obvious pattern takes the form of a rule: Each term is twice the size of the preceding term. This rule is recursive and can be expressed in the difference equation

$$a_{n+1} = 2a_n.$$

But the difference equation alone does not determine a sequence, we also need an initial term. Thinking of the 1 as being in position 0 leads to the data table

position	0	1	2	3	4
term	1	2	4	8	16

.

But if we think of the 1 as being in position 1, we get a slightly different table,

position	1	2	3	4	5
term	1	2	4	8	16

.

Another pattern is revealed if we recognize that the terms of the sequence are all powers of 2, and write them using exponents. Then we can rewrite the tables above as

position	0	1	2	3	4
term	2^0	2^1	2^2	2^3	2^4

and

position	1	2	3	4	5
term	2^0	2^1	2^2	2^3	2^4

.

For the first of these new data tables, we see a simple direct rule: each term can be computed by using the position number as an exponent applied to 2. This can be expressed in the functional equation

$$a_n = 2^n.$$

The second new data table shows a very similar pattern, but this time each exponent is one less than the corresponding position number. A valid functional equation for this table is

$$a_n = 2^{n-1}.$$

In this example we see several valid descriptions of a single number sequence, using tables and functional equations, as well as difference equations combined with initial terms. The point to understand is that there is not just one correct way to describe a number sequence. In the homework exercises, you are free to choose a description that you find most convenient or appropriate. It is not important that your description be the same as another student's. What is important is that you produce correct descriptions of given sequences.

1.3 Exercises

Reading Comprehension. Answer each of the following with a short essay. Give examples, where appropriate, but also answer the questions in written sentences.

(1) Write an essay of about half a page on difference equations and functional equations. Your answer should include the definition for each kind of equation, an example of each, and describe the differences between the two as well as the advantages of each. Also, give an example of a sequence which is described by both a difference and a functional equation.

(2) @Explain why difference equations are often used in combination with an initial term or initial value, but functional equations are not.

(3) @In this section we have seen that a recursive rule for a sequence can be formulated using a difference equation with variable position numbers. If properly formulated, would such an equation satisfy the random stranger test all of the time, some of the time, or none of the time? What if the random stranger is algebraically literate? Justify your answer.

(4) @In this section we have also seen that a direct rule for a sequence can be formulated using a functional equation with a variable position number. If properly formulated, would such an equation satisfy the random stranger test all of the time, some of the time, or none of the time? What if the random stranger is algebraically literate? Justify your answer.

(5) In this chapter we have represented patterns verbally, visually, and algebraically. Write an essay comparing and contrasting these methods. Do all three methods convey the same information? Should all three methods be used for each pattern? Are there situations in which one method is superior?

Math Skills.

(6) @If possible, determine the requested terms of each sequence described below. If it is not possible, explain why.

 a. @$a_n = (-1)^n$; compute a_{10}
 b. $b_{n+1} = b_n + 2 + 3n$, $b_0 = 12$; compute b_3
 c. @$c_n = 5n - 8$; compute c_7
 d. $d_n = 3d_{n-1} - 7$; compute d_2
 e. @$u_{n+2} = 2u_{n+1} - u_n$, $u_0 = 100$, and $u_1 = 75$; compute u_2, u_3, and u_4.

1.3. Exercises

(7) @Each part below shows equations for the first several terms of a number sequence. For each part, using a variable n, state one equation that represents all the given equations. Two solutions to part a are provided as a guide.

    a. $\quad a_2 = a_1 + 5 \qquad$ **Solution** 1: $a_n = a_{n-1} + 5$
          $a_3 = a_2 + 5 \qquad$ **Solution** 2: $a_{n+1} = a_n + 5$
          $a_4 = a_3 + 5$

    b. $\quad b(1) = b(0) + 3$
          $b(2) = b(1) + 3$
          $b(3) = b(2) + 3$

    c. @ $\quad c(1) = 1 \cdot 2$
          $c(2) = 2 \cdot 3$
          $c(3) = 3 \cdot 4$
          $c(4) = 4 \cdot 5$

    d. @ $\quad d_2 = d_1 + 1$
          $d_3 = d_2 + 2$
          $d_4 = d_3 + 3$

    e. $\quad u_1 = 5(11)$
          $u_2 = 5(11^2)$
          $u_3 = 5(11^3)$
          $u_4 = 5(11^4)$

    f. @ $\quad v_4 = v_1 + v_3$
          $v_5 = v_2 + v_4$
          $v_6 = v_3 + v_5$

    g. $\quad w_2 = 2w_1$
          $w_3 = 3w_2$
          $w_4 = 4w_3$

(8) There is a pattern in the following number sequence.

$$600, \ 500, \ 450, \ 425, \cdots$$

Each number in the sequence is found by adding 200 to half the preceding number. For example, add 200 to half of the original 600 to get 500, then add 200 to half of 500 to get 450, and so on. Write a difference equation that expresses this pattern.

(9) @Each part states a rule describing a number sequence. For each one indicate whether the rule is recursive or direct. For each recursive rule, find a difference equation. For each direct rule, find a functional equation.

    a. @Each term of the sequence is found by adding 5 to the preceding term.
    b. @Each term of the sequence is found by cubing the position number.
    c. Each term of the sequence is found by multiplying the preceding term by $\frac{1}{2}$.
    d. Each term of the sequence is found by multiplying the position number by 7 and adding 3 to the result.
    e. Each term of the sequence is found by subtracting 5 from the preceding term and multiplying the result by 2.
    f. Each term of the sequence is found by multiplying its position number by the preceding term.

(10) @In a through f below several terms of a number sequence are shown. For each sequence formulate a functional equation, or a difference equation with initial term, or if possible, both.

   a. @81, 27, 9, 3, 1, $\cdots$
   b. 20, 22, 24, 26, 28, $\cdots$
   c. @1, 4, 9, 16, 25, 36, $\cdots$
   d. 1, 10, 100, 1000, $\cdots$
   e. @$\frac{1}{2}, \frac{1}{4}, \frac{1}{6}, \frac{1}{8}, \frac{1}{10}, \cdots$
   f. $\frac{1}{2}, \frac{2}{3}, \frac{3}{4}, \frac{4}{5}, \cdots$

(11) @In exercise 8 in Section 1.1 the sequence 4, 10, 16, 22, $\cdots$ was discussed in relation to the pictures below.

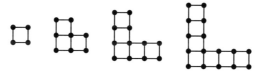

Each picture can be considered to be a collection of points (shown as dots) and lines connecting the points, where each line starts at one point and ends at another. Count the lines in the pictures to find the terms of the sequence.

In completing the instructions below, refer to the terms of the sequence as a_1, a_2, and so on. For example, $a_1 = 4$, $a_2 = 10$.

   a. @Find a recursive pattern in this sequence, and describe the pattern verbally.
   b. @Write a difference equation for the pattern you described in part a.
   c. @Find a direct pattern for the sequence, and describe the pattern verbally.
   d. @Write a functional equation for the pattern you found in part c.
   e. @Find a_8 for this sequence and draw the corresponding picture. Use the picture to check whether your calculation gave the correct answer.

(12) Consider the sequence 800, 790, 780, 770, 760, $\cdots$.

   a. Write a difference equation and initial term for this sequence.
   b. If we think of 800 as a_0, verify that $a_n = 800 - 10n$ is a functional equation for the sequence.
   c. Which equation is more easily used to write a verbal description of a pattern in the sequence? Explain.
   d. You can find a_{100} for this sequence using either equation—they should result in the same answer. Which equation is easier to use? Explain.

(13) @Consider the sequence 144, 48, 16, $\frac{16}{3}, \frac{16}{9}, \frac{16}{27}, \cdots$

   a. @Write a difference equation and initial term for this sequence.
   b. @If we think of 144 as a_0, verify that $a_n = 144/3^n$ is a functional equation for the sequence.
   c. Which equation is more easily used to write a verbal description of a pattern in the sequence? Explain.
   d. You can find a_{100} for this sequence using either equation—they should result in the same answer. Which equation is easier to use? Explain.

1.3. Exercises

Problems in Context.

(14) Mold Growth. A biologist observes the way a patch of mold grows in a glass dish. Using image processing software, she can obtain an estimate of the size of the patch (in square inches) each day. She sees that the data follow a pattern: each day the patch is 1.24 times larger than it was the preceding day. On the first day the patch measured 0.8 square inches. The next day it was $0.992 = 1.24 \times 0.8$ square inches. The day after that it was $1.23 \approx 1.24 \times 0.992$ square inches, and so on.

   a. Determine how large the patch will be on the fifth day.

   b. Write out a difference equation and initial term for the pattern.

   c. You now know the first several terms of the sequence. Eventually the mold will fill the dish. Extend the sequence to predict when (to the nearest day) the patch will first fill the dish assuming the dish can contain 7 square inches of mold.

   d. Under different conditions we would expect the mold to grow at a different rate but still to grow by a multiplicative rule. If the size of the patch were 0.8 square inches on the first day, 0.89 on the second day, and 0.97 on the third day, what number should each term be multiplied by to determine the next term? Explain your method.

(15) @Tree Farm. A large Christmas tree farm acquires several hundred additional acres from a neighboring property. The owner is developing a model for bringing the new acreage into production. The following simplifying assumptions are included in the model:

   - The growth cycle of each tree is 8 years, from planting to harvest.
   - Each year 2,600 trees will be harvested.
   - Each year 5,000 trees will be planted.
   - On the average, 70% of the newly planted trees are *viable*, meaning they will survive to harvest.
   - At the start of the model, there are approximately 25,000 trees at various stages of maturity. It is estimated that 90% of those trees are viable.

   Define a sequence $t_0, t_1, t_2, \cdots$, where t_n represents the number of viable trees on the farm, n years after the start of the model.

   a. @Based on the stated assumptions, calculate the values of t_n for $n = 0, 1, 2, 3,$ and 4. Take into account the fact that viable trees are harvested each year, and that not all of the trees planted every year are viable.

   b. @Based on your calculations formulate a difference equation and initial term for t_n. (This should be possible using methods of this section.)

   c. @Based on your calculations, also find a functional equation for t_n. (We have not yet seen a method for this, but it is possible if you can find the right sort of pattern.)

   d. @The assumptions involve simplifications that are not actually observed in the real world. Identify as many of these simplifications as you can, explaining your reasoning.

e. @The assumptions indicate that 70% of newly planted trees are viable, but that for the existing trees on the farm at the start of the model, 90% are viable. What might account for the difference between these two different viability percentages?

The next four problems all concern the diagram in Figure 1.12. It shows the first four pictures of a pattern that can be continued indefinitely. By counting various parts of each picture in the pattern, we are led naturally to number sequences.

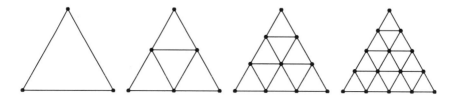

Figure 1.12. The first four pictures in a pattern that can be continued indefinitely.

(16) Counting Triangles. Each picture in Figure 1.12 can be thought of as a combination of a number of smaller triangles. In particular, consider triangles formed by three adjacent vertices, or dots. These are triangles whose interiors do not contain any other visible lines. In the first picture there is just one triangle. In the second, there are four such triangles. In the third, there are nine such triangles. In this way, we obtain a sequence 1, 4, 9, $\cdots$, which will be referred to in this problem as the *triangle sequence*.

   a. If we label the terms of the triangle sequence with the letter t, then the first three terms are $t_1 = 1, t_2 = 4$, and $t_3 = 9$. What is t_4?

   b. Find and describe a pattern that governs the triangle sequence. Is your pattern recursive?

   c. Use your pattern to predict the values of t_5 and t_6. Then draw the next two pictures in Figure 1.12, and find t_5 and t_6 by counting. Did you get the same results?

   d. Find a difference equation, or a functional equation, or if possible, both, for the triangle sequence. [Hint: for a difference equation, it may useful to know that the nth odd number is given by $2n - 1$.]

(17) @Counting Vertices. The dots in each picture in Figure 1.12 are called *vertices*. Counting the number of vertices in the successive pictures produces a sequence v_1, v_2, v_3, and so on, referred to in this exercise as the *vertex sequence*. There are three vertices in the first picture, so $v_1 = 3$. There are 6 vertices in the second picture, so $v_2 = 6$.

   a. @Find v_3 and v_4.

   b. @Find and describe a pattern that governs the vertex sequence. Is your pattern recursive?

1.3. Exercises

c. @Use your pattern to predict the values of v_5 and v_6. Then draw the next two pictures in Figure 1.12, and find v_5 and v_6 by counting. Did you get the same results?

d. @Find a difference equation, or a functional equation, or if possible, both, for the vertex sequence.

(18) **Counting Edges.** Consider the line segments connecting adjacent vertices in each picture. These are called *edges*. Counting the number of edges in the successive pictures produces a sequence e_1, e_2, e_3, and so on, referred to in this exercise as the *edge sequence*. There are three edges in the first picture, so $e_1 = 3$. There are 9 edges in the second picture, so $e_2 = 9$.

a. Find e_3 and e_4.

b. Find and describe a pattern that governs the edge sequence. Is your pattern recursive?

c. Use your pattern to predict the values of e_5 and e_6. Then draw the next two pictures in Figure 1.12, and find e_5 and e_6 by counting. Did you get the same results?

d. Find a difference equation, or a functional equation, or if possible, both, for the edge sequence.

(19) Find another sequence associated with the pictures in Figure 1.12, by counting something in each picture. Give your sequence a name, and analyze it following the same outline as in the preceding three problems.

(20) @**Finance.** Savings accounts often offer a fixed annual interest rate. The balance in a savings account at the end of each year can be considered as one term of a number sequence. For example if $1,000 is invested it would be natural to consider that as the balance after zero years and write $b_0 = 1000$. If the annual interest rate is 2%, then after one year the balance in the account will be $b_1 = 1,000 + 2\%$ of $1,000 = 1,000 + 0.02 \cdot 1,000 = 1,020$. Similarly, after two years we will have $b_2 = 1,020 + 0.02 \cdot 1,020 = 1,040.40$ and so on.

a. @Write an equation, either difference or functional, for b_n.

b. @What assumptions are inherent in the model for b_n?

c. @This model can be extended to account for additional investments. If we deposit two hundred dollars each year, the balance will grow faster. To avoid confusing this model with the one discussed in parts a and b, let us refer to the terms of the new sequence as c_0, c_1, and so on. The first few terms are below. Write a difference equation for c_n.

$c_0 = 1,000.00$
$c_1 = 1,000.00 + 0.02 \cdot 1,000.00 + 200 = 1,220.00$
$c_2 = 1,220.00 + 0.02 \cdot 1,220.00 + 200 = 1,444.40$
$c_3 = 1,444.40 + 0.02 \cdot 1,444.40 + 200 = 1,673.288$

(21) @**Revisiting Descriptions.** Each of the following items provides a brief description of a sequence considered in a problem in context in Section 1.1. The original problems asked students to find patterns for the sequences, and in an exercise in Section 1.2 you were asked to write verbal descriptions of the patterns. Now find

a difference equation or a functional equation representing each of your descriptions.

Note: although a brief description of each problem context is provided here, it may be helpful to reread the detailed descriptions given with the original exercises, and to review your findings for them in both Section 1.1 and Section 1.2.

   a. @Coin Triangles: the number of coins needed to form triangles of increasing side length. The triangular numbers are 1, 3, 6, 10, $\cdots$.
   b. @Ball Pyramids: the number of balls needed to form triangular based pyramids of increasing side length. The tetrahedral numbers are 1, 4, 10, 20, $\cdots$.
   c. @Coin Tossing: the number of possible outcomes when a coin is tossed once, twice, three times, etc. The first several terms for this sequence are 2, 4, 8, 16, 32.

2

Arithmetic Growth Models

Recognizing and analyzing patterns is an important part of applying mathematics to real world problems. Accordingly, Chapter 1 focused on number patterns, and particularly, patterns described by simple rules. Now we begin a systematic study of various types of rules. This chapter will look at sequences such as

$$3, 7, 11, 15, 19, 23, 27, \cdots$$

in which each term represents a constant increase (or decrease) from the preceding term. These sequences are said to exhibit *arithmetic growth*. We will work out general properties of their difference equations, functional equations, and graphs, and will see how such sequences can be applied.

Later chapters will look at other types of rules, exploring the properties of corresponding sequences. As we proceed, the rules and properties will become increasingly complicated, and the methods discussed in Chapter 1 will be indispensable. In contrast, most or all of the ideas we develop in this chapter can be understood on the basis of common sense and basic quantitative reasoning. The goal is for students to become familiar with the methods of Chapter 1 in very comfortable surroundings. With this in mind, readers should make it a point to practice using difference equations, functional equations, subscript or parenthesis notation, and the terminology of recursive and direct patterns, even if they do not seem necessary to understand examples or complete homework exercises.

Many applications concern changes that occur over time. These often involve streams of data values that occur at regular intervals, e.g., every hour or every day or every month. In such a context, the concept of arithmetic growth can be stated succinctly as follows:

> **Arithmetic Growth Assumption:** Under the assumption of Arithmetic Growth, equal periods of time result in equal increases of the variable.

For example, in Chapter 1 we considered the average concentration of carbon dioxide (CO_2) for the entire atmosphere on a yearly basis (see page 2). We noticed that each year the concentration increased by approximately the same amount, 2 parts per million. Based on this observation we can formulate a *model* for CO_2 concentration with the assumption that *average atmospheric concentration increases by 2 parts per million each year*. That is an arithmetic growth assumption; it leads to an arithmetic growth model. The model only approximates reality, because in our data the annual increases are only approximately equal to 2 parts per million.

2.1 Properties of Arithmetic Growth

Arithmetic growth models all share several common features. For example, they have similar difference equations, graphs, and functional equations. The common features are implications of arithmetic growth, so knowing about them can help us decide whether an arithmetic growth assumption is appropriate. On the other hand, they are also aspects of the model that we can apply once arithmetic growth is assumed. For both reasons, in this section we focus on common features of arithmetic growth models.

Recognizing Arithmetic Growth Sequences. There is an extended version of a data table that makes it clear at a glance whether or not a sequence exhibits arithmetic growth. To illustrate it, consider the sequence $2, 8, 14, 20, 26, 32, \cdots$. Considering the initial term to be a_0, we can list the position numbers and terms in Table 2.1.

Table 2.1. A data table with a column of differences.

n	a_n	Difference
0	2	
		6
1	8	
		6
2	14	
		6
3	20	
		6
4	26	
		6
5	32	

The table also has an extra column for the differences between successive terms. From $a_0 = 2$ to $a_1 = 8$, there is an increase of 6. In other words, the *difference* $a_1 - a_0 = 6$, so we enter that in the difference column between the lines for $n = 0$ and $n = 1$. In the same way, each entry in the difference column is found by subtracting an entry of the a_n column from the entry directly below it. This difference is entered in the table cell that touches the two entries that were subtracted.

In an arithmetic growth sequence each successive term increases by the same amount over the preceding term, so all of the differences are the same. This is immediately apparent in the table. Because the difference column entries are all equal, we can see at a glance that this sequence is an instance of arithmetic growth.

In contrast, let us construct an extended data table for this sequence: 5.1, 7.7, 10.5, 13.0, 15.5, 18.1, defining the starting term as b_0. See Table 2.2.

Because the entries in the difference column are not all the same, this sequence is not an example of arithmetic growth. In this case, because the differences are all pretty close in value, we might consider the sequence to be approximately following an

2.1. Properties of Arithmetic Growth

Table 2.2. The entries in the difference column are not all equal for this sequence. Therefore it is not an arithmetic growth sequence.

n	b_n	Difference
0	5.1	
		2.6
1	7.7	
		2.8
2	10.5	
		2.5
3	13.0	
		2.5
4	15.5	
		2.6
5	18.1	

arithmetic growth pattern. But the rule is: a sequence is an arithmetic growth sequence if and only if all the entries in the difference column are identical.

Difference Equations. For the CO_2 model, if the starting concentration is 381 and we assume an annual increase of 2 parts per million, the resulting sequence is

$$381, 383, 385, 387, \cdots.$$

Introducing c as a label, we can represent the terms using subscript notation. We make the starting position number 0, so that $c_0 = 381, c_1 = 383$, and so on. Then the sequence follows the difference equation

$$c_{n+1} = c_n + 2.$$

This is a typical arithmetic growth difference equation.

This same difference equation is valid whether we call the starting position number 0, 1, or any other whole number. Thus, if we say $c_1 = 381$, $c_2 = 383$, and so on, it will still be true that $c_{n+1} = c_n + 2$. The literal meaning of the equation is *each term is two more than the preceding term*, and that holds for the sequence $381, 383, 385, 387, \cdots$ regardless of the position number attached to 381.

Let us consider some other examples of arithmetic growth. In a study of a flu epidemic, the equation

$$p_{n+1} = p_n + 500$$

might be used to indicate that the number of people who have been infected is going up by 500 per month. Here p_n represents the number of people who have been infected by the start of month n. To use this equation we need an initial value. Taking $p_0 = 1000$ would indicate that at the start time for the model, 1000 people had already been infected.

Similarly, the equation

$$f_{n+1} = f_n + 10$$

could be used to represent the fine for an overdue book at the library where fines increase by 10 cents per day. If the book is returned n days after the due date, the fine will be f_n. In this case, the way we have described the problem implies that $f_0 = 0$, because if the book is returned *zero* days past the due date, there will be no fine.

For a somewhat different example, a sequence g_n could represent the amount (in grams) of available fuel after n months of operation in a model of a satellite propulsion system. The equation

$$g_{n+1} = g_n - 4.35$$

would indicate a consumption of 4.35 grams of fuel each month. Again we need an initial value to use the difference equation. We take $g_0 = 3{,}100$, indicating the amount of fuel available at the start time for the model.[1] In this case the variable is shrinking rather than growing, but because the amount of change is constant we still consider this to be arithmetic *growth*. The successive terms can be thought of as growing smaller. This is also referred to as *negative growth*, or *decay*.

Comparing the difference equations for all of these examples,

$$\begin{aligned} c_{n+1} &= c_n + 2, \\ p_{n+1} &= p_n + 500 \\ f_{n+1} &= f_n + 10 \end{aligned}$$

and

$$g_{n+1} = g_n - 4.35,$$

we see a common form. Each difference equation says that any term of the sequence is obtained by adding a constant to the preceding term. Even the equation for g_{n+1} can be thought of as adding a constant, if we write it in the form

$$g_{n+1} = g_n + (-4.35).$$

All of these examples are instances of a general form that applies in all arithmetic growth models. The general form is given as follows.

Arithmetic Growth Difference Equation: Every arithmetic growth sequence follows a difference equation of the form

$$a_{n+1} = a_n + d, \qquad (2.1)$$

where d represents a constant added amount, also called the *common difference*.

This serves as a template for difference equations in all arithmetic growth models. We use a as a generic label, understanding that in any specific application of arithmetic growth, some other label might take a's place. Similarly, we use d as a place holder for whatever added amount appears in any specific case.

Looked at in this way, (2.1) defines a whole family of closely related difference equations, specifically, the arithmetic growth difference equations. Notice that d is a special kind of variable. In any specific arithmetic growth model, d will be replaced by an actual number, and will remain constant throughout our analysis of the model. The value of d only changes if we modify the model. Thus, d is useful for understanding how different models are related to each other, rather than for analyzing a specific model. Such a variable is referred to as a *parameter*. Parameters will arise in later chapters in connection with other families of difference equations. They are also familiar in algebra where they are used in families of equations. For example, in the generic linear equation $y = mx + b$, m and b are parameters. Like the d in (2.1), they must be replaced with numbers to obtain the equation of a particular line.

The letter d should be thought of as standing for the word *difference*. In any arithmetic growth sequence, the numerical difference between one term and the next is a constant. The parameter d is equal to this constant *difference*.

---

[1] The difference equation and initial value for this model are based on data from the Algerian satellite ALSAT-2A. See [**34**].

2.1. Properties of Arithmetic Growth

The assumption of arithmetic growth, as described verbally in the box on page 39, leads to a difference equation of the form of (2.1). Actually, the arithmetic growth assumption implies more than the difference equation. In the CO_2 example, the difference equation says that the concentration of CO_2 increases by the same amount every year. But the verbal statement would also indicate constant increases every month, or week, or day. This broader idea of arithmetic growth will be taken up in greater detail later. The point of emphasis here is that if you agree to use an arithmetic growth model, as described verbally, then you are led to the difference equation (2.1).

This illustrates an important idea—using a simple assumption about how terms of a number sequence vary to devise a difference equation. Here it is the arithmetic growth assumption (equal growth occurs in equal periods of time) that leads to (2.1). In later chapters we will consider other sorts of growth assumptions and study the difference equations they inspire.

Numerical and Graphical Properties. One way to study a sequence in a model is to systematically compute the values of terms. Note though that these computations are part of the *model*, as opposed to the observed data on which the model is based. Usually, additional data values are *collected*, not computed.

For the CO_2 example, we have already found the terms 381, 383, 385, 387, $\cdots$. These can be incorporated in the data table

n	0	1	2	3	4	5
c_n	381	383	385	387	389	391

which reflects our decision to label the starting term c_0. Often questions about a model can be answered by reference to such a table. For the CO_2 model, we might like to know how long it will take the concentration to reach 400, or what the concentration will be in a future year, say 2025.

Both questions can be answered by extending the table systematically term by term. For the first question, we continue until we find a term c_n that is at or above 400. For the second question, we have to remember that c_0 is the value for 2006. Then 2025 is 19 years later, and we have to continue the table until we reach c_{19}.

Systematic computation of this sort is referred to as a *numerical* approach or method. This can involve computing only a few terms, or hundreds or thousands. As long as we are working directly with numerical values, we are using a numerical approach.

What can we observe from numerical methods in arithmetic growth models? For one thing, the terms of the sequence increase (or decrease) with perfect regularity. This is, after all, the underlying assumption of arithmetic growth. As soon as we see the first two terms, 381 and 383, and observe an increase of 2, we know that the increases for all successive terms will be the same.

In fact, the terms of an arithmetic growth sequence are completely determined by two quantities, the starting term a_0 and the constant difference d. If you are told that $a_0 = 100$ and $d = 7$, you can immediately compute sequence terms of 100, 107, 114, 121, and so on.

Computing a large number of terms by hand can be tedious. Fortunately, calculators and computers make the computation almost effortless. This is true whenever we are working with a difference equation, but is especially true for arithmetic growth sequences. See the supplementary technology guide [23] for further discussion of numerical computation.

Numerical Method Example. In the model for satellite fuel, how long will it take to use up all the fuel? To analyze this, we produce a data table as shown below.

Table 2.3. Using a numerical method to determine how long the fuel will last in the satellite model.

n	g_n
0	3,100.00
1	3,095.65
2	3,091.30
3	3,086.95
4	3,082.60
⋮	⋮
710	11.50
711	7.15
712	2.80
713	−1.55

We have left out all the lines between $n = 4$ and $n = 710$ to save space, but they were all computed as part of our numerical method. The last few lines show that after 712 months only 2.80 grams of fuel would remain, according to the model. The last of the fuel would then be used up in the following month. Thus, we predict that the fuel would be used up after 713 months, or about 59.4 years.

Graphs. Another aspect of arithmetic growth sequences is that their graphs are straight lines. Specifically, when we graph the terms of the sequence, we see the individual points are all in a line. Often, to emphasize this fact, we display the line as part of the graph, with or without the individual data points.

Figures 2.1–2.3 show graphs for three of the sequences we have been discussing.

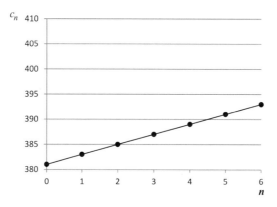

Figure 2.1. Graph for the CO_2 sequence.

The first graph depicts c_n (CO_2 concentration). It shows individual points as well as the line on which they fall. For f_n (library fine amounts) the second graph shows only the individual points. In contrast, for g_n (amount of fuel on a satellite) the graph

2.1. Properties of Arithmetic Growth

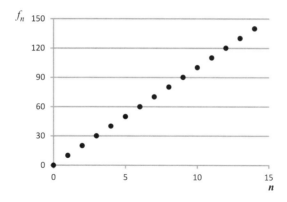

Figure 2.2. Graph for the library fine sequence.

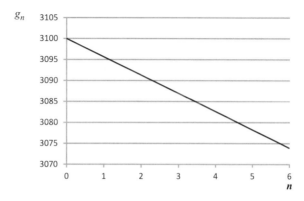

Figure 2.3. Graph for the satellite fuel sequence.

does not display any individual points. The points were used to find the line, but only the line appears on the graph. These are all valid formats. Which one is best in a particular application depends on what you wish to emphasize.

The fact that the individual points line up is a direct consequence of the arithmetic growth assumption. Each successive term increases or decreases by the same amount, d, so each successive point of the graph rises or falls by the same amount, again d. Moreover, the value of the parameter d is reflected in the steepness of the line. If d is positive, the line slopes up to the right, whereas for negative d the line slopes down to the right. The bigger d is, the more steeply the line slopes. Later we will discuss this in greater detail. Because the graphs always involve straight lines, arithmetic growth models are sometimes referred to as *linear* models.

On the other hand, many phenomena have graphs that are not straight lines. A famous example from economics is the Laffer curve (Figure 2.4). It shows one conception of how government revenue (i.e., income) varies with the tax rate. We emphasize that the shape of the curve is meant to be figurative, rather than a quantitatively accurate plot of revenue versus tax rate. The point is the general appearance of an arch, and definitely not a straight line, inferred from the fact that a tax rate of either zero or 100% will result in zero revenue. In more general terms, the Laffer curve warns not to use a linear model for the growth of revenue as the tax rate is increased. Indeed, looking

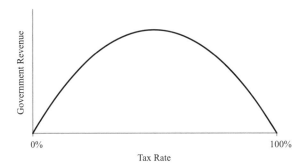

Figure 2.4. The Laffer curve. A tax rate of 0% means no taxes are due, so government revenue is 0. At the other extreme, a 100% tax rate means no one keeps any earnings. Assuming a rational person would not work under those conditions, government revenue would be 0 in this case as well. For intermediate tax rates, we know that government revenue is positive. These considerations suggest a curve of the sort shown, though this graph should not be taken as an accurate depiction of the actual revenue curve. See [**19**]. There is much to be said about this topic, including arguments refuting the entire concept of a Laffer curve. See [**27**, pp. 130-136].

only at the right side of the curve, some economists argued that decreasing the tax rate could increase revenue.

Specifics of tax policy aside, our goal here is to recognize that the straight line graph of arithmetic growth models is a special characteristic, and indicates that these models only apply in certain contexts. Phenomena and models with graphs other than straight lines are called *nonlinear*. They will be considered in later chapters. Now we continue our discussion of arithmetic growth models, where the graphs *are* linear.

Graphical Methods. Earlier we used a numerical method to predict how long the fuel in the satellite model would last. This question (and similar ones) can also be answered using a graphical method. It requires that we reformulate the question in terms of the graph, and then try to find the answer visually. For the question at hand, we want to know when g_n will equal zero. The points on the graph all have the form (n, g_n), and we particularly want to find one of the form $(n, 0)$. Such a point would lie on a horizontal line at a height of 0. In a standard xy graph, that would be the x-axis. But in our graph (Figure 2.3), the lowest horizontal line is at a level of 3070, so the x-axis doesn't even appear in the figure.

A more complete picture is provided by Figure 2.5, where the horizontal axis extends to $n = 750$, and corresponds to a g_n value of zero. In particular, the value of g_n equals 0 at the point where the sequence's graph crosses the axis. That appears to occur where $n = 720$, approximately.

As this example shows, solving a problem graphically requires that the point or points of interest appear in the graph. Moreover, the accuracy of answers obtained graphically is limited by the graph's resolution. In Figure 2.5, for example, we can estimate values on the n-axis to the nearest multiple of 25, but it is not possible to accurately read the values to the nearest whole number.

2.1. Properties of Arithmetic Growth

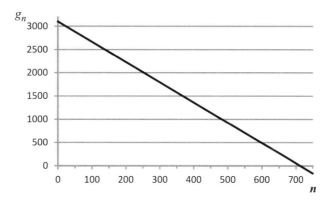

Figure 2.5. Extended graph for the satellite fuel sequence.

These are limitations of graphical methods when applied to static images in print and on line.[2] Fortunately, with graphical computer and calculator applications, the resolution and extent of graphs can be adjusted interactively. For example, by entering the satellite model difference equation and initial value in a graphing calculator, the user can dynamically extend or shrink the range shown on each axis, zooming out for the big picture to reveal where features of interest occur, and zooming in to locate those features with great accuracy. These ideas are explored more fully in the supplementary technology guide [23].

In general, a graphical method always involves visually inspecting a graph. For static graphs, the accuracy with which the graph can be read must limit the accuracy of the answers we obtain. But even when that is true, graphical methods are often helpful to get an approximate answer. In addition, expressing a question in terms of some feature of a graph provides a different view point, and this often contributes to the overall understanding of a model. We will see this later in the chapter when we consider linear functions and equations. But it will take on even greater significance in later chapters when we study non-linear phenomena.

Functional Equations. Do you recall the distinction between functional equations and difference equations discussed in Chapter 1? The difference equation is used *recursively*; the initial term is used to find the second term, which in turn is used to find the third term, and so on. To find the 100th term in the sequence, we have to compute all of the 99 terms that precede it. A functional equation is direct. We can use it to find the 100th term in a single computation. For this reason, a functional equation is very useful.

Every arithmetic growth model has a functional equation. For the CO_2 model, the equation is

$$c_n = 381 + 2n.$$

How was this equation found? Here is one approach: use the difference equation to compute the first several terms of the sequence, but without carrying out any of the

---

[2]In this context, *static* means fixed or unalterable. Computer graphics that can be modified by the user are called *dynamic*.

operations, as in the following equations.

$$c_0 = 381$$
$$c_1 = 381 + 2$$
$$c_2 = 381 + 2 + 2$$
$$c_3 = 381 + 2 + 2 + 2.$$

For each succeeding term of the sequence, we simply added 2 to the right side of the preceding equation. But notice that we didn't actually perform those additions. We just wrote them down with + signs. This reveals a simple pattern. Following that pattern, we could write the next line as

$$c_4 = 381 + 2 + 2 + 2 + 2.$$

But more importantly, we can jump ahead and predict an equation for any term, for example c_7 without working out equations for c_5 and c_6. The number of added 2's on each line is the same as the subscript on the left side of the equation, so for c_7 we write seven 2's,

$$c_7 = 381 + 2 + 2 + 2 + 2 + 2 + 2 + 2.$$

This is going to get unwieldy quickly, so let us combine the added 2's to obtain

$$c_7 = 381 + 2 \cdot 7.$$

With the same approach the earlier equations can be rewritten as

$$c_0 = 381$$
$$c_1 = 381 + 2 \cdot 1$$
$$c_2 = 381 + 2 \cdot 2$$
$$c_3 = 381 + 2 \cdot 3$$
$$c_4 = 381 + 2 \cdot 4.$$

Now the pattern is so simple, we can immediately jump to

$$c_{19} = 381 + 2 \cdot 19$$

or

$$c_{42} = 381 + 2 \cdot 42$$

or even

$$c_{1000} = 381 + 2 \cdot 1000.$$

Such an equation could be written for any term of the sequence. With algebra we express this idea economically with the single equation

$$c_n = 381 + 2n,$$

which we understand to be valid when n is replaced by any whole number.

At this point we have a strong guess about the functional equation for the sequence. But all we have to justify it is an observed pattern, and sometimes patterns are misleading, as we saw in Chapter 1. Fortunately, there is an alternative way to think about the functional equation. Suppose we wish to predict the value of c_{14}. With $c_0 = 381$ as the starting value, we can interpret c_{14} as the value after 14 years. Since the CO_2 concentration increases by 2 every year, after 14 years there will be an increase of $2 \cdot 14$. This shows that $c_{14} = 381 + 2 \cdot 14$. And similar reasoning applies for any number of years.

2.1. Properties of Arithmetic Growth

This leads us to the same functional equation we found before: $c_n = 381 + 2n$. But this time we reached the equation by reasoning about the computations, rather than by merely extending an observed pattern.

Throughout this book we will use patterns to understand functional equations for models. Proper mathematical analysis requires finding logical reasons to validate an observed pattern. For the case of arithmetic growth, because the pattern is so simple, the logical validation is reasonably simple. As we progress to more complex models in later chapters, logical validation of functional equations will become correspondingly complicated. In some instances we will merely hint at a necessary validation, or even omit it altogether. But the reader should understand the difference between observing a pattern and validating it. Furthermore, be assured that the authors have formulated proper validations for all the patterns we use, even though those validations are not all presented.

Using either patterns or logical reasoning, we can find functional equations for the library fine and satellite fuel examples. The table below shows the respective initial terms and the difference and functional equations for each example, for comparison purposes.

Example:	CO_2	Library Fine	Satellite Fuel
Initial Term	$c_0 = 381$	$f_0 = 0$	$g_0 = 3{,}100$
Difference Equation	$c_{n+1} = c_n + 2$	$f_{n+1} = f_n + 10$	$g_{n+1} = g_n - 4.35$
Functional Equation	$c_n = 381 + 2n$	$f_n = 0 + 10n$	$g_n = 3{,}100 - 4.35n$

The functional equations for all of these examples can again be united in a single generic equation, just as we found for difference equations on page 42.

> **Arithmetic Growth Functional Equation:** If an arithmetic growth sequence follows the difference equation $a_{n+1} = a_n + d$, then its functional equation is given by
> $$a_n = a_0 + dn. \tag{2.2}$$

As before, a is a generic label for the sequence.[3]

For the functional equation, we have an additional parameter, a_0. In any application of arithmetic growth, such as the three examples represented in the table above, we may use a different letter in place of a, and specific numbers will take the place of d and a_0 in the functional equation.

Like the functional equation in the CO_2 model, the generic functional equation can be found in two ways. The examples in the table reveal a pattern, and that alone is enough to suggest the generic functional equation. Going further, the generic functional equation can also be derived logically by similar reasoning as used earlier, but with symbolic parameters d and a_0 rather than specific numbers. Although this is more abstract than the earlier argument, it demonstrates that the generic functional equation is valid for *all* possible parameter values, and for all whole numbers n.

Functional Equations and Numerical Methods. Earlier, we used a numerical method to predict when the fuel would run out in the satellite model. There we used the difference equation, and systematically computed all the values of g_n until we found

---

[3] In (2.2) note that dn means $d \times n$, not d_n; this n is not a subscript.

the point where negative terms first appear. Now we have the functional equation, $g_n = 3{,}100 - 4.35n$, providing another possible approach: systematic trial and error. Will the fuel last one year? Compute $g(12) = 3{,}100 - 4.35 \cdot 12 = 3047.8$ to see that yes, there will still be fuel after a year. Will it last for ten years? That would be 120 months, and $g(120) = 3{,}100 - 4.35 \cdot 120 = 2578$. So there will still be fuel. Will it last for 100 years? This time we find $g(1{,}200) = 3{,}100 - 4.35 \cdot 1{,}200$, which is negative. We cannot actually have a negative amount of fuel, but this shows that the fuel will not last for 100 years. In this way, we can systematically refine our questions, getting closer and closer to the point where the fuel runs out. The table below shows how the entire process might play out. In the n column are our successive guesses about how long the fuel might last, while the entries in the g_n column are computed from the functional equation.

Table 2.4. Systematic trial and error to discover when the satellite fuel will run out.

n	g_n
12	3,047.8
120	2,578.0
1,200	−2,120.0
600	490.0
800	−380.0
700	55.0
750	−162.0
712	2.8
713	−1.5

For this example systematic trial and error requires much less calculation than the earlier approach using the difference equation. But functional equations may be difficult or impossible to derive in some models, so that trial and error is not possible. Both approaches are instances of numerical methods because they work directly with the numerical values of the terms of the sequence.

Functional Equations and Theoretical Methods. We have seen both numerical and graphical methods for answering questions about a model. There is another important approach, referred to in this book as a *theoretical method*.[4] To illustrate, we consider again the question: when will the fuel run out in the satellite model? The answer is the point in the sequence at which $g_n = 0$. This equation, combined with the functional equation for g_n, leads to

$$3{,}100 - 4.35n = 0.$$

Now we can use algebra to solve the equation, and thus determine the unknown n, as follows. First, we add $4.35n$ to both sides of the equation, obtaining

$$3{,}100 = 4.35n.$$

Next, exchange the two sides of the equation,

$$4.35n = 3{,}100.$$

---

[4] Some authors refer to this as a symbolic method.

2.1. Exercises

Finally, dividing both sides of the equation by 4.35 shows that

$$n = 3{,}100/4.35,$$

which is approximately 712.6. Thus, according to the model the satellite fuel will last for between 712 and 713 months. Dividing by 12 we see that is 59 years and between 4 and 5 months.

Although we used numbers in this analysis, it differs from a *numerical* approach because we also used algebra. But the concept of a theoretical method encompasses more than just using algebra to solve an equation, as indicated in the following points.

First, theoretical methods can involve many types of analysis, including algebraic manipulation, solution of equations and inequalities, geometry, statistics, and more advanced types of mathematics. The most common use of theoretical methods in this book involves solving functional equations algebraically, but you will also see instances where these methods are used to derive properties of models. We have already seen one example, the derivation of a functional equation for the CO_2 model. Second, theoretical methods often help us understand models on a deeper level than the answers we obtain using numerical and graphical methods, because they reveal relationships among variables in a model. Finally, numerical, graphical, and theoretical methods can and should be used in combination, each contributing part of our understanding of a model. This idea will be developed further in Section 2.2.

Profile for Arithmetic Growth. We have considered quite a few aspects of arithmetic growth models. In later chapters we will consider other kinds of models. One of our goals in doing so is to see how different kinds of models compare, what sorts of properties they can have, and how to recognize when they are appropriate for applications. To facilitate comparisons we will develop a profile of each type of model. Here is part of the profile for arithmetic growth models.

Table 2.5. Profile for arithmetic growth sequences.

Verbal Description:	Each term increases (or decreases) from the preceding term by a constant amount
Parameters:	Initial term a_0, constant difference d
Difference Equation:	$a_{n+1} = a_n + d$
Functional Equation:	$a_n = a_0 + dn$
Graph:	Straight line

2.1 Exercises

Several problems require the creation of graphs. It is recommended that these graphs be produced using computer software (such as Excel) or an online graphing application, and printed for inclusion with your solutions. Alternatively, graphs may be produced by hand on graph paper, but if so, it is important to plot points with the greatest possible accuracy.

Reading Comprehension.

(1) Write a short essay on the topic of arithmetic growth. Tell what features are shared by all arithmetic growth models, and give details about graphical properties, difference equations, and functional equations of these models. How can you tell whether a particular application might (or might not) be an appropriate place to use an arithmetic growth model?

(2) Why are arithmetic growth models often referred to as linear models? Think of your answer as an explanation to help a fellow student who is struggling with this topic. Your answer should be as specific as possible and include at least one example.

(3) Find a recent newspaper or magazine article that uses either a numerical, graphical, or theoretical approach to draw a conclusion about a model. Explain the example you have chosen briefly, using your own words. It does not necessarily need to be an arithmetic growth model.

(4) @What are parameters and how do they differ from variables? Give a description and examples.

Mathematical Skills.

(5) @For each of the following items a sequence is given either numerically or graphically. Determine whether or not the sequence is arithmetic. If not, explain why not. If so, give the values for the constant difference, d, and the initial term, a_0.

   a. @3, 5, 7, 9, 11, 13, $\cdots$
   b. 7.1, 8.3, 9.5, 10.7, 11.9, 13.1, $\cdots$
   c. @1, 4, 16, 64, 256, $\cdots$
   d. $\frac{1}{2}, \frac{1}{4}, \frac{1}{6}, \frac{1}{8}, \frac{1}{10}, \cdots$
   e. @150, 145, 140, 135, 130, 125, $\cdots$
   f. @

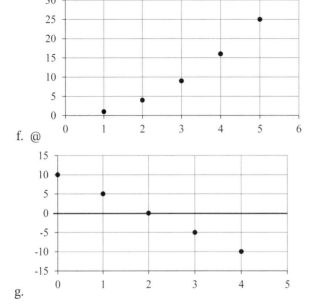

   g.

2.1. Exercises

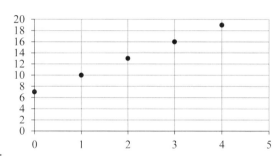

h.

(6) @For each functional equation below, find the corresponding difference equation and the initial term.

a. @$a_n = 15 - 3n$

b. $b_n = 7 + \frac{1}{3}n$

c. $c_n = 20 - 5n$

d. $d_n = 20 + 5n$

e. @$e_n = 10n - 3$

f. $f_n = 10n + 3$

(7) @Each part below gives a difference equation and the value of an initial term for a sequence. Find the functional equation for each sequence.

a. @$a_{n+1} = a_n + 2;\ a_0 = 1$

b. $b_{n+1} = b_n + 6.2;\ b_0 = 5$

c. $c_{n+1} = c_n + \frac{1}{5};\ c_0 = -312$

d. $d_{n+1} = d_n - 1.3;\ d_0 = 100$

e. @$e_{n+1} = e_n - \frac{1}{2};\ e_0 = 70$

f. $f_n = f_{n-1} - 10;\ f_1 = 23$

g. $g_n = g_{n-1} + 0.8;\ g_1 = 11.3$

(8) Graph each of the sequences, $a_n, \cdots, g_n$, defined in the previous problem. Your graphs should include at least five clearly labeled points. Graphs may be produced carefully by hand or printed after being created with computer software. See the technology guide [23] for more information.

(9) @For each of the following items a sequence is given either numerically or graphically. For each sequence find functional and difference equations.

a. $20, 24, 28, 32, 36 \cdots$

b. @$\frac{1}{4}, \frac{1}{3}, \frac{5}{12}, \frac{1}{2}, \frac{7}{12}, \cdots$

c. $-18, -28, -38, -48, -50, \cdots$

d. $1.1, 2.4, 3.7, 5, 6.3, \cdots$

e.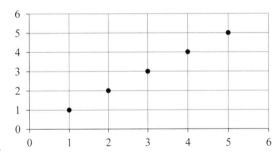

f.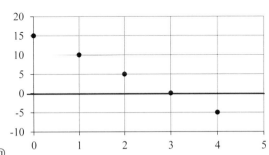

(10) @In each part below find specified terms for an arithmetic growth sequence.

   a. @The initial value is $a_0 = 5$ and the sequence begins 5, 8, 11, 14, 17, $\cdots$. Find the terms a_6 and a_{400} in the sequence.

   b. The initial value is $b_0 = 10$ and the sequence begins 10, 15.6, 21.2, 26.8, 32.4, $\cdots$. Find the terms b_9 and b_{150} in the sequence.

   c. @Each term decreases from the preceding term by 12. The initial value is $c_0 = 325$. Find c_7 and c_{100}.

   d. Each term increases from the preceding term by 3.4. The initial value is $d_0 = 9$. Find d_{10} and d_{200}.

   e. @Find e_9 and e_{250} for sequence e_n whose graph is shown here.

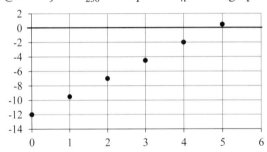

2.1. Exercises

f. Find f_7 and f_{300} for sequence f_n whose graph is shown here

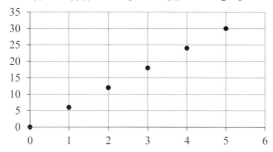

(11) Each equation below defines a sequence. For each,

  i. compute the value of the sequence for $n = 0, 1, 2$, and $n = 10$,
  ii. find a difference equation, and
  iii. graph the sequence. Your graph should include at least five clearly marked points.

  a. $a_n = 10n - 2$
  b. $b_n = 2 - 3n$
  c. $c_n = 5 - n$
  d. $d_n = 1 + 2.7n$

Problems In Context.

(12) @Flu Epidemic. In this section one of the examples of arithmetic growth is

$$p_{n+1} = p_n + 500; \quad p_0 = 1,000,$$

where p_n represents the number of people who have been infected with flu by the start of month n.

  a. @Graph at least the first 5 terms of the sequence and leave room to extend the graph several more terms.
  b. @Use a graphical approach to determine how long it will be before at least 9300 people are infected with flu.
  c. @Use a numerical approach to determine how long it will be before at least 9300 people are infected with flu and compare the answer to the one you found in the previous part of this problem.

(13) Ebola Outbreak. In a 2014 outbreak of the Ebola virus in West Africa, one report included the statement "At current rates of transmission, each infected person is passing the disease to about two more people"[9]. In trying to understand this situation, a student reasons as follows. Suppose that there are 100 people infected initially. Each of those infects two more people, so then there will be 200 new patients (they are in generation 2), and a total of 300 patients. Now each of the 200 new patients infects two more, for a new set of 400 patients (generation 3) and a total of 700. Continuing in this way, we can track the number of newly infected people in each generation finding 100, 200, 400, 800, ⋯, or we can track the total number of infected people after each new generation, finding 100, 300, 700, 1500, ⋯. Would

an arithmetic growth model be appropriate for either of these sequences? If so, find appropriate difference and functional equations, identifying all your variables. If not, explain why not.

(14) @Diving Pressure. As a diver descends into the ocean ambient pressure builds at a rate of about 1 atmosphere (atm) for every 10 meters (m) of depth. A diver will experience 1 atm at the surface, 2 atm at a depth of 10 m, 3 atm at a depth of 20 m, and so on.

   a. @Let d_n be the depth in meters at which a diver experiences n atms of pressure. Compute d_n for $n = 1, 2, 3,$ and 4.
   b. @Graph the sequence from part a.
   c. @Explain why this situation can be accurately modeled by an arithmetic growth sequence. Are there any limitations we should consider? For example, does it make sense to allow $n = 0$?
   d. @Find a difference and a functional equation for d_n.
   e. @A diver has a camera that can withstand pressures up to 25 atm. How deep can the camera be safely taken?
   f. @A team of divers descends to a depth of 27 meters. What pressure will they experience?

(15) Simple Interest. A student borrows $5,000 from his aunt. He promises to repay the loan as soon as possible. It is agreed that he will pay the original $5,000, plus simple interest, in one lump sum payment. The interest will be calculated at one quarter percent per month. Let p_n be the amount of money the student will have to pay the aunt if he makes the payment after n months. For example, if he makes the payment after a year, that is 12 months and the amount to be repaid would be p_{12}. At a quarter percent per month, the interest will be 3 percent in 12 months. Now 3 percent of $5,000$ is $0.03 \times 5,000 = 150$. So if the loan is paid back after 12 months, the amount to be repaid is $\$5,150$, the original loan amount plus the $150 in interest. That is p_{12}. In contrast, p_6 is what the student has to pay if he makes the payment after 6 months.

   a. What is p_0? Does that make sense?
   b. What is the difference equation for p_n?
   c. What is the functional equation for p_n?
   d. Use the functional equation to determine how much must be repaid if the payment is made after 18 months.
   e. Eventually, the student repaid his Aunt at a cost of $6,000. How long was that after he borrowed the original $5,000? That is, find n so that $p_n = \$6,000$.

Digging Deeper.

(16) @When we know the initial term a_0 and the common difference d for an arithmetic growth sequence, the functional equation $a_n = a_0 + dn$ can be immediately formulated. In this problem you will be asked to find functional equations using other information. Find a functional equation for each part below. Show your work or explain your reasoning.

a. $@a_1 = 12$ and $d = 4$
   b. $b_3 = 8$ and $b_{n+1} = b_n + 1.4$
   c. $c_{n+1} = c_n - \frac{1}{2}$ and $c_3 = 12$
   d. $@e_2 = 8$ and $e_6 = 52$
   e. The first term of the sequence is $f_1 = 6$. There is no f_0 term. The second term is $f_2 = 0$.

(17) @Two equations involving a single variable are called equivalent if they are both true for exactly the same values of the variable. For example $x + 1 = 6$ and $10/x = 2$ are equivalent because they are both true when x is 5, and for no other values of x. However $x(x + 3) = 5x$ and $x + 3 = 5$ are not equivalent because the first is true for both $x = 0$ and $x = 2$, whereas the second is only true for $x = 2$.
For each part below an equation is given, and an operation to transform the equation. In each case, tell whether the original equation and the result of the operation are equivalent.

   a. $@x - 2 = 7$; add 2 to both sides
   b. $3x - 4 = 8$; divide both sides by 3
   c. $@(x + 1)(2x) = (x + 1)(x - 3)$; divide both sides by $(x + 1)$
   d. $5 + 6x = -(x + 1)$; multiply by -1 on the right side
   e. $4x = 11$; take the absolute value of both sides.
   f. $3x + 6 = 4x + 8$; factor the right side

(18) For each of the following, write an equation for n as a function of a_n. For example, for part a, since $a_n = 15 - 3n$, we have $3n = 15 - a_n$ so $n = 5 - a_n/3$. Thus we have expressed n as a function of a_n.

   a. $a_n = 15 - 3n$ b. $a_n = 15 + 3n$
   c. $a_n = -3n + 20$ d. $a_n = 3n + 20$
   e. $a_n = 20 - 5n$ f. $a_n = 20 + 5n$
   g. $a_n = 20n - 3$ h. $a_n = 20n + 3$

2.2 Applications of Arithmetic Growth

In Section 2.1 we considered mathematical properties of arithmetic growth models, such as difference equations, functional equations, and graphs. Now we turn to applications.

The applications we have in mind begin with a problem situation or context, about which we wish to answer one or more questions. Our overall strategy is to introduce a mathematical framework, called a mathematical model, that resembles the real context. Questions about our problem situation are translated into questions about the model, which we answer (if possible) using mathematical operations and analysis. The answers are mathematically correct conclusions about the model.

We recognize that the model is not an exact portrayal of the problem situation, and that the conclusions of the model may only provide approximate answers to our original questions. Indeed, validating and/or improving models are important facets of the mathematical modeling approach. But we will not delve into them here. Our focus will be on how to formulate models and use them to answer questions about a problem context.

A Modeling Outline. Many mathematical operations are expressed in terms of an *algorithm*, which is a set of prescribed steps to be performed in a specified order. Examples include procedures for addition, subtraction, multiplication, and division of numbers, and algebraic manipulations such as expanding $(x+1)^2$ to x^2+2x+1. Conceptually, an algorithm is supposed to be clear-cut and mechanical—at each step one knows exactly what to do, with no uncertainty and no need to exercise judgment.

The process for adding two numbers illustrates the idea of an algorithm. To add 1,768 and 421, we first write them like so.

$$\begin{array}{r} 1768 \\ +421 \\ \hline \end{array}$$

Then we work from right to left, adding digits in each vertical column, recording the result below the line. The actions we take at each step may depend on the results of preceding steps—sometimes we have to carry, sometimes we don't, but at every step we know exactly what to do. And we know that when all of the steps have been carried out in the prescribed manner, we will have the desired result. That is how an algorithm works.

There is no algorithm for developing and using mathematical models. Even in the restricted context of this section, where arithmetic growth sequences are our sole concern, applications will require judgment and insight. The information given and questions posed will vary from problem to problem. It will be up to the student to choose appropriate notation, and identify or adopt assumptions, questions, and methods of analysis.

Our primary goal for this section is to explain the modeling process, preparing students to develop their own models for exercises at the end of the section. Listed below are common steps that usually occur in developing and using models. We emphasize that this list should be viewed as a rough outline, and not as a step-by-step algorithm. Most problems will involve all the steps in some form, but in a way that varies from one problem to another. Nor should the order of the steps be rigidly followed. For example, graphs and tables are mentioned in the sixth step of the outline, but they may be useful in the preceding steps. Also, sometimes working on a later step of the outline prompts you to revise an earlier step. The point of the outline is to help you organize your work and to highlight important components of the modeling process.

Modeling Outline:

(1) Understand the general characteristics of the problem situation, including given information and questions to be answered.

(2) Define variables.

(3) Identify assumptions.

(4) Find values of parameters.

(5) Formulate difference and functional equations.

(6) Where appropriate, create diagrams, tables, and graphs representing important aspects of the model.

(7) Formulate specific questions about the model.

2.2. Applications of Arithmetic Growth

(8) Find answers to the questions.

(9) Translate conclusions about the model into conclusions about the problem situation.

(10) Reflection and reconsideration. Are the results reasonable? Are the assumptions? Were there any errors? Are there ways to improve the model?

We proceed to consider several examples.

Example 1: Time of Death. According to the Writer's Medical and Forensics Lab [36], body temperature can be used to estimate time of death. Quoting the source,

> Normal body temperature during life is 98.6 degrees F. After death, the body loses heat progressively until it equilibrates with that of the surrounding medium. The rate of this heat loss is approximately 1.5 degrees per hour until the environmental temperature is attained, then it remains stable.

The expression *degrees F* means *degrees Fahrenheit*. This is also represented symbolically, as in 98.6°F.

Suppose that a dead body is discovered in a hotel room at 9 PM. The temperature in the room is 68°F. Upon discovery, the temperature of the corpse is found to be 84°F. Find the time of death.

Solution.

General Characteristics. The problem poses a single question: when did death occur? This asks us to find a particular time, so time will be one variable. We are also given information about the temperature of the corpse, so temperature will also be a variable. We can formulate a model by understanding the given context in terms of a sequence of data values. One way to do this is to imagine a sequence of hourly temperature readings, starting with the temperature at the time of death. We can conceptualize the situation in this way, even though the problem does not say that such data have been collected. We note that the problem statement includes two additional facts: normal body temperature during life, and rate of heat loss after death.

Define Variables. Choosing the letter T (for temperature), we denote our sequence of temperatures $T_0, T_1, T_2, \cdots$, where T_0 is the temperature at the time of death, T_1 is the temperature one hour later, and so on. In this context, the position number n represents elapsed time, in hours, from the time of death. For reference purposes, we incorporate this information into a formal definition of variables:

Let T_n be the temperature of the body, in degrees F, n hours after death.

Identify Assumptions. By the end of the course you will have learned about several different kinds of models, and part of each modeling problem will be to decide which kind of model to use. But at this point of the course we have only seen arithmetic growth models, so that is the kind of model we will use here. With that in mind, we note three assumptions:

(1) T_n is an arithmetic growth sequence,

(2) body temperature is initially 98.6°, and

(3) body temperature decreases by 1.5° each hour.

We are also given the temperature of the room, but we shall see that this fact does not enter into the model. As you work on a problem, it may not be obvious what facts are relevant, and there is no harm in including something that later is not needed. So a fourth assumption about room temperature could be included in the list above. But your solution will be easier to follow if you create a final draft that excludes references to unnecessary information.

Parameters. We know that two parameters are included in an arithmetic growth model, the initial value T_0 and the difference d. These are both given directly in our list of assumptions. The initial temperature is 98.6, the temperature of the body when death occurred. And a decrease of 1.5° every hour means that $d = -1.5$. If these facts are not at first apparent, a good strategy is to work out the first several terms of the sequence, as follows. By definition, T_0 is the temperature at the time of death. What is that? Right up to the moment of death the victim is alive, and so has a temperature of 98.6. Proceeding, T_1 is the temperature an hour later. But we know the temperature decreases by 1.5° each hour. So we must have $T_1 = 98.6 - 1.5 = 97.1$. By similar reasoning, succeeding terms are $T_2 = 97.1 - 1.5 = 95.6$, $T_3 = 95.6 - 1.5 = 94.1$, and $T_4 = 94.1 - 1.5 = 92.6$. Computing these values should help you recognize arithmetic growth at work, and identify the parameters.

Difference and Functional Equations. The difference and functional equations for this model are

$$T_{n+1} = T_n - 1.5$$

and

$$T_n = 98.6 - 1.5n,$$

respectively. The difference equation formalizes the recursive pattern of repeated 1.5 degree reductions in temperature. The functional equation provides a means for direct (as opposed to recursive) determination of T_n for any n. Because they offer alternative ways to find the same thing, we can use the equations together to check for errors.

Diagrams, Tables, Graphs. We display below a table of values and a graph for the sequence T_n. These were created using the methods discussed in Section 2.1. Tables and graphs offer additional perspectives on a model, sometimes revealing aspects that are not easily recognized working with equations alone. They provide yet another alternative way to answer specific questions about the model. Also, they should generally be included as part of the documentation of a model, because they help communicate what you have done to a broader audience. For all of these reasons, you should create tables and graphs for your models, even if you do not find them strictly necessary to find specific answers.

2.2. Applications of Arithmetic Growth

n	T_n
0	98.6
1	97.1
2	95.6
3	94.1
4	92.6
5	91.1
6	89.6

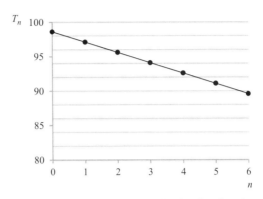

Tables and graphs may be created at any convenient point in the development of the model. For example, as soon as the first several terms of the sequence were computed, the table and graph could have been produced. On the other hand, in the process of answering specific questions about the model, it might prove worthwhile to extend the table and graph to additional data points.

Questions About the Model. The original problem statement asks us to find the time of death. How do we restate this as a question about the model? Observe that in the model, the start time is the time of death, and that is not known. However, we do know the temperature of the body was 84° when it was discovered. Is that the value of T_n for some n? Suppose, for example, we find that $T_8 = 84$. That would tell us that the discovery of the body occurred 8 hours after death. Notice how the definitions of variables makes this clear: T_8 is the temperature of the body 8 hours after death. And since we are given the time at which the body was discovered, we can also determine the time of death. In essence, we can change the original question to *How long had the victim been dead when the corpse was discovered?* That translates directly to a question about the model: *For what n does $T_n = 84$?*

Find Answers. One approach to finding the answer is to look at the table and graph already produced. Is there an entry in the table or a point on the graph where $T_n = 84$? No. But we see immediately that extending the table and graph for a few additional terms should reveal the desired information. Accordingly, we create the following revised table and graph.

n	T_n
0	98.6
1	97.1
2	95.6
3	94.1
4	92.6
5	91.1
6	89.6
7	88.1
8	86.6
9	85.1
10	83.6

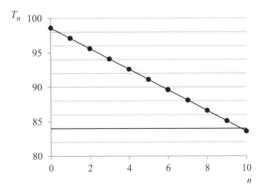

Referring to the table, we see that $T_9 = 85.1$ and $T_{10} = 83.6$, so $T_n = 84$ should occur somewhere between $n = 9$ and $n = 10$. The graph portrays the same information visually. Note that the line corresponding to a T value of 84 has been highlighted.

For a more exact answer, we combine the functional equation with $T_n = 84$ to obtain
$$98.6 - 1.5n = 84.$$
Solving this equation for n we find
$$n = (98.6 - 84)/1.5 = 9.73$$
to two decimal places. An exact answer is $n = 146/15 = 9\frac{11}{15}$.

Translate Conclusions. We have found an exact answer to the question *In the model, for what n does $T_n = 84$?* This indicates a time of death 9.73 hours before 9 PM, or at approximately 11:15 AM. Or, using the exact answer, we can convert the fractional part 11/15 from hours to minutes, observing that 1/15 of an hour is 4 minutes. Thus, death occurred 9 hours and 44 minutes before 9 PM, and hence at 11:16 AM.

Reflection and Reconsideration. We have found an answer, but is it correct? Is it reasonable? Perhaps we made a mathematical error. Or, even if the mathematical steps were performed perfectly, it may be that the assumptions in the model are questionable. Turning a critical eye on your own work is an important aspect of mathematical modeling.

We can be fairly confident that the mathematical steps were correctly done, because we actually found answers three different ways. Both the table and graph indicated that n should be between 9 and 10, and by visual inspection, the graph suggests a value of n more than 9.5. These observations are consistent with the exact answer we found using algebra. At least we can see that our different views of the model are consistent.

But we can also check the conclusion by a different approach. Supposing that the death did occur at 11:16 AM, what temperature would we expect to find at 9 PM? We know that the temperature should decrease by 1.5° every hour, or by 3° every two hours. So, with a temperature of 98.6 at 11:16, we expect a temperature of 95.6 at 1:16, 92.6 at 3:16, 89.6 at 5:16, and so on. Continuing in this way, we predict a temperature of 83.6 at 9:16. This is consistent with the given information, and indicates that our answer is in the right ballpark.

Because we know that our model is only an approximate representation of the true situation, we should not expect an answer that is exactly correct. Just on the basis of common sense, we would expect there to be some uncertainty about the time the body was discovered. So even if the model is reasonably accurate, the exact time of death will probably be somewhat different than what the model indicates. Also, in establishing the time of death, the investigators probably do not require accuracy down to the second. Both of these factors influence how we interpret the results. In particular, we found an approximate solution of $n = 9.73$ as well as an exact figure of $n = 9\frac{11}{15}$. The exact value may be preferred for aesthetic reasons, but in practical terms, 9.73 is just as believable. Similarly, it is as reasonable to round the time of death to 11:15 as it is to report the exact value of 11:16. The one minute difference is almost certainly insignificant.

2.2. Applications of Arithmetic Growth

Another point worth considering is the use of a fractional value for n. In our original formulation of the model, n is a position number, and hence represents a whole number. On the other hand, n is also interpreted as a number of hours, and so fractional values make sense. But the derivation of the arithmetic growth functional equation is based on n representing a whole number. Does the equation remain valid for values of n that are not whole numbers?

In the context of the current problem, this is essentially asking whether the temperature decrease each hour is evenly distributed during the hour. In other words, if the temperature decreases by 1.5° in an hour, can we further assume a decrease of 0.75° every half hour? Or of 0.5° every 20 minutes? Adopting this plausible assumption, using the functional equation with fractional values of n is justified. We will discuss this issue at greater length later, in connection with continuous and discrete variables. As we will see, although permitting fractional values of n is valid in this problem, it will not always be so.

Finally, we might also question the assumptions of the model. In the context of a textbook exercise, we can legitimately accept the given information as correct. But in an actual application in the real world, assumptions are always subject to discussion and criticism. For the problem under discussion, two obvious questions are whether the body will really cool by the same amount every hour, and if so, whether the 1.5 degree figure is correct. For the first question, notice that the model does not agree with the verbal problem description in one important respect: in the model the body continues to cool even after it reaches the environmental temperature. Everyday experience (in other contexts) tells us that the body will cool down to the temperature of the surrounding room, but not lower. Indeed, the quoted passage from the Medical and Forensics Lab states as much. Yet in the model, with $n = 40$ we find $T_{40} = 98.6 - 40 \cdot 1.5 = 38.6$, much colder than the room termperature of 68. The model may still be accurate for the time period of the problem context, but we should be aware that it cannot be valid in the long term.

Regarding the second question, we again rely on our everyday experience. We know that a warm object will cool more quickly in a refrigerator than in a heated room. Analogously, in the context of the model, we expect a body's cooling rate to depend on the temperature of its surroundings. Since the model does not take the surrounding temperature into account, we should question the accuracy of the 1.5 degree figure. One response would be to investigate how our conclusions might change under different assumptions. For example, if the body cools by 2° per hour rather than 1.5°, how much earlier or later would we place the time of death?

Deeply analyzing questions of this sort is beyond the scope of our discussion here. They are mentioned to illustrate the meaning of the last step of the outline, and to enrich the reader's understanding of the methods of mathematical modeling.

Many students will wonder how much reflection is reasonable in homework exercises or on examinations. Certainly it is worthwhile to check for mathematical errors and consider whether your answers are reasonable. For more conceptual considerations, such as whether the assumptions of the model are valid, students should not expect to find definitive conclusions. However, they should be prepared to indicate what assumptions might be questioned, and why. In the context of this specific example, there is another assumption that might reasonably be questioned, and the reader is challenged to identify it. A question about this assumption will appear in the exercises at the end of the section.

As the preceding discussion shows, working through an applied modeling problem involves many considerations, requires some judgment and insight, and probably does not proceed in a linear fashion. As you proceed you may wish to modify earlier steps, for example. At the end you may discover inconsistencies that reveal errors. For all of these reasons, it is often helpful to work out a problem in a rough form, making changes as necessary, and then write up a final draft of the solution at the end. This not only will result in clear and coherent solutions for evaluation on assigned work, but will also reinforce your understanding of how mathematical modeling is actually used in the real world.

Homework Solution. In the preceding discussion our goal has been to expose the rationale for our approach. The reader is encouraged to think in similar terms in solving problems in this section. However, it is not always necessary to include as much explanation in your solutions as has been provided above. In Figure 2.6 we display a sample solution for the time of death problem, as a student might include in a homework assignment. The table and graph were produced using computer software, and attached to the page. If you are studying this text as part of a course, consult the instructor for guidelines on writing up solutions to applied problems.

Example 2: College Tuition. Table 2.6 shows the annual in-state tuition at the College of William and Mary, for each year from 2001 to 2012. Develop an arithmetic growth model based on this information, and estimate the expected tuition in 2015 and in 2020. If a child is 5 years old in 2015, how much will four years tuition at William and Mary be when the child reaches the age of 18, according to the model?

Table 2.6. In-state tuition charges for the College of William and Mary, in thousands of dollars.[5]

Year	Tuition	Year	Tuition
2001	4.780	2007	9.164
2002	5.528	2008	10.426
2003	6.430	2009	10.800
2004	7.096	2010	12.188
2005	7.778	2011	13.132
2006	8.490	2012	13.570

Solution.

General Characteristics. In an arithmetic growth model there will be a sequence of values. We can think of the annual tuition figures as terms of such a sequence: $4.780, 5.528, 6.430, \cdots$. The problem statement directs us to develop an arithmetic growth model. For the model to fit the data exactly, each term should increase by the same amount over the preceding term. To investigate whether this is true, we add

---
[5]Data source is reference [55].

2.2. Applications of Arithmetic Growth

a column to the table (see Table 2.7) showing how much the tuition increases for each year.

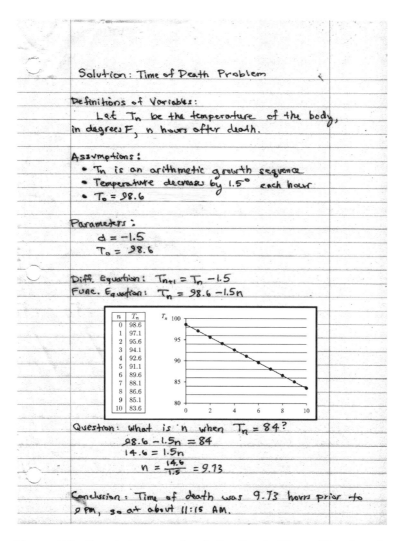

Figure 2.6. Sample homework solution for the Time of Death problem.

We see that the yearly increases are not all the same, and that there is a fair amount of variation. This tells us that the model will not fit the data exactly. But it may still be the case that an arithmetic growth model can approximate the given data.

Here, it will be worthwhile to examine a graph of the data, as shown in Figure 2.7. We have included a straight line joining the first and last data points, merely as a visual reference, and it does appear that the data points are all pretty close to this line. Thus, we will proceed to develop an arithmetic growth model. The points of the model will fall exactly on a straight line, and will be close to, but not exactly equal to, the original data points.

66　　　　　　　　　　　　　　　　　　　　　　Chapter 2. Arithmetic Growth Models

Table 2.7. Tuition data with yearly increases, in thousands of dollars.

Year	Tuition	Increase
2001	4.780	
		0.748
2002	5.528	
		0.902
2003	6.430	
		0.666
2004	7.096	
		0.682
2005	7.778	
		0.712
2006	8.490	
		0.674
2007	9.164	
		1.262
2008	10.426	
		0.374
2009	10.800	
		1.388
2010	12.188	
		0.944
2011	13.132	
		0.438
2012	13.570	

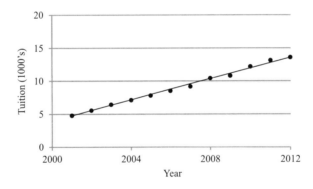

Figure 2.7. Graph of the tuition data.

Define Variables. We again choose the letter T (for tuition in this example), so that T_n is the tuition in year n, in units of thousands of dollars. We emphasize here that T_n is a term in the sequence in our *model*, not the actual data value given in the table.

It will be convenient to take the starting term to be T_1, so that T_1 is the tuition in 2001, T_2 the tuition in 2002, and so on. Then it will be particularly simple to translate between a year and the corresponding position number, e.g., T_{17} corresponds to 2017.

As in the earlier example, position number n represents elapsed time, but now in units of years. Mimicking the approach of the preceding example, we state the variable definitions formally, for future reference.

Let T_n represent in-state tuition, in thousands of dollars, at the College of William and Mary for year n, where year 1 is 2001.

2.2. Applications of Arithmetic Growth

Identify Assumptions. As specified in the directions, we assume that the model sequence $T_1, T_2, T_3, \cdots$ is an arithmetic growth sequence.

Parameters. As always for an arithmetic growth model, we wish to identify the initial value T_0 and the common difference d. We would like the terms T_1 through T_{12} to approximate the corresponding data values very closely. Since it is not obvious how to find the parameters that are the best possible, we start with a simple but plausible approach and hope that it produces reasonably accurate results. First, we compute the parameter d as the average of all of the differences in Table 2.7. Rounded to three decimal places, that gives us $d = 0.799$. Second, we choose T_0 so that T_1 equals the first data value, 4.780. Using our arithmetic growth assumption, T_1 must be 0.799 units higher than T_0. That means $T_0 = T_1 - 0.799 = 4.780 - 0.799 = 3.981$.

Difference and Functional Equations. Now that we have found the parameters, we can substitute them into the standard arithmetic growth difference and functional equations, to find

$$T_{n+1} = T_n + 0.799$$

and

$$T_n = 3.981 + 0.799n.$$

These are the difference and functional equations for our model.

Diagrams, Tables, Graphs. A table of values and a graph are shown below for our sequence T_n. For later reference we also include in the table the original data values and the errors, that is, the differences between each term of T_n and the corresponding data value. To be more precise, each error is found by subtracting a data value from the corresponding term T_n of the model. Thus, nonzero errors can be either positive or negative. A positive error indicates that the model value is too high; negative indicates that the model value is too low.

In the graph the original data values appear as dots, while the model is represented by a straight line. The individual terms T_n are not shown on this line, because they would be too close to many of the original data points. Comparing the dots with the line, we obtain a visual portrayal of the errors. Where a data point is below the line, the corresponding error is positive; a data point above the line indicates the error is negative. And the vertical distance between each data point and the line is the absolute value of the error.

Notice that the worst error is 0.411. For this point the model over-estimates the tuition by $411, out of $9,575. For about half the points, the model is off by less than $100.

n	Data	T_n	Error
1	4.780	4.780	0.000
2	5.528	5.579	0.051
3	6.430	6.378	−0.052
4	7.096	7.177	0.081
5	7.778	7.976	0.198
6	8.490	8.775	0.285
7	9.164	9.574	0.410
8	10.426	10.373	−0.053
9	10.800	11.172	0.372
10	12.188	11.971	−0.217
11	13.132	12.770	−0.362
12	13.570	13.569	−0.001

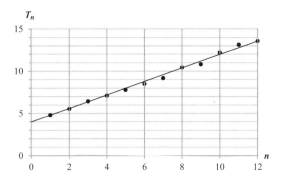

Questions About the Model. The problem statement asks for tuition estimates for 2015 and 2020. As we have defined our variables, this corresponds to finding T_{15} and T_{20}. We are also asked for the four-year tuition total at age 18 for a child age 5 in 2015. Since age 18 is 13 years older than age 5, we want the tuition total for 2028 through 2031. Thus, we would like to compute $T_{28} + T_{29} + T_{30} + T_{31}$.

Find Answers. Using the functional equation we compute the following values:

n	15	20	28	29	30	31
T_n	15.966	19.961	26.353	27.152	27.951	28.750

This gives us T_{15} and T_{20} directly. We can also use the values in the table to compute

$$T_{28} + T_{29} + T_{30} + T_{31} = 110.206.$$

Translate Conclusions. The interpretation of these results is as follows. T_{15} represents the tuition for 2015. Thus, the model predicts tuition will be $15,966 for 2015, and similarly, $19,961 for 2020. The sum of T_{28} through T_{31} indicates a four year tuition total of $110,206 at age 18 for the child who is 5 in 2015.

Reflection and Reconsideration. This application is very similar to the prior example, in terms of potential sources of error. We can again review our findings using graphical and numerical methods, checking that all of the methods produce consistent results. We can also question whether future tuition increases are likely to follow the same trends as the past twelve years. Many possible factors can influence tuition at a public university, including the general health of the economy and political forces in state government. Accordingly, we should not be surprised to see sudden changes that render our arithmetic growth model incorrect. In contrast, for the prior example, the goal was to model a physical process. We expect the rate of cooling of an inert body to depend on surrounding conditions, but otherwise to follow a predictable and consistent course of events. Thus, two bodies that start at the same temperature and are subjected to the same conditions should cool off at the same rate. Also, the validity of

2.2. Applications of Arithmetic Growth

an arithmetic growth model is asserted by a forensic expert, presumably based on observations in many cases. These factors should give us greater confidence in the model for that example.

As observed earlier, the data presented in this problem closely approximate an arithmetic growth sequence. Our model produces annual tuition projections that are very close to the actual data for the first 12 terms. So we might be fairly confident that the model will continue to be accurate for the first few years after 2012. However, using the model to project as much as 13 years into the future should not be counted on to produce accurate results.

One aspect of the reflection phase is gaining new insights about a particular problem or the methods employed. For this example it is striking that the model reproduces the first and last data points almost exactly. And if we had defined d as the *exact* average of the first differences, without rounding, the model would have reproduced the first and last data values *exactly*. Is this just a coincidence?

No. The same thing will happen in any arithmetic growth model defined in the same way. That is, with any given set of data, if we define a_0 to equal the first data point exactly, and define d to be the average of all the differences between consecutive data values, then the model will agree exactly with the last data point as well. This assertion can be justified with a little thought, taking into account how an average is computed, and the interested reader is invited to think it out. Our main purpose in mentioning it here, though, is to illustrate the discovery of new ideas during the reflection phase of modeling.

Example 3: Making T-shirts. An established business manufactures various items of clothing. For this problem we focus on a single factory set up to produce t-shirts, and on the cost of production for a single typical day. With the existing equipment and facilities, it is possible to make at most 125,000 shirts per day. But depending on other variables (such as sales demand) the company might decide to make fewer shirts, or even suspend operations for a while and make no shirts at all. Our goal is to develop a model that shows how the costs vary as we increase the number of shirts produced. The board of directors has tasked us to answer two specific questions:

(1) What will it cost to produce 90,000 shirts?

(2) If the budget is limited to $60,000, how many shirts can be produced?

We distinguish between two different types of cost. First, there are costs that do not depend on how many shirts are made. They include costs of the building that are incurred whether or not any shirts are produced, such as the cost of insurance, taxes, security, and maintenance. These are called *fixed* costs. Company analysts have determined that fixed costs for a typical day amount to about $17,500.

Second, there are costs that do depend on the number of shirts made, such as the cost of fabric, wages for the workers who make the shirts, and energy costs to operate equipment in the factory. These are called *variable* costs, and are calculated on a per-shirt basis. The analysts have calculated that current variable costs total about $0.40 per shirt.[6]

---

[6]Variable cost and factory capacity are based on a 2010 report about a factory in Bangladesh. See [**43**].

Solution.

General Characteristics. Does it make sense to use a sequence model in this problem? To get an idea of the situation, let us look at some examples. If the company is idle, and produces no shirts, then there are only fixed costs, totaling $17,500. If one shirt is produced, the variable cost will be $0.40, so the total cost will be $17,500.40. For two shirts, the variable costs will be $0.80, so the total cost in this case is $17,500.80. Continuing in this fashion, we can find the total cost to produce 3, 4, or 5 shirts. However, this approach is unrealistic. Can we expect a factory capable of producing 125,000 shirts to produce just one or two? In addition, mathematically and conceptually, it will be inconvenient to work with quantities as small as a single shirt.

Accordingly, let us represent the quantity of shirts in units or lots of 10,000. Then the fixed costs will be $17,500 and variable costs will be $4,000 per lot. Arguing as before, we find the cost to produce nothing will be $17,500, to produce one lot will be $21,500, two lots will be $25,500, and so on. Continuing in this fashion we obtain Table 2.8.

Table 2.8. Cost (in dollars) to produce shirts in lots of 10,000.

Number of Lots	Total Cost ($)
0	17,500
1	21,500
2	25,500
3	29,500
4	33,500
5	37,500

We make two observations. First, arranged in this way our results form a data table for a sequence. The entries in the first column are position numbers, and those in the second column are the terms. Second, this sequence exhibits arithmetic growth. Each time we increase the number of lots by 1, the total cost increases by a constant amount, $4,000. We conclude that an arithmetic growth sequence should be a good model for this context.

Define Variables. Just as we elected to represent shirts in lots of 10,000, so too it will be convenient to represent monetary amounts in units of $1,000. We use the letter c for *cost* for the terms of the sequence. Accordingly, we define our variables as follows:

> Let c_n represent the cost to produce n lots of 10,000 shirts each, and express c_n in units of thousands of dollars.

Identify Assumptions. We assume that c_n is an arithmetic growth sequence. This is consistent with the table of values, reproduced below in units of thousands of dollars.

In graphing those values, we note that the individual points line up, further justifying the use of an arithmetic growth model (see Figure 2.8). Going beyond what is in the table or the graph, the problem context tells us that with each additional lot, c_n will increase by 4. That implies that the sequence terms do in fact grow arithmetically.

2.2. Applications of Arithmetic Growth

Table 2.9. c_n is the cost in thousands of dollars to produce n lots of 10,000 shirts each.

n	c_n
0	17.5
1	21.5
2	25.5
3	29.5
4	33.5
5	37.5

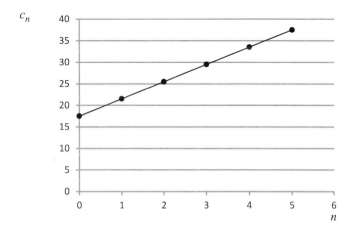

Figure 2.8. Individual points from the data table lie on a straight line.

We also assume that $n \leq 12.5$ because the factory can only produce 125,000 shirts per day.

Parameters. The arithmetic growth parameters are $c_0 = 17.5$ and $d = 4$.

Difference and Functional Equations. The difference and functional equations are
$$c_{n+1} = c_n + 4$$
and
$$c_n = 17.5 + 4n.$$

Diagrams, Tables, Graphs. We have already produced a table and a graph for this sequence. They show how the model behaves up to $n = 5$. This corresponds to 50,000 shirts, far less than the factory's 125,000 shirt capacity. To answer the questions posed in the problem statement, it might be informative to extend both the table and the graph to $n = 13$.

Questions About the Model. In the problem statement, question 1 asks for the cost of producing 90,000 shirts. That corresponds to 9 lots of 10,000, so the question becomes *In the model, what is c_9?* The second question asks how many shirts can be produced with a budget of $60,000. In the context of the model this becomes *For what value of n does $c_n = 60$?*

Find Answers. For question 1, applying the functional equation, with $n = 9$, we find
$$c_9 = 17.5 + 9 \cdot 4 = 53.5.$$

For question 2 we again use the functional equation. This time we set c_n equal to 60 and leave n as an unknown. That gives us
$$60 = 17.5 + 4n.$$
Using algebra to solve for n results in
$$n = (60 - 17.5)/4 = 10.625.$$

Translate Conclusions. For the first question, $c_9 = 53.5$ means a cost of $\$53,500$ to produce $90,000$ shirts. For the second question, $n = 10.625$. But n is the number of $10,000$ shirt lots. Therefore, the total number of shirts is $10.625 \cdot 10,000 = 106,250$ shirts.

Reflection and Reconsideration. For this problem we do not have specific data to work with. We have no way to assess the validity of the fixed and variable costs described in the problem statement. Within the context of the given information, an arithmetic growth model is appropriate.

Are our answers to the questions in the problem statement reasonable? Returning to the original data, to make $125,000$ shirts at 0.40 dollars per shirt would cost $\$50,000$, and with fixed costs of $\$17,500$, the total would be $\$67,500$. That is what it costs to operate the factory at maximum capacity. To produce only $90,000$ would cost less, so our figure of $\$53,500$ is at least in a reasonable range.

We can also verify our conclusions using alternate methods. For example, in Figure 2.9 we show an extended graph for the model. We can answer the stated questions approximately by reading this graph. It shows that c_9 is about 53, and that c_n reaches 60 for n around 10.5. These agree well with our earlier results of 53.5 and 10.625. Extending the data table provides another alternate method, but we leave that to the reader.

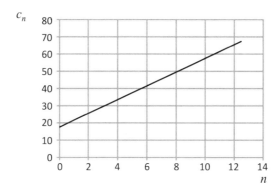

Figure 2.9. Extended graph for the t-shirt cost model.

Because n is in units of $10,000$, it makes sense to consider fractional values of n. For example, a single shirt is one ten-thousandth of a lot, so increasing the number of shirts by one corresponds to increasing n by 0.0001. This means that a four decimal place figure, such as 5.6871, corresponds to a whole number of shirts. On the other

2.2. Applications of Arithmetic Growth

hand, extending n to additional decimal places would be questionable, since it would require quantifying fractional parts of a single shirt.

Does the functional equation still apply when we allow non-whole number values of n? The given information specifies the variable cost per single shirt, which we used in the model to establish the cost of each lot of 10,000. This implies that the functional equation will remain valid for non-whole numbers n having 4 or fewer decimal digits.

The experience of working through this example shows that an arithmetic growth model can be applied in any analogous problem. As long as we agree to analyze costs in terms of a fixed component and a variable (per item) component, an arithmetic growth model will be feasible.

Additional Considerations. The three preceding examples illustrate how arithmetic growth models are developed and used. We have seen common features of the models, such as the use of difference and functional equations. We have also seen that each problem is a little different from the others, and that judgment and understanding are needed.

We conclude this section with some observations about functions and variables. Our goal is to recognize considerations from the examples that arise frequently in arithmetic growth models.

Functions. In any arithmetic growth model, an important role is played by the functional equation $a_n = a_0 + dn$. When the parameters a_0 and d are replaced with specific numbers, two variables remain, n and a_n. As we have seen, given a value for either variable, the functional equation allows us to determine the other.

In some cases the goal is to find a particular term a_n of the sequence. Thus, a value for n is given and we want to know the value of the corresponding a_n. These are called *find-a_n questions*. For example,

> What will the tuition be in 2020?

and

> What will it cost to produce 90,000 t-shirts?

are both find-a_n questions. In the context of the model, each question reduces to finding a_n for a specific n.

In contrast, when we are given the value of a_n and asked to find n, that is referred to as a *find-n question*. Both

> How long after death would it take for body temperature to reach 85°?

and

> How many t-shirts can be manufactured for $60,000?

are find-n questions.

In many models position numbers represent times, and a find-n question asks when something occurs. Thus, the first of the two preceding questions might be restated as *When does body temperature reach 85°?* In a similar way, whenever n represents time, a find-n question is also a find-*when* question. But keep in mind that n won't always indicate a time. In the t-shirt model, n indicates how many shirts are to be produced.

The distinction between find-n and find-a_n questions reflects an important aspect of working with functions. To illustrate, we momentarily put aside the idea of number sequences, and look at a familiar equation of the sort that is studied in algebra and precalculus classes:

$$y = x^2 + \frac{5}{x}. \tag{2.3}$$

Notice that the two variables do not play comparable roles here. If we replace x with a specific value, such as $x = 8$, we can immediately calculate y. The reverse is not true. If we replace y by 20 for example, a corresponding value of x cannot be immediately calculated. Indeed, it is not immediately obvious whether such a value even exists, or if one does exist, how to find it.

We use special terminology to emphasize the distinct roles played by x and y in (2.3). We say the equation defines y as a function of x, meaning that we can immediately compute y as soon as a value of x is given. Another way to say the same thing is this: y is determined by x.

The simplest way to use the equation is first to *choose* a value of x then *compute* the value of y. In choosing x we don't need to know y, so the choice of x is independent of y, and we call x the *independent* variable. When we compute the value of y, the result depends on the value of x, so we call y the *dependent* variable.

Notice that the dependent variable appears by itself on one side of the equation, whereas the independent variable appears in a more complicated algebraic combination on the other side of the equation. That might seem backward if you consider being *by itself* to be an aspect of *independence*. But that is incorrect. The terms *dependent* and *independent* refer to how we *use* each variable, not to the visual appearance of the variable in the equation. And while we can identify which variable is which based on appearance, we have to do so carefully. It is incorrect to say, *y is all by itself and therefore is independent*. The correct analysis is *y is all by itself, and therefore dependent, because the equation tells us how the value of y depends on the other variables*.

Starting with a numerical value for x and computing the corresponding y is referred to as *function evaluation*. We can often reverse the process, first choosing y and then trying to find corresponding values of x. This is referred to as *function inversion*.

This terminology can be used to distinguish between find-a_n questions and find-n questions using a functional equation. The functional equation expresses a_n as a function of n, because a_n appears by itself on one side of the equation while n appears in a more complicated algebraic combination on the other side. In other words, n is the independent variable and a_n is the dependent variable. To answer a find-a_n question, we are given the value of the independent variable (n) and want to find the dependent variable. This is a direct computation, and an instance of function evaluation. On the other hand, to answer a find-n question, we have to reverse the roles of the variables. This is an instance of function inversion, and generally involves more than direct calculation.

Continuous and Discrete Variables. When we work with a sequence in a model, we understand the position number n to be a whole number. It makes sense to refer to the third term a_3 or the fourth term a_4, but there is no three-and-a-half-th term. On the other hand, we have also seen that the position number can have a second interpretation that remains valid when n is not a whole number. For example, in the time of death model, we identify n with elapsed time, so that T_3 is the temperature

2.2. Applications of Arithmetic Growth

at a time 3 hours after death. In that context it makes sense to interpret $T_{3.5}$ as the temperature 3.5 hours after death. And as we argued in discussing this model, it is reasonable to assume that the hourly decrease in temperature, 1.5°, occurs uniformly during each hour. Thus, in half an hour the body should cool by 0.75°, so $T_{3.5}$ is 0.75° cooler than T_3. With this assumption, the functional equation gives correct values for T_n even when n is not a whole number.

But that sort of logic is not always valid. Consider the tuition model, where T_3 is the tuition for 2003 and T_4 is the tuition for 2004. To be more specific, T_3 is the tuition for the academic year starting in the fall of 2003. But what might $T_{3.2}$ mean? Since there is no academic year corresponding to 3.2, it is not clear what meaning, if any, $T_{3.2}$ could have. In this case, the model only makes sense for whole number values of n.

This leads us to the distinction between continuous and discrete variables. In the first example, where n can assume any numerical value, we say that n is a *continuous variable*. In the second example the value of n is restricted to be a whole number, and we say n is a *discrete variable*. Discrete variables need not be restricted to whole number values. For the t-shirt model n represents the number of 10,000 shirt lots produced by a factory. As discussed earlier, that means n can be a non-whole number with up to four decimal places. In other words, the only permitted values of n are $0, 0.0001, 0.0002, 0.0003, \cdots, 0.9999, 1.0000, 1.0001, \cdots$. This is another instance of a discrete variable n. Significantly, n has a specific first value, second value, third value, and so on, with no values in between. This is the key idea of a discrete variable: the permitted values are separate from one another. In contrast, for any value of a continuous variable, there is no *next* value. In the time of death model, we understand that T_2 is the temperature for a time of 2 hours. We can also conceive of the temperature after $2.1, 2.01$ or 2.001 hours. Thus we can understand the meaning of n equal to $2.1, 2.01, 2.001$, or any other decimal of the form $2.00\cdots01$. In this instance it is not possible to list the permitted values of n. Visually, they make up a continuous set of points on a number line, with no intervening gaps.

Interpreting n as a discrete variable has significant implications for find-a_n and find-n questions. To illustrate, let us consider the tuition model, recalling the functional equation

$$T_n = 3.981 + 0.799n.$$

Although the right-hand side of this equation can be computed for any value of n, in the model n only makes sense as a whole number. Accordingly, we should only ask find-T_n questions for whole number values of n. Although we can ask *What is T_n after 3.2 years?*, and compute $T_{3.2} = 3.981 + 0.799 \cdot 3.2 = 6.5378$, neither the question nor the answer is meaningful: there is no academic year corresponding to $n = 3.2$, so how can there be a tuition figure for $n = 3.2$? In general, for the tuition model, find-T_n questions are meaningful only when n is a whole number.

The situation for find-n questions is similar. To see why that is so, consider first $T_7 = 9.574$ and $T_8 = 10.373$. Because n can only be a whole number, it cannot have any value between 7 and 8. Likewise, the tuition cannot have any value between 9.574 and 10.373. We might ask *For what n is $T_n = 10$?* But there is no meaningful solution. Although the equation

$$3.981 + 0.799n = 10$$

is algebraically solvable, the solution $n = 7.533\cdots$ is not a whole number, and so is not an acceptable value for n in this model. Consequently, there is no solution to the question *In what year is the tuition equal to $10,000?*. On the other hand, suppose we ask *What is the first year that tuition is $10,000 or higher?* This time there is a perfectly valid answer: $n = 8$. Before $n = 8$ the tuitions are all less than $10,000, and for $n \geq 8$ the tuitions are all more than $10,000. This illustrates that questions about a model may need to be carefully formulated when n is a discrete variable.

The distinction between defining n to be continuous or discrete can be understood in relation to the graph of a_n. When n is discrete, the graph consists of individual points, separated by spaces, because the possible values of n are separated on the horizontal axis. This is shown in the graph for the library fine sequence (Figure 2.2 on page 45). In the graph, n is assumed to be a whole number of days, and each point of the graph shows the fine for a book returned some specific number of days late.

In contrast, we have seen that n has a meaningful interpretation as a continuous variable in the time of death problem. Conceptually, graphing points for all possible T_n will produce a continuous line. It will have no gaps because the possible n values cover the horizontal axis with no gaps. This is illustrated in the graph on page 61. The dots on the graph highlight the values of T_n for whole numbers n. But the other points on the line are also meaningful. They show how the temperatures T_n vary for values of n between consecutive whole numbers.

This graphical distinction between continuous and discrete variables is useful conceptually, but it is not always observed in practice. Sometimes, for convenience or emphasis we might draw a discrete model's graph as a continuous line, as we did in the graph of the tuition model on page 68. There, as discussed earlier, n is only meaningful as a whole number. Accordingly, on the graph, only the points of the line with whole number n-coordinates are meaningful. Nevertheless, we represented the model with a solid line to facilitate a visual comparison of the model and the data.

As a final variation on these ideas, let us look again at the library fine model, where f_n is the fine when a book is returned n days past the due date. Earlier we considered this as an example where n should be interpreted as a discrete variable. But now we ask, is there a way to interpret this when n is not a whole number? Here is one approach. Let us assume that the library has an after-hours book return facility, so that books can be returned at any time. And for simplicity, also assume that the library closes at the same time every day. If a book is returned before closing time on the date it is due, there is no fine. After that if the book is returned any time in the next 24 hours, that is, before closing time on the next day, the fine is 10 cents. If the book is returned any time in the next 24 hours, the fine is 20 cents. Thinking in this way, we can define n to represent the elapsed time, in days, between when the book was due and when it was returned. And because the book can be returned at any time, 24 hours per day, n need not be restricted to whole numbers. Defining f_n as the fine for a book returned n days past its due date makes as much sense when $n = 3.12$ as it does when $n = 3$.

Although our definition of f_n remains valid for fractional values of n, the functional equation, $f_n = 10n$, does not. For $n = 1.5$, this becomes $f_{1.5} = 15$. But that is not the fine the library charges for a book that is returned one and a half days late. On the first day after the due date, the fine would be ten cents any time up to closing. Then the fine jumps to 20 cents, where it remains for the next 24 hours. But a fine of 15 cents would never be incurred. The library rules do not assess a partial fine for a half day, or any other fraction. So in this model, we can permit non-whole number values of n conceptually, but we cannot use the functional equation unless n is a whole number.

2.2. Applications of Arithmetic Growth

We can recognize the same thing visually, by creating a graph based on the analysis in the preceding paragraph (Figure 2.10). The result is a series of horizontal lines, each

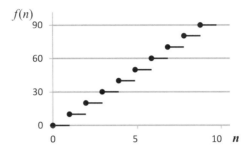

Figure 2.10. Graph for library fines when n is a continuous variable.

of which represents one 24-hour period. During the first 24 hours the fine is zero; for the next 24 hours the fine is 10 cents, and so on. The points corresponding to whole number values of n are depicted as dots. These are the same points that we originally saw in Figure 2.2.

For values of n between whole numbers, the correct values of $f(n)$ do not fill in the line connecting the dots. This shows that the correct graph of $f(n)$ is not a straight line, and therefore cannot be the graph of the functional equation $f(n) = 10n$. It also illustrates a general graphical fact. In an arithmetic growth model, when we use the functional equation with values of n that are not whole numbers, that is equivalent to assuming we can connect the individual points of the graph with a straight line. The library model shows that this is not always correct.

In most of the models you will encounter in this book it will be possible to interpret n as a continuous variable, and the functional equations will be equally valid for whole number and fractional values of n. However, as the examples above show, there are exceptions. In some cases, like the tuition model, it just will not make sense to consider n as a continuous variable. In others, like the library fine model, even though we can interpret n meaningfully as a continuous variable, the functional equation will not be valid unless n is a whole number. Unfortunately, there is no simple rule for deciding which situation holds in a particular model. Therefore, when you are developing and analyzing models, you should give some thought to whether it makes sense to treat n as a continuous variable.

Which is better, discrete or continuous? Throughout the book, we will study situations where the models are most easily formulated in terms of sequences and difference equations. That means we will begin with the idea of n as a whole number. But in most of these cases we will also derive a functional equation within which we will want to interpret n as a continuous variable. If that is feasible in the model context, formulating and answering find-a_n and find-n questions is simplified. If not, we have to be more careful analyzing the model.

2.2 Exercises
Reading Comprehension.

(1) There are various methods for analyzing data to create a model and using the model to make predictions. The methods presented in this section were categorized as numerical, graphical, and theoretical. Compare and contrast these types of methods. Your answer may be organized in table or paragraph form. For each method your answer should include at least a description, an example, some advantages, and some disadvantages.

(2) This section includes a list of common steps used to create a model. Which of these steps seems the most obvious? Which seems the most challenging? Briefly explain your choices.

(3) @In the time of death example the reader was challenged to identify a questionable assumption included in the model (see page 64). What is the questionable assumption, why might it be incorrect, and how might that affect the conclusions from the model? [Hint: Suppose death resulted from a severe bacterial infection.]

(4) @Suppose $a_n = 2n + 3$ and we want to find a_3. Is that a case of function evaluation or function inversion? Briefly justify your answer.

(5) @Questions about continuous and discrete variables.

   a. Describe continuous and discrete variables in your own words.

   b. @In a fish population model, p_n is the number of fish after n years, in units of thousands. Thus, the equation $p_2 = 3$ means that there are 3,000 fish after 2 years. In this model, would it make sense for n to be a continuous variable? A discrete variable? What about p_n?

   c. Suppose d represents an amount of money, in units of dollars, taking values such as 0.35, 199.99, and 1,000,000. Is d a continuous or discrete variable? Explain.

   d. Re-visit your descriptions for part a. Did they capture everything you thought about in parts b and c? If not, revise your descriptions.

Math Skills.

(6) @In each part a sequence is given, either numerically or graphically. For each one, tell whether the sequence is exactly, approximately, or not at all an example of arithmetic growth. Briefly explain your answer.

   a. @1, 4, 9, 16, 25, $\cdots$

   b. 7.7, 17, 26.3, 35.6, 44.9, $\cdots$

   c. @5, 8, 10, 12, 15, 17, $\cdots$

2.2. Exercises

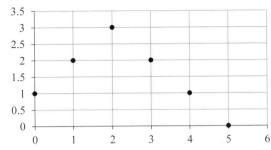
d.

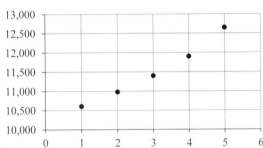
e.

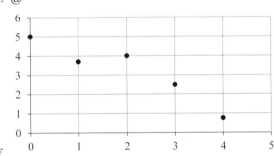
f.

(7) @For each equation below find n such that the statement is satisfied. Use either a numerical or theoretical approach. State which method you used.

   a. @$a_{n+1} = a_n + 6$ and $a_0 = 10$, find n such that $a_n = 34$
   b. @$b_n = 52 - 3n$, find n such that $b_n = 7$
   c. $c_n = 2(5 + 3n)$, find n such that $c_n = 40$
   d. $d_{n+1} = d_n - 8$ and $d_0 = 100$, find n such that $d_n = 20$
   e. $e_n = 1 + 7n$, find n such that $e_n = 57$

Problems in Context.

(8) @Battery Power. A weather balloon carries a battery-powered radio transmitter which sends weather data back to the ground. When the balloon is sent up, the battery carries a charge of 30 units. It uses up 2.4 units of charge per hour. Let q_n represent the charge on the battery n hours after the balloon is sent up.

   a. @Using a numerical method, find q_1, q_2, and q_3.
   b. @What is the difference equation for q_n?
   c. @What is the functional equation for q_n?
   d. @What will the charge be 4 hours after launch?

e. @The radio transmitter cannot continue to work once the charge on the battery falls below 4 units. How many hours will that take?

(9) **Wetlands.** In 2011 the US Department of Interior released a five-year survey on wetlands and found that from 2004 to 2009 America's wetlands declined slightly. The report says, "The net wetland loss was estimated to be 62,300 acres between 2004 and 2009, bringing the nation's total wetlands acreage to just over 110 million acres in the continental United States, excluding Alaska and Hawaii" [**52**]. Let's suppose that in one county the number of acres of wetlands has been decreasing by about 200 acres per year for the past several years and there were 36,000 acres of wetlands in 2014. An interested ecologist uses the notation w_n for the number of acres of wetlands n years after 2014. Develop an arithmetic growth model for this problem as follows:

  a. Formulate the arithmetic growth difference equation for w_n.
  b. Formulate the initial value and any restrictions on n.
  c. Create a graph for the model. Your graph may be produced with technology and printed or created by hand using graph paper and plotting the points as accurately as possible.
  d. Formulate the functional equation for w_n.
  e. Use the model to predict how many acres of wetlands will remain in the county in the year 2050.
  f. Use the model to determine when there will be only 25,000 acres of wetlands in the county.
  g. Reflect on your conclusions. Do they seem plausible? What other considerations should be taken into account? What other questions could this model be used to answer? How might we want to extend the model?

(10) **@Housing Cost.** An economist is studying the cost of housing. In one county, the data show that despite some ups and downs the average price of a new home increased by about $2,000 per year from 2004 to 2010, starting at $135,000 in 2004. The economist uses the notation h_n for the average price of a new home n years after 2004. Develop an arithmetic growth model for this problem by completing the steps below.

  a. @Formulate a difference equation for h_n. State the corresponding initial value and any restrictions on n.
  b. @Formulate a functional equation for h_n.
  c. @Create a graph for the model.
  d. @Use the model to predict the average home price in the year 2025.
  e. @Use the model to predict when the average home price will reach $250,000.
  f. @Reflect on your conclusions. Do they seem plausible? What other considerations should be taken into account? Does it seem plausible that the price of a new home would increase consistently every year? What other questions could this model be used to answer? How might we want to extend the model?

(11) **Auto Theft.** According to the San Antonio, Texas, Uniform Crime Report there were 5,893 cases of auto theft in 2011 and 6,577 cases of theft in 2013 [**46**]. Using c_n to represent the number of car thefts n years after 2011, develop an arithmetic growth model for this situation.

2.2. Exercises

a. What is the difference equation and initial value for the model?

b. What is the functional equation for this model?

c. According to the model, how many car thefts would have occurred in 2014?

d. According to the model, when will the number of car thefts in a year first exceed 10,000?

e. Reflect on your conclusions. What other considerations should be taken into account? What other questions could this model be used to answer? How might we want to extend the model?

(12) @Cooling System Malfunction. On board a UFO, far out in space, the cooling system for the hyper-drive develops a malfunction. The engineers studying the problem estimate that temperature is rising at 2.8 degrees per hour. The temperature was 148 degrees at 9 AM Standard Galactic Time on Sunday. If the temperature reaches 200 degrees, the reactor will have to be shut down. Applying appropriate steps from the modeling outline on page 58, develop an arithmetic growth model for this situation and use your model to answer the following questions. You may use any combination of graphical, numerical, and theoretical methods.

a. @What will the temperature be by 4 PM Standard Galactic Time on Sunday?

b. @According to the model, when will the reactor have to be shut down?

c. Reflect on your conclusions. Do they seem plausible in the UFO context? What other considerations should be taken into account? What other questions could this model be used to answer? How might we want to extend the model?

(13) Spread of Disease. A scientist studying the spread of a new disease in a small town decides to use an arithmetic growth model. She estimates that 3,700 people have the disease at the start of her study, and that there are 45 new cases each day.

a. Applying appropriate steps from the modeling outline on page 58, develop an arithmetic growth model for this situation, and use your model to predict when the number who are or have been infected will reach 10,000.

b. The scientist has also found that about 3 percent of the people who get the disease require treatment with a special medicine. Before the outbreak, the local hospital had a limited supply of the medicine, amounting to 500 doses. This was provided to patients as necessary both before, and during, the study. According to the model, how long after the start of the study will the special medicine last?

c. Suppose that the small town in the study is isolated—very few people arrive or leave. Given what you know about the way diseases spread, do you think an arithmetic growth model is reasonable? Consider both predictions made over a short period of time, and those over much longer periods of time.

d. How would your answer to the previous question change if the town were in a popular tourist area? Would it matter how long the tourists stay in the town?

Digging Deeper.

(14) @At the end of Example 2: College Tuition, on page 69, the assertion is made that, "with any given set of data, if we define a_0 to equal the first data point exactly, and define d to be the average of all the differences between consecutive data values, then the model will agree exactly with the last data point as well." Think about this assertion, try a few examples of your own, and convince yourself it is a true statement. Write an explanation of your work that would help elucidate the ideas for a classmate struggling with them. If possible, use algebra to show that the statement is true.

(15) Consider again the "Spread of Disease" exercise about the epidemic in a small town. Modify the previous discussion as follows. Suppose that the spread of the disease is increasing, and that the researcher observes 45 new cases on the first day of the study, 50 new cases on the next day of the study, and 55 new cases the day after that. She models the number of *new* cases per day using an arithmetic growth model. Here, let c_n be the number of new cases of the disease in day n of the study. So $c_1 = 45$, $c_2 = 50$, and so on. Create a model for the number of new cases each day, then use numerical methods to figure out how many days the hospital's supply of medicine will last.

(16) @A modified version of the library fine model was discussed on page 76. In that discussion, it was determined that the amount of the fine could be calculated using $f(n) = 10n$ when n is a whole number. The value of the expression $10n$ is defined for fractional values of n, but as shown in Figure 2.10, the graph for the amount of the library fine is a collection of line segments, not a single line. Therefore the equation $f(n) = 10n$ is not valid when n is not a whole number. If we decide to use the equation anyway, how large an impact might that have on the results? That is, what is the largest possible error that can arise in predicting a library fine if the equation $f(n) = 10n$ is used for a value of n that is not a whole number?

2.3 Linear Functions and Equations

In Section 2.1 we found that the graphs of arithmetic growth models always appear as straight lines. We also found arithmetic growth models always have functional equations that can be expressed in a form similar to

$$g_n = 3{,}100 - 4.35n. \qquad (2.4)$$

This is an example of a *linear equation*, so called because the graph is a straight line. It is a variant of the familiar form $y = mx + b$ studied throughout the mathematics curriculum. In this section we will study linear equations in depth, covering ideas connected with functions, graphs, and solving equations.

Recall that (2.4) is the functional equation we found in the satellite fuel example. This equation expresses g_n as a function of n because it permits us to compute a value of g_n as soon as we know n. To be more specific, we refer to this as a *linear* function because the functional equation is a linear equation.

A word is in order here about terminology, and in particular about the terms *equation, expression,* and *function*. It is a common error to use the word *equation* for just

2.3. Linear Functions and Equations

about anything that can be written down with algebraic symbols. Many students refer to all of the following as equations,

$$y = 3x + 4 \qquad 3x + 2 \leq 5y \qquad 7x^3 + 5xy - 2$$

though that is only correct for the first. The middle example should properly be called an inequality, while the third is an expression. An expression, having no equal sign or inequality, can be thought of as a recipe for computing something. Indeed, if you replace every variable in an expression with a number, and perform the indicated operations, you end up with a number. Looking at a particular linear example, $y = 3x+4$ should be called a linear *equation,* while $3x + 4$ should be called a linear *expression.*

Both equations and expressions are closely connected to the concept of *function.* Throughout this book, the convention will be to use *expression* and *function* more or less interchangeably, and to use *equation* to emphasize the idea that there is an equal sign present. To illustrate this distinction using the earlier examples, $y = 3x + 4$ will be called a linear *equation,* while $3x + 4$ will be called a linear *function.*

It is significant in the examples above that once we establish a numerical value for x, there is only one correct corresponding value of y. Stated more succinctly, each x corresponds to a *unique* y.[7] This is the defining characteristic of the function concept in theoretical mathematics. An equation like those above, with one variable (y) equal to an expression involving a second variable (x), almost always has the uniqueness property, because substituting numbers for variables in an algebraic expression almost always produces a unique value. An exception is an equation such as

$$y = \pm\sqrt{x^2 + 1},$$

where the $\pm$ sign indicates that y may assume two different values. Although there are other exceptions in more advanced areas of mathematics, we will not encounter any in this book. As a general rule, any equation of the form

$$y = \text{expression involving only } x$$

defines y as a function of x, so long as no $\pm$ sign appears in the expression.

So far, linear equations and functions have been discussed only in connection with specific examples. Let us proceed to a definition of the concept of linearity. Conceptually, this can be approached in several ways. Our first approach is based on the mathematical operations used to perform a calculation. When we use (2.4) to compute a_n, what operations are performed? The variable n is multiplied by one constant, and added to another. Generalizing slightly leads to the following first definition of linear function.

> **Operational Definition of Linearity:** Linear functions and linear equations involve just two types of operations:
>
> (1) addition or subtraction, and
>
> (2) multiplication by a constant or division by a constant.

For example, consider the operations that have to be performed in

$$z = 3(x - 7y) + w/5.$$

---

[7]The mathematical meaning of unique is *one and only one.* In contrast, in every day usage *unique* is sometimes understood to mean *distinct,* i.e., all different. For a function, there can be only one y for each x, but there is no requirement that different x's have different y's.

In each multiplication at least one of the two factors is a number, and the only division that appears is division by 5, a number. There are also several additions and subtractions. This is a linear equation. It defines z as a linear function of x, y, and w. In contrast, the following examples are not linear equations:

$$xy + 2 = z$$
$$x^2 + y^2 = z^2$$
$$y = -2 + \sqrt{4 - 3x}$$
$$y = \frac{3x + 4}{4x - 3}$$

In each of these there are operations that are neither addition or subtraction, nor multiplication by a constant or division by a constant. For example, in the second equation there appears an x^2. This is an abbreviation for $x \cdot x$ and so requires multiplication of two non-constant quantities. You should try to decide why each of the other equations is not a linear equation.

The preceding definition establishes one conceptual meaning of linear functions and equations. It is also useful to have a rule with which we can distinguish a linear equation at a glance. In this book we will usually be concerned with a linear function depending on a single variable. For such linear functions, we have a second, more algebraic, definition.

> **Algebraic Definition of Linearity:** We say y is a linear function of x if it can be expressed in the form:
> $$y = (\text{a constant}) \cdot x + (\text{a constant}). \qquad (2.5)$$

We call this the *standard form* of a linear equation, though it is also commonly referred to as the *slope-intercept form*. Any equation that can be expressed in the standard form is linear.

There are no restrictions on the values of the constants. They can be positive, negative, or zero; integers, fractions, or decimal expressions. They can even be numbers like $\sqrt{2}$ or π, which have specific values, but cannot be represented by a finite or repeating decimal.

Here are some examples. The equation

$$y = 1.3x + 8$$

expresses y as a linear function of x. It has the exact form of (2.5) with 1.3 as the first constant and 8 as the second. The choice of specific letters to use as variables is not significant in defining a linear function. Thus

$$w = 1.3t + 8$$

defines w as a linear function of t. Similarly, the satellite model functional equation

$$g_n = 3{,}100 - 4.35n$$

is not in the standard form. But rewriting it as

$$g_n = -4.35n + 3{,}100$$

shows that g_n is a linear function of n.

2.3. Linear Functions and Equations

It is permitted for the two constants in (2.5) to be equal, so
$$y = 1.3x + 1.3$$
is a linear equation. What about this equation:
$$y = x?$$
This is algebraically equivalent to
$$y = 1 \cdot x + 0$$
so even though $y = x$ is not in exactly the same form as (2.5), it can be expressed in that form, and is therefore a linear equation.

This last example illustrates an important point. Algebra permits us to express a given equation in many different forms.[8] In some cases, the form in which an equation appears makes it difficult to recognize whether it is linear. Consider
$$y = (5x - 1)(2x + 3) - 10(x^2 + 3x).$$
This is certainly not in the same form as (2.5). Can it be expressed in the same form, by using algebra? It is hard to tell just by looking at the equation. With a few steps of algebra, the equation can be expressed as
$$y = -17x - 3,$$
so the equation is linear. But that is not at all obvious from the original form of the equation.

The algebraic definition of linearity provides a standard form for a linear equation. In whatever form an equation originally appears, if we can put it into the standard form, we know it is linear. The validity of this idea depends on the two forms of the equation being perfectly *equivalent*. But what does that mean? The answer depends on what happens when we replace the variables with numbers, because we think of variables as representing unknown numbers. With that in mind,
$$y = x$$
and
$$y^2 = x^2$$
are not equivalent. To see this, consider replacing x by 2 and y by -2. Then the first equation becomes a false statement, but the second is true. This shows that the two equations do not produce compatible results for all possible values of the variables. Conversely, two equations or expressions are equivalent if they both produce the same result whenever we replace the variables with numbers.

The preceding example shows that some care is required in changing an equation from one form to another. In particular, squaring both sides does not necessarily produce an equivalent equation. Later we will go into more detail about valid algebraic operations for changing an equation to a different form. We will also see that there are several different standard forms for linear equations, and why each is useful.

Straight lines in the xy plane are represented by linear equations. Sometimes we are interested in lines that are defined graphically or geometrically. For example, in Figure 2.11 equilateral triangle OAB stands on the x-axis with point $O = (0, 0)$ and

---

[8]In fact, an equation can be expressed in infinitely many different forms. For example, $y = 2x$ can be expressed as $y = 2(x - 1) + 2$, or $y = 2(x - 2) + 4$, or $y = 2(x - 3.04) + 6.08$, etc.

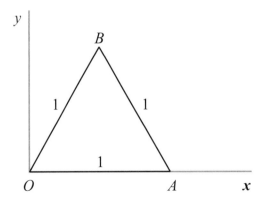

Figure 2.11. The sides of the triangle are represented by linear equations.

point $A = (1, 0)$. Each side of the triangle is part of a straight line, and so has a linear equation. Methods for finding the equations will be presented later. Without going into the details now, the equation of side AB is $y = -\sqrt{3}\,x + \sqrt{3}$. For side OB the equation is $y = \sqrt{3}\,x$, which can also be written in the standard form as $y = \sqrt{3}\,x + 0$. As these equations show, constants like $\sqrt{3}$ can arise in quite simple geometric arrangements. The base of the triangle is part of the x-axis. That has the equation $y = 0$. Can this be written in the standard form? Yes: $y = 0 \cdot x + 0$.

Algebra and Solving Linear Equations. We turn now to methods for solving linear equations. As we saw in Section 2.3, variables in models can be either continuous or discrete. For greatest simplicity, except where noted otherwise we assume the variables in the following discussion are all continuous. The methods we consider can still be applied for discrete variables, but the conclusions must then be interpreted carefully.

Consider again our satellite fuel model, where the functional equation

$$g_n = 3{,}100 - 4.35n$$

relates a satellite's fuel supply g_n to the number of months n the satellite has been in operation. We can use this equation directly to find g_n for any given value of n. That is called evaluating the expression $3{,}100 - 4.35n$ for a particular n. We can also assign a value to g_n and ask for the corresponding n. For example, to find when the fuel will be half gone, we can set g_n equal to 1,550, thus obtaining the equation

$$1{,}550 = 3{,}100 - 4.35n.$$

Here the task is to find the value (or values) of n for which this becomes a true statement. This is referred to as *solving* the equation.

As we discussed earlier, there are several methods that can be used to solve an equation. For example, we can use a numerical approach such as systematic trial and error, or a graphical approach to find at least an approximate answer visually. But generally the most efficient and exact approach is to use algebra. We touched on this topic briefly on page 50. Let us now look at the algebraic process of solving equations in more detail.

2.3. Linear Functions and Equations

For the particular case of linear equations, there is an algebraic method that always applies. Usually a linear equation has exactly one answer, although it is also possible for there to be no solutions, or infinitely many. The algebraic method tells us which of these cases holds, and leads us to the solutions, if any.

As already mentioned, we usually apply algebra to replace one equation (or expression) with an equivalent one. For linear equations, the necessary operations can be categorized as follows. You can

- algebraically rearrange either side,
- add or subtract the same thing on both sides, or
- multiply or divide both sides by the same thing (but not zero.[9])

In the first category we include operations like factoring, reducing fractions, and using the distributive law to remove parentheses. It has been proven that the operations in the list always change a given equation into an equivalent equation.[10]

Using these operations, our goal is to reduce a given equation to the simple form *variable* = *number*. Here is a somewhat involved example:

$$2(3x - 5) + 6 = 2 + 4x - 2(x - 4)$$
$$6x - 10 + 6 = 2 + 4x - 2x + 8$$
$$6x - 4 = 2x + 10$$
$$6x - 4 - 2x = 2x + 10 - 2x$$
$$4x - 4 = 10$$
$$4x - 4 + 4 = 10 + 4$$
$$4x = 14$$
$$4x/4 = 14/4$$
$$x = 7/2.$$

In this example, each step has been done separately to emphasize what operations are being performed. You should take a few minutes to study the equations, figuring out what is done at each step.

When you solve an equation for yourself, you may be able to do several steps at once. That can shorten the process significantly.

No Solution; Every Number a Solution. It is possible to have linear equations without any solution, or for which every number is a solution. Although these equations rarely occur in practice, it is not always obvious when they do occur. For example, the following equation looks like a typical linear equation

$$3x + 2 = 5 + 3(x - 4).$$

A few steps of algebra lead to the equation

$$3x + 2 = 3x - 7.$$

---

[9]The exclusion of zero must be kept in mind when multiplying or dividing by a variable expression which might equal zero for some values of the variables. For example, dividing both sides of $x^2 = 2x$ by x is not valid when $x = 0$, a possibility we have to consider separately as a special case.

[10]Recall that for other operations, including squaring both sides of an equation, the new equation and the original can fail to be equivalent. This in turn can lead to spurious solutions or hide legitimate ones. That won't happen if we only use operations in the list.

This is an impossible situation and is never true: no matter what x is, the left-hand side will be greater than $3x$, while the right-hand side will be less than $3x$. So the original equation has no solution. On the other hand, this equation

$$2 - 4x = 4(3 - x) - 10$$

is true for every value of x. Try a few. If you make $x = 0$ the equation becomes $2 = 12 - 10$, which is certainly true. What do you get if $x = 1$? $x = -1$? In all these cases the resulting equation is true. Now no number of examples can *prove* that this equation is true for every x. However, this conclusion *can* be reached by using algebra. Applying the same kind of steps as in the preceding examples, we can reach the equation $2 - 4x = 2 - 4x$. Since this is true for all values of x, so is the original equation. As already stated, equations of this type rarely actually occur in practice, although they do appear once in a while, especially as a result of an error. It is important to understand what they mean when they do occur. Then, if an error has been made, it will be easier to spot.

More than One Variable. So far the examples have all involved just one variable. The same kinds of operations can be applied in equations with more than one variable. In that case, it is often useful to simplify the equation to a form with one of the variables isolated on one side of the equation, and all other terms on the other side of the equation. As an example, we will again use the satellite model

$$g_n = 3{,}100 - 4.35n.$$

Using algebra, we can isolate n on one side of the equation:

$$g_n = -4.35n + 3{,}100$$
$$g_n - 3{,}100 = -4.35n$$
$$\frac{g_n - 3{,}100}{-4.35} = n$$
$$n = -\frac{g_n - 3{,}100}{4.35}$$

This process is described as solving for n in terms of g_n. Notice that the final equation gives n as a function of g_n, because as soon as the value of g_n is substituted, the value of n can be immediately computed. Solving the equation for n in this way is sometimes described as *inverting* the original equation, the one giving g_n as a function of n. The new equation for n as a function of g_n is useful for answering find-n questions. For example, to find n when the fuel has been reduced to 1,000 grams, compute

$$n = -\frac{1{,}000 - 3{,}100}{4.35} = 482.76$$

(to two decimal places).

In the preceding example, it makes the most sense to think of n as a continuous variable. In this case, we usually use parentheses rather than subscript notation. Thus, the functional equation for g_n becomes

$$g(n) = 3{,}100 - 4.35n$$

and the inverse equation can be written

$$n(g) = -\frac{g - 3{,}100}{4.35}.$$

2.3. Linear Functions and Equations

This notation emphasizes how one variable depends on another. Thus, in the first equation, we write $g(n)$ to stress that g depends on n. Conversely, we write $n(g)$ in the second equation because there n depends on g. When this kind of emphasis is not necessary, we can also write the equations as

$$g = 3{,}100 - 4.35n$$

and

$$n = -\frac{g - 3{,}100}{4.35}.$$

Here it is implicit that g depends on n in the first equation, and n depends on g in the second.

Graphs of Linear Equations. An equation with two variables, like those in the preceding example, can be represented by a graph. For a linear equation the graph will always be a straight line. Geometric features of the line, such as where it crosses an axis, are closely connected to algebraic features of the equation. Understanding the connections allows us to pass easily from a given graph to its equation, or from an equation to the graph. We will explore these connections presently, focusing on three standard forms for linear equations. But first we look at two geometric features of graphs.

Intercepts. The variables in an equation dictate the axes that appear in the graph. By convention, if the variables are x and y, the vertical axis is the y-axis and the horizontal axis is the x-axis. Points where a line or curve cross the axes have special significance. Those on the x-axis are called *x-intercepts*, and those on the y-axis are called *y-intercepts*.

Because an intercept lies on an axis, one of its coordinates must be zero. Specifically, an x-intercept has $y = 0$ and a y-intercept has $x = 0$. Visually, such points are easily identified on a graph. This is illustrated in Figure 2.12. The curve on the left has a y-intercept at point B and x-intercepts at points A, C, and D. The graph on the right is a straight line. It has one y-intercept at B and one x-intercept at A.

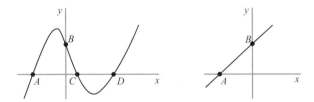

Figure 2.12. Marked points showing x- and y-intercepts for a curve and a straight line.

Intercepts are also algebraically significant, especially if we think in terms of functions. For example, suppose $y = f(x)$ is a function of x. Then the x-intercepts are points where the function has a value of zero. To find them we have to solve the equation $f(x) = 0$. In a number sequence model this corresponds to a find-n question.

Similarly, the y-intercept occurs when we take $x = 0$. It can be found by direct evaluation of the function equation. In a number sequence model, the y-intercept is the same as the initial value a_0, and answers a find-a_n question.

Although intercepts are of interest for any sort of curve, here we are particularly concerned with linear equations and straight lines. For example, we know from the algebraic definition of linearity that $y = 2x + 3$ is a linear equation, and hence the graph is a straight line. The y-intercept is 3, as we can verify by substituting 0 for x. In general, for an equation in the form

$$y = (\text{a constant}) \cdot x + (\text{a constant}) \tag{2.6}$$

the second constant is the y-intercept. This is illustrated in Figure 2.13 for the lines $y = 3x + 5$ and $y = 3x - 5$.

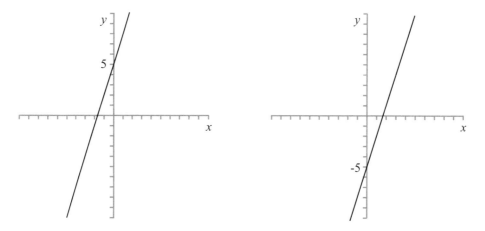

Figure 2.13. Graph for $y = 3x + 5$ on the left and $y = 3x - 5$ on the right.

Slope. The second geometric feature we consider is referred to as the slope of a straight line. The slope quantifies the steepness of the line. More specifically, the numerical value of the slope tells how much the line moves up or down for every unit you move to the right. For the equation $y = 3x + 5$, the slope is 3. This indicates that the line moves up 3 units for every unit you move to the right. For the line $y = -3.2x + 5$, the slope is -3.2 so the line moves *down* 3.2 units for each unit to the right. Both of these are shown in Figure 2.14. Whenever we express the equation of a line in the form of (2.6), the first constant is the slope.

The slope can be computed between any two points on a line. For example, consider the points $(1, 3)$ and $(6, 7)$. As shown in Figure 2.15, we can go from the first point to the second by traveling up 4 units (referred to as the *rise*) and then to the right 5 units (referred to as the *run*). The slope is the ratio, 4/5, often expressed verbally as *the rise over the run*. This can be computed in a single formula as

$$\frac{\text{rise}}{\text{run}} = \frac{7-3}{6-1} = \frac{4}{5}. \tag{2.7}$$

The slope of a line is usually represented by the letter m, as in the familiar equation $y = mx + b$. Extending the prior example, we can compute the slope between two points (x_1, y_1) and (x_2, y_2) using

$$m = \frac{y_2 - y_1}{x_2 - x_1},$$

2.3. Linear Functions and Equations

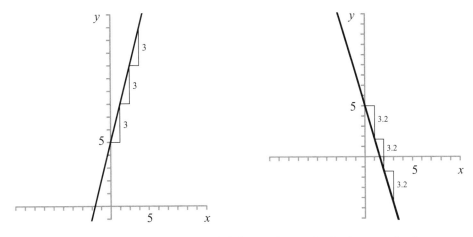

Figure 2.14. The line on the left has a slope of 3; it rises 3 units for every unit moved to the right. The line on the right has a slope of -3.2; it falls 3.2 units for every unit moved to the right.

which is known as the *slope formula*.[11] It is not valid if $x_1 = x_2$ because division by zero is undefined. Consequently, the slope of a vertical line cannot be computed, and is also said to be undefined.

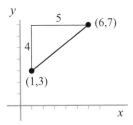

Figure 2.15. The slope between two points is *the rise over the run*.

Moving from one point to another, the rise can be positive, negative, or zero. It is negative when the second point is below the first: moving between the points we experience a negative rise, or a fall. Similarly, the run can be positive, negative, or zero, with a positive value for motion to the right and a negative value for motion to the left. Depending on the signs of the rise and the run, the slope can be positive, negative, zero, or, as noted before, undefined when the run is zero. A line with positive slope extends upward to the right. A line with negative slope extends downward to the right. A slope of 0 indicates a horizontal line. An undefined slope occurs for a vertical line. As an example of a negative slope, consider the line from $(1, 6)$ to $(4, 2)$. Using the slope formula we find $m = -4/3$, which is negative. If you graph these two points you can verify that they define a line extending downward to the right.

---

[11] The use of subscripts in the slope formula is similar to our earlier use for numbering the terms of a sequence. In this case, because we are dealing with two points, there will be two x-coordinates and two y-coordinates. The subscript notation helps us remember the meaning of each variable in the formula: x_1 is the x-coordinate of the first point, y_2 is the y-coordinate of the second point, etc.

When we compute the slope using two points, it does not matter which we take as the first point. As an illustration, let us recompute the slope we found in (2.7), with the order of the points reversed. That is, take $(6, 7)$ as the start point and $(1, 3)$ as the end point. Then the slope is given by

$$\frac{y_2 - y_1}{x_2 - x_1} = \frac{3 - 7}{1 - 6} = \frac{-4}{-5} = \frac{4}{5}.$$

Notice that the signs changed for both the numerator and denominator of the fraction. By the rules of equivalent fractions, we get the same answer as before. Verbally, going up 4 and right 5 creates the same slope as going *down* 4 and *left* 5.

Slopes can be interpreted in more than one way. For the preceding example, where the slope is 4/5, we can say that the line rises by 4/5 of a unit for every unit moved to the right. That is consistent with how we defined slope. But it is also true that the line rises 4 units for every 5 units to the right, just as it does between $(1, 3)$ and $(6, 7)$. In fact, something similar holds for any other fraction equal to 4/5. For example, since $12/15 = 4/5$, we can say that the line rises 12 units for every 15 units moved to the right, as shown in Figure 2.16. Interpreting a slope in this way provides a helpful link between the geometric meaning and the numerical value expressed as a fraction. The verbal expression of the fraction 4/5 as *four over five*, can be considered as directions for moving along the line: go *UP four, OVER five*.

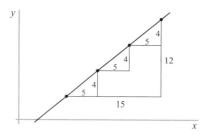

Figure 2.16. The slope of the line is $4/5 = 12/15$. The line rises 4 units for every 5 units moved to the right, and also rises 12 units for every 15 units to the right.

The Slope-Intercept Form. We proceed now to the first of three special forms for a linear equation, namely, the slope-intercept form.

Slope-Intercept Equation of a Line:
$$y = mx + b. \qquad (2.8)$$

Actually, this is just the standard form of a linear equation by another name. In

$$y = (\text{a constant}) \cdot x + (\text{a constant})$$

representing the constants as parameters m and b produces (2.8).

As explained in the preceding discussion, m is the slope of the line, and b is the y-intercept. That is why (2.8) is called the slope-intercept form of the equation. Knowing the slope and intercept immediately leads to a visual understanding of the graph. Begin at the point b on the y-axis, and proceed on a diagonal moving up m units for every

2.3. Linear Functions and Equations

unit you move to the right.[12] Indeed, associating b with *begin* and m with *move* is one way to remember which parameter is which.

You can generate a sequence of points following the recipe *up m over* 1 repeatedly. For a numerical example, if the equation is $y = 3x + 5$, begin at 5 on the y-axis, and generate the points by moving up 3 and over 1 several times. This is the pattern we saw earlier in Figure 2.14. As a second example, for $y = \frac{4}{5}x - 3$, we can use the fractional form of the slope as illustrated in Figure 2.16. Thus, we would begin at -3 on the y-axis, and construct points by moving up 4 and over 5 repeatedly.

There is a close connection between this approach to graphing a linear equation and an arithmetic growth sequence, as revealed by the following example. Consider a sequence that starts with $a_0 = 5$, and proceeds through repeated additions of $d = 3$. In a list of terms, $5, 8, 11, 14, 17, \cdots$, each term is 3 more than the preceding term. Therefore, when we graph the individual terms of the sequence, we begin with 5 on the y-axis, and each new point is one unit to the right and 3 units above the preceding point. But this is exactly the same process we used to graph the line $y = 3x + 5$. In this way we see that the y-intercept of the equation is the same as the initial term of the sequence, and the slope of the equation is the same as the added amount of the sequence. In other words, the linear equation $y = mx + b$ and the arithmetic growth sequence $a_0, a_0 + d, a_0 + 2d, \cdots$ are essentially the same, with $a_0 = b$ and $d = m$. The same conclusion is reached when we put the functional equation $a_n = a_0 + dn$ into slope-intercept form $a_n = dn + a_0$. This expresses a_n as a linear function of n, with slope d and y-intercept a_0.

We have seen how the slope-intercept form of a linear equation provides an immediate visualization of the graph. It is also useful for the reverse problem: determine the equation given the graph. For example, what is the equation of the slanting line shown in Figure 2.17? It appears in the graph that the y-intercept is 2. From that point, we

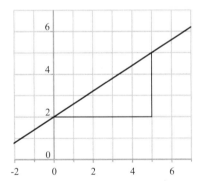

Figure 2.17. What is the equation of the slanting line?

can reach another point of the line by moving 5 units to the right and 3 units up, as indicated in the figure. But that is the same as moving up 3 units and over 5 units, so the slope is 3/5. The equation can therefore be immediately written down as $y = \frac{3}{5}x + 2$.

---

[12]This is literally true when $m \geq 0$. It is also true for $m < 0$ if we consider a negative upward movement as going down. For example, if $m = -1$, we interpret *go up* -1 *units* to mean *go down* 1 *unit*.

The Point-Slope Form. One limitation of the slope-intercept form is that it is sometimes inconvenient to show the y-intercept of a line in the graph. An example is shown in Figure 2.18. Notice that the axes have not been drawn in the usual locations. That

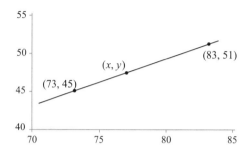

Figure 2.18. The axes in this graph are not where $x = 0$ or $y = 0$, and the y-intercept of the line cannot be read off the graph.

is, they do not meet at the 0 point of each axis. This is often done when the points to be displayed are in a cluster far away from $(0, 0)$. Remember that our notion of y-intercept assumes that x is 0 on the vertical axis. Therefore, on this graph it is an error to think of the y-intercept as the point where the line crosses the vertical axis. Instead, the true y-intercept, and the y-axis itself, would actually be far to the left of the points shown in the figure.

It is still possible to determine the equation of the line in a simple way. First, using the two points shown, $(73, 45)$ and $(83, 51)$, the slope is easily computed. Either observe that the graph goes over 10 (from 73 to 83) and up 6 (from 45 to 51), and compute the slope as the rise over the run, or use the slope formula

$$\frac{y_2 - y_1}{x_2 - x_1} = \frac{51 - 45}{83 - 73}.$$

Either way, we find the slope as 6/10 or 0.6. Now imagine another point on the line at an unknown location (x, y). Let us recompute the slope, going from $(73, 45)$ to (x, y). This time the slope is $(y - 45)/(x - 73)$. Since we already know the slope is 0.6, we conclude that

$$\frac{y - 45}{x - 73} = .6$$

or

$$y - 45 = .6(x - 73).$$

This is referred to as the *point-slope* form of a linear equation. It allows one to write down the equation for a line as soon as the slope and one point of the line are known. In this case, we know the slope is 0.6 and one point of the line is $(73, 45)$.

Although we depended heavily on a graph to derive the point-slope form, it can be applied without referring to a graph. This is shown in the following example.

Problem: A straight line passes through the points $(2, 10)$ and $(11, 3)$. Find an equation of the line.

Solution: Using the given points, we compute the line's slope as

$$\frac{3 - 10}{11 - 2} = -\frac{7}{9}.$$

2.3. Linear Functions and Equations

Applying the point-slope form, we can immediately write

$$y - 10 = -\frac{7}{9}(x - 2).$$

Note that the solution gives us a valid equation for the line, without any need to create or refer to a graph. Also, we used $(2, 10)$ as one known point of the line, so its coordinates appear directly in the equation. It would have been equally correct to use the point $(11, 3)$, in which case we would have found

$$y - 3 = -\frac{7}{9}(x - 11).$$

The two equations are equivalent, and can be simplified to the same slope-intercept form.

The general form of the point-slope equation is usually written in the following standard form.

Point-Slope Equation of a Line:

$$y - y_0 = m(x - x_0). \tag{2.9}$$

In this equation, x_0, y_0, and m are parameters. As usual m represents slope, and (x_0, y_0) represents any specific point on the line. When we use the point-slope equation there will be numerical values for these parameters, just as in the example m is $-7/9$ and (x_0, y_0) is $(2, 10)$ or $(11, 3)$ (or any other point of the line). The variables x and y should be thought of as representing many different points all on the line.

Like the slope-intercept form, the point-slope form can be used to visualize the graph given the equation. For example, in $y - 5 = 0.12(x - 4)$ we can recognize immediately that the point $(4, 5)$ is on the line, and the slope is 0.12. As before, we can generate a series of points, this time by starting at $(4, 5)$ and repeatedly going up 0.12 and over 1.

The Two-Intercept Form. There is a third form of a linear equation that is especially useful if we know where the line crosses both axes. In Figure 2.19 the line shown crosses the x-axis at 5 and the y-axis at 3. From this information the following equation can be written down immediately:

$$\frac{x}{5} + \frac{y}{3} = 1 \tag{2.10}$$

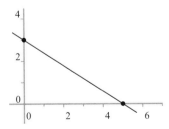

Figure 2.19. Using both x- and y-intercepts to find the equation.

This equation can be derived from the point-slope form, taking $(5, 0)$ and $(0, 3)$ as two points of the line. However, this is not obvious at a glance, and requires a few steps of algebra. The following more elegant justification requires just a few mental calculations, and avoids algebra altogether.

First, since (2.10) is clearly a linear equation, its graph is a straight line. Second, by substitution, we can see that (5, 0) and (0, 3) satisfy the equation, and so are points on the line. Thus, (2.10) represents the line with intercepts 5 and 3, as claimed.

This example uses what is called the *two-intercept form* of a linear equation. The general expression of this form can be given as follows.

> **Two-Intercept Equation of a Line:**
> $$\frac{x}{a} + \frac{y}{b} = 1. \tag{2.11}$$

Here, the letters a and b are parameters. In using this form, a should be replaced by the x-intercept of the line and b should be replaced by the y-intercept of the line. Note: there is no two-intercept form for a line that goes through the point $(0,0)$. For such a line, the x- and y-intercepts are both 0. With $a = 0$ and $b = 0$, (2.11) would not make sense. Similarly, there is no two-intercept form for a horizontal or vertical line. (Why?)

The two-intercept form is often applicable for analyzing mixtures with two ingredients. Here is a typical example. A horse trainer wishes to make a grain supplement of corn and oats, in order to add 100 grams of protein to the diet of a mare with a new foal. The trainer knows from the information on the packages that using corn alone, 3.50 quarts would be needed, while using oats alone, 2.58 quarts (really, $10\frac{1}{3}$ cups) would be needed. Clearly, various mixtures of corn and oats could also be used. The problem is to determine how the amounts of corn and oats must be chosen.

To put this into a more mathematical framework, let us use variables: x for the amount of corn and y for the amount of oats, both in units of quarts.[13] Each valid mixture can be represented by an ordered pair (x, y). The information above tells us that $(x, y) = (3.50, 0)$ is one such pair, and $(x, y) = (0, 2.58)$ is another. Conceptually, if we could find all the valid pairs, they would make up some sort of curve. The two points we already know would then be the x- and y-intercepts. If we assume that the curve is actually a straight line, then we can immediately write down its equation using the two-intercept form:

$$\frac{x}{3.50} + \frac{y}{2.58} = 1.$$

Once we have this equation, we can manipulate it algebraically. For example, solving for y leads to

$$y = 2.58\left(1 - \frac{x}{3.50}\right).$$

Then, for any given amount of corn (x) we can immediately compute the necessary amount of oats (y).

The point of this example is to illustrate why the two-intercept form is often useful in mixture problems. With the information given we could have shown linearity (rather than assuming it), but elected not to do so to simplify the discussion. Mixture problems will be considered in greater detail in the next section.

---

[13] We usually try to select letters that remind us what each variable represents, such as T for temperature and c for costs. In this example the obvious letters would be c for corn and o for oats. However, the letter o is too easily confused with zero, so we decided to use the generic variables x and y.

2.3 Exercises

Reading Comprehension.

(1) @Explain the difference between a linear function and a linear equation.

(2) In the reading there are two different definitions of linear function: one that emphasizes the way the function appears, and one that concerns the operations that are used in computation. Write versions of these definitions in your own words.

(3) On page 87 there is a very detailed example of solving a linear equation. Explain how each equation after the first was obtained from the preceding equation. For example, the second equation was obtained by expanding $2(3x - 5)$ to $6x - 10$ on the left, and $-2(x - 4)$ to $-2x + 8$ on the right. Both of these steps are applications of the distributive law.

(4) @Suppose in the equation $T = 0.01h + 72$, the variable h represents height above the ground in feet, and the variable T represents temperature in degrees Fahrenheit at height h. Why might it be useful to solve the equation for h in terms of T? How could this be accomplished?

(5) @Slope concept. Write brief essay answers to the following. It may be helpful to include examples in your answers.

   a. @Give a definition of slope in your own words.
   b. How is slope related to the graph of a linear function?
   c. @When using two points to compute slope, why does it not matter which one is used as the first point?
   d. How is slope a kind of rate?

(6) @Intercept concept. Write brief essay answers to the following. It may be helpful to include examples in your answers.

   a. @What is an intercept?
   b. @How many intercepts might the graph of a linear function have?
   c. How are intercepts related to find-n and find-a_n questions?
   d. What else do we know about the intercepts for the graph of a linear equation?

(7) In this section you read about three different forms for linear equations. Write out the three forms using parameters (like m and b, or m, x_0, and y_0). For each equation, explain in words what each parameter tells you about the graph. Give a specific example of each equation, using numbers in place of the parameters, and show the graph of each example.

(8) @Explain how a linear equation with one variable can have no solution. Explain how a linear equation can have an infinite number of solutions. Include examples in your answer.

(9) Explain what it means to *invert* a function. Give an example.

Math Skills.

(10) @You know from the reading that the three equations below are not linear. Give a specific reason to show that each is nonlinear. For example, the reading gives $z^2 = x^2 + y^2$ as an equation that is not linear, citing the following specific reason: There appears an x^2, which means $x \cdot x$, whereas linear equations avoid multiplying two variables together.

  a. $2 + xy = z$
  b. @$y = -2 + \sqrt{4 - 3x}$
  c. $y = \frac{3x+4}{4x-3}$

(11) @Determine if each of the following equations is linear. If so, find any solutions. If not, explain how you know it is nonlinear. There is an example of such a reason in the previous problem.

  a. @$3x - 4 = 5 + 2x$
  b. $\frac{3}{x} = 7$
  c. @$3x - 2 = 7x + 4$
  d. $\sqrt{17}\, x = \sqrt{68}$
  e. @$5 = \sqrt{x}$
  f. $\frac{x}{2} - 1 = 6$
  g. @$0.2x + 17 = 0.5 + x$
  h. $2x(x - 5) = 168$
  i. @$4(x - 5) + 5(x - 4) = 3(x - 6) + 6(x - 3)$

(12) @If a function or equation is linear, it can be put in the form $y = mx + b$ or $f(x) = mx + b$. Determine if each of the following is linear. If so, put it in the form above. If not, explain why it is nonlinear.

  a. @$f(x) = 1.2(x - 5) + 9.4$
  b. $3x - 4 = 5 + 2y$
  c. @$f(x) = 6 + \sqrt{9x}$
  d. $3(y - 5) + 3 = 9x + 6(2 - x)$
  e. @$x(1.3 + x) - y = x^2 + 7x - 12.4$
  f. $f(x) = x - 2(x + 7)$
  g. @$2(x - 3) = 3(y - 2)$
  h. $f(x) = \frac{2+x}{x}$
  i. @$4(2 - x) + 5(y - 3) = x + 1$

(13) @Find any and all solutions for each of the following linear equations. Remember there may be no solutions, exactly one solution, or it may be that every number is a solution.

  a. @$3(x - 5) + 3 = 9x + 6(2 - x)$
  b. $2(x - 3) = 3(x - 2)$
  c. @$4(4 - x) + 5(x - 3) = x + 1$
  d. $2a - 7 = 3(a + 3) - (16 + a)$

2.3. Exercises

e. $@x + 12 = 2x - 1$

f. $2a + 3 - 0.2a = 1.8a$

(14) @For each of the following linear equations you are given the value of one variable. Use it to find the other unknown value.

a. $@y = 7x - 4, x = 2$

b. $f(n) = -3(n - 1) + 5, n = 1$

c. $@a_n = 12 - n, n = 5$

d. $a_n = 2n - 3, a_n = 13$

(15) @How would you use the following information to construct a graph of the line? You are not being asked to construct the graph, but rather to provide instructions.

a. @The slope is $m = 3$ and the y-intercept is $b = 1$.

b. The point $(-1, 2)$ is on the line and the slope is $m = \frac{3}{7}$.

c. The x-intercept is 2 and the y-intercept is -1.3.

(16) @Find the equation for the line shown in each of the following graphs. Note: pay attention to the scale on each axis.

a. @
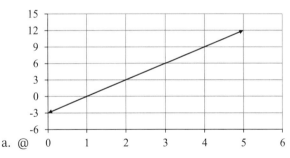

b.
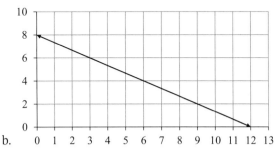

c.
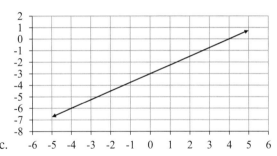

(17) @For each equation, tell what form the equation is in (point-slope, slope-intercept, or two-intercept form), and describe how to create a quick graph.

   a. @$y = 3x - 4$
   b. @$y - 2 = 0.5(x - 3)$
   c. $y = x + 2$
   d. $y - 5 = 2x$
   e. @$x/2 + y/4 = 1$
   f. $y = 2.3x$

(18) @For each of the following descriptions, find an equation for the line. Use the form (point-slope, slope-intercept, or two-intercept) that most directly fits the given information.

   a. @A line with an x-intercept of 5 and a y-intercept of 3.
   b. A line with an x-intercept of -4 and a y-intercept of 1.
   c. @A line with a slope of 4 and a y-intercept of -6.
   d. A line with a slope of $\frac{7}{6}$ and a y-intercept of 3.
   e. @A line that passes through $(4, 7)$ and has a slope of $-\frac{1}{3}$.
   f. A line that passes through $(-1, 5)$ and has a slope of 2.

(19) On page 85 we saw that $y = (5x - 1)(2x + 3) - 10(x^2 + 3x)$ is equivalent to $y = -17x - 3$. Show the algebra to get from the first equation to the second.

Digging Deeper.

(20) @In the reading we saw that $y^2 = x^2$ has different solutions than $y = x$ so the two equations are not equivalent. What about $y^3 = x^3$? Is that equivalent to $y = x$? What about $y^4 = x^4$? Is there a pattern? What do you expect of higher powers of x and y?

(21) We know that the two-intercept form is $\frac{x}{a} + \frac{y}{b} = 1$. Is the 1 important? What happens if we use $\frac{x}{3} + \frac{y}{5} = 2$? Are the intercepts $(3, 0)$ and $(0, 5)$?

(22) @In this problem we will look more closely at the relationship between the two-intercept form of a line, $\frac{x}{a} + \frac{y}{b} = 1$, and the slope-intercept form $y = mx + b$. Before we go on, note that in both forms b represents the y-intercept and recall that $m = \frac{\text{rise}}{\text{run}}$ for any pair of points on the line.

   a. @Let's start with an example. Write an equation in slope-intercept form for the line shown in the graph.

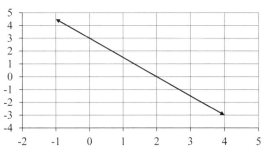

2.4. Applying Linear Functions and Equations

b. In case you have not already done so, take a moment to notice that intercepts are (2, 0) and (0, 3). We know immediately the two-intercept form is $\frac{x}{2} + \frac{y}{3} = 1$. Solve this equation for y, that is convert it algebraically to slope-intercept form. Compare it to the equation you found based on the graph.

c. @Now think about the steps you took above. Will this process work for other lines? To get a more general result, repeat the process for a line with intercepts $(a, 0)$ and $(0, b)$. That is, use those points to write the equation of the line in two-intercept form. Once you've done that, algebraically manipulate the equation to convert it to slope-intercept form. Separately, use the points $(a, 0)$ and $(0, b)$ to determine the slope of the line. Compare the results of your calculations.

d. @Will this process work for all lines? Are there any exceptions? If so, what are they?

(23) A straight line has x-intercept at a and y-intercept at b. Use the slope-intercept equation or the point-slope equation to find an equation for the line. Then, using algebra, rearrange your equation to the form $x/a + y/b = 1$.

(24) @In this section we considered again the satellite model g(n). We found the inverse equation by solving for n. The new equation made it easy to calculate the number of days the satellite could travel, n, with g grams of fuel. Let's consider applying this method in other examples.

a. @Recall the CO_2 example from Section 2.1 where $c(n)$ represents the concentration of CO_2 in the atmosphere n years after 2006. We found the functional equation $c(n) = 381 + 2n$ (see page 48). Invert the equation and explain what the new equation means.

b. @Can we apply the same method to the library fine example from section 2.1, where the fine on a book returned n days late is given by $f(n) = 10n$? If the method is appropriate, what will the new equation tell us? If not, why not?

(25) Review Figure 2.11 on page 86, which depicts an equilateral triangle. It can be shown that B has coordinates $(1/2, \sqrt{3}/2)$. (This can be done by drawing a line perpendicular to the x-axis connecting the x-axis and B then using the Pythagorean theorem on the resulting right triangle.) Using the coordinates of B, find the equations of the lines making the two sides of the triangle that meet at B, and verify that your equations agree with the equations given in the reading following the figure.

2.4 Applying Linear Functions and Equations

As we have seen, linear equations arise out of arithmetic growth models. In that context we are concerned with terms of a sequence, and the equations we find relate the terms to their position numbers. Applications of this approach were considered in Section 2.2. Now we consider situations in which we do not have a sequence, but rather have two variable quantities that we wish to relate to one another. Our focus will be on cases where their relationship can be expressed in the form of a linear equation, so that the methods of the preceding section apply. Usually, we will eventually wish to express one variable as a linear function of the other.

In a given context, how do we know that the variables of interest really are linearly related? In some cases we don't! Developing a model always involves simplifying assumptions. In particular, for a model with variables x and y, we might *assume* that y is a linear function of x. This is sometimes referred to as a *linearity assumption*. Of course, we want our model to be a close approximation to the real world problem. So a linearity assumption is not always appropriate. If we have data for our two variables, and if the graph shows points that are close to lining up, that is one indication that a linear model will be useful. In other cases it may be possible to justify a linearity assumption based on some conceptual understanding of how our variables are related. As a third possibility, linearity is sometimes assumed purely for convenience, or because we do not have enough information about the situation to develop a more sophisticated model. In these cases, it is especially important to remember that any conclusions we derive are only valid if the linearity assumption is a reasonably accurate reflection of the real world context.

We proceed to a series of examples to illustrate these ideas in greater detail. Keep in mind that homework and examination questions may include information that indicates a linear model, either directly or indirectly. As you study the examples, look for the characteristics that indicate linearity.

Example 1: Constant Rate Assumption. The original discussion of the satellite model indicated that 4.35 grams of fuel are used each month. This is a representative figure based on published data for a specific satellite [**34**]. In actual fact the amount of fuel used varies—some months it is more than 4.35 grams and some months it is less. In developing a model, we made the simplifying assumption that the fuel used in a month is constant, and that led us to an arithmetic growth difference equation. Now we reconsider the situation putting aside the methods of sequences and difference equations, and focusing instead on the concept of a constant rate of fuel consumption. As we will see, this is tantamount to assuming a linear relationship between available fuel and time.

Rates and Slopes. The key idea is this: the concepts of *rate* and *slope* are essentially the same. A rate is a comparison between two changing quantities. Common examples are gas mileage (miles per gallon), speed (miles per hour), and pay rate (dollars per hour). Whenever the word *per* occurs in this way, a rate is being described.[14] For the satellite model, saying that 4.35 grams of fuel are used up every month is the same as saying the *rate* of fuel consumption is 4.35 grams per month.

Now consider a variant on the satellite fuel model, where there are 60 grams of fuel available at time 2, and 16.5 grams at time 12. This information can be displayed in a graph showing fuel in grams on the vertical axis and time in months on the horizontal axis (Figure 2.20). Between these two points is a span of $12 - 2 = 10$ months, during which the fuel supply falls by $60 - 16.5 = 43.5$ grams. A decrease of 43.5 grams in 10 months amounts to a rate of $43.5/10 = 4.35$ grams per month, assuming an equal decrease for each month. But this is identical to the slope calculation between the two points on the graph. The change in fuel supply is the rise, the time span is the run,

---

[14] In a similar way, we can think of the word *percent* as meaning "per hundred". So, when a store advertises a 10% markdown, that means the price will be reduced by 10 dollars per hundred. Viewed in this way, a percentage is a kind of rate.

2.4. Applying Linear Functions and Equations

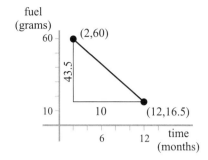

Figure 2.20. The slope between two points can be viewed as a rate.

and the rate is rise/run = slope. In fact, using the slope formula for our two points we would find
$$\frac{y_2 - y_1}{x_2 - x_1} = \frac{16.5 - 60}{12 - 2} = \frac{-43.5}{10}.$$
This is the same calculation we did to find the rate, except that the slope formula automatically determines the correct sign, positive for an increase and negative for a decrease.

In other applications the variables on the x- and y-axes will have other units. But whatever the units on the axes are, a slope with numerical value of m means a change of m y-units occurs for each advance of one x-unit. In other words, the slope is a rate: m y-units per x-unit.

Constant Rate Implies Linearity. Understanding the connection between rates and slopes has a special significance. Assuming a constant rate is the same as assuming a constant slope between any two points on the graph, which in turn implies that the graph is a straight line. So a constant rate assumption is equivalent to a linearity assumption.

Returning to the satellite fuel model, we originally assumed a constant fuel consumption rate of 4.35 grams per month. However, because months do not all have the same length, it will be more convenient to express time in units of years.[15] In 12 months the satellite will use up $4.35 \cdot 12 = 52.2$ grams of fuel. Let us now develop a model assuming a constant fuel consumption rate of 52.2 grams per year. That means we are assuming that the amount of fuel is a linear function of time. And because the amount of fuel is decreasing over time, the slope is -52.2. As in our work with difference equation models, it is important to include a careful statement defining the variables and any assumptions in our model, as follows.

> In this model we consider a linear function $g(t)$ representing the amount of fuel on board a satellite at time t. Time is measured in units of years, with $t = 0$ at the start of January 1, 2015. This is called the *start time*. The fuel is measured in units of grams. At the start time, there are 3,100 grams of fuel. Thus, $g(0) = 3,100$, and $g(t)$ is the amount of fuel t years later. We also assume that fuel is consumed at a constant rate of 52.2 grams per year.

---

[15]Admittedly, years do not all have the same length either, but leap years are rare enough to justify overlooking them: one extra day every four years, corresponding to approximately $4.35/30 = .145$ grams of fuel, is not likely to be significant.

The linearity assumption shows that $g(t)$ is given by an equation of the form

$$g(t) = mt + b,$$

where the parameters m and b are to be determined from given information. We know that m is the slope of g's graph, and as already discussed, that must be -52.2. We also know that b is the intercept on the vertical axis, that is, $b = g(0) = 3{,}100$. Thus, our equation becomes

$$g(t) = -52.2t + 3{,}100.$$

Once we have the equation, we can answer find-g questions by direct calculation. For example, to find the amount of fuel remaining after 5 years we calculate $g(5) = -52.2 \cdot 5 + 3{,}100 = 2{,}839$ grams. We can also answer find-t questions, such as *When will the fuel be half gone?* Since half of the initial supply is $3{,}100/2 = 1{,}550$ grams, this translates to the question

For what value of t is $g(t) = 1{,}550$?

and leads to the equation

$$-52.2t + 3{,}100 = 1{,}550.$$

Using the methods of Section 2.3, we can solve this equation to find $t = 29.69$ years, to two decimal places. As a check, we compute

$$g(29.69) = 1{,}550.182.$$

Finally, because the start time for the model is at the beginning of 2015, the answer $t = 29.69$ means about 2/3 of the way through 2044.

This example illustrates two important general principles concerning two related variables. Representing the variables as x and y, we can formulate the principles as follows.

- Assuming a constant rate of change is the same as assuming a linear relationship between x and y.

- If the constant rate is m and if $y = b$ when $x = 0$, then the relationship is modeled by $y = mx + b$.

As an example of applying this, let us consider the pressure exerted on a scuba diver by the surrounding water. One commonly used model assumes that the pressure is one atmosphere (atm) when the diver is at the surface, and increases by one atm for every 10 meters of depth (see [7]). That translates to a constant rate of $1/10 = 0.1$ atm per meter. We recognize that this is a linear model, and has equation

$$p = md + b,$$

where p is pressure in atm, d is depth in meters, m is the constant rate of 0.1, and b is the pressure at a depth of 0. Since zero depth means at the surface, $b = 1$. Hence the equation becomes

$$p = 0.1d + 1.$$

Thus, using the two principles, we quickly derive an equation for this model.

2.4. Applying Linear Functions and Equations

Example 2: Structural Linearity. In the preceding example we began by assuming a constant rate, then found that implies a constant slope, and hence a linear equation. Now we will see a case where linearity is not simply *assumed*, but rather, is a logical consequence of the problem structure. We reconsider the following problem, first discussed in Section 2.3.

> A horse trainer wishes to make a grain supplement of corn and oats, in order to add 100 grams of protein to the diet of a mare with a new foal. The trainer knows from the information on the packages that using corn alone, 3.50 quarts would be needed, while using oats alone, 2.58 quarts (really, $10\frac{1}{3}$ cups) would be needed. Clearly, various mixtures of corn and oats could also be used. The problem is to determine how the amounts of corn and oats must be chosen.

Let us again use variables x and y, respectively, for the amount of corn and the amount of oats in an acceptable mixture, that is, one that contains 100 grams of protein. For example, the given information tells us that 3.5 quarts of corn with 0 quarts of oats is an acceptable mixture. For this mixture we have $x = 3.5$ and $y = 0$, and we can represent the mixture by the point $(3.5, 0)$ on a graph with x- and y-axes. See Figure 2.21. Using similar logic, we find that 0 quarts of corn plus 2.58 quarts of oats is

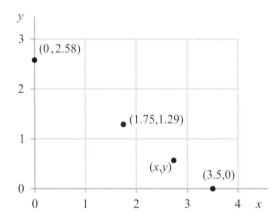

Figure 2.21. Acceptable mixtures of corn and oats are represented by points (x, y).

an acceptable mixture, represented by the point $(x, y) = (0, 2.58)$. As another example, we can reason that $(x, y) = (1.75, 1.29)$ is also an acceptable mixture. We know that 3.5 quarts of corn have 100 grams of protein, so $3.5/2 = 1.75$ quarts will provide 50 grams of protein. Similarly, we see that $2.58/2 = 1.29$ quarts of oats will also provide 50 grams of protein. Therefore, combining 1.75 quarts of corn with 1.29 quarts of oats provides 100 grams of protein, as desired.

These are just three possible mixtures of corn and oats that provide 100 grams of protein. But there are an infinite number of acceptable mixtures, and each contributes a point to our graph. Our goal is to see that all of these points line up in a straight line.

To reach the goal, we begin with a simple assumption about how nutrients are distributed in grain. In a one quart sample of corn, there will be a certain amount of protein. Practically this amount is a constant—any one quart sample will contain the same amount of protein as any other one quart sample. We assume that whatever variation exists is small enough to be disregarded.

How much protein is in a quart of corn? For the moment, let us call this unknown amount p. Then there will be p grams of protein in a quart of corn, $2p$ grams in 2 quarts, $3p$ grams in 3 quarts, and so on. In particular, 3.5 quarts of corn must contain $3.5p$ grams of protein. But the problem statement tells us that 3.5 quarts of corn contain 100 grams of protein. Therefore $3.5p = 100$, and we deduce that $p = 100/3.5$ or about 28.57. The exact numerical amount is not important. In a similar way we can work out how many grams of protein are in a quart of oats. We call this q, and find $q = 100/2.58$. To two decimal places that is 38.76, but again, the exact numerical value is not important.

Now consider one possible combination of corn and oats, represented by the point (x, y). The x quarts of corn contribute $x \cdot p = x \cdot 100/3.5$ grams of protein and the y quarts of corn contribute $y \cdot q = y \cdot 100/2.58$ grams of protein. The point will be on our line if and only if the total protein in the combination is 100 grams. This condition is expressed by the equation

$$x\frac{100}{3.5} + y\frac{100}{2.58} = 100.$$

So, from a basic understanding of the problem context we are led by logic to an equation for the two variables. We recognize it as a linear equation. In this example, we have derived a linear model directly from the problem structure. To emphasize this result, we say that this problem is *structurally* linear.

Using algebra the equation can be expressed as

$$\frac{x}{3.50} + \frac{y}{2.58} = 1.$$

This is the same two-intercept form we found in our previous discussion of the model. Similar logic applies in many kinds of mixture problems, showing that they can be modeled with linear equations. Here are a few examples.

- A company obtains coal from two different suppliers. One charges $ 60.58 per short ton, and the other charges $ 46.02. For a total of $ 5000, the company can purchase x tons from the first supplier and y tons from the second. How are x and y related?

- The heat provided by a fuel (such as coal) can be measured in BTUs. The coal provided by the first supplier in the preceding item provides 12,500 BTUs per ton. For the second supplier it is 11,800 BTUs per ton. To obtain a total of 600,000 BTUs, a company will buy x tons from the first supplier and y tons from the second. How are x and y related?

- A farmer has $ 80,000 to spend on production of corn and soybeans. If she grows only corn, she can afford to cultivate 210 acres. If she grows only soybeans, she can afford to cultivate 408 acres. But that is not an option because she only owns 350 acres. If she wants to cultivate x acres of corn and y acres of soybeans, how are x and y related?

- A consumer budgets 60 euros each month to spend on movies and used books. Each movie costs 10 euros and each used book costs 5 euros. If the consumer sees x movies and buys y books, how are x and y related?

2.4. Applying Linear Functions and Equations

Problems of this sort are probably structurally linear. That is, with the same logic as in the grain mixture problem, it is probably possible to demonstrate the correctness of a linearity assumption. If so, the two intercept equation is likely to be a convenient means for finding an equation for the model.

Example 3: Fitting a Line to Data. We have seen one example where linearity is simply assumed, and another where it arises out of our understanding of the problem context. Now we will look at an example where available data suggest adopting a linearity assumption.

The problem context is a proposed fundraising activity by a student organization. The members intend to sell canned soft drinks between classes. They know from their economics classes that the number of drinks that can be sold (referred to as the *demand*) depends on the price per drink. To investigate this, they conduct a survey to estimate the demand on a typical day at various prices. Their results are displayed in Figure 2.22, with prices shown on the horizontal axis and demand on the vertical axis.[16] For example, the survey results indicate that at a price of eighty cents each the students

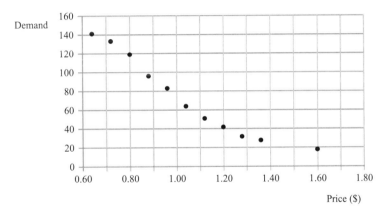

Figure 2.22. Demand data for selling soft drinks. Each individual point, marked as a dot, shows how many soft drinks can be sold at a certain price.

can expect to sell 119 drinks. Accordingly, there is a dot at the point $(p, d) = (0.80, 119)$, where p represents price per drink and d represents demand.

Clearly, the points on the graph do not fall on a straight line. But one can imagine drawing a line that comes close to the points. In Figure 2.23 two possible lines are shown, labeled L and M. In advanced mathematics and statistics, there are sophisticated methods for finding the best possible line, subject to a careful definition of *best*, but those methods are beyond the scope of this book. Instead, for illustrative purposes, we imagine drawing a line with a ruler, and consider how to find its equation. You can carry out this process working with a printed copy of the graph. Lay a ruler on the graph, adjust it by eye to get as close as possible to the data points, and then draw a line.

---

[16] These data come from an actual survey conducted by American University students in 1994, but have been adjusted for inflation.

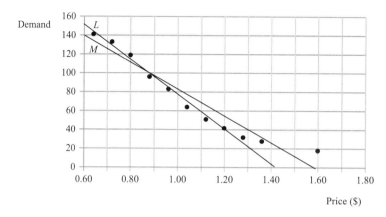

Figure 2.23. Demand data with two illustrative lines. Line M attempts to stay as close as possible to all the data points. Line L is closer than line M to almost all the data points, but is far away from the last two data points.

This involves judgment and visualization, and there is no one right answer. Indeed, we show two possible lines in Figure 2.23 to emphasize the variation that is possible. In drawing line M we tried to get as close as possible to all the data points. For line L, we disregarded the final two data points and drew a line that comes as close as possible to the remaining points. A rationale for doing this is that we do not expect to charge a price as high as in the last two data points, so it makes sense to concentrate on the part of the demand curve that is likely to apply in our market analysis. For now, our focus will be on finding an equation for one of our lines. We will see in Chapter 3 how this demand model can be used to design a pricing strategy.

For the sake of discussion, let us agree to adopt line L for our model. How can we find its equation? We try to identify two points on the line as accurately as possible, and then use the point-slope form of a linear equation. If possible, we would like to find two points where the line passes right through an intersection of grid lines, or comes very close to such an intersection. These do not have to be data points. For example, line L comes very close to $(1.10, 60)$, $(1.20, 40)$, and $(1.30, 20)$. We can pick any of these as representative points on the line, and in fact, all three of these points lie on a single line that passes through diagonally adjacent corners of the grid. However, extending this line to the left, observe that it will eventually reach the point $(0.60, 160)$, which is noticeably removed from line L. This illustrates an important idea. When we identify points that appear to be on or very close to the line, there is almost always some inaccuracy. Its effects can be minimized by choosing representative points that are far apart. So, to find the equation of the line, we can use any of $(1.10, 60)$, $(1.20, 40)$, and $(1.30, 20)$ as a first point, but should choose a reasonably distant second point.

Let us agree to choose $(1.20, 40)$ for our first point. Unfortunately, L doesn't seem to pass close to any distant grid line intersections. So for a second point, we estimate the intersection between L and the left-hand border of the graph to be $(0.60, 152)$.

Now we find the equation of the line through $(0.60, 152)$ and $(1.20, 40)$. The slope is given by

$$m = \frac{40 - 152}{1.20 - 0.60} = -\frac{112}{0.60} = -\frac{560}{3}.$$

2.4. Applying Linear Functions and Equations

Combining this with the point $(1.20, 40)$, the point-slope equation[17] is

$$d - 40 = -\frac{560}{3}(p - 1.20).$$

This leads to

$$d = -\frac{560}{3}(p - 1.20) + 40,$$

expressing d as a linear function of p.

This completes the development of our linear demand model. We began with a set of data, decided the points fall close enough to a line to justify a linear model, drew a line that comes close to the data points, and then found its equation.

An important part of developing a model is to reflect on whether it produces reasonable results. For this example, we view our equation as modeling how price and demand are related, but know that the model is not exactly correct. For instance, taking the price as 0.80, the equation says

$$d = -\frac{560}{3}(0.80 - 1.20) + 40.$$

That works out to 114.67 to two decimal places. In the context of the problem, we should restrict the demand to whole number values. We do not expect to sell exactly 114.67 drinks. The equation gives us an approximate prediction of the demand, and we can further refine that prediction by rounding up to the nearest whole number, 115. Is that accurate? The actual survey data result is 119, so our model is off by 4 drinks out of 119. Whether that is accurate enough to be useful depends on the goals of the analysis and how much accuracy is required in the final answer. But remember that this model is for sales on a typical day. We expect that we will sell more drinks some days and fewer drinks on other days. Common sense suggests that sales results will vary by more than 4 drinks day by day. So an error of 4 in the model does not appear to be unreasonable.

Example 4: Modeling Outline. In Section 2.2 we presented an outline for developing arithmetic growth models based on difference equations (see page 58). Recall that the outline is not intended to be followed rigidly. It provides a list of possible steps in developing and applying a model, but variations are to be expected in deciding what steps to do and in what order.

With a few adjustments, the same outline applies when developing linear models without any reference to difference equations. Indeed, in the preceding examples most of the steps of the outline have been discussed in one form or another. Let us look at one more example, this time highlighting how the development reflects the modeling outline.

Cell Phone Company Example. A cellular telephone company has equipment that can service 100 thousand customers. In 2010 they had 70 thousand customers. Over the last few years the number of customers has been increasing by about 4,500 per year. Assuming that the growth continues at the same pace, what will the situation be in the year 2020? 2030? How long will it be before additional equipment will be needed?

---

[17] In the standard point-slope equation we have changed x to p, the variable on the horizontal axis, and y to d, the variable on the vertical axis.

General Characteristics. The problem statement primarily concerns the number of customers of the company. We understand that this is growing over time. The questions in the problem either ask for the number of customers for a specified year, or ask when there will be a specified number of customers. Therefore, we should include variables representing the number of customers and time. The problem statement also says to assume a constant growth rate of 4,500 customers per year.

Define Variables. We choose variable names that remind of us of their meanings, c for the number of customers and t for the time, in years. For convenience, we can express c in units of thousands. Should t be defined as a discrete or continuous variable? Notice that the given information specifies a single number for the customers in an entire year. We can say that $c = 70$ for $t = 2010$, but is that an average over the year? Or a snapshot at one day of the year, e.g., on January 1? Or perhaps it is the peak number of customers during the year. The given information doesn't make this clear, so it is uncertain how to interpret a value of c for a fractional part of a year. Accordingly, we will restrict t to whole number values, at least initially. With these ideas in mind, we can define our variables as follows:

We define t to be a whole number representing the year, and c is the number of customers, in thousands, for that year.

Assumptions. We assume that c increases by 4.5 each year. (Remember that c is expressed in units of thousands.) We are given that $c = 70$ when $t = 2010$.

Step 4: Graphical Representation of Given Information. We represent the two variables on a graph, with t on the horizontal axis and c on the vertical axis. See Figure 2.24. We plot points of the form (t, c) where c is the number of customers for year t.

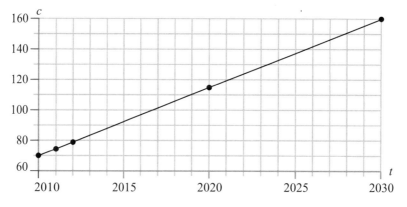

Figure 2.24. Graph of cell phone customers by year. If there are c customers for year t the point (t, c) appears on the graph.

For example, there will be a point at $(2010, 70)$. The assumption that c increases by 4.5 per year indicates a constant rate, and in terms of the graph, that means a constant slope of 4.5. For example, there will be points at $(2011, 74.5)$ and $(2012, 79)$ as shown on the graph. The constant rate assumption implies linearity, so the points all fall on a straight line, also shown on the graph.

2.4. Applying Linear Functions and Equations

For a linear model, either variable can be selected for the horizontal axis. Usually when one variable represents time, it is assigned to the horizontal axis, as we have done here. This is the same as treating time as the independent variable. One reason to do so is that we often conceptualize how something varies over time. That generally meets the uniquenss requirement for a function, because an observable variable (such as temperature) cannot have two different values at the same time. Rates are also more commonly expressed using time as the independent variable. In the present problem, it is very natural to say that the company is gaining so many customers per year. But if we make the number of customers the independent variable, the rate would be in units of years per customer. That can be understood—.01 years per customer would indicate a waiting of time of a hundredth of a year to gain each additional customer—but it is much less natural.

Find an Equation. We have already indicated that the graph is a line with slope 4.5 and that (2010, 70) in on the graph. Therefore, according to the point-slope form, we have the equation

$$c - 70 = 4.5(t - 2010).$$

We can express c as a function of t by rewriting the equation in the form

$$c = 4.5(t - 2010) + 70. \tag{2.12}$$

This can be used to find c when we have a given value of t.

We can also express t as a function of c by solving the first equation for t. That leads to

$$t = \frac{c - 70}{4.5} + 2010. \tag{2.13}$$

This can be used to find t when we have a given value of c.

Formulate Questions. The first questions ask what the situation will be in 2020 and 2030. We will interpret those to ask what $c(2020)$ and $c(2030)$ are. The final question asks how long it will be before new equipment is needed. Since it is given that the existing equipment can serve 100,000 customers, we interpret this to ask when $c(t)$ will be 100.

Answer Questions and Translate to Original Context. Using (2.12) with $t = 2020$,

$$c = 4.5(2020 - 2010) + 70 = 115.$$

Thus, according to the model, there will be 115,000 customers in 2020. By similar reasoning, we find that there will be 160,000 customers in 2030.

For the remaining question, we use (2.13), with $c = 100$. That produces

$$t = \frac{100 - 70}{4.5} + 2010 = 2016.666\cdots.$$

However, we have defined t to be a discrete variable taking only integer values. So, according to the model, the existing equipment will be sufficient through 2016, but will not be sufficient for 2017.

As a double-check on this answer, we can compute $c(2016)$ and $c(2017)$, using (2.12). We find $c(2016) = 97$ and $c(2017) = 101.5$, confirming the earlier answer.

Reflect on the Answers. Are the conclusions of the model reasonable? If the number of customers increases by 4,500 per year, then in ten years the increase will be 45 thousand. Thus, with 70 thousand in 2010, we expect $70 + 45 = 115$ thousand in 2020, and $115 + 45 = 160$ thousand in 2030. That is what the model predicts as well. Also, the predicted value of 115 thousand in 2020 exceeds the firm's current operational capacity. This shows that new equipment will be needed sometime between 2010 and 2020. Therefore the model's prediction that new equipment will be needed during 2016 is reasonable. We can also see from the graph that the conclusions of the model are in the right numerical range.

We should also ask if the assumptions are valid. We have no way of knowing whether the annual increase in the number of customers will remain constant. It might grow more rapidly, if the popularity of the system grows by word of mouth. The more customers there are, the more opportunities to recruit new customers. On the other hand, the total number of customers is limited by the size of the population. When c gets close to that limit, it becomes unreasonable to expect constant annual increases to continue. From the given information, it is impossible to know whether either of these possibilities will occur. But it is worthwhile to consider conditions under which constant growth might be unlikely to persist.

Proportional Reasoning. Having now seen several examples of the use of linear models, we turn to an important aspect of such models, a concept referred to as *proportional reasoning*. To illustrate this idea, we consider again the soft drink demand model. Suppose that, from the original survey data, you are given only the following table.

Price	Demand
0.80	120
1.20	40

What demand would you expect at a price of $1.00? Take a minute now to answer this question just using common sense.

Did you predict a demand of 80 drinks at a price of $1.00 per drink? Many people do, arguing as follows:

> *The demand decreased by $120 - 40 = 80$ drinks when the price was raised by 40 cents. If we only raise the price by half as much, that is by 20 cents, then the demand should only decrease by half as much, 40 drinks. Thus at a price of $1.00 per drink the demand should be $120 - 40 = 80$. In other words, since a price of $1.00 is half way between $0.80 and $1.20, the corresponding demand should be half way between 120 and 40.*

This kind of logic is called *proportional reasoning*. It is very commonly used. Here it is based on the assumption that the ratio *(demand decrease)/(price increase)* gives the same value when the price increase is 40 cents as it does when the price increase is 20 cents, or anything else. More generally, when we are considering the way that two related quantities vary, proportional reasoning assumes that changes in each quantity are always in the same proportion.

2.4. Applying Linear Functions and Equations

As an example of another aspect of proportional reasoning, consider the following question: Based on the table above, what would you predict as the demand when the price is set at $1.30? That is a price increase of 10 cents compared to the second row of the table. We have already seen that a price increase of 40 cents leads to a decrease of 80 in the demand. A 10 cent increase is one fourth as much, so we expect the decrease in demand to also be one fourth as much, or 20 drinks. So, demand will decline from 40 drinks to 20 drinks if we raise the price from $1.20 to $1.30.

The connection of proportional reasoning to linear models is quite direct: a linearity assumption will always lead to the same results as using proportional reasoning. Indeed, the ratio of corresponding changes in two variables is the same as the slope, so assuming that this ratio has a fixed value is exactly the same as assuming a constant slope.

The idea of proportional reasoning is fairly clear in the example above, because the numbers are simple. Many people can look at the table and immediately determine that a price of $1.00 should produce a demand of 80. But based on the original survey data, we should have had the following table.

Price	Demand
0.80	119
1.20	42

Given this information, and asked to predict the demand for a price of $0.87, almost no one could see the answer at a glance. In fact, many people would probably have difficulty finding an answer at all. Thus, the idea of proportional reasoning can be much more difficult to apply than it is to understand.

This is where the ability to find an equation is so useful. Knowing that proportional reasoning is the same as assuming linearity, we can apply the methods of this section to find an equation for demand as a function of price. Based on the figures in the second table, we would find

$$\text{demand} = 119 - \frac{77}{0.40}(\text{price} - 0.80).$$

This equation allows immediate computation of demand for *any* price. From this perspective, the methods we have covered for formulating and applying linear models can be viewed as a reliable way to use proportional reasoning.

Conversely, we should recognize that proportional reasoning should only be applied in cases where linearity is an appropriate assumption. For a problem where the data definitely do *not* follow a straight line, a linear model will not provide accurate results, and proportional reasoning should not be used. We will see many examples of such problems in succeeding chapters.

A Common Error. As already mentioned, for most people, proportional reasoning is easy to use when the numbers work out nicely. In these cases it is easy to obtain reasonable answers just using common sense, without any need for the linear equation methods presented in this chapter. Unfortunately, using common sense to formulate models can also lead to dramatic errors. Here is an illustration.

Returning to the information in the first table, let us again consider the connection between price and demand for soft drinks. We saw that it is easy to predict the demand at certain prices, like $1.00, because the numbers work out simply. To predict the demand at other prices, say $0.87, it would be very handy to have an equation giving the demand d as a function of the price p. We saw that raising prices by 10 cents reduces demand by 20. Proportional reasoning then indicates that each one cent increase in the price will result in a two drink reduction in demand. How can that be expressed as an equation?

Many students will write as an answer

$$p + 0.01 = d - 2$$

and will interpret it loosely to mean *an addition of $0.01 to price produces a subtraction of 2 from demand*. This is wrong. Can you see why?

The problem is that = is being misinterpreted. It does not mean *produces*. It means *is the same number as*. Reread the equation in that light. It says that the price, plus one cent more, is the same as the number of drinks minus 2. This just isn't true. In the first place, the two sides of the equations are in different units, cents on the left and drinks on the right, and thus cannot be equal. Second, we know that at a price of $0.80 the demand is 120 drinks. That is when $p = 0.80$ we know $d = 120$. Using proportional reasoning again, at a price of 0.81 the demand would be 118. Here the $p + 0.01$ is 0.81, and the $d - 2$ is 118. These numbers go together, but they are certainly not equal. To repeat: $p + 0.01 \neq d - 2$.

This example illustrates the difficulty of translating English into algebraic statements. English words and phrases have many meanings and shades of nuance, whereas an algebraic equation has just one correct meaning. Loosely translating *produces* to the = symbol is the wrong approach.

So what is the right approach? The answer is to use all your tools for working with linear equations. First, recognize that by using proportional reasoning we are really adopting a linear model, meaning a linear equation for p and d. We know two data points, because $d = 120$ when $p = 0.80$ and $d = 40$ when $p = 1.20$. To find the equation for the straight line joining these data points, we can calculate a slope and use the point-slope form for a line. That leads to the equation

$$d - 120 = -200(p - 0.80).$$

This is the correct equation. In it, it is not easy to find the numbers mentioned in the statement:

For each $0.40 cent increase in price there is an 80 drink decrease in demand.

Most readers of this book will naturally understand the ideas underlying linear relationships and proportional reasoning. They can apply the ideas in a common sense way, without equations or algebra, when simple numbers are involved. In contrast, the discussion in Section 2.3 might appear needlessly complicated and involved, delving into the fine details of slopes, intercepts, parameters, and several different standard forms for a linear equation. The point of the preceding example is to illustrate why the algebraic tools are sometimes necessary.

To repeat a familiar theme, the models to be considered in later chapters are more complicated than linear models, and require more complicated forms of analysis. But in broad terms, our methods will be the same ones applied in this chapter: identifying and understanding numerical, graphical, and algebraic properties. In this way, the treatment here of linear models will provide a foundation for the material to follow.

2.4 Exercises

Reading Comprehension.

(1) Consider how we can recognize when a linear model might be applicable. Review the examples in this section. For each, some justification for using a linear model was given. Make a list of characteristics which indicate a linear model should be used.

(2) Explain how the slope of a line can be thought of as a rate. Include at least one example with units.

(3) @A wildlife biologist is studying the health of rainbow trout in the Pacific Northwest's Spokane River. Her data include, among other things, the length and weight of each fish she examines. On the average, she finds that a 300 mm trout weighs about 290 grams, and a 400 mm trout weighs about 630 grams.[18]

  a. @Assume that proportional reasoning is valid, and give an example of a predicted weight for trout of some other length than those mentioned, explaining your reasoning.

  b. @Assume that proportional reasoning is *not* valid, and explain what conclusions can be drawn from the numerical data and a graph of the data.

Problems In Context.

(4) @Problems from the Reading. On page 106, following the discussion of mixture problems and the two-intercept form, several scenarios were described. Construct a model and answer the question for each one. Note that your solution should include both the model and the specific answer to each question. In constructing your model, you are free to use any form of linear equation that you prefer.

  a. @A company obtains coal from two different suppliers. One charges $60.58 per short ton, and the other charges $46.02. For a total of $5,000, the company can purchase x tons from the first supplier and y tons from the second. The second supplier has only 70 short tons available. If the company purchases 70 short tons from the second supplier, how many should they purchase from the first supplier?

  b. The heat provided by a fuel (such as coal) can be measured in British thermal units (BTUs). The coal provided by the first supplier in part a provides 12,500 BTUs per ton. For the second supplier it is 11,800 BTUs per ton. To obtain a total of 600,000 BTUs, our company will buy x tons from the first supplier and y tons from the second. (We are no longer concerned with the purchase price.) If 21 short tons are purchased from the first supplier, how many short tons should be purchased from the second supplier?

  c. @A farmer has $80,000 to spend on production of corn and soybeans. If she grows only corn, she can afford to cultivate 210 acres. If she grows only soybeans, she can afford to cultivate 408 acres. But that is not an option because she only owns 350 acres. If she cultivates 100 acres of corn, how many acres of soybeans should she cultivate? In order to cultivate the soybeans she must

---

[18]Data from [**44**]

have both enough land and enough money. What if she cultivates 50 acres of corn? Note that you must define variables for this model.

d. A consumer budgets 60 euros each month to spend on movies and used books. Each movie costs 10 euros and each used book costs 5 euros. If the consumer wishes to buy as many books as the number of movies she sees, how many books should she buy? Note that you must define variable for this model.

(5) **Manufacturing Backpacks.** A company manufactures backpacks. The total variable cost to make each backpack, including materials and labor, is $23. In addition, the company has fixed expenses of $12,000 per month for items such as rent and insurance; these expenses do not depend on the number of backpacks made. Using this information, develop a linear model for monthly total costs as a function of the number of backpacks made each month. Write a short report about your model, defining the variables you use, indicating your assumptions, showing the equation for costs, explaining your reasoning, and reflecting on your conclusions.

(6) **@Paper Airplanes.** Abigail is playing with a paper airplane. When she throws it from shoulder height, four feet for Abigail, the plane takes 8 seconds to reach the ground. Assuming a linear model, how long will the plane take to reach the ground if Abigail throws it while standing on an 18 inch tall stool? Your answer should include your model, a brief explanation of your reasoning, and reflection on your conclusion.

(7) **@Temperature and Pressure.** When a sealed container is heated, the pressure inside the container rises. This is tested in a laboratory experiment with a sealed air tank. When the tank is at 70 degrees Fahrenheit, the pressure inside is 15 pounds per square inch (psi). After the tank has been heated to 247 degrees Fahrenheit the pressure inside is 20 psi.

   a. @Assuming the relationship is linear, find an equation relating the temperature and pressure.

   b. @Use your equation to predict what the pressure would be at a temperature of 150 degrees.

   c. @In your equation, is temperature a function of pressure, is pressure a function of temperature, or is neither a function of the other? How can you tell?

(8) **Corn and Lima Bean Mixture.** A dietitian is developing various mixtures of corn and lima beans to provide a predetermined amount of fiber. Let C stand for the amount of corn in an acceptable mixture, and let B stand for the amount of beans. If she uses corn alone, 2.5 cups are required (so $C = 2.5$ when $B = 0$). Similarly, if she uses just lima beans 1.8 cups are needed ($B = 1.8$ when $C = 0$).

   a. Using this information, find an equation relating B and C.

   b. Based on your equation, how many cups of beans are needed if 1 cup of corn is used?

   c. In your equation, is the amount of corn a function of the amount of lima beans, is the amount of lima beans a function of the amount of corn, or is neither a function of the other? How can you tell?

2.4. Exercises

(9) @Electric Company Study. The electric company in a small town wants to project future growth. The company has collected the data in the table below showing how many customers they had in past years. The column headed n represents the time in years starting from $n = 0$ in 1996.

Year	n	Number of Customers
1998	2	23,717
2000	4	24,313
2002	6	24,781
2004	8	25,103
2006	10	25,673
2008	12	25,965
2010	14	26,690
2012	16	26,909
2014	18	27,529

The electric company analysts developed the following equation to model demand: $C = 22{,}850 + 250n$ where C is the number of customers n years after 1996. The graph below shows both the data and the model.

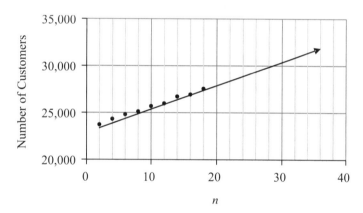

a. @What are the implicit assumptions of the model the analysts have developed?

b. @For each of the assumptions you listed, determine if the assumption is reasonable. If so, what, if any, future events might invalidate the assumption? If not, briefly explain why not.

c. @The company can presently generate enough electricity to serve a total of 30,000 customers. Use the analysts' model to estimate when the number of customers will be as high as the company can serve with their current generating capabilities. Write a brief explanation of how you arrived at your answer.

d. @What might the company do over the next several years to determine whether the projections of the model should be relied on?

(10) Temperature Conversion. On the Fahrenheit temperature scale, water freezes at 32 degrees and boils at 212 degrees. The Centigrade scale is defined so that water freezes at 0 and boils at 100 degrees.

a. There is a linear equation that can be used to convert a Centigrade temperature, C, to Fahrenheit, F. Another way of saying this is that the equation gives F as a function of C. Find this equation. For a hint, see the footnote.[19]

b. Is it possible for a temperature to be the same number using both Fahrenheit and Centigrade? If so, how could that temperature be found? If not, why not?

(11) @Diving. As discussed in an example, scuba divers are subject to the effects of increasing pressure as they go deeper and deeper into the ocean. Using units of feet for depth and pounds per square inch (psi) for pressure, divers assume a pressure of 15 psi at the surface, and an increase of 15 psi for every 33 feet of depth. Thus, at a depth of 33 feet the pressure is 30 psi.

a. @Write an equation for pressure as a function of depth.

b. @What is the pressure at a depth of 100 feet?

c. @Suppose an underwater camera can withstand pressures of up to 1,000 psi. How deep into the ocean can the camera safely go?

(12) Air Pollution. In a large city, air pollution increases during the day, as auto emissions and other types of pollution enter the atmosphere. One day in Smogville, the pollution level was 20 parts per million at 8 in the morning, and had increased to 80 parts per million by noon. Develop a linear model for the pollution level for that day as a function of time. The Smogville Air Quality Management Division is required to publish an unhealthy air alert on any day when the pollution level reaches 150 parts per million. If the linear model is valid from 8 in the morning to 6 at night, will it be necessary to publish an alert?

(13) @Demand. A small boutique will offer a new line of handmade bags for the summer. The owner is working with the artist on pricing the bags. The bags will all be the same price and that price will be between $50 and $125. After doing some market research, the owner has concluded that if she charges $125 for each bag she will sell only 10 bags over the summer and that for every $5 less she charges for a bag, she will have 2 more sales. That is if she charges $120, she expects to sell 12 bags. Similarly at $115, she expects to sell 14 bags. When she next talks with the artist, she will find out how many bags the artist plans to make. She would like to be able to suggest a retail price immediately. In preparation, she wants to incorporate the information she has into a formula relating the number of bags and the suggested retail price. Determine if the situation can be modeled accurately by a linear model. If so, write an equation to model the situation. If not, explain why not.

Digging Deeper.

(14) @The admission at Frank's Funtime Fantasy Land is $11.50 for children under 18 and $25.00 for anyone age 18 or older. On a Saturday, 4,400 people enter the park and $67,839.50 is collected. How many children and how many adults attended?

---

[19] Make a graph showing data points of the form (C, F), where C is the Centigrade temperature corresponding to a Fahrenheit temperature of F. This graph will be a straight line. You are looking for the equation of this line.

2.4. Exercises

(15) Anthony and Claudia Jean are selling homemade brownies and iced cookies. The brownies are $1.75 each and cookies are $2 each. They have prepared 6 dozen brownies and 13 dozen cookies for today. At the end of the day all the left over food is donated to a local charity. While they did not keep a record of how many of each item was sold or donated, Anthony and Claudia Jean know that they made $415.75. Can they now determine from the information they have how many of each item they sold? If so, how? If not, why not?

(16) @Tabatha reads a lot of books and reads at a rate 10 lines of text every minute.

   a. @How many seconds does it take Tabatha to read one line of text?
   b. @Write *three* equations that describe the number of lines Tabatha reads, L. One which relates L to the number of minutes, m, that Tabitha spends reading. A second which relates L to the number of seconds, s, that Tabitha spends reading. A third which relates L to the number of hours, h, that Tabitha spends reading.
   c. @How long would it take her to read *Book Ten* of John Milton's *Paradise Lost*, which is 1,104 lines long?

(17) Tommy wants to rent a car for one week to take a trip to New York and back. He calls two car rental companies to get prices. Arnie's Alright Automobiles rents a Honda Civic for $120 a week plus $0.11 per mile. Carl's Comfortable Cars has the same vehicle for $70 per week and $0.20 per mile. Tommy must drive 634 miles to New York.

   a. For Tommy's trip, which car offers the better deal?
   b. Is there a driving distance for which the two rental options would cost the same amount? If so, find it. If not, explain why.

3

Quadratic Growth

Chapter 2 showed that all arithmetic growth models have several features in common, including the forms of the difference and functional equations, the shapes of the graphs, and methods of application. Indeed, it is in recognition of these commonalities that all arithmetic growth models are considered to be part of the same *family* of models. In this chapter we will extend this idea by studying a second family of models, namely, the quadratic growth models.[1]

The organization of this chapter is similar to that of Chapter 2. In Section 3.1 we will define and discuss the properties of quadratic growth sequences. Applications of these sequences are covered in Section 3.2. A broader discussion of quadratic functions and equations will follow in Section 3.3, and Section 3.4 will concern one specific application of quadratic functions and equations.

3.1 Properties of Quadratic Growth

Definitions. To introduce the concept of quadratic growth, we revisit an example from Chapter 1, the Added Multiples of Three sequence

$$5, 8, 14, 23, 35, 50, \cdots$$

(see page 16). We again use t as a label for the terms of this sequence, and define $t_0 = 5, t_1 = 8$, and so on. This is a little different from our earlier discussion of the sequence in which the initial term was defined to be t_1. Taking the initial term as t_0 will lead to a more convenient form of the functional equation.

This sequence is not an instance of arithmetic growth, because the successive differences are not all the same. We can see this clearly in Table 3.1, which shows position numbers, terms, and the differences between the successive terms.

---

[1] The term *quadratic* is connected with the algebraic methods that apply to these models. Historically, these methods first began to develop around 2000 BC. They were studied in ancient Babylonia, Egypt, Greece, China, India, and the Islamic Empire, generally in connection with problems about squares and rectangles, also called quadrilaterals. As used today, the algebraic methods include quadratic equations and the quadratic formula, topics you may recall studying in an earlier course.

Table 3.1. Added multiples of three sequence with differences.

n	t_n	Difference
0	5	
		3
1	8	
		6
2	14	
		9
3	23	
		12
4	35	
		15
5	50	

Although the differences are not constant, they do follow a simple pattern. In fact, the differences form an arithmetic growth sequence. This idea is so important that it is worth repeating: the added multiples of three sequence does not exhibit arithmetic growth, but the sequence of differences between successive terms *does*. This is the defining characteristic of quadratic growth.

Because the sequence of differences will be important throughout this and succeeding chapters, let us formalize the terminology. Given a sequence such as

$$4, 9, 16, 25, 36, 49, \cdots,$$

subtracting each number from the one that follows generates a new sequence, in this case,

$$5, 7, 9, 11, 13, \cdots.$$

This defines the *differences* for the original sequence. In symbols, the sequence of differences can be displayed as

$$t_1 - t_0, t_2 - t_1, t_3 - t_2, \cdots,$$

with the nth term given by

$$\text{difference}_n = t_{n+1} - t_n.$$

Positive differences represent increases and negative differences represent decreases. For the example, from 4 to 9 there is an increase of 5, from 9 to 16 the increase is 7, etc. For the sequence

$$25, 13, 7, 4, 2, 1, \cdots$$

the differences are

$$-12, -6, -3, -2, -1, \cdots.$$

The initial difference is given by $13 - 25 = -12$, which we can think of as an increase of -12 or a decrease of 12.

Using this terminology, we now state a concise definition of quadratic growth.

> **Verbal Definition of Quadratic Growth:** In a quadratic growth sequence the differences are not all the same; they follow a pattern of arithmetic growth.

Read that statement carefully! The terms themselves don't exhibit arithmetic growth —the *differences* do.

3.1. Properties of Quadratic Growth

Second Differences. There is another way to describe quadratic growth sequences, using a concept called *second differences*. Second differences are the differences of the differences. We have already seen how you can start with a sequence and compute the differences. Take the differences of these differences in the same way, and that gives what are called the second differences of the original sequence. For example, suppose we begin with the sequence 6, 12, 20, 30, 42, 56. Then the differences are computed to be 6, 8, 10, 12, 14. Now look at the differences for these numbers—from 6 to 8, then from 8 to 10, 10 to 12, and so on. Those are the second differences of the original terms. And in this example, the second differences are all the same. They all equal 2. This is illustrated schematically in Figure 3.1.

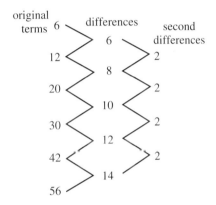

Figure 3.1. Constant second differences.

For any quadratic growth sequence, the differences follow a pattern of arithmetic growth. That means that the differences of these differences must be a fixed nonzero constant. That is the second way to think about quadratic growth.

> **Constant Second Difference Definition of Quadratic Growth:** In a quadratic growth sequence, the second differences are constant and not zero.

This idea applies to the example in Figure 3.1. For the sequence 6, 12, 20, 30, 42, 56, ⋯ all of the second differences are equal, and nonzero, so the sequence is an example of quadratic growth. On the other hand, in an exercise in Section 1.1 we were introduced to the tetrahedral numbers 1, 4, 10, 20, 35, 56, ⋯, for which the first differences are 3, 6, 10, 15, 21, ⋯ and the second differences are 3, 4, 5, 6, ⋯. Because the second differences are *not* all equal, this is not a quadratic growth sequence.

Tables with First and Second Differences. The format in Figure 3.1 is intended to remind us how differences and second differences are computed, and is convenient for working out examples by hand. A somewhat more formal alternative appears in Table 3.2, which shows first and second differences for the tetrahedral numbers. This is the format we will use in showing first and second differences from this point forward.

Creating a table of this sort often reveals a pattern. It may be as simple as constant second differences, which would indicate a quadratic growth sequence. Or there

Table 3.2. Tetrahedral numbers with first and second differences.

n	a_n	1st Diffs	2nd Diffs
0	1		
		3	
1	4		3
		6	
2	10		4
		10	
3	20		5
		15	
4	35		6
		21	
5	56		

may be some other pattern, which is what we see in Table 3.2: the second differences are consecutive whole numbers, starting with 3. As a particularly surprising example, consider the sequence 6, 12, 24, 48, 96, 192, $\cdots$, in which each term is twice the preceding term. First and second differences for this sequence are shown in Table 3.3, where we see that the first difference column reproduces the orginal terms of the sequence, and likewise for the second differences column. By the end of this book, you will have learned to recognize a number of such patterns, and seen how they can be used to analyze sequences.

Table 3.3. Doubling sequence with first and second differences.

n	a_n	1st Diffs	2nd Diffs
0	6		
		6	
1	12		6
		12	
2	24		12
		24	
3	48		24
		48	
4	96		48
		96	
5	192		

With the terminology of first and second differences, we can express the definitions of arithmetic and quadratic growth a little more succinctly. For arithmetic growth sequences the first differences are constant. For quadratic growth sequences the first differences are not constant but the second differences are. Notice that when the first differences are constant, the second differences are all zero. That is why we require nonzero second differences in our description of quadratic growth. Otherwise, every arithmetic growth sequence would also be classified as a quadratic growth sequence, in a trivial way, with constant second differences equal to zero.

The idea of constant second differences is useful in finding additional terms of a quadratic growth sequence. For example, in Figure 3.1, the next entry in the differences column would have to be 16 in order to continue the pattern of constant second differences. That implies that the next entry in the original terms column should be 56 + 16 = 72. This reveals a recursive process for extending the sequence, and so foreshadows the use of a difference equation. We turn now to a more systematic look at the difference equations for quadratic growth models.

3.1. Properties of Quadratic Growth

Difference Equations. We have already seen a difference equation for the added multiples of three sequence, based on a pattern (see page 28). We began with some specific equations like these,

$$t_1 = t_0 + 3 \cdot 1$$
$$t_2 = t_1 + 3 \cdot 2$$
$$t_3 = t_2 + 3 \cdot 3$$
$$t_4 = t_3 + 3 \cdot 4,$$

which show how each of the first few terms is related to the preceding term. Following the pattern of these equations, we can write a corresponding equation for any term of the sequence. As a particular instance,

$$t_{16} = t_{15} + 3 \cdot 16.$$

This leads to the difference equation

$$t_{n+1} = t_n + 3(n+1). \tag{3.1}$$

However, we now know more about arithmetic growth, and so can obtain the same result in a more direct way. We know that the sequence of differences,

$$3, 6, 9, 12, 15, \cdots,$$

is an instance of arithmetic growth, with an initial term of 3 and a common difference also equal to 3. Therefore any one of the differences can be computed as $3 + 3n$ for the corresponding position number n. But we also know that this is the difference between t_n and t_{n+1}. Together, these observations show that

$$t_{n+1} - t_n = 3 + 3n,$$

leading to

$$t_{n+1} = t_n + 3 + 3n.$$

That is equivalent to (3.1).

For any quadratic growth sequence similar developments can be followed. We can use patterns to find a difference equation, or we can apply our knowledge of arithmetic growth to obtain an equation for the nth term of the sequence of differences. By either method, we derive a difference equation of the form

$$a_{n+1} = a_n + (\text{a constant}) + (\text{a constant}) \cdot n \tag{3.2}$$

The constants are parameters, meaning that in any specific quadratic growth sequence they will be replaced with particular numbers. Note that the second constant cannot be zero. Otherwise, the difference equation would become

$$a_{n+1} = a_n + (\text{a constant}),$$

making the original sequence an instance of arithmetic growth.

If we use d and e to stand for the constants in (3.2), we obtain the following general form.

> **Quadratic Growth Difference Equation:** A quadratic growth sequence always satisfies an equation of the form
>
> $$a_{n+1} = a_n + d + en, \tag{3.3}$$
>
> where d and e are parameters, and $e \neq 0$.

The difference equation for a quadratic growth sequence can be expressed in many different forms. We refer to (3.3) as the *standard form*.

We can also write the equation as

$$a_{n+1} - a_n = d + en.$$

In terms of differences this says

$$\text{difference}_n = d + en.$$

The condition $e \neq 0$ means that $d + en$ is not a constant. Thus we arrive at an alternate way to think about quadratic growth:

> **Alternate Verbal Description of Quadratic Growth:** A quadratic growth sequence is one in which the difference between a_{n+1} and a_n is a (nonconstant) linear function of n.

This is closely related to our first definition (page 122), recognizing that terms in an arithmetic growth sequence are given by a linear function of n.

Using the Difference Equation. Let us see how we can use the difference equation, and check that it is correct. For the added multiples of three sequence, we found the difference equation

$$t_{n+1} = t_n + 3 + 3n.$$

Substituting 0 for n, we obtain

$$t_1 = t_0 + 3 + 3 \cdot 0.$$

Using the fact that $t_0 = 5$, the equation becomes $t_1 = 8$, which we know is correct.

Next, replace n with 1. The difference equation becomes

$$t_2 = t_1 + 3 + 3 \cdot 1,$$

or more simply,

$$t_2 = t_1 + 6.$$

We just verified that $t_1 = 8$, so now $t_2 = 14$, again correct.

We can continue in this fashion, verifying that the difference equation produces the correct values for t_n up to $n = 5$. We can also find new terms of the sequence according to the same rule. Thus, with $n = 5$, the difference equation becomes

$$t_6 = t_5 + 3 + 3 \cdot 5 = 50 + 18 = 68.$$

In similar fashion we can use the difference equation to continue the sequence for as many terms as we wish.

Finding d and e. There is a simple way to find the difference equation for a quadratic growth sequence from a table showing the first and second differences. As an example we consider the sequence $12, 21, 28, 33, 36, 37, \cdots$, and use the label b with $b_0 = 12, b_1 = 21$, etc. It is not immediately obvious whether this is a quadratic growth sequence, so we make Table 3.4. The constant second differences tell us that this is a quadratic growth sequence. The parameters d and e are the first entries in the 1*st Diffs* and 2*nd Diffs* columns, respectively. Thus $d = 9, e = -2$, and the difference equation for this sequence is

$$b_{n+1} = b_n + 9 - 2n.$$

3.1. Properties of Quadratic Growth

Table 3.4. A sequence table with first and second differences. The second differences are all equal to -2 so this is a quadratic growth sequence. The parameter d is the first entry in the *1st Diffs* column, or 9. The parameter e is the first entry in the *2nd Diffs* column, or -2.

n	b_n	1st Diffs	2nd Diffs
0	12	9	
1	21	7	-2
2	28	5	-2
3	33	3	-2
4	36	1	-2
5	37		

Why does this method work? Focus just on the *1st Diffs* column. That is an arithmetic growth sequence, with initial term 9 and common difference -2. This shows that the nth entry of the column is

$$\text{difference}_n = 9 - 2n,$$

and therefore

$$b_{n+1} - b_n = 9 - 2n.$$

This is equivalent to the preceding difference equation.

This procedure for finding the standard difference equation for a quadratic growth sequence is worth highlighting for future reference.

> **Finding Quadratic Growth Parameters d and e:** In any quadratic growth sequence, there is a difference equation of the form
>
> $$a_{n+1} = a_n + d + en,$$
>
> where d and e are parameters that represent constant numerical values, and $e \neq 0$. When the first several terms of the sequence, starting with a_0, are entered in a data table with first and second differences, the parameter d is the first entry of the column of first differences, and the parameter e is the first entry in the column of second differences.

We emphasize the ease with which a quadratic growth difference equation can be found. As in the example, for any sequence we can form a table with first and second differences. If the second differences are constant, but not the first differences, that shows our sequence is an instance of quadratic growth. Defining d and e as the top entries in the columns for first and second differences, we obtain the difference equation. Then we can use the difference equation to extend the sequence for additional terms.

For the b_n example, using $b_5 = 37$, we proceed to compute

$$b_6 = b_5 + 9 - 2 \cdot 5 = 37 - 1 = 36$$
$$b_7 = b_6 + 9 - 2 \cdot 6 = 36 - 3 = 33$$
$$b_8 = b_7 + 9 - 2 \cdot 7 = 33 - 5 = 28.$$

To verify that these terms are the correct continuation of the pattern of quadratic growth, use them to extend Table 3.4 to $n = 8$. You will find that the *1st* and *2nd Diffs* columns continue with the same pattern as for the preceding terms.

Although we now have a ready means to find a quadratic growth difference equation, it is the functional equation that will be most useful in applications. As we will see, the standard difference equation is the starting point in finding quadratic growth functional equations. Before turning to that topic, though, we take a brief look at graphs of quadratic sequences.

Graphs. Quadratic growth sequences can be graphed using the same methods we have discussed in previous chapters (see page 17). In Figure 3.2 and Figure 3.3, we show graphs for the added multiples of three sequence t_n and the example sequence b_n discussed above. As usual, each term of the sequence is represented with a dot. In the first graph we have also drawn straight lines between the dots, to emphasize the appearance of an upward bending arc. For the second graph no lines have been included. Both are valid approaches to graphing a sequence. The decision to include or omit such lines is a matter of personal preference.

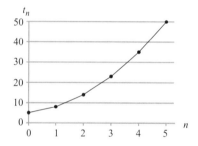

Figure 3.2. Graph for the added multiples of three sequence. The straight lines connecting the dots are provided for emphasis only. Only the dots correspond to terms of the sequence.

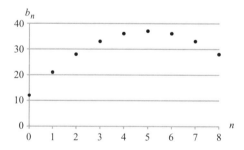

Figure 3.3. Graph for the example sequence b_n. No connecting lines have been drawn between the dots.

The shapes revealed in Figure 3.2 and Figure 3.3 are characteristic of quadratic growth. In each case the dots lie on a special shaped curve called a parabola. The main point of emphasis here is that these graphs are definitely not straight lines. This is a visual representation of the distinction between arithmetic and quadratic growth, and has several important consequences. For one thing, proportional reasoning is not

3.1. Properties of Quadratic Growth

valid for quadratic growth models. For another, quadratic growth models can increase at first, but later decrease, as shown in Figure 3.3, or vice versa. In these cases the graph will have a highest point or a lowest point, and such a point has special significance in the context of a model. We saw this same idea in connection with the Laffer curve on page 46. There, the inverted U shape of the curve was meant to be taken in a figurative sense, showing that raising taxes can raise government revenue in some cases but lower it in others. Now we see that this same shape actually arises in quadratic growth models.

All of these observations highlight important differences between models with straight line graphs, and those whose graphs are not straight lines. Models of the latter type are referred to as *nonlinear*, to emphasize the distinction. Quadratic growth models are always nonlinear. But we can say much more than that. As we will see later in the chapter, the graphs of quadratic growth models have specific features that allow us to predict their shapes quite accurately.

In order to analyze the graphs of quadratic growth models, we need to know about their functional equations. We turn to that topic next.

Functional Equations. We turn now to functional equations. Remember that difference equations are *recursive*, they let you compute the next term of a sequence once you know the preceding term. The usual way to compute a particular term using the difference equation is step by step: start out with a_0, then compute a_1, then a_2, then a_3, and so on, until you reach the one you want. In contrast, functional equations are direct. If you want to compute the 25th term, you simply substitute 25 for n and immediately compute a_{25}.

To highlight this distinction, consider again the added multiples of three sequence, t_n. We have found a difference equation

$$t_{n+1} = t_n + 3 + 3n$$

and know how to find terms recursively. But most readers will be unable at this point to predict the value of t_{100} directly.

As we are about to see, quadratic growth sequences all have functional equations that share a common form. Moreover, we can find functional equations by recognizing a pattern in the systematic application of difference equations. To illustrate this idea we will work out a functional equation for the b_n example considered earlier.

Recall that the initial term is $b_0 = 12$, and that the difference equation is

$$b_{n+1} = b_n + 9 - 2n.$$

Starting with b_0 we can systematically work out b_1, then b_2, and so on. For example,

$$b_1 = 12 + 9 - 2 \cdot 0.$$

We leave the answer in this form, showing the operations that go into the calculation, instead of proceeding to a final answer of 21. Our goal is to identify a pattern produced by repeated application of similar operations, and for that we need to be able to see what the operations are.

Proceeding, to get b_2, take the value we just found for b_1 and add $9 - 2 \cdot 1$. That gives the equation

$$b_2 = 12 + 9 - 2 \cdot 0 + 9 - 2 \cdot 1.$$

Similarly, $b_3 = b_2 + 9 - 2 \cdot 2$, so take what we already have for b_2 and add $9 - 2 \cdot 2$:
$$b_3 = 12 + 9 - 2 \cdot 0 + 9 - 2 \cdot 1 + 9 - 2 \cdot 2.$$

To reveal the pattern more clearly, we rewrite this equation keeping all of the 9's together, and also keeping all the subtracted terms together. This is what it looks like:
$$b_3 = 12 + 9 + 9 + 9 - 2 \cdot 0 - 2 \cdot 1 - 2 \cdot 2.$$

This process can be continued as long as we wish. The following pattern results:

$$\begin{aligned}
b_0 &= 12 \\
b_1 &= 12 + 9 & &- 2 \cdot 0 \\
b_2 &= 12 + 9 + 9 & &- 2 \cdot 0 - 2 \cdot 1 \\
b_3 &= 12 + 9 + 9 + 9 & &- 2 \cdot 0 - 2 \cdot 1 - 2 \cdot 2 \\
b_4 &= 12 + 9 + 9 + 9 + 9 & &- 2 \cdot 0 - 2 \cdot 1 - 2 \cdot 2 - 2 \cdot 3.
\end{aligned}$$

This pattern can be expressed a little more compactly by combining all the 9's together and all subtracted terms together. Then the equations above become

$$\begin{aligned}
b_0 &= 12 \\
b_1 &= 12 + 9 \cdot 1 & &- 2(0) \\
b_2 &= 12 + 9 \cdot 2 & &- 2(0 + 1) \\
b_3 &= 12 + 9 \cdot 3 & &- 2(0 + 1 + 2) \\
b_4 &= 12 + 9 \cdot 4 & &- 2(0 + 1 + 2 + 3).
\end{aligned}$$

Now the pattern should be clear. We can use it continue the list for several more equations, obtaining

$$\begin{aligned}
b_5 &= 12 + 9 \cdot 5 & &- 2(0 + 1 + 2 + 3 + 4) \\
b_6 &= 12 + 9 \cdot 6 & &- 2(0 + 1 + 2 + 3 + 4 + 5) \\
b_7 &= 12 + 9 \cdot 7 & &- 2(0 + 1 + 2 + 3 + 4 + 5 + 6).
\end{aligned}$$

In fact, we can use the pattern to write an equation for any b_n, without finding all the preceding equations. For example, the predicted equation for b_{10} would be
$$b_{10} = 12 + 9 \cdot 10 - 2(0 + 1 + 2 + 3 + 4 + 5 + 6 + 7 + 8 + 9).$$

In particular, we can now compute the numerical value $b_{10} = 12$, without finding the preceding values. Thus we have found a *direct* pattern for computing terms of the sequence.

At this point, we have merely extended an observed pattern. And while patterns do not always continue, in this case the predicted equations are correct. This can be verified using the difference equation, similarly to the way we derived the first four equations.

Notice that the equation for b_0 does not follow the pattern. That is because b_0 is the initial term. The pattern does give correct equations for every b_n with $n \geq 1$, just as we saw for $n = 10$.

An Amazing Shortcut. Although we now have a direct method for calculating b_n, it is not very convenient to use. Suppose you wanted to compute b_{50}. The pattern says to start with $12 + 9 \cdot 50$, and that is easy enough. But next we must subtract 2 times the total of all the whole numbers from 0 through 49.[2] Just writing this down is awkward,

---

[2] Check the pattern carefully: the last number added in the parenthesis is one less than the subscript on the left side of the equation, so for b_{50}, the last number in the parentheses is 49.

3.1. Properties of Quadratic Growth

and nobody would want to actually do all that addition. Fortunately, there is a shortcut that can be applied to problems of this type.[3]

Here is how the shortcut works. First, we can ignore the starting 0, which contributes nothing to the total. To add up all the numbers from 1 to 49, multiply 49 by the next whole number, 50, and divide by 2. The answer is $49 \cdot 50/2 = 1{,}225$ or equivalently, $\frac{49 \cdot 50}{2} = 1{,}225$. (The two forms, $49 \cdot 50/2$ and $\frac{49 \cdot 50}{2}$ will be used interchangeably.) Similarly, to add up all the numbers from 1 to 100, just multiply 100 by 101, and divide by 2: $100 \cdot 101/2 = 5{,}050$. You can see that with a calculator this shortcut is easy to use even with very large numbers. The total of all the whole numbers from 1 to 1,000 is $1{,}000 \cdot 1{,}001/2 = 500{,}500$.

This shortcut is an amazing simplification. It almost seems too good to be true. In fact, how do we know it *is* true? Of course, it is easy to check whether it gives the right answer for small numbers. We can add up $1 + 2 + 3 + 4$ directly to find the answer 10, and see that the shortcut answer $4 \cdot 5/2$ is correct. But how do we know that it works for gigantic numbers like 1,000 or 10,000 or a million? One answer is that a deductive proof (similar to what many students study in high school geometry) is known. The main concept involved in this proof will be presented near the end of the section. For the present, you are asked to accept that the shortcut works as described so that it can be used in the continued study of quadratic growth functional equations.

Before returning to that subject, one additional digression is justified. The idea of adding up many, many numbers is sometimes awkward to describe. We have referred before to adding up all the whole numbers from 1 to 1,000, for example. But it is simply not practical to write out all 1,000 numbers with plus signs.

For this reason we introduce a shorthand notation that is commonly used with sums. It uses three dots to indicate continuation of a pattern. In this approach, rather than writing $1 + 2 + 3 + 4 + 5 + 6 + 7$ out in full, the shorter $1 + 2 + 3 + \cdots + 7$ is used. Similarly, $1 + 2 + 3 + \cdots + 100$ is understood to mean adding up all the whole numbers from 1 to 100. The first few terms are included to show what the pattern is, and the last number tells where to stop adding. So $2 + 4 + 6 + \cdots + 100$ means to add all the *even* whole numbers from 2 to 100. Using this shorthand notation, the shortcut for adding whole numbers can be written

$$1 + 2 + 3 + \cdots + n = \frac{n(n+1)}{2}.$$

Or, in a form that is more compatible with quadratic growth functional equations,

$$0 + 1 + 2 + 3 + \cdots + n = \frac{n(n+1)}{2}. \tag{3.4}$$

Notice that this equation makes sense for $n = 0$, whereas the preceding version does not.

Now we can turn again to the discussion of functional equations in quadratic growth.

---

[3]The shortcut has been known for centuries, and has been discovered independently by many mathematicians from different cultures. The prominent 19th century German mathematician Karl F. Gauss is supposed to have figured it out at the age of 11 when his mathematics class was assigned to add all of the whole numbers from 1 to 100.

A Functional Equation for b_n. Remember that we were studying the pattern of values for b_n. We had worked out the following equations:

$$\begin{aligned}
b_0 &= 12 \\
b_1 &= 12 + 9 \cdot 1 - 2(0) \\
b_2 &= 12 + 9 \cdot 2 - 2(0 + 1) \\
b_3 &= 12 + 9 \cdot 3 - 2(0 + 1 + 2) \\
b_4 &= 12 + 9 \cdot 4 - 2(0 + 1 + 2 + 3) \\
b_5 &= 12 + 9 \cdot 5 - 2(0 + 1 + 2 + 3 + 4) \\
b_6 &= 12 + 9 \cdot 6 - 2(0 + 1 + 2 + 3 + 4 + 5) \\
b_7 &= 12 + 9 \cdot 7 - 2(0 + 1 + 2 + 3 + 4 + 5 + 6).
\end{aligned}$$

Now we can apply the shortcut (3.4) for adding whole numbers. For example, in the last equation, instead of $0 + 1 + 2 + 3 + 4 + 5 + 6$ we can write $6 \cdot 7/2$. Applying the shortcut in a similar fashion to each of the last few equations in the set above, we get

$$\begin{aligned}
b_4 &= 12 + 9 \cdot 4 - 2 \cdot \frac{3 \cdot 4}{2} \\
b_5 &= 12 + 9 \cdot 5 - 2 \cdot \frac{4 \cdot 5}{2} \\
b_6 &= 12 + 9 \cdot 6 - 2 \cdot \frac{5 \cdot 6}{2} \\
b_7 &= 12 + 9 \cdot 7 - 2 \cdot \frac{6 \cdot 7}{2}.
\end{aligned}$$

There is again an easily recognized pattern here. You will have no trouble guessing that $b_{73} = 12 + 9 \cdot 73 - 2(72)(73)/2$. Using the variable n, the pattern can be expressed in the general form

$$b_n = 12 + 9n - 2 \cdot \frac{(n-1)n}{2}. \tag{3.5}$$

That is a functional equation for b_n. Using algebra, we can simplify it a little to

$$b_n = 12 + 9n - (n-1)n. \tag{3.6}$$

With this functional equation we can compute any term of the sequence. For example, to find b_{15} we set $n = 15$ and compute

$$b_{15} = 12 + 9 \cdot 15 - 14 \cdot 15 = -63.$$

We can also verify that it gives correct answers for the terms we already know. In particular, substituting $n = 0$ we find

$$b_0 = 12 + 9 \cdot 0 - (-1)(0) = 12.$$

Interestingly, even though the pattern we found does not apply for the initial term, the equation still gives the correct value. Indeed, the functional equation gives the correct value of b_n for every whole number n.

Another Quadratic Growth Functional Equation Example. The process we used for the b_n sequence can be followed for any quadratic growth sequence. In fact, given the first several terms of any sequence, we can examine the first and second differences to determine whether quadratic growth is in evidence. If it is, we can find the difference equation, and then use that systematically to derive a functional equation. Here is an example.

3.1. Properties of Quadratic Growth

Suppose we wish to find a functional equation for the sequence

$$5, 27, 59, 101, 153, 215, \cdots.$$

We compute the differences and the second differences as shown in the following table.

n	a_n	1st Diffs	2nd Diffs
0	5		
		22	
1	27		10
		32	
2	59		10
		42	
3	101		10
		52	
4	153		10
		62	
5	215		

The second differences are all equal to 10. This indicates that the terms belong to a quadratic growth sequence.[4] What is more, as in an earlier example we can read the parameters $d = 22$ and $e = 10$ out of the table, thus finding the difference equation

$$a_{n+1} = a_n + 22 + 10n.$$

We also observe that $a_0 = 5$.

Following the same process as before, we now use the difference equation systematically with n equal to 0, 1, 2, 3, and so on. With a little simplification, we find the pattern

$$\begin{aligned}
a_0 &= 5 \\
a_1 &= 5 + 22 \cdot 1 + 10(0) \\
a_2 &= 5 + 22 \cdot 2 + 10(0 + 1) \\
a_3 &= 5 + 22 \cdot 3 + 10(0 + 1 + 2) \\
a_4 &= 5 + 22 \cdot 4 + 10(0 + 1 + 2 + 3) \\
a_5 &= 5 + 22 \cdot 5 + 10(0 + 1 + 2 + 3 + 4) \\
a_6 &= 5 + 22 \cdot 6 + 10(0 + 1 + 2 + 3 + 4 + 5) \\
a_7 &= 5 + 22 \cdot 7 + 10(0 + 1 + 2 + 3 + 4 + 5 + 6).
\end{aligned}$$

Applying the shortcut (3.4), we rewrite the equations as

$$\begin{aligned}
a_0 &= 5 \\
a_1 &= 5 + 22 \cdot 1 + 10(0)(1)/2 \\
a_2 &= 5 + 22 \cdot 2 + 10(1)(2)/2 \\
a_3 &= 5 + 22 \cdot 3 + 10(2)(3)/2 \\
a_4 &= 5 + 22 \cdot 4 + 10(3)(4)/2 \\
a_5 &= 5 + 22 \cdot 5 + 10(4)(5)/2 \\
a_6 &= 5 + 22 \cdot 6 + 10(5)(6)/2 \\
a_7 &= 5 + 22 \cdot 7 + 10(6)(7)/2.
\end{aligned}$$

Based on the pattern of these equations, we formulate the general equation

$$a_n = 5 + 22n + 10 \cdot \frac{(n-1)n}{2}.$$

---

[4]That is hardly surprising in an example at this point in the chapter, but in some other context, if you are trying to decide whether a set of data might follow a simple model, looking at differences in this way is a good start. If the differences themselves are constant, you will know that an arithmetic growth model is correct. If the differences are not constant but the second differences are constant, that indicates a quadratic growth model. If no differences are constant, some other type of model may be required.

If you check this with the first few values of n, you will see that it correctly gives the terms $5, 27, 59, 101, 153, 215$.

A General Quadratic Growth Functional Equation. As the two examples illustrate, we can use the difference equation for any quadratic growth sequence to find a functional equation, and the form of the equation will always be the same:

$$a_n = \text{(a constant)} + \text{(a constant)} \cdot n + \text{(a constant)} \cdot (n-1)n/2.$$

Furthermore, each constant in this equation is a known parameter. The first constant is a_0, and the second and third constants are, respectively, the parameters d and e from the difference equation. Thus, we have

$$a_n = a_0 + dn + e(n-1)n/2. \tag{3.7}$$

This gives a functional equation in a parametric form that can be used with any quadratic growth sequence. For future reference we summarize this result below.

> **Quadratic Growth Functional Equation:** If a quadratic growth sequence has an initial value of a_0 and obeys the difference equation $a_{n+1} = a_n + d + en$, then a functional equation is
> $$a_n = a_0 + dn + e(n-1)n/2.$$
> Here, d is the initial first difference, $a_1 - a_0$, and e is the common second difference.

As a final example, let us use this result to find a functional equation for the added multiples of three sequence. We have already seen that its difference equation is

$$t_{n+1} = t_n + 3 + 3n,$$

so we know that d and e both equal 3. We also know that $t_0 = 5$. Substituting these values into the general quadratic growth functional equation leads to

$$t_n = 5 + 3n + 3(n-1)n/2. \tag{3.8}$$

To make sure we have made no errors, we can apply this equation with a few small values of n. For $n = 2$ the equation becomes

$$t_2 = 5 + 3 \cdot 2 + 3(2-1)2/2 = 5 + 6 + 3 = 14,$$

which is correct. Similarly, with $n = 5$ we find

$$t_5 = 5 + 3 \cdot 5 + 3(5-1)5/2 = 5 + 15 + 30 = 50,$$

again producing the correct value.

It is remarkable how easily we can find functional equations for quadratic growth sequences. Initially, although we understood a clear pattern for the terms of the t_n sequence, we had no simple way to find t_{100}. But now we can immediately compute

$$t_{100} = 5 + 3 \cdot 100 + 3(99)100/2 = 5 + 300 + 14{,}850 = 15{,}155.$$

3.1. Properties of Quadratic Growth

Find-n Questions. The computation of t_{100} illustrates the most direct way to use a functional equation. It answers what we have called a find-a_n question. But we have also seen the importance of find-n questions. For example, we might ask how far you must go in the added multiples of three sequence to reach a term of one million. In symbols, that is asking to find n such that

$$t_n = 1{,}000{,}000.$$

Here we take a brief look at methods for answering such questions. In Section 3.2 we will see applications of quadratic growth where find-n questions arise naturally.

The problem of finding n so that $t_n = 1{,}000{,}000$ can be approached numerically, graphically, or algebraically. Using a numerical approach, we can apply (3.8) in a process of systematic trial and error. We have seen that $t_{100} = 15{,}155$, which is far too small. Let's try $n = 1{,}000$. We have

$$t_{1000} = 5 + 3 \cdot 1000 + 3(999)1000/2 = 1{,}501{,}505,$$

which is too large. For a third trial,

$$t_{500} = 5 + 3 \cdot 500 + 3(499)500/2 = 375{,}755.$$

Continuing in this fashion, we can keep refining our choices of n. Eventually we will see that $t_{815} = 997{,}565$ and $t_{816} = 1{,}000{,}013$. These results show that there is no term of the sequence that is exactly equal to one million, and that the first term that exceeds one million is t_{816}.

We can also discover this result graphically. Changing n to x and t_n to y, the functional equation becomes

$$y = 5 + 3x + 3(x-1)x/2.$$

The condition $t_n = 1{,}000{,}000$ likewise becomes

$$y = 1{,}000{,}000.$$

A point where the graphs of both equations meet is the graphical equivalent of the condition that $t_n = 1{,}000{,}000$. Using a graphing calculator or a graphing application on a computer, and interactively adjusting the scale of the graph, the point or points of intersection can be found (Figure 3.4).

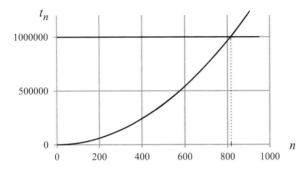

Figure 3.4. The condition $t_n = 1{,}000{,}000$ occurs where the curve $y = 5 + 3x + 3(x-1)x/2$ intersects the horizontal line $y = 1{,}000{,}000$.

To approach this same problem algebraically, set $t_n = 1{,}000{,}000$ in (3.8) to obtain

$$1{,}000{,}000 = 5 + 3n + 3(n-1)n/2.$$

Applying the distributive law to the factor $(n-1)$ and combining like terms leads to

$$(3/2)n^2 + (3/2)n - 999{,}995 = 0,$$

which is a *quadratic equation*. It can be solved using the quadratic formula, a familiar tool from algebra and precalculus courses. We will consider these matters at length in section Section 3.3, but if you recall how to use the quadratic formula, feel free to apply it in the exercises for this section.

Proving the Amazing Shortcut. The patterns that permitted us to find functional equations for quadratic growth models were based on the shortcut for adding consecutive whole numbers:

$$1 + 2 + \cdots + n = n(n+1)/2.$$

When the shortcut was introduced, the reader was asked to accept its validity without much justification. Now we will see that the shortcut is valid for any number of terms, and why.

The key idea involves adding up the terms of the sum in a different order. To illustrate this idea, here is the sum for $n = 10$ ordered two ways:

$$1 + 2 + 3 + 4 + 5 + 6 + 7 + 8 + 9 + 10 = (1+10) + (2+9) + (3+8) + (4+7) + (5+6).$$

On the right there are 5 parenthetical groups, each with a sum of 11. Note that the number of groups is $n/2$ because each group has two terms in it. The sum of each group is $n + 1$. So the sum is equal to $(n/2)(n+1) = 5 \cdot 11 = 55$. This agrees with the shortcut formula $n(n+1)/2$. And what we observe in this example will occur for any even n.

For an odd number of terms, things are a little bit different. For $n = 9$ group like so:

$$1 + 2 + 3 + 4 + 5 + 6 + 7 + 8 + 9 = (1+8) + (2+7) + (3+6) + (4+5) + 9.$$

This time there are $5 = (n+1)/2$ groups (one of which is just a single number) and the total of each group is $9 = n$. The overall total is thus $9 \cdot 5 = n \cdot (n+1)/2$. This, too, agrees with the shortcut formula $n(n+1)/2$. And as before, what we observe in this example will occur for any odd n.

Putting both arguments together shows that the shortcut is valid for any n, even or odd.

Profile for Quadratic Growth. At the end of Section 2.1, we presented a profile of arithmetic growth. We should update that profile to incorporate the concept of differences covered in this chapter. The updated version is:

Table 3.5. Updated profile for arithmetic growth sequences.

Verbal Description:	Each term increases (or decreases) from the preceding term by a constant amount
Identifying Characteristic:	Constant first differences
Parameters:	Initial term a_0, constant difference d
Difference Equation:	$a_{n+1} = a_n + d$
Functional Equation:	$a_n = a_0 + dn$
Graph:	Straight line

To summarize the results of this section, here is an analogous profile for quadratic growth:

Table 3.6. Profile for quadratic growth sequences.

Verbal Description:	First differences form an arithmetic growth sequence
Identifying Characteristic:	Constant second differences (nonzero)
Parameters:	Initial term a_0, initial first difference d, constant second difference $e \neq 0$
Difference Equation:	$a_{n+1} = a_n + d + en$
Functional Equation:	$a_n = a_0 + dn + e(n-1)n/2$
Graph:	Points lie on a parabolic curve

3.1 Exercises

Reading Comprehension. Answer each of the following with a short essay. Give examples, where appropriate, but also answer the questions in written sentences.

(1) What is a quadratic growth sequence?

(2) @In the equation $a_{n+1} = a_n + d + en$, d and e are called parameters. Explain what parameters are and how they differ from variables (such as n).

(3) Describe three differences between quadratic growth models and arithmetic growth models.

(4) @It is stated in the reading that quadratic growth sequences are nonlinear. In what way does the functional equation for a quadratic growth sequence reveal this nonlinearity? In what way does the graph?

(5) Explain what *second differences* are, and why they are significant in the context of quadratic growth sequences.

Mathematical Skills.

(6) @For each of these sequences construct a table of first and second differences. Use your table to determine the next 3 terms in each sequence.

a. @1, 6, 9, 10, 9, 6, 1 ⋯
b. @1, 2, 9, 28, 65, 126, 217 ⋯
c. −10.3, −5.3, 9.7, 34.7, 69.7, 114.7, 169.7 ⋯

(7) Use the functional equation $a_n = 60 - 5n + 4(n-1)n/2$ to determine the first five terms in the sequence. Then construct a table of first and second differences to verify that the resulting sequence is quadratic, has initial first difference $d = -5$, and constant second difference $e = 4$.

(8) Find the second differences of the sequence 7, 14, 28, 56, 112, ⋯. This sequence is not quadratic. What do you notice about the second differences?[5]

(9) @In a quadratic growth model, the initial term is $b_0 = 6$, and the difference equation is $b_{n+1} = b_n + 0.5 + 0.2n$.

   a. @Use the difference equation to determine b_1, b_2, b_3, and b_4.
   b. @Make a graph of the sequence, including at least 6 points on the graph. (You may do this by hand, or use a graphing calculator or computer.)
   c. @What are the parameters d and e for this sequence?
   d. @Find the functional equation for b_n. Use it to check the answer you got for b_4, and also to find b_{10}.

(10) In a quadratic growth model, the initial term is $z_0 = 10$, and the difference equation is $z_{n+1} = z_n + 2 - n$.

   a. Use the difference equation to determine z_1, z_2, z_3, and z_4.
   b. Make a graph for the z_n's. Include at least 6 points on the graph. (You may do this by hand, or use a graphing calculator or computer.)
   c. What are the parameters d and e?
   d. Find the functional equation for z_n. Use it to check the answer you got for z_4, and also to find z_{14}.

(11) The first differences of a sequence are shown below. Use them to determine the functional equation of the sequence.

n	a_n	Difference
0	7	
		4
1	11	
		7
2	18	
		10
3	28	
		13
4	41	
		16
5	57	

(12) For a sequence $a_0, a_1, a_2, \cdots$, the first differences follow this arithmetic rule:

$$\text{difference}_n = 1.2 + 7.5n.$$

The initial term of the sequence is $a_0 = 20$.

---

[5] The same thing was observed on page 124 for the sequence 6, 12, 24, 48, 96, 192, ⋯. This idea is explored more in the Digging Deeper section of these exercises.

3.1. Exercises

a. Determine a_1, a_2, a_3, a_4, and a_5.

b. Find the functional equation for a_n.

(13) @A quadratic growth sequence has initial term $a_0 = 105$ and difference equation $a_{n+1} = a_n - 1 + 4n$.

   a. @Find a_{20}, showing work to justify your answer.

   b. @For what n does $a_n = 1700$? You may use numerical or graphical methods, as described on page 135. Alternatively, you are free to use any method you know for solving a quadratic equation. In either case, show work justifying your answer.

(14) @A quadratic growth sequence has initial term $a_0 = 10$, $a_1 = 10.5$, and $a_2 = 13.4$.

   a. @Find a_{16}, showing work to justify your answer.

   b. @For what n does $a_n = 476$? You may use numerical or graphical methods, as described on page 135. Alternatively, you are free to use any method you know for solving a quadratic equation. In either case, show work justifying your answer.

(15) A quadratic growth sequence has initial term $a_0 = 82$ and difference equation $a_{n+1} = a_n + 3 - 5n$.

   a. Find a_{13}, showing work to justify your answer.

   b. For what n does a_n first equal or fall below 0? You may use numerical or graphical methods, as described on page 135. Alternatively, you are free to use any method you know for solving a quadratic equation. In either case, show work justifying your answer.

(16) A quadratic growth sequence has initial term $a_0 = 8$, $a_1 = 5$, and $a_2 = 4$.

   a. Find a_{50}, showing work to justify your answer.

   b. For what n does a_n first equal or exceed 150? You may use numerical or graphical methods, as described on page 135. Alternatively, you are free to use any method you know for solving a quadratic equation. In either case, show work justifying your answer.

(17) @For each of the following, determine whether it is an arithmetic growth sequence, a quadratic growth sequence, or neither, giving a reason for your answer. If it is arithmetic or quadratic growth, find the difference and functional equations.

   a. @3, 5, 8, 12, 17, 23 $\cdots$

   b. @12, 13, 17, 26, 42, 67 $\cdots$

   c. 3.85, 3.54, 3.23, 2.92, 2.61, 2.30, $\cdots$

   d. 5, 16, 23, 26, 25, 20, $\cdots$

(18) @For each of the following, determine whether it is an arithmetic growth sequence, a quadratic growth sequence, or neither, giving a reason for your answer.

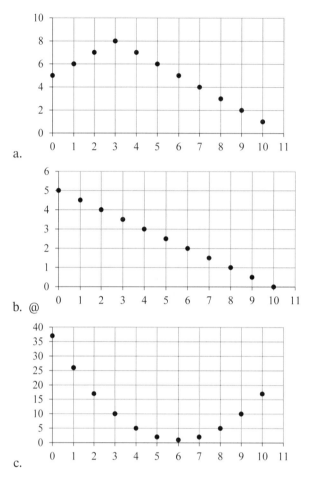

a.

b. @

c.

(19) @If you add up all the whole numbers from 1 to 50, what is the total?

(20) If you add up all the whole numbers from 1 to 1,000, what is the total?

(21) Compute $1 + 2 + 3 + \cdots + 25$.

Problems in Context.

(22) @Triangle Edges. The figure below shows four large equilateral trianges, in which all but the first have been subdivided into grids of smaller triangles.

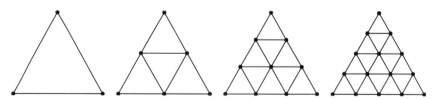

In each triangle, we refer to any line joining two adjacent dots as an edge. There are $e_1 = 3$ edges in the first triangle, $e_2 = 9$ in the second, $e_3 = 18$ in the third, and $e_4 = 30$ in the fourth. In general, let e_n be the total number of edges in a subdivided triangle whose bottom side has been divided into n edges.

3.1. Exercises

   a. @Find a pattern in the e sequence.
   b. @According to your pattern what are e_5 and e_6?
   c. By drawing an appropriate diagram, verify that your predicted value of e_5 is correct.
   d. @How many edges are in a subdivided triangle whose bottom side has been divided into 30 edges?

(23) @Pieces of Gold. Quadratica the Wise once solved a problem for a wealthy merchant. Told by the merchant to name her own reward, Quadratica pointed to the merchant's chess board and asked for one gold piece to be put on the first square, 2 gold pieces to be put on the second square, 3 on the third, and so on until all 64 squares on the chessboard were filled. How many pieces of gold were then on the chessboard?

(24) Total Hits. A film student posts a video on YouTube that attracts quite a following. There were 100 hits on the first day, 140 hits the second, 180 hits the third. Tracking the popularity of the video, the student records the values of $H_1, H_2, H_3, \cdots$, where H_n is the total hits for days one through n. Thus, $H_2 = 100 + 140 = 240$ total hits for days one and two. Similarly there were $H_3 = 100 + 140 + 180 = 420$ total hits for days one through three. Assuming the number of hits each day continues to grow according to the same pattern as for the first three days, find H_n for the next several values of n, and show that they define a quadratic growth sequence. Find the functional equation and use it to predict when the total number of hits will surpass one million.

(25) Sum of Odd Numbers. What is the sum of the first n odd whole numbers? [Hints: (1) Work out several examples. The sum of the first 3 odd whole numbers is $9 = 1 + 3 + 5$. Find the sum of the first 4, then the first 5, and so on. Use your results to define a sequence. What kind of sequence is it? (2) A reasonable initial term of the sequence would be "the sum of the first 0 odd whole numbers." What value makes sense as the initial term of the sequence?]

Digging Deeper.

(26) @Tetrahedral Numbers. In a Problem in Context (Ball Pyramids) in Section 1.1 the tetrahedral numbers 1, 4, 10, 20, 35, 56, 84, $\cdots$ were revealed in a problem about pyramids of balls. The sequence was re-visited in this section and shown not to be quadratic. Still, there is a tantalizing pattern to explore.

   a. @Based on the pattern of second differences, find the next three terms after 84.
   b. @Show that the first differences of the tetrahedral numbers forms a quadratic growth sequence.
   c. @Find a difference equation for the tetrahedral numbers.

(27) On page 124 we saw that the sequence 6, 12, 24, 48, 96, 192 $\cdots$ has a surprising property: **the differences are the same as the original terms**. Find other sequences with this property. Find a general rule that all such sequences must obey. For example, will every doubling sequence have this property? Will other non-doubling sequences have the property?

(28) @Suppose a quadratic growth sequence has $a_0 = 18$, $a_1 = 24$, and $a_4 = 82$.

   a. @What are a_2 and a_3?

   b. @You now have enough information to construct a functional equation for this sequence. Are 3 terms of a quadratic growth sequence always enough to find a functional equation? Why or why not?

(29) We have seen a shortcut for sums of the form $1 + 2 + 3 + \cdots + 50$. But what if we add multiples of 7 rather than consecutive whole numbers?

   a. Can you find a shortcut for sums of the form $7 + 14 + 21 + \cdots + 84$?

   b. Can you find a shortcut for sums of multiples of any number?

(30) @We have seen a shortcut for sums of the form $1 + 2 + 3 + \cdots + 50$. But what if we start with something other than 1? Can you find a shortcut for sums of the form $23 + 24 + 25 + \cdots + 68$?

3.2 Applications of Quadratic Growth

In this section we will look at three situations in which quadratic growth models arise. In the first we are led to a quadratic growth model by properties of observed data. For example, we may see a pattern of equal or nearly equal second differences in a sequence of data values. In the second situation we arrive at a quadratic growth difference equation using a method of recursive analysis. This approach can often be applied in problem contexts involving networks. The third instance of quadratic growth arises when it is of interest to add up successive data values in an arithmetic growth model.

Data Properties Suggesting Quadratic Growth. We saw in the preceding section that a quadratic growth sequence is characterized by first differences exhibiting arithmetic growth, or by constant second differences. It is rare to observe these conditions exactly in real data. But if they hold approximately, that is a clue that a quadratic growth model might provide a good approximation. If the conditions do not hold even approximately, properties of the data still might suggest using a quadratic model in some cases. We will see examples of both situations below, starting with a case where the first differences are closely approximated by an arithmetic growth sequence.

Motion in Freefall. In an early 17th century study of motion, Galileo investigated falling objects and balls rolling down ramps.[6] A modern version of his experiments might be formulated as follows. Lay out a measuring tape along the path of motion, and every second record the moving body's position along the tape. So, for a rolling ball, the starting point on the tape would be at 0. After one second it might be at 10 on the tape, after another second it would be at 40, then 90, and so on. These successive positions on the tape, 0, 10, 40, 90, and so forth, form a data sequence.

---

[6]For more on this subject, see [33].

3.2. Applications of Quadratic Growth

Using this and related methods, Galileo made many discoveries. Here is how he described one in which he seemed to take special pleasure:

> I have discovered by experiment some properties of [motion] which are worth knowing and which have not hitherto been either observed or demonstrated ... for so far as I know, no one has yet pointed out that the distances traversed, during equal intervals of time, by a body falling from rest, stand to one another in the same ratio as the odd numbers beginning with unity. [**26**, p. 153]

The distances referred to here are actually the first differences of the data described above. If the successive positions are 0, 10, 40, 90, and so on, then the distance traveled in the first second would be $10 - 0 = 10$. Similarly, the distance traveled in the second second would be $40 - 10 = 30$, and in the third second, $90 - 40 = 50$. So Galileo's discovery says that the first differences in the motion data grew in proportion to the odd numbers. If the first difference was one foot, then the succeeding differences would be 3 feet, 5 feet, 7 feet, etc. Or, if the first difference was 10 feet, then the succeeding differences would be 30 feet, 50 feet, 70 feet, and so on. Whatever he found for the difference of the first two data values, the difference of the next two would be three times as great, then 5, then 7, and would continue in the pattern of odd numbers. Of course, Galileo's experiments did not produce exact measurements of times and distances. Even if we repeat the experiment today, with modern equipment, we would not expect the differences to come out exactly in the pattern of odd numbers. But they would be close enough to that pattern to allow us to recognize it.

Galileo's discovery implies quadratic growth of the data sequence, with the first differences exhibiting arithmetic growth and the second differences constant. For example, if the ball travels 10 feet in the first second, then the first differences are approximately 10, 30, 50, 70, and so on, and the second differences would all be about 20. Similarly, if the ball travels 2 feet in the first second, the differences in succeeding seconds will be 6, 10, 14, etc. (the doubles of the odd numbers) and the second differences will all be 4. These observations tell us that quadratic growth will be a good choice for a model. In fact, quadratic models are used in physics today for the motion of falling bodies.

This example illustrates the use of a quadratic model as a matter of choice. If the data have second differences that are approximately constant, the analyst may choose to adopt a model in which the second differences are truly constant. That means choosing a quadratic growth model. The model will be an approximation to the real data, and if the approximation is accurate enough for the intended purposes, then the assumption of constant second differences is justified. So, for these examples, we might say that quadratic growth is just one possible way to approximate a real problem.

A Reel Application. We turn next to a more contemporary example involving material wound onto or off of a reel at a constant rate. For example, magnetic tape was once a common means of recording and retrieving data for computers, audio, and video equipment. The tape passed across a reader at a constant speed, winding off of one reel and onto another. A similar arrangement is still in use today in certain manufacturing operations. For example, a strip of paper towel material is drawn at a constant speed through a perforating machine before being wound onto a cardboard tube. Other examples include making cords of various kinds: twine, yarn, rope, as well as wire and cable. Part of the goal in modeling any of these operations is to predict how quickly the winding material fills its spool.

In one specific instance, McKelvey analyzed the operation of a reel-to-reel tape recorder [38]. The machine included a numerical counter intended for indexing material on the tape. McKelvey wished to develop a model relating the number displayed on the counter to elapsed time.

The nature of this relationship is not obvious. For the machine under consideration, the displayed counter number increases in direct proportion to the number of rotations completed by one tape reel, so that one complete rotation always advances the counter by an equal amount. But the length of time for each rotation is *not* constant. When the reel is full, each rotation moves a much longer section of tape than when the reel is nearly empty, and so takes a longer time.

In order to investigate the relationship between elapsed time and the displayed counter numbers, McKelvey collected a series of data points using the following procedure. For the initial data point both the counter and the time were set to 0. Then he simultaneously started the tape recorder and a stop watch, and noted down the times when the counter reached 100, 200, 300, and so on. His data are shown in Table 3.7, along with the first and second differences.

Table 3.7. Tape recorder time readings, with first and second differences. Stopwatch times were collected for counter readings of 100, 200, 300, etc.. Thus, for each data point, the counter reading is equal to $100n$. In the original data, times were recorded in minutes and seconds, but they have all been converted to seconds here.

n	Time in seconds	1st Diffs	2nd Diffs
0	0		
		242	
1	242		43
		285	
2	527		42
		327	
3	854		39
		366	
4	1220		40
		406	
5	1626		40
		446	
6	2072		40
		486	
7	2558		

Noticing that the second differences are nearly constant, we can formulate a quadratic growth model. We will define our variables as follows. First, n is the counter reading, in units of 100's. Thus, $n = 3$ means a counter reading of 300. Second, for each n, we define T_n to be the corresponding stop watch reading, in seconds. Our model will lead to a functional equation for T_n, with which we will be able to convert any value of n (and hence, any counter reading) to a time.

To construct our quadratic growth sequence, we next select values for the parameters $T_0, d,$ and e introduced in Section 3.1.[7] As a first attempt, let us take $T_0 = 0$ and $d = 242$, the first entries in the second and third columns of the table, and define e as

---
[7] These appear in the profile for quadratic growth sequences on page 137.

3.2. Applications of Quadratic Growth

the average of all the second differences, approximately 40.667. Thus, we obtain the difference and functional equations

$$T_{n+1} = T_n + 242 + 40.667n$$

and

$$T_n = 242n + 40.667n(n-1)/2.$$

How closely does our model approximate the actual data? In this example, with only eight data points, we can easily compare each T_n with the corresponding data value, as in the following table.

Data	0.0	242.0	527.0	854.0	1220.0	1626.0	2072.0	2558.0
T_n	0.0	242.0	524.7	848.0	1212.0	1616.7	2062.0	2548.0
Error	0.0	0.0	−2.3	−6.0	−8.0	−9.3	−10.0	−10.0

The final row shows the errors—that is, the amount by which each term T_n is incorrect. These range from 0 for the first two data values to −10 for the last two. The negative sign indicates that the model value is too low.[8] Overall, these results show that the model is pretty accurate. An error of ten seconds or less is not very much when compared to correct values of around 2000.

On the other hand, we notice a definite pattern in the errors. They show that the model consistently underestimates the observed data, and that the amount of the underestimate gets progressively worse. So we should be able to get better results by adjusting our parameters. The goal is to increase the value of T_n, especially for the larger values of n. With a little experimentation, the authors found that defining $T_0 = -4$, $d = 246$, and $e = 40$ produces the following improved results.

Data	0	242	527	854	1220	1626	2072	2558
T_n	−4	242	528	854	1220	1626	2072	2558
Error	−4	0	1	0	0	0	0	0

Now the model exactly reproduces six of the eight data points, and is off by four seconds for one of the remaining points and by one second in the other. This is nearly perfect agreement between the data and the model, providing strong evidence of a quadratic relationship in the underlying mechanics of the tape recorder.

This close agreement is shown visually in Figure 3.5. In the figure the dots represent our original data points and the curve represents the functional equation derived in our model. As already stated, the model and the data are in perfect agreement for all but two of the points. And for those two, the differences between the model and the data are too small to detect visually.

In discussing the Galileo example, we commented that a quadratic growth sequence is just one possible choice in developing a model. But the close agreement we find in the tape model is so striking, it is hard to imagine a better alternative approach. To illustrate this idea, in Figure 3.6, we have plotted the original data, along with a straight line that comes fairly close to all the data points. The line represents what we

---

[8]We have computed each error by subtracting the correct data value from the value the model produces. In some references, errors are defined by subtracting in the opposite order, that is, the values produced by the model are subtracted from the corresponding data values. These two approaches differ only in the sign attached to each error. Thus, in our approach, a positive error indicates a model value that is too high, while for the alternative a positive error indicates a model value that is too low.

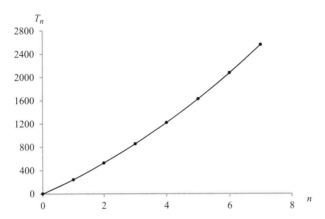

Figure 3.5. The tape recorder data with our revised quadratic growth model. The data values are represented by dots, and the model by the curve. To the unaided eye, it appears that the curve goes right through all of the data points. The model sequence does not produce the data values exactly for two points, but the errors are too small to observe in the figure.

might have obtained had we chosen to model the data with an arithmetic growth sequence. It provides a fairly accurate approximation to the original data. But when this graph is compared with Figure 3.5, it is clear that an arithmetic growth model cannot really match the data, while the quadratic growth model appears to be a perfect fit.

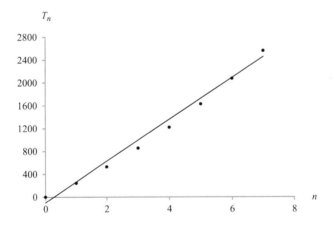

Figure 3.6. The tape recorder data with a possible arithmetic growth model. The original data values are represented by dots. The model is shown in the form of a straight line. We can judge the accuracy of the model by looking at how far the dots stray away from the line.

For this example, the quadratic model and the data are in such close agreement that there would be little point in trying to refine the parameter values further. In contrast, in some applications such close agreement is impossible to obtain. As we have indicated previously, there are known mathematical techniques for finding the best possible model in such a situation, though they are beyond the scope of this book.

3.2. Applications of Quadratic Growth

The next example illustrates how such models are used, without going into the details of how their parameters are determined.

North Polar Ice Cap. In the preceding examples, nearly constant second differences prompted us to adopt quadratic growth models. But data can suggest using a quadratic growth model even when second differences do not appear to be nearly constant. To illustrate this point, we consider a model for the shrinking north polar ice cap, as presented by Witt [**56**].

Our data consist of yearly figures for the extent (in millions of square kilometers) of the north polar ice cap in the month of September each year from 1979 to 2012 [**41**]. See Table 3.8 and Figure 3.7.

Table 3.8. North polar sea ice extent in September, each year from 1979 through 2012. Extent is given in units of millions of square kilometers.

Year	Extent	Year	Extent	Year	Extent	Year	Extent	Year	Extent
1979	7.20	1986	7.54	1993	6.50	2000	6.32	2007	4.30
1980	7.85	1987	7.48	1994	7.18	2001	6.75	2008	4.73
1981	7.25	1988	7.49	1995	6.13	2002	5.96	2009	5.39
1982	7.45	1989	7.04	1996	7.88	2003	6.15	2010	4.93
1983	7.52	1990	6.24	1997	6.74	2004	6.05	2011	4.63
1984	7.17	1991	6.55	1998	6.56	2005	5.57	2012	3.63
1985	6.93	1992	7.55	1999	6.24	2006	5.92		

In Figure 3.7, we have included a straight line representing a possible arithmetic growth model. The data points are scattered fairly widely around the line, in contrast to Figure 3.5 where the data points appear to lie almost exactly on the model curve. Here we recognize that there is a good deal of variability in the extent of sea ice from one year to the next, and our goal is not to predict the exact amount of ice in any particular year. Rather, we would like a description of the underlying trend. These data points do not appear to be scattered at random. There is a visible downward aspect as we proceed from older to more recent values. The line is meant to represent that downward trend, and so is referred to as a *trend line*. In adopting a linear model, we are implicitly assuming that the spread of data points about the trend line in the future will be similar to what we see in the figure. If so, even though we cannot predict the exact amount of ice in any specific year, we *can* predict that the amount of ice will fall within a certain range.

This is illustrated in Figure 3.7 by a shaded area centered on the trend line. For the extant data, nearly all of the points lie in the shaded area, and we expect that to remain true in the future. Applying this in a specific example, we would expect the data point for 2020 to fall within the shaded region. Therefore, based on the figure, we predict that the sea ice extent for 2020 will be between 3.5 and 5.1 million square kilometers.

For this type of analysis, where we look for a trend in a scattered set of points, there are a great many possible lines that might be drawn. On the basis of a casual visual inspection, it would be difficult to single out one specific model. This is where the ability to find the best possible model is useful. The mathematics behind choosing

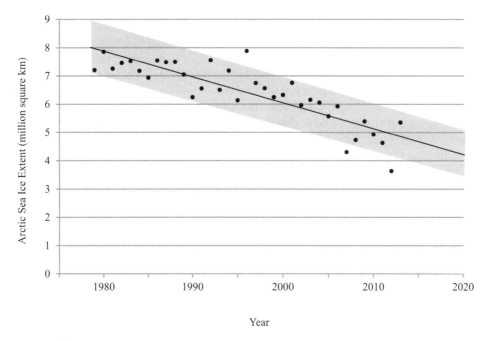

Figure 3.7. September north polar sea ice extent, 1979–2012, shown as dots. The straight line represents the best possible arithmetic growth model. Nearly all of the data points lie in the shaded area, and we suppose that future observations of sea ice extent will continue this pattern.

the best parameters is a bit beyond the scope of the present discussion, but the main idea is roughly to choose a model so that the average vertical distance from the dots to the line is as small as possible. That is the model represented by the line in Figure 3.7.

But why should we choose the best *straight line*? Or equivalently, why should we choose the best arithmetic growth model? Perhaps some other type of model would be better. And indeed, it is not difficult to visualize an arching trend *curve* (instead of a trend line) as in Figure 3.8. This suggests considering a quadratic growth model, even though an analysis of the second differences shows they are nowhere near to being constant.

Just as in the case of a linear model, there are known procedures for defining a quadratic model with the best possible parameters. As before, that means roughly that the average vertical distance from the data points to the graph of the model is as small as possible. However, with a quadratic model, the graph is a parabolic trend curve, not a straight trend line.

The procedures for finding the best quadratic model are pre-programmed in many popular graphing calculators and computer spreadsheet applications. Carrying out this analysis with the sea ice data produces the quadratic model shown in Figure 3.8. As before, the data points are shown as dots, and the quadratic model is shown as a curve. There is also a shaded area indicating how the data points are spread about the curve.

To the eye, the quadratic model does seem to track the cluster of data points more accurately than the linear model. That is not conclusive proof that the quadratic model

3.2. Applications of Quadratic Growth

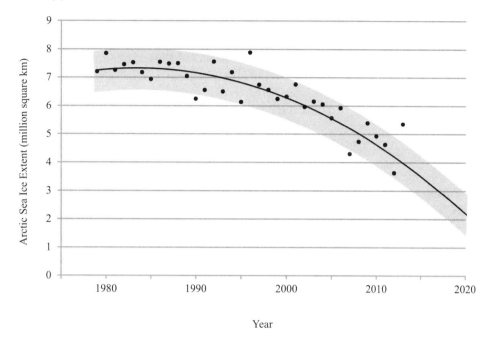

Figure 3.8. The sea ice data with a possible quadratic growth model.

is correct, but without further information, it suggests that the quadratic model is probably more credible than the linear model.

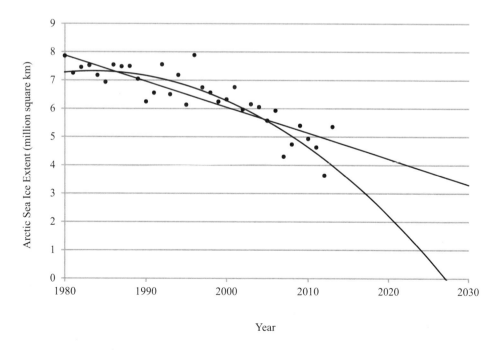

Figure 3.9. Sea ice extent data with linear and quadratic models.

The two models differ greatly in predicting future levels of September sea ice extent. To highlight the difference, we have included both models in a single graph, with the time line extended to 2030. See Figure 3.9. Because the parabolic curve bends away from the line, the further we look into the future, the wider is the separation between the two models. The quadratic model predicts that sea ice could completely disappear in the summer by 2025, whereas the linear model predicts that sea ice extent should still be in a range between 2 and 5 million square kilometers as late as 2030. Remember that we do not predict future levels of sea ice to follow either model curve exactly, but rather to range within an area centered on the curve. Even so, it is interesting to compare where each curve touches the horizontal axis. For the quadratic model that appears to be around 2027; for the linear model the line will not reach the axis until 2066.

As you can see, the two models lead to widely different predictions on the future course of summer sea ice disappearance. Here we are looking at extremely simple models and using relatively unsophisticated methods. But what we have seen hints at the importance and difficulty of validating accurate models for climate change phenomena. At the very least, we can see how a quadratic model might be considered as one possibility, even when the data are far from following the constant second difference rule. And we can see how dramatic an impact the nature of the model can have on what it predicts for the future.

In all of the examples discussed so far, a quadratic model was selected based on a set of data. But quadratic models can be suggested by other considerations. In some applications, the structure of the problem context leads naturally to the selection of a quadratic model. In particular, it may be possible to work out a difference equation based on a logical analysis of the context. If the difference equation has the right algebraic form, we can recognize it as an instance of quadratic growth. We will see several examples of that next.

Recursive Analysis and Network Problems. Quadratic growth difference equations arise naturally in a variety of problems featuring networks. A network can be thought of as a collection of objects connected together in some way. They may be cities connected by roads or airline routes, telephones connected by phone lines, computer processors connected by data pathways, or satellites connected by radio links. As an illustration, we will consider here a network of computers connected together to exchange email. A very naive approach would be to connect each pair of computers directly with a dedicated physical link, say a telephone line.[9] That might require a lot of phone lines for large numbers of computers. But how many? Suppose there are 100 computers, how many phone lines are needed? What if there are 1,000 computers? This problem can be analyzed using difference equations, and we will see that the type of equation that is developed is a quadratic growth difference equation.

---

[9]This is not a realistic way to connect large networks of computers, but it is an easily visualized example that illustrates the idea of quadratic growth in network problems. A more realistic problem is simply to enumerate the number of different pairs of computers. This might be of interest in analyzing the email traffic within the network. As a first step in monitoring the flow of data it would be natural to keep track of how much mail passed from any one computer to each of the others. This can be recorded for each pair of computers. How many pairs are there? Analyzing that question is mathematically identical to counting how many wires are needed to connect each pair of computers.

3.2. Applications of Quadratic Growth

It is easy enough to count the number of connections needed when the number of computers is small. For two computers, only one telephone line is needed. For three computers, three lines are needed. How many are needed for four computers? A good way to find the answer is by drawing a diagram. Represent each computer with a dot, and draw lines so that every dot is connected to each of the others. (One possible configuration is shown on p. 152, but don't look at it until after you have worked it out for yourself.) Once you have a correct diagram, you can count the lines to see how many are needed to connect four computers.

For larger numbers of computers, this process of counting soon becomes impractical. Difference equations offer an alternative. The main idea is this: it is much easier to count the number of *new* telephone lines required to add one new computer to the network, than it is to count the total number of phone lines. If there are 10 computers already on the network and one new computer is to be connected, how many new lines are needed? The new computer just needs to be connected to each of the existing ones, 10 in all. Then every old computer is connected to the new computer, and of course all the old computers were already connected to each other. To repeat the conclusion: with 10 computers on the network, 10 new lines are needed to add one computer.

This reasoning can be applied to any size network. If there are 50 computers on a network, and one computer is added, how many new telephone lines are needed? Fifty: the new computer has to be connected to each of the fifty existing computers. Similarly, if there are 100 computers on the network, adding one more computer requires 100 new lines. And in general, if there are n computers on the network, adding one more computer requires n new lines.

By focusing on how the number of phone lines increases as one new computer is added, we are finding just the kind of relationship needed to formulate a difference equation. With that as a goal, let us define a sequence, as follows:

If n computers are connected in a network, t_n is the number of telephone lines required to provide a direct connection joining every pair of computers.

Our direct counting above showed that $t_2 = 1$, $t_3 = 3$, and $t_4 = 6$.

Now let us reconsider the analysis of adding one computer to an existing network. With 10 computers already connected, the number of existing lines will be t_{10}. We have seen that adding one computer will require 10 new lines, for a total of $t_{10} + 10$. But that is the number of lines for a network of 11 computers. So we conclude

$$t_{11} = t_{10} + 10.$$

Similar reasoning for an existing network of 100 computers shows that

$$t_{101} = t_{100} + 100.$$

In the general case, if we start with a network of n computers and t_n lines, and then add one more computer, we will have to add n new lines, for a total of $t_n + n$. That is the number of lines for $n + 1$ computers, showing that

$$t_{n+1} = t_n + n. \tag{3.9}$$

In this way we derive a difference equation for our sequence.

From the algebraic form of the difference equation, we recognize that t_n is a quadratic growth sequence. In fact, the preceding equation is of the form

$$t_{n+1} = t_n + d + en$$

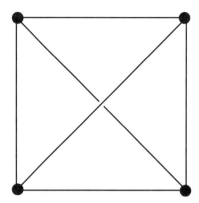

Figure 3.10. 6 Lines needed for 4 computers.

with $d = 0$ and $e = 1$. And now that we know the values of d and e, we can immediately substitute them into the quadratic growth functional equation (see page 134) to find

$$t_n = t_0 + dn + en(n-1)/2$$
$$= t_0 + 0 \cdot n + 1 \cdot n(n-1)/2$$
$$= t_0 + n(n-1)/2.$$

To complete this equation, we need to find t_0. With 0 computers, it is clear that we don't need any phone lines, so it would be reasonable to say that $t_0 = 0$. That would give the equation

$$t_n = n(n-1)/2 \qquad (3.10)$$

Can that really be right? What we did to find the functional equation almost seems too easy. Just to be on the safe side, let's check whether the equation correctly produces the terms of the sequence we have already found, $t_2 = 1$, $t_3 = 3$, and $t_4 = 6$. Substituting $n = 2$ into (3.10) produces $t_2 = 2 \cdot 1/2 = 1$, and that is the right answer. The reader is invited to verify that the equation also gives correct values for t_3 and t_4. This shows that the choice of $t_0 = 0$ really is correct for our difference equation, and that (3.10) correctly gives the functional equation for this problem.

Now we can answer some of the questions we raised before. If there are 100 computers in the network, how many telephone lines would be needed to connect each pair directly? The answer is $t_{100} = 99 \cdot 100/2 = 4{,}950$. What if there are 1,000 computers? Using common sense, you might guess that with 10 times as many computers you would need 10 times as many lines. But the functional equation gives $t_{1{,}000} = 999 \cdot 1{,}000/2 = 499{,}500$. That is more than 100 times more telephone lines! This shows one of the reasons that mathematical models are important. In many situations, relying on *common sense* can lead to unreliable conclusions. In this example, we have seen how a difference equation approach leads indisputably to a quadratic growth model. Our knowledge of these models gives us a functional equation, and leads to conclusions that we might never have expected just using *common sense*. Actually, what is referred to as *common sense* in this paragraph is nothing more than proportional reasoning. It is only the idea that doubling one variable should double the other variable that is flawed, not common sense in general. Perhaps it would be more correct to say that *unschooled* common sense can lead to errors. In any case, as this example

3.2. Applications of Quadratic Growth

illustrates, in quadratic growth models proportional reasoning is not appropriate. This point will be discussed further in Section 3.3.

The questions we have answered so far involve setting a value for n and predicting the result for t_n. That is a find-t_n question. We may also be interested in find-n questions, where t_n is given but we don't know n. As an example, suppose we only have 2,000 telephone lines available to us, and we want to know how many computers can be connected. That means we know t_n is 2,000 and want to find n, leading to the equation

$$2{,}000 = n(n-1)/2.$$

Although we can investigate this equation using numerical and graphical methods, it cannot be solved algebraically using the methods of the previous chapter; it is not a *linear* equation. It is called a *quadratic* equation, and will be studied in detail in the next section. In particular, we will see that quadratic equations can always be solved algebraically using the quadratic formula. Then we will be able to answer find-n questions algebraically in all quadratic growth models.

The computer network problem is a dramatic example of a powerful tool for developing difference equation models. We directly discovered a difference equation for our sequence, merely by thinking about how each new t_n could be obtained from the preceding term. We will refer to this as a *recursive analysis*, because it looks at a recursive rule for generating successive terms. In the example the recursive analysis produced a difference equation we recognized and had studied, leading effortlessly to a functional equation as well.

Let us see another example of this approach. Although it is a fictitious situation, and of no practical significance, it illustrates the recursive analysis approach in a concrete way.

The Handshake Problem. At Quaint City High School (QCHS), it is the custom to hold an annual Senior Ball in honor of the graduating class. The students elect a King and Queen of the Ball, and follow a curious handshaking custom. To open the Ball, the King and Queen shake hands three times. Each of the guests (that is, the other students) must shake hands twice with the King and twice with the Queen. Also, every guest must shake hands with each of the other guests exactly one time. The Hand Shake Problem asks: *If there are a total of n guests, how many handshakes occur?*

The situation in this problem can again be viewed as a kind of network. Represent the King, Queen, and their guests as large dots. Represent each handshake as a line joining the parties who are shaking hands. The resulting diagram with three guests is shown in Figure 3.11. The diagram indicates that 18 handshakes occur when only 3 guests attend the ball. Do you agree?

To analyze this situation, let us define a sequence, as follows:

If n guests attend the QCHS Senior Ball, h_n is the number of handshakes that must occur.

As in the computer network example, we can create network diagrams for small values of n, and so compute the corresponding values of h_n. In fact, we have already seen that $h_3 = 18$. In a similar way we can find $h_2 = 12$. With $n = 0$ or 1, we can count handshakes without even drawing a diagram. Remember that n is the number of guests. If $n = 0$, only the King and Queen attend the ball, and custom dictates that they shake

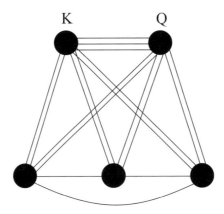

Figure 3.11. Network diagram for a ball with three guests. The King and Queen dots are indicated with the letters K and Q. The other dots represent guests. Do you agree that 18 handshakes will occur?

hands three times. So $h_0 = 3$. For $n = 1$ there will be one guest, who must shake twice with the King and twice with the Queen. That makes 4 handshakes. But the King and Queen shake hands 3 times. So in total, $h_1 = 7$. For future reference, we display these results in the following table.

n	0	1	2	3
h_n	3	7	12	18

Next we make a recursive analysis. Suppose that there are 10 guests. All the required handshakes have been duly performed, and everyone is having a fine time. Then an 11th guest arrives. It is Johnny Comelately. He must shake hands twice with the King, twice with the Queen, and once with each of the ten guests who arrived on time. So Johnny's arrival results in $4 + 10$ additional handshakes. We conclude that

$$h_{11} = h_{10} + 4 + 10.$$

It would also be correct to write $h_{11} = h_{10} + 14$, but as we will see, that would obscure the pattern we are trying to find.

Now repeat the preceding analysis, but supposing this time that 50 guests showed up on time, and that Johnny is the 51st. He must still shake twice each with the King and Queen. But this time he must shake the hand of the 50 guests who arrived on time. Consequently, Johnny's arrival in this case precipitates $4 + 50$ new handshakes, and we write

$$h_{51} = h_{50} + 4 + 50.$$

Based on these two cases, we can now pass to the general recursive step. With n guests at the Ball, who have already completed h_n handshakes, one additional guest arrives. He or she must shake twice with the King, twice with the Queen, and once with each of the original n guests. That requires $4 + n$ additional handshakes, and leads to the equation

$$h_{n+1} = h_n + 4 + n.$$

This is a difference equation for the sequence h_n, obtained through recursive analysis.

3.2. Applications of Quadratic Growth

As before, we recognize that h_n must be a quadratic growth sequence, because it conforms to a quadratic growth difference equation. We can also see, by inspection, that $d = 4$ and $e = 1$. Proceeding, from the standard quadratic growth functional equation, we now derive

$$h_n = h_0 + 4n + n(n-1)/2.$$

But we already determined that $h_0 = 3$. Thus, our functional equation becomes

$$h_n = 3 + 4n + n(n-1)/2.$$

Is this correct? Have we made any errors? Let us apply the equation with $n = 3$. We get

$$h_3 = 3 + 4 \cdot 3 + 3(2)/2 = 18,$$

which is the same thing we found earlier with a network diagram. For $n = 2$ the functional equation says

$$h_2 = 3 + 4 \cdot 2 + 2(1)/2 = 12,$$

which again agrees with our earlier results.

And now we can find the number of handshakes required with *any* number of guests. For example, if the King and Queen play host to 142 of their classmates, the total number of handshakes will be

$$h_{142} = 3 + 4 \cdot 142 + 142(141)/2 = 10{,}582.$$

That's a lot of handshakes!

The two preceding examples demonstrate the idea of a recursive analysis. It may be surprising that we can work out how one data value is related to the next, even when we don't know exactly what the data values are. But as the examples show, this approach can be highly effective, leading directly to an exact difference equation.

For the network problems we have considered here, the recursive analyses led to difference equations in which we recognized the algebraic form of quadratic growth. Thus, the use of a quadratic growth model was dictated by each problem context. Accordingly, we consider these to be instances of *structural quadratic growth*. In other words, quadratic growth is implicit in the structure of each of these problems.

Next we will consider another instance of structural quadratic growth. In this new situation, we will use a model for one aspect of a problem to find a model for a related aspect. This time, it will be the decision to use an arithmetic growth model for the first aspect that will force us to use a quadratic model for the second.

Sums of Arithmetic Growth Models. In each difference equation model we have studied so far, the focus has been on a single sequence of numbers, a_0, a_1, a_2, etc. In some situations it is also of interest to calculate a running total for these figures. Here is a specific example.

The House Construction Problem. The development plan for Quaint City includes a limited number of lots for single family houses. Over several years, the planning commission keeps track of the number of new homes constructed, as shown in Table 3.9. At the start of 2009 there were 60,000 available lots. The commission would like to project the number of available lots in 2020, and predict when there will be no lots remaining.

Table 3.9. New home construction figures.

Year	2009	2010	2011	2012	2013	2014
Houses Built	3,000	3,800	4,600	5,400	6,200	7,000
Running Total	3,000	6,800	11,400	16,800	23,000	30,000

In developing a model for this problem, we define n to be a year number, starting with $n = 0$ in 2009, and define h_n to be the number of houses built in year n. Thus h_0, h_1, h_2, and so on are the entries in the second row of the table.

The table also shows a running total for the annual construction figures. This keeps track of the total houses built since the start of 2009. For 2009, the running total is the same as the number of houses built, 3,000. In 2010, there were 3,800 new homes built. That means the total for 2009 and 2010 combined was $3{,}000 + 3{,}800 = 6{,}800$. That figure appears in the table at the bottom of the column for 2010. The same is true for the other columns. So, for 2013 the entry in the running total row is 23,000, and that indicates that 23,000 homes were built from 2009 through 2013.

Notice that the running total for each year can be found by adding that year's number of new houses to the running total through the previous year. For instance, the running total through 2012 was 16,800. In 2013 an additional 6,200 new homes were built. That brought the running total of homes up to $16{,}800 + 6{,}200 = 23{,}000$.

The running total figures are referred to as the *sums* for the original data sequence. We will denote these running totals or sums in this example by s_0, s_1, s_2, etc. In general, s_n is the sum of housing figures for years 0 through n. Put another way, s_n is the total of h_0, h_1, h_2, and so on up to h_n.

Using the variables n, h_n, and s_n, we can reconstruct the table as follows.

Table 3.10. Reformulated new home construction table.

n	0	1	2	3	4	5
h_n	3,000	3,800	4,600	5,400	6,200	7,000
s_n	3,000	6,800	11,400	16,800	23,000	30,000

It is easy to see how a model for s_n can be used to answer the commission's questions. The number of lots remaining after n years will be 60,000 minus s_n. And when s_n reaches 60,000, there will be no lots left. But what does all this have to do with quadratic growth? The answer is: if the original data follow an arithmetic growth model, then the running totals follow a quadratic growth model. Let's see why.

We can formulate a difference equation for s_n in two steps. First, we will relate s_n and h_n. Second, we will develop a *functional* equation for h_n. Combining these two steps will give the desired difference equation for s_n by itself.

For the first step, we use the observation made earlier: the total for any year is the number of houses built that year plus the total up to the previous year. Let us look at that in terms of s's and h's. The total for years 0 through 3 is s_3. It can be found by adding the previous year's total, s_2, to the number of houses built in year 3. That is expressed in the form of an equation as

$$s_3 = s_2 + h_3.$$

3.2. Applications of Quadratic Growth

Similarly, the total for years 0 through 2, s_2, is the total for the previous year s_1 plus the number of houses built in year 2, h_2. That is,

$$s_2 = s_1 + h_2.$$

The examples reveal a pattern that can be expressed by the single equation

$$s_{n+1} = s_n + h_{n+1}.$$

Next, just look at the h_n data. You should recognize that these figures follow an arithmetic growth pattern. The first year the figure is 3,000, and the figures increase by 800 each year. You should also remember how to express that as both a difference equation, $h_{n+1} = h_n + 800$, and as a functional equation, $h_n = 3,000 + 800n$. Combining both of these, we obtain a functional equation for h_{n+1}:

$$h_{n+1} = h_n + 800 = 3,000 + 800n + 800 = 3,800 + 800n.$$

Now we will use this equation for h_{n+1} in the earlier difference equation for s_{n+1}, as follows

$$s_{n+1} = s_n + h_{n+1}$$
$$= s_n + 3,800 + 800n.$$

Do you recognize the form of this difference equation? It is another instance of quadratic growth.[10] This time, the parameters are $d = 3,800$ and $e = 800$. Observe that the starting value for s_0 is the same as h_0, namely, 3,000. Accordingly, we can immediately write out a functional equation for s_n.

$$s_n = s_0 + dn + e(n-1)n/2$$
$$= 3,000 + 3,800n + 800(n-1)n/2.$$

Just to double check this result, use the equation to compute

$$s_5 = 3,000 + 3,800 \cdot 5 + 800(4)5/2 = 30,000.$$

This agrees with the number in the bottom row of Table 3.10 for $n = 5$, as it should.

Although we found the functional equation in this example by using a formula that works for all quadratic growth models, we could also use patterns to find the formula. We know that s_n can be found by adding up the h_n values from h_0 through h_n. We can list the first several h_n values using the functional equation $h_n = 3,000 + 800n$, and then add them up. Here is one way to organize the calculations:

$$\begin{aligned}
h_0 &= 3,000 \\
h_1 &= 3,000 + 800 \cdot 1 \\
h_2 &= 3,000 + 800 \cdot 2 \\
h_3 &= 3,000 + 800 \cdot 3 \\
+\ h_4 &= 3,000 + 800 \cdot 4 \\
\hline
h_0 + h_1 + \cdots + h_4 &= 5 \cdot 3,000 + 800(1 + 2 + 3 + 4) \\
&= 5 \cdot 3,000 + 800(4)(5)/2.
\end{aligned}$$

---

[10] Since the sequence s_n is defined as the sums for the sequence h_n, it should come as no surprise that the h_n values are the differences of the s_n sequence. Similarly, the differences for h_n are the same as the second differences for s_n. If h_n is an arithmetic growth model, its differences are constant. That means that the second differences for s_n are also constant, and that shows in another way that s_n is given by a quadratic growth model.

Note that at the final step we used the shortcut for adding the whole numbers from 1 up to 4.

The preceding calculation shows that

$$s_4 = 5 \cdot 3{,}000 + 800(4)(5)/2.$$

How would this change if you computed s_7 instead of s_4? The reader is invited to predict the final result. Then, to verify the prediction, write down the equations for h_0 through h_7 in the same format shown above, and add them all up. You should end up with

$$s_7 = 8 \cdot 3{,}000 + 800(7)(8)/2.$$

These results suggest a general pattern, leading to the equation

$$s_n = (n+1)3{,}000 + 800(n)(n+1)/2.$$

This equation is not in the same form as the one we found earlier, but it gives the same results. As an illustration, here is the way s_5 is computed with our new equation.

$$s_5 = 6 \cdot 3{,}000 + 800(5)(6)/2 = 30{,}000.$$

That is the same result we found earlier.

This example introduces the idea of the sums of a sequence. Sums of sequences are often of interest in models. Now we know that adopting an arithmetic growth sequence for a problem automatically leads to a quadratic growth sequence for the sums.

This completes our presentation of three situations that lead to quadratic growth. As a final example in this section, we will develop models for world oil consumption and reserves.

The Oil Reserves Problem. It is generally accepted that petroleum is a non-renewable resource. There is a certain amount available, and once that is gone, other sources of energy will have to be relied upon. When can we expect the existing oil to run out?

As a first step in analyzing this problem, we look at recent figures on world oil consumption, as published by British Petroleum [5]. The data are reproduced in Table 3.11.[11] We consider the entries in the second column as terms in a sequence,

Table 3.11. Annual world oil consumption, in billions of barrels.

Year	World Oil Consumption
2003	29.2788
2004	30.3981
2005	30.8019
2006	31.1436
2007	31.6652
2008	31.5298
2009	31.0655
2010	32.0473
2011	32.4609
2012	32.9147
2013	33.3358

---

[11] In the original source, consumption was listed on a per-day basis. Those figures were multiplied by 365 (or 366 in leap years) to obtain the data here.

3.2. Applications of Quadratic Growth

$C_0, C_1, C_2,$ and so on. In other words, C_n represents the world oil consumption in year n, where 2003 is year 0. A graph for the data values with a superimposed linear model is given in Figure 3.12.

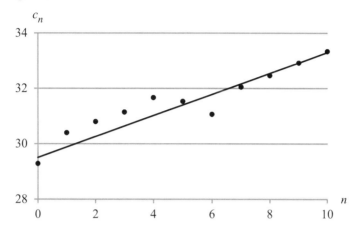

Figure 3.12. Graph for the data sequence C_n, the world oil consumption in year n with $n = 0$ in 2003. The vertical scale is in units of billions of barrels. The straight line represents a linear model for the data.

The data points present a roughly linear appearance, except for the period from $n = 4$ to $n = 6$. That corresponds to 2007 − 2009, and probably reflects the global economic recession of that period. However the rate of increase in consumption before and after the recession is pretty constant, as reflected in the slopes among the first 5 points and the slopes among the last 5 points. Accordingly, let us adopt an arithmetic growth model for the data. Based on a visual inspection and some trial and error, the authors found the straight line shown in the figure, given by the functional equation

$$c_n = 29.5 + 0.38n. \tag{3.11}$$

Here, we are using a lower case c to distinguish between the original data values and our approximating model. For each n, C_n is the actual consumption figure from the table, and c_n is the corresponding value defined by (3.11).

This is not a very sophisticated model. For example, it does not account for changing economic conditions. Also, the consumption data includes fuel from sources other than petroleum, such as natural gas and ethanol. So we do not expect this model to be highly reliable in predicting future consumption trends. But it will serve for illustrative purposes.

A model for annual consumption says nothing directly about the depletion of oil reserves. For that we need to study how the cumulative consumption of oil grows over time. Accordingly, let us consider the running totals for our annual consumption model.

Formally, we define

$$s_n = c_0 + c_1 + \cdots + c_n.$$

Then s_n will model the cumulative world oil consumption for years 0 through n. We can compute the first several values of c_n using (3.11), and then obtain the first few s_n in a running total column. See Table 3.12.

Table 3.12. Terms of c_n and s_n. The c_n values are from (3.11); the s_n values are running totals for the c_n values.

n	c_n	s_n
0	29.50	29.50
1	29.88	59.38
2	30.26	89.64
3	30.64	120.28

We know that s_n will be a quadratic growth sequence because the terms are running totals for an arithmetic growth sequence. Therefore, the functional equation must be of the form
$$s_n = s_0 + dn + en(n-1)/2.$$
Proceeding, observe that $s_0 = c_0 = 29.5$ in our model. One way to find d and e is by computing first and second differences for the s_n sequence. Using the entries from the third column of Table 3.12, we display these differences in Table 3.13.

Table 3.13. Terms of the sequence s_n, with first and second differences.

n	s_n	1st Diffs	2nd Diffs
0	29.50		
		29.88	
1	59.38		0.38
		30.26	
2	89.64		0.38
		30.64	
3	120.28		

As expected, the second differences are constant, and we see that $e = 0.38$. The first entry in the column of first differences reveals that $d = 29.88$. Thus, the functional equation for s_n is
$$s_n = 29.5 + 29.88n + 0.38n(n-1)/2.$$
We can check that this gives the same values for s_n as those shown in Table 3.12.

Although constructing a table as above is a valid way to find the parameters for our quadratic growth sequence, an algebraic approach is more efficient. We will again begin by finding the difference equation. As a first step, we consider an example: computing s_5. We know that
$$s_5 = c_0 + c_1 + \cdots + c_5 = s_4 + c_5.$$
Applying (3.11) shows that
$$c_5 = 29.5 + 0.38 \cdot 5.$$
Thus we can write
$$s_5 = s_4 + 29.5 + 0.38 \cdot 5.$$
This relates s_5 to s_4. We can view this as an instance of a difference equation with s_4 playing the role of s_n and s_5 the role of s_{n+1}. In fact, carrying out the same steps starting with the computation of s_{n+1} leads to
$$s_{n+1} = s_n + 29.5 + 0.38(n+1).$$

3.2. Applications of Quadratic Growth

This is almost in the standard form for a quadratic growth difference equation. With two more algebraic steps,

$$s_{n+1} = s_n + 29.5 + 0.38n + 0.38$$
$$= s_n + 29.88 + 0.38n,$$

we obtain the standard form, and identify the parameters $d = 29.88$ and $e = 0.38$, as before.

The same logic also shows, more generally, that for an arithmetic growth sequence with functional equation

$$a_n = a_0 + en, \tag{3.12}$$

the sequence of sums will have *difference* equation

$$s_{n+1} = s_n + a_1 + en$$

and hence the functional equation

$$s_n = a_0 + a_1 n + en(n-1)/2. \tag{3.13}$$

Note that we are using e instead of d for the added constant in (3.12) to avoid conflicting meanings for the parameter d.

Here we see again that algebra helps us find and verify general results. In any model that involves running totals of an arithmetic growth sequence, we can apply (3.13) to immediately derive the functional equation for s_n. In the exercises, readers are encouraged to apply both approaches. Fill in a few lines in a table for s_n and compute the first and second differences to obtain the parameters d and e. Then double check your results by applying (3.13).

Regardless of the method by which it is found, the functional equation for s_n can be used to answer questions about our model. In particular, we can project when the oil will run out. The British Petroleum report on oil consumption also provides figures on oil reserves. According to the report, as of the end of 2013 there were 1,687.9 billion barrels of proven oil reserves. Using our model, we can ask when those known reserves will be used up. We note first that 2013 is year 10, and that the total consumption between 2013 and year n is therefore $s_n - s_{10} = s_n - 345.4$. We want to know when this will equal the known reserves of 1,687.9. That is, we want

$$s_n - 345.4 = 1,687.9$$

or more simply

$$s_n = 1,687.9 + 345.4 = 2,033.3.$$

Thus we obtain the equation

$$29.5 + 29.88n + 0.38n(n-1)/2 = 2,033.3.$$

Using a numerical approach, we can systematically generate the values of s_n until the desired level is reached. Proceeding in this fashion reveals that $s_{50} = 1,989.0$ and $s_{51} = 2,037.9$. Therefore, the oil reserves known in 2013 would be used up during year 50. Since year 0 is 2003, year 50 will be 2053.

Have we proven that the oil will run out in about 35 years? No. We do not know that the arithmetic growth model for consumption will be accurate in the future. And it is almost certain that additional oil reserves will be discovered. But what we can say is this: if world consumption continues to grow at roughly the same rate in the future as observed over the past decade, then the oil reserves known by the end of 2013 will be used up in about 35 years. That is typical of the kinds of predictions models provide.

This concludes the discussion of situations in which quadratic growth arises in applications. We have seen that a quadratic model can be adopted because of properties of a data set, for example constant (or nearly constant) second differences. In such cases, the use of a quadratic model is a matter of choice. In contrast, we have also seen how recursive analysis can lead directly to a quadratic growth difference equation, particularly in contexts involving networks. Finally, we considered sums of sequences, an idea that is useful in many kinds of models. For arithmetic growth sequences, the sums are always quadratic growth sequences.

3.2 Exercises

Reading Comprehension. Answer each of the following with a short essay. Give examples, where appropriate, but also answer the questions in written sentences.

(1) @Under what conditions might one decide to adopt a quadratic growth model for a stream of data values $a_0, a_1, a_2 \cdots$? That is, what properties of the data indicate that a quadratic model might be appropriate?

(2) Explain what is meant by *structural quadratic growth*. Include at least one example of a sequence with structural quadratic growth and at least one example without.

(3) @Is proportional reasoning appropriate with quadratic growth models? Give an example to support your answer (either yes or no).

(4) Explain the concept of *sums of an arithmetic growth sequence*. Give an example illustrating why this is of interest, and how it can be useful.

(5) Give two examples of quadratic growth models from the reading, one for which n can be interpreted to be a continuous variable, and the other for which n only makes sense as a discrete variable. Explain in each case why n is interpreted as it is.

(6) Explain what is meant by *recursive analysis*.

Math Skills.

(7) @In the computer network model we found the functional equation $t_n = n(n-1)/2$. Find the lowest n for which t_n is above 3,000. Use graphical and numerical methods. (See Figure 3.13.)

(8) In the Handshake Problem, we found the functional equation $h_n = 3 + 4n + n(n-1)/2$. Find the n for which h_n is closest to 1,000. Use graphical and numerical methods. (See Figure 3.14.)

(9) @Let a_n represent the nth term of the sequence 5, 8, 11, 14, 17, $\cdots$, where we consider 5 to be a_0. Now consider the sequence of sums $s_n = a_0 + a_1 + \cdots + a_n$. Numerically, we see that the sum sequence is 5, 5 + 8, 5 + 8 + 11, $\cdots$. In this problem you will use two approaches to find a functional equation for s_n then use the equation to find a particular sum.

3.2. Exercises

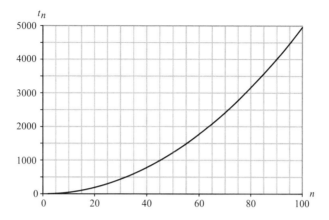

Figure 3.13. Graph for computer network problem.

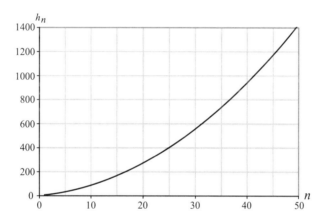

Figure 3.14. Graph for handshake problem.

a. @Make a table of values for s_n. Include the first and second differences in your table. The first differences should be the sequence a_n. Use the first and second differences to find a functional equation for s_n.

b. @Use (3.13) (see page 161) to obtain a functional equation for s_n.

c. @Find the sum $5 + 8 + 11 + 14 + 17 + \cdots + 155$.

(10) Let a_n represent the nth term of the sequence 10, 10.7, 11.4, 12.1, 12.8, $\cdots$, where we consider 10 to be a_0. Now consider the sequence of sums $s_n = a_0 + a_1 + \cdots + a_n$. Numerically, we see that the sum sequence is 10, 10 + 10.7, 10 + 10.7 + 11.4, $\cdots$. In this problem you will use two approaches to find a functional equation for s_n then use the equation to find a particular sum.

a. Make a table of values for s_n. Include the first and second differences in your table. The first differences should be the sequence a_n. Use the first and second differences to find a functional equation for s_n.

b. Use (3.13) (see page 161) to obtain a functional equation for s_n.

c. Find the sum $10 + 10.7 + 11.4 + 12.1 + \cdots + 40.1$.

Problems in Context.

(11) @Non-Terrestrial Gravity Data. Ten-year old boy genius, Johnny, built his own rocket ship and flew to the moon. Near the moon Johnny came to a complete stop and let gravity pull his ship down. While falling, Johnny recorded the following pieces of information.

Time	Distance
1	0.81
2	3.25
3	7.30
4	12.96
5	20.23

Each entry in the first column of the table represents an elapsed time in units of seconds, measured from the start of the ship's fall. Each entry in the second column represents the total distance the ship has fallen, in units of meters, up to the corresponding time value. For example, the table shows that after falling for 3 seconds, the ship had traveled a distance of 7.30 meters.

  a. @Analyze the data. That is, define variables, determine what kind a sequence this is, find a difference equation, and find a functional equation to model the data.

  b. Make a table of error values for your model. Recall that we find error by subtracting the correct data value from the value that the model produces.

  c. @If Johnny stopped 1,000 meters above the lunar surface, when will he hit the surface (assuming he doesn't try to avoid it).

(12) Sales Data. Angela, the maker of Fabulous Pants, is considering her recent sales.

Month	Sales (in dollars)
August	600
September	890
October	1,160
November	1,410
December	1,640

  a. Analyze the data. That is, define variables, determine what kind a sequence this is, find a difference equation, and find a functional equation to model the data.

  b. If sales continue like this, when will the sales for one month first reach or exceed $2,850?

(13) @Computer Network. Here is a modified version of the computer network problem. At a large research facility, all communications between computers are provided by hard wired connections, for security reasons. The facility has a cloud computing network, with two servers and a variable number of user computers. They are all connected as follows. For every pair of users there is a direct wire connection. In addition, each user has two direct connections to each of the servers, to provide rapid and reliable access to such services as printing and data storage and

3.2. Exercises

retrieval. Finally, the two servers must constantly exchange information so that each has a complete copy of the data stored on the other. To accommodate this exchange, the servers are connected with four direct lines.

   a. @Draw a diagram of this situation with three users. How many connection wires are needed?
   b. @Repeat the preceding question with two users, then one user, and finally with no users.
   c. @Let T_n be the number of connection wires needed when there are n users. Use a recursive analysis to find a difference equation for T_n, and verify that it is consistent with your answers to the preceding questions.
   d. @Find a functional equation for T_n.
   e. @How many connection wires are needed if there are 100 users?
   f. @If the facility has 25,000 connection wires, what is the maximum number of users that can be included in the network?

(14) **Antibiotic Interaction.** A drug company is planning to test how several different antibiotics interact. The idea is to use every antibiotic in a test with every other antibiotic. In each test, a patient will be given a combination of two pills. To provide a comparison, the researchers will also do a test of each drug by itself. The researchers do not want the doctors and patients to know which tests involve two antibiotics, and which just involve one, because that might influence the results. So they decide to use two kinds of placebos (phony pills that contain no medicine). One is red and one is blue.

Your task is to figure out how many tests need to be run. Each test involves two pills, either two antibiotics, an antibiotic and a red pill, or an antibiotic and a blue pill. Every possible combination of antibiotics will be tested, and each antibiotic will be tested in combination with each of the phony pills. Let t_n stand for the number of tests that will be needed if there are n different antibiotics.

For example, t_3 is the number of tests necessary if there are three antibiotics. The first antibiotic must be tested with each of the other 2 antibiotics as well as with each of the 2 placebos (for a total of 4 tests). The second antibiotic has already been tested with one antibiotic so must only be tested with 1 other antibiotic and each of the 2 placebos (for a total of 3 more tests). Finally, the last antibiotic has already been tested with the other two antibiotics and so must only be tested with the 2 placebos (for a total of 2 more tests). We conclude that $t_3 = 9$.

   a. Find t_4 and explain your answer.
   b. Suppose the tests for 15 antibiotics were already planned. Then the director of research decided that a 16th antibiotic should be included. How many **new** tests will need to be added? Explain.
   c. Use the answer to the preceding question to write an equation that relates t_{15} and t_{16}.
   d. Find a difference and a functional equation for t_n. [Hint: first make a table of values for t_0 to t_4.]
   e. How many tests are needed if there are 50 antibiotics?

(15) @Soccer Tournament. In a round-robin soccer tournament, every team must play each of the other teams once. How many games must be scheduled if there are 10 teams? If there are 20 teams? If there are n teams? [Hint: This can be thought of as a network problem and recursive analysis can be applied.]

(16) Selling More Each Year. An electronics store began selling home computers in 1990. In 1990 they sold 25 computers. In the next year they sold 35 computers. The year after that they sold 45 computers, and so on. Assuming that the number sold each year follows an arithmetic growth model, the store manager predicts that $s_n = 25 + 10n$ computers will be sold in year n (where 1990 is year 0). Let T_n be the total sales for years 0 through n.

    a. Compute directly from the given information the total sales for 1990 through 1993.

    b. Use your knowledge of quadratic growth models to find a functional equation for T_n.

    c. The store manager estimates that the total market for home computer sales is limited to 5,000 homes in the store's area. Because of competition with other stores, the store is only expected to be able to sell a total of 1,000 computers. According to the manager's equation for total sales over n years, how many years will it take the store to sell this many computers? Does this seem like a reasonable prediction? Why or why not?

(17) @Logging More Each Year. A lumber company owns 50,000 acres of forest land. In the first year of operation they log 1,200 acres. The next year they log 1,400 acres. The next year they log 1,600 acres. Assuming that the number of acres logged each year follows an arithmetic growth law, how many acres total will have been logged after 10 years? How many years will it take to log the entire 50,000 acres?

(18) More Carbon Dioxide Each Year. A researcher developed a linear model for the number of motor vehicles registered in the US. She derived the equation

$$M_n = 2.75 \cdot n + 193.06,$$

where M_n is the number of vehicles in the US, in millions, in year n, with $n = 0$ in 1990. For example, the model says that in 2015 there will be

$$M_{25} = 2.75 \cdot 25 + 193.06 = 261.81$$

million vehicles in the US.

Assume that each vehicle generates about 6 tons of carbon dioxide per year. According to your linear model, how many tons of carbon dioxide will be added to the atmosphere between 2015 and 2050 by automobile exhaust? [Hint: determine how many tons will be added in 2015, in 2016, in 2017 and so on. Show this is an arithmetic growth sequence, and use it to analyze the problem.]

(19) Auto Fuel Economy. A graph containing the following data values was posted by a user on an internet forum dedicated to automobile fuel economy [39].

Speed	50	55	60	65	70
MPG	60.1	57.3	53.6	48.9	43.1

3.2. Exercises

These are fuel economy figures in units of miles per gallon for a 2012 Toyota Prius at several different speeds, shown in miles per hour. We can treat the second row of the table as a data sequence, with $m_0 = 60.1$, $m_1 = 57.3$, and so forth.

a. Find the first and second differences for these values, and show that this is approximately a quadratic growth sequence.

b. Create a quadratic growth model for the data, using a trial-and-error approach to choose parameters for which the model approximates the data as closely as possible. Provide a graph for your results, showing both the original data values and the model.

c. Provide a table showing each original data value, the corresponding value from the model, and the difference between the two (that is, the model *error*).

d. Write a description of your model, including definitions for the variables, and difference and functional equations.

e. Use your model to predict the fuel economy at a speed of 90 miles per hour, explaining your reasoning.

f. Reflect on your model. Are your answers reasonable? When you use different approaches (graphical, numerical, algebraic) do you get consistent results? Does the model make sense in general terms? Is it credible for all values of n, and if not, over what range of n values *is* it credible?

Digging Deeper.

(20) @In this section we discussed a pattern discovered by Galileo. Galileo's discovery involves distances traveled by a falling body over a series of equal periods of time, every second, or every two seconds, or every 5 seconds, or some other similar arrangement. According to the pattern, however far the body falls in the first time period, it will fall 3 times that initial distance in the second time period, 5 times the initial distance in the third time period, 7 times in the fourth time period, and so on. Now consider a diver about to jump off a high platform. In the first one-quarter of a second the diver falls 1 foot.

a. @How long will it take for the diver to reach the water if the platform is 16 feet high?

b. @How far does the diver fall in the last quarter-second before hitting the water?

c. What is the diver's speed in feet per second for the last quarter-second? [Hint: since you know how many feet the diver covered in a quarter-second, four times that amount would be covered in a full second, traveling at the same speed. So the speed in feet per second is four times the number of feet traveled in the last quarter-second.]

d. @Now repeat the same calculations for a platform 64 feet high. That is four times higher than the platform in the first part of the problem. Does the diver fall four times longer? Is the diver going four times as fast in the last quarter-second?

(21) We have seen how quadratic growth is, in a way, an extension of the idea of arithmetic growth. This problem explores the idea of extending quadratic growth in a similar way.

a. On page 124 we discussed the sequence of tetrahedral numbers, 1, 4, 10, 20, 35, 56, 84, ⋯. Show that this sequence has constant *third* differences. We will call such sequences cubic growth sequences.

b. Show that the first differences of the tetrahedral numbers form a quadratic growth sequence. Use this idea to find a difference equation for the tetrahedral numbers. (These first two steps were Digging Deeper exercises in section 3.1. If you completed those problems, you may want to refer to your work before moving on.) Then formulate a conjecture about the general form of difference equations for cubic growth sequences.

c. Suppose $a_0, a_1, a_2, \cdots$ is a quadratic growth sequence. Consider the sums $s_0 = a_0, s_1 = a_0 + a_1, s_2 = a_0 + a_1 + a_2, \cdots$. Show that the sum sequence is a cubic growth sequence.

d. We saw that for a sequence with constant first differences, the functional equation is linear. In fact, the functional equation is

$$a_n = a_0 + dn = a_0 + d\frac{n}{1},$$

and a_0 and d are the top entries in the a_n and differences columns. Similarly, for a sequence with constant second differences, the functional equation is quadratic. In this case, the functional equation is

$$a_n = a_0 + d\frac{n}{1} + e\frac{n(n-1)}{1 \cdot 2},$$

where a_0, d, and e are the first entries in a_n, differences, and second differences columns.

Based on these observations, guess what the form should be for the functional equation for a cubic growth sequence, assuming that a_0, d, e, and f are the first entries in a_n, differences, second differences, and third differences columns. Use your guess to formulate a functional equation for the tetrahedral numbers. Is your equation correct? That is, does it give the correct values for the tetrahedral numbers we have already found: 1, 4, 10, 20, 35, 56, 84, ⋯?

3.3 Quadratic Functions and Equations

In Chapter 2 the functional equations of arithmetic growth sequences led us to study linear functions and equations, an important topic in its own right. In the same way, the functional equations for quadratic growth models lead us to study quadratic functions and equations, the subject of this section. We will take a systematic look at algebraic forms, graphical features, methods for manipulating and solving quadratic equations, and related topics. To get started, we will develop appropriate definitions.

Definitions of Quadratic Functions and Equations. Let us review an example. One of the first cases we considered was the sequence

$$5, 27, 59, 101, 153, 215, \ldots$$

(see page 133). We found its difference equation

$$a_{n+1} = a_n + 22 + 10n$$

3.3. Quadratic Functions and Equations

and functional equation
$$a_n = 5 + 22n + 10 \cdot \frac{(n-1)n}{2}$$

The functional equation can be algebraically simplified as follows:
$$\begin{aligned} a_n &= 5 + 22n + 10 \cdot \frac{(n-1)n}{2} \\ &= 5 + 22n + 5(n-1)(n) \\ &= 5 + 22n + 5(n^2 - n) \\ &= 5 + 22n + 5n^2 - 5n \\ &= 5 + 17n + 5n^2 \\ &= 5n^2 + 17n + 5. \end{aligned}$$

This is typical of quadratic growth functional equations. In fact, because quadratic growth functional equations can always be expressed in the standard form
$$a_n = a_0 + dn + e \cdot \frac{(n-1)n}{2}$$
we can apply similar algebraic steps as above to obtain a similar result. Thus, every quadratic growth functional equation can be put in the form
$$a_n = \text{(a constant)} \cdot n^2 + \text{(a constant)} \cdot n + \text{(a constant)},$$
and the first constant on the right is equal to $e/2$. This is a quadratic equation, and defines a_n to be a quadratic function of the variable n. However, in this section we will use the more traditional x and y variables to discuss functions. Accordingly, we have the following.

> **Definition of Quadratic Functions and Equations.** We say y is a quadratic function of x if it can be expressed in the form.
> $$y = \text{(a constant)} \cdot x^2 + \text{(a constant)} \cdot x + \text{(a constant)}, \quad (3.14)$$
> where the first constant is not zero. We call this the *standard form* of a quadratic equation. Any equation that can be expressed in the standard form is quadratic.

It is customary to represent the constants in the standard form as parameters a, b, and c. Thus, the standard quadratic equation becomes
$$y = ax^2 + bx + c, \quad (3.15)$$
with $a \neq 0$. Because the powers of x appear from highest to lowest, this standard form is also referred to as *descending form*. The parameters a, b, and c are also referred to as *coefficients*.

The standard quadratic, (3.15), is very similar to the standard linear equation
$$y = mx + b.$$

The choice of letters a, b, and c for the quadratic equation and m and b for the linear equation is dictated by tradition. Perhaps it would make more sense to write the standard quadratic equation
$$y = ax^2 + mx + b$$

to emphasize the structure of a linear function with an additional term ax^2. But that would run contrary to the practice in mathematics textbooks over many decades. So keep in mind that the letters are placeholders for numerical constants, and which letters are used has little significance. In particular, the b in the quadratic equation should not be confused with the b in the linear equation.

Note the wording of the definition. It says a function is quadratic if it *can be* expressed in the standard form. Functions that appear in some other form might still be quadratic. Indeed, the following are all quadratic functions:

$$3x^2 + 4x - 5 \qquad (3x-2)(4x-5) \qquad (3x+4)^2 \qquad \frac{x}{2} + \frac{3x^2}{4}$$

Each can be rewritten in the standard form using algebra. For instance, $(3x-2)(4x-5)$ can be rewritten as $12x^2 - 23x + 10$. Similarly, since $\frac{3x^2}{4}$ is the same as $\frac{3}{4}x^2 = 0.75x^2$, the last example above can be expressed as $0.75x^2 + 0.5x + 0$.

As these examples show, it sometimes requires some algebra just to recognize whether or not a function is a quadratic. Very often, quadratic functions appear in an application in some form other than the one in the definition. For instance, we found quadratic growth functional equations in a form like $1 - 2n + 3(n-1)n/2$, which is quadratic but not in the standard form. Moreover, as in the linear case, quadratic functions can be expressed in a variety of different forms which are especially useful for different purposes. One form is most convenient if you are trying to graph a quadratic equation, while another is more helpful if you are trying to find the x-intercepts of the graph. For all of these reasons, algebra plays a fundamental role in much of what we will cover in this section. With that in mind, we turn next to some of the algebraic properties of quadratic functions.

Algebraic Considerations. In many earlier discussions, the idea of changing one algebraic form into another one has been used. Let us take a closer look at this idea.

It is possible to write down two expressions that are different in appearance, but which always lead to the same numerical results. A very simple example is provided by the expressions $3x + 4x$ and $7x$. No matter what number is used for x, these two expressions produce the same result, and they are therefore said to be *equivalent*.

Two equivalent expressions always define a single function. Consider the equation $y = 3x + 4x$. This defines y as a function of x. As soon as a value is defined for x, y can be immediately computed from the equation. In the same way, $y = 7x$ defines y as a function of x. Both equations define the *same* function. When x is assigned a value, the two equations produce the same result for y. The steps that are taken to reach that y differ for the two equations, but the result is the same. In the case of the computer network example of Section 3.2, $n(n-1)/2$ and $0.5n^2 - 0.5n$ are equally valid ways to express the number of telephone lines as a function of the number of computers.

How can you tell whether two expressions define the same function? One approach is to substitute some numerical values for the variables. The two expressions cannot define the same function if they produce different results. Unfortunately, if the results agree, this test is not conclusive. Consider $x^3 + 6x + 2$ and $7x + 2$. If you replace x by 1, each expression gives a result of 9. If you replace x by -1, the result is -5 in both cases. And if x is replaced by 0, both of the results equal 2. All of these might lead you to suspect that the two expressions define the same function. But these three cases of agreement do not prove equality. As mentioned, it only takes one case of

3.3. Quadratic Functions and Equations

disagreement to show that the two expressions define different functions. We can find just such a case in this example by taking $x = 2$. That leads to $2^3 + 6 \cdot 2 + 2 = 22$ from the first expression and $7 \cdot 2 + 2 = 16$ from the second. These are different results, so the expressions define different functions.

There is another method for determining whether two expressions define the same function: use rules of algebra to transform one into the other. For example, one rule of algebra states that a pattern of the form $A(B + C)$ can always be replaced by $AB + AC$, and vice versa. This is known as the distributive law, and it is used frequently. It is what justifies "expanding" $4(3 + x)$ to $12 + 4x$. Similarly, the distributive law shows that $p(500 - 4p) = p \cdot 500 - p \cdot 4p$ or, more simply, $500p - 4p^2$. Used in the reverse direction, the distributive law justifies combining like terms, as in replacing $3x + 4x$ with $7x$. This is also called factoring.

Combining both applications of the distributive law, the expression $(2x-5)(3x+1)$ can be rewritten as follows:

$$\begin{aligned}(2x - 5)(3x + 1) &= 2x(3x + 1) - 5(3x + 1) \\ &= 2x \cdot 3x + 2x \cdot 1 - 5 \cdot 3x - 5 \cdot 1 \\ &= 6x^2 + 2x - 15x - 5 \\ &= 6x^2 - 13x - 5\end{aligned}$$

Checking for Errors. Although in general no number of examples can be relied on to prove that two functions are equal, quadratic functions of a single variable are a special case. For these functions, it is only necessary to check three values of the variable. If the functions agree for three different values of the variable, then they are actually the same function. This can be very useful in checking for algebra errors. Consider the example above, which starts with $(2x-5)(3x+1)$. After doing the algebra, a result of $6x^2 - 13x - 5$ is reached. If no errors have been made, $(2x - 5)(3x + 1)$ and $6x^2 - 13x - 5$ should produce the same result no matter what value is used for x. To check this, it is enough to try any three values of x, say 0, 1, and 2. With $x = 0$, each expression results in -5. With $x = 1$, we find $(2 \cdot 1 - 5)(3 \cdot 1 + 1) = (-3)(4)$ in the first expression and $6 \cdot 1^2 - 13 \cdot 1 - 5 = -12$ in the second. Similarly, using $x = 2$ in each expression produces the same result, -7. Because these are quadratic functions, three examples provide conclusive proof that the two expressions are equal, so no errors were made in the algebra. In practice, you will not want to check three examples every time you use algebra. It is a good idea to check at least one value though. That will usually reveal an error, if there is one. It is not a conclusive test, as checking three examples is. It is possible for an error to go undetected when you test just one example, but you have to be pretty unlucky for that to happen.

Second Differences. We first encountered quadratic functions in our study of quadratic growth sequences, which we recognize from their constant second differences. In the more general study of quadratic functions, the idea of constant second differences still holds, in the following sense. Suppose $f(x)$ is a quadratic function. Create a table of values for x and $f(x)$, selecting uniformly spaced x values. The corresponding $f(x)$ values will have constant second differences.

For example, let $f(x) = 3x^2 - 5x + 2$, and construct a table with x values starting at $x = -2$ and increasing in steps of three: $-2, 1, 4, 7, 10, 13$. These x's and the corresponding $f(x)$'s appear in the table below.

x	-2	1	4	7	10	13
$f(x)$	24	0	30	114	252	444

If you compute the first and second differences of the sequence of $f(x)$ values, you will see that all the second differences equal 54. Similarly, if you space the x values one half unit apart, you find the following table:

x	-2.00	-1.50	-1.00	-0.50	0.00	0.50
$f(x)$	24.00	16.25	10.00	5.25	2.00	0.25

The second differences for the second row of this table are again constant. This time they all equal 1.5.

There is an algebraic explanation for this phenomenon. In each case, the x values form an arithmetic growth sequence. For example, in the second table, we can generate the x values by starting at -2 and repeatedly adding 0.5. Using a position number n, starting at 0, we can therefore express the nth x value as $-2 + 0.5n$. Also, the sequence of $f(x)$ values can be denoted by $a_0 = f(-2)$, $a_1 = f(-1.5)$, and so on. Now consider the difference between two consecutive terms, say a_n and a_{n+1}. We compute

$$a_n = f(-2 + 0.5n) = 3(-2 + 0.5n)^2 - 5(-2 + 0.5n) + 2$$

and

$$\begin{aligned} a_{n+1} &= f(-2 + 0.5n + 0.5) \\ &= f(-1.5 + 0.5n) \\ &= 3(-1.5 + 0.5n)^2 - 5(-1.5 + 0.5n) + 2. \end{aligned}$$

By subtracting these, we obtain an expression for $a_{n+1} - a_n$. The algebra to work that out is a bit tedious, but can be completed using nothing beyond the methods we have already seen. Without filling in all of the intermediate steps, let us look at the final answer:

$$a_{n+1} - a_n = -7.75 + 1.5n.$$

After one final rearrangement,

$$a_{n+1} = a_n - 7.75 + 1.5n$$

we recognize this as a quadratic growth difference equation. Therefore the f values in the table form a quadratic growth sequence, and must have constant second differences. Moreover, by analyzing the algebraic steps that were omitted above, one can see that a similar result will apply in general, for any quadratic function $f(x)$, and any arithmetic growth sequence of x values in our table.

To summarize these findings in a slightly different form: if $f(x)$ is any quadratic function, and if $a_0, a_1, a_2, \cdots$ is an arithmetic growth sequence, then $f(a_0), f(a_1), f(a_2), \cdots$ will be a quadratic growth sequence.

Graphs. On a graph with two axes, you are already familiar with the idea of graphing linear functions and equations. Quadratics are graphed in the same way. As an example, consider the function $x^2 + 3x + 5$. If you make $x = 3$, then you obtain a result of $3^2 + 3 \cdot 3 + 5 = 23$. That corresponds to one point on the graph, $(3, 23)$. If a point on the graph is labeled (x, y), then the y must be what you get by using x in the function.

3.3. Quadratic Functions and Equations

That means $y = x^2 + 3x + 5$. For this reason, the *equation* $y = x^2 + 3x + 5$ always goes with the graph of the *function* $x^2 + 3x + 5$.

If you graph a variety of quadratic functions, you will see some patterns emerge. The graphs are always in the shape of a $\cup$ or a $\cap$. The $\cup$ or $\cap$ is symmetric. That is, there is a vertical center line that divides it exactly in half, so that each side is the mirror image of the other. The center line is also called the *axis of symmetry*. It intersects a $\cup$ at its lowest point, and intersects a $\cap$ at its highest point. These qualitative properties are displayed in several graphs in Figure 3.15.

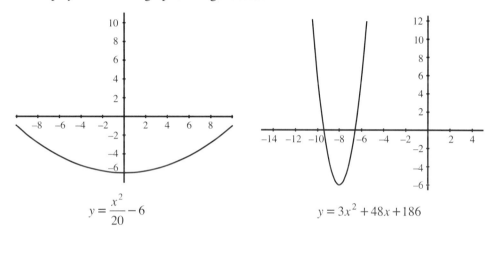

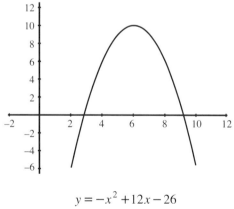

Figure 3.15. Sample quadratic function graphs.

The graph of a quadratic function makes a special shape called a *parabola*. Many other kinds of functions have graphs that are roughly in the shape of a $\cup$ or $\cap$, but only graphs of quadratic functions are called parabolas.[12] A parabola with the $\cup$ shape is said to open upward; the $\cap$ shape is said to open downward.

---

[12] These curves were defined and studied by ancient Greek mathematicians, who found many properties that distinguish parabolas from other $\cup$ and $\cap$ shaped curves. For example, when a mirror is in the shape of a parabola, light rays parallel to the center line and incident on the "inside" of the parabola all reflect through a common point, called the focus. This is not true for all $\cup$ or $\cap$ shaped curves, just for parabolas.

That characteristic parabolic shape provides a graphical explanation of the behavior of quadratic functions. As you are aware, graphs of linear functions are straight lines, and using proportional reasoning amounts to assuming that some graph follows a straight line. In contrast, no part of a parabolic graph follows a straight line for very long. Even on the steepest parts of the sample graphs, if you draw a straight line along the side of the curve, the parabola will soon curve away from the line. This indicates that a quadratic function ultimately increases or decreases much more rapidly than any straight-line function. Because their graphs do not behave like straight lines, quadratic models are referred to as *nonlinear*; they are also said to possess the attribute of *nonlinearity*.

These ideas can be related to the computer network example. Recall that we found a network with 100 computers would need approximately 5,000 phone lines, while a network with 1000 computers would need almost 500,000 lines (page 152). Comparing these results, there is a ten-fold increase in the number of computers. Using proportional reasoning, we would expect the required number of telephone lines also to increase ten-fold, to 50,000. But that is a drastic underestimate of the true figure of nearly 500,000. This is a reflection of the nonlinearity of the model. It occurs because proportional reasoning assumes that the graph follows a straight line, whereas the true graph is a parabola, and curves up more steeply than a line.

In general, any quadratic function with a positive x^2 term eventually increases much more rapidly than a linear function. This can be made more specific. As a rule of thumb, for a quadratic function, a proportional change in x must be *squared* to estimate the corresponding change in y. Thus, doubling x should be expected to produce a four-fold increase in y, because $2^2 = 4$. Similarly, if we triple x we should expect a nine-fold increase in y. And, as in the computer example, increasing x ten-fold results in increasing y one hundred-fold.

Beyond the idea of nonlinearity, we can determine very precise information about the graph of a quadratic function from its coefficients in the standard form ax^2+bx+c. Several specific instances are detailed in the following paragraphs. They are illustrated for the example of $y = 3x^2 - 12x + 6$, as shown in Figure 3.16.

$\cup$ **Or** $\cap$. When a, the coefficient of x^2, is positive the graph opens upward ($\cup$); when negative, the graph opens downward ($\cap$). Thus, for the equation $y = 3x^2 - 12x + 6$ we immediately recognize that the graph will have the $\cup$ shape. This is indeed the case, as shown in Figure 3.16.

The figure illustrates the $\cup$ versus $\cap$ rule in one example, but how do we know it applies in general? Here is one justification. Suppose that a is positive. Then for large positive values of x, ax^2 will be positive, and much greater than the bx and c terms. For such values of x, y will be positive. Therefore, as the curve is traced to the right, it must eventually rise above and stay above the x-axis. And that can only happen on a $\cup$ shaped graph. A similar argument applies when a is negative.

y-**intercept.** The c coefficient, for example the 6 in $y = 3x^2 - 12x + 6$, is not multiplied by a variable, and so is called the *constant* coefficient. It gives the y-intercept of the curve. Notice that the y-intercept appears in the same position in the $mx + b$ form for the equation of a line. In both cases, the constant coefficient is the value of y when x is set to 0. But $x = 0$ only for a point on the y-axis. Thus, in both the linear and quadratic cases, the constant coefficient tells us immediately where the curve crosses the y-axis.

3.3. Quadratic Functions and Equations

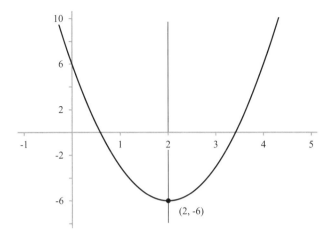

Figure 3.16. Graph of $y = 3x^2 - 12x + 6$, with center line and vertex.

Center Line and Vertex. Next, look at the center line (or axis of symmetry) that cuts through the exact middle of the curve. It can be determined exactly by marking the point at $-b/(2a)$ on the x-axis. For the example under discussion, $-b/(2a) = -(-12)/6 = 2$. The center line for this example crosses the x-axis at $x = 2$.

One important aspect of the location of the center line is that it determines the highest point on a downward opening parabola ($\cap$) and the lowest point on an upward opening parabola ($\cup$). In either case, that extreme point on the curve is called the *vertex* of the parabola. The x for the vertex is the same as the x for the center line: $-b/(2a)$. By substituting this for x in the equation of the curve, we can find the y, and hence the exact location of the vertex. In the example we have been working with, $-b/(2a) = 2$. With $x = 2$, we find $y = 3 \cdot 2^2 - 12 \cdot 2 + 6 = 12 - 24 + 6 = -6$. This shows that $(2, -6)$ is the vertex. It is also the lowest point on the curve.

We have now seen how the parameters from the standard form $ax^2 + bx + c$ easily lead to four characteristics of the graph. Presently we will show that the parameters can also be used to find a fifth aspect of the graph, the x-intercepts. But first we pause to comment briefly on the significance of such results.

Today graphing calculators and computers can be used to obtain high quality graphs almost effortlessly. For this reason, the information about the graph given above may not seem very important. However, each of these characteristics is connected with the overall behavior of the function, and provides a bridge between its numerical and graphical properties. By studying both the graphs and the way aspects of the graphs can be found using the coefficients, you will develop a richer understanding of quadratic functions. This knowledge is also useful in checking for errors. A graph obtained from a computer or calculator should show the features we can predict from the coefficients. Otherwise, some error has been made.

x-intercepts. There is one more aspect of the graph of a quadratic function that can be determined from the coefficients—the x-intercepts. These are the points (if any) where the graph crosses the x-axis. Whereas the y-intercept answers the question *What happens when $x = 0$?* the x-intercepts answer the inverse question *Is there an x for which $y = 0$?* This distinction is closely related to one made earlier between find-a_n

questions and find-n questions. Given a specific example, such as $3x^2 - 12x + 6$, we can easily see what happens when $x = 0$: direct substitution produces $3 \cdot 0^2 - 12 \cdot 0 + 6 = 6$. In contrast, it is not so easy to find an x for which $3x^2 - 12x + 6 = 0$.

The distinction may be more meaningful in terms of an application. Consider a model for world oil reserves, where x represents a number of years (starting from a reference point, such as 2003) and y represents the amount of oil available at time x. In this case, the y-intercept is the amount of oil at time 0. That is, it represents the world oil supply at the start time for the model. In contrast, an x-intercept represents a *time* when $y = 0$. This is a prediction of how long a time remains before the oil is used up.

Algebraically, the x-intercepts for a quadratic function $ax^2 + bx + c$ are the solutions to the equation $ax^2 + bx + c = 0$. They are the values of x, if any, which lead to a result of 0 when substituted into the function. These numbers are also referred to as the *roots* or *zeros* of the function. As we have seen in examples, numerical and graphical methods can be used to answer find-n questions, and these same methods can be used to find roots, at least approximately. Now we will consider algebraic methods for finding roots of quadratic equations.

Solving Equations. Some readers who have studied quadratic equations before may recall that roots of the equation $ax^2 + bx + c = 0$ are given by an algebraic expression called the *quadratic formula*. In one sense, this makes solving quadratic equations almost effortless. One need only enter the coefficients into the formula, and compute the result. But the formula itself is somewhat complicated, and it is not at all obvious why such a formula is valid. Accordingly, we will devote the next few pages to deriving the quadratic formula. One reason for doing so is to show how and why the formula holds. Another is to show how someone might discover this formula.

We will begin by considering a very special kind of quadratic equation, exemplified by $x^2 = 7$, and employing a numerical approach. With a calculator, you can try various guesses for x. If you substitute 2.6 for x, the result is $x^2 = 6.76$, which is too low. Trying 2.7 leads to 7.29 which is too high. So the solution must be between 2.6 and 2.7. Continuing with this approach we can get better and better approximations to an answer. For example, 2.6457 is too low and 2.6458 is too high. Clearly, there can be no exact answer in the form of a finite decimal between 2.6 and 2.7. Just imagine multiplying such a decimal by itself by hand. For example, consider the multiplication

$$\begin{array}{r} 2.64575 \\ \underline{\times 2.64575} \end{array}$$

The very first thing you multiply is 5×5, and the result, 25, gives you the last decimal digit of the answer, namely 5. For any finite decimal expression of this sort, with nonzero digits to the right of the decimal point, it is impossible to multiply the expression by itself and end up with all 0's following the decimal point. In particular, it is impossible to end up with the whole number 7. So there is no exact decimal solution for $x^2 = 7$. A similar argument shows that there is no exact fractional solution. If you replace x with a fraction in lowest terms, say 66/25, then x^2 will still be a fraction. It can't work out to be exactly 7. Intuitively speaking, in $\frac{66 \cdot 66}{25 \cdot 25}$ there is nothing in the numerator to cancel the 25's in the denominator, so the result cannot be a whole number.

Although $x^2 = 7$ has no exact solution as a fraction or finite decimal, conceptually an exact solution must exist. We can see this visually by comparing the graphs of $y = x^2$

3.3. Quadratic Functions and Equations

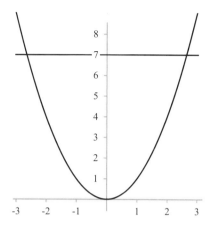

Figure 3.17. Intersecting graphs. The parabola is the graph of $y = x^2$. The horizontal line is the graph of $y = 7$. If $P = (x, y)$ is a point of intersection then x and y must satisfy both equations. Therefore $x^2 = y = 7$, showing that x is an exact solution of the equation $x^2 = 7$.

and $y = 7$, as in Figure 3.17. The horizontal line and the parabola clearly intersect at two points, and the x-coordinate of each intersection point is an exact solution of $x^2 = 7$. Therefore there *are* exact solutions, but they cannot be represented by fractions or finite decimals. For this reason, the exact solutions are said to be *irrational*.

As shown in the figure, there are two exact irrational solutions to $x^2 = 7$. One is approximately 2.6457, as we saw earlier. The other is the negative of the first. There is a mathematical notation for these exact solutions. The positive solution is written $\sqrt{7}$ (pronounced *the square root of 7*), and the negative solution is then $-\sqrt{7}$. In general, the $\sqrt{}$ of a positive number z means the positive solution of the quadratic equation $x^2 = z$. So $\sqrt{13}$ means the positive solution of $x^2 = 13$. Put another way, $\sqrt{13}$ is the positive number which, when multiplied by itself, gives 13. There is only one solution to $x^2 = 0$, namely 0, so $\sqrt{0} = 0$.

There are many algebraic properties of $\sqrt{}$. You have probably studied them in a previous course, and they will not be systematically described here. We will use the properties as needed. If you need to review the properties of $\sqrt{}$, consult any college algebra text. Your instructor can help you locate one.

There is one aspect of square roots that we will review, namely, that there are no real square roots of negative numbers.[13] Consider, for example, $\sqrt{-3}$. This is supposed to be a solution of the equation $x^2 = -3$. But there is no such real number x. Whatever number you use for x, multiplying it by itself to form x^2 produces a positive result, or zero. The result is never negative, and in particular, cannot equal -3.

There is a graphical way to reach the same result. Rewrite the equation in the form $x^2 + 3 = 0$. The solutions to this equation would be x-intercepts on the graph of $y = x^2 + 3$, as mentioned in the discussion of x-intercepts (see page 176). The coefficients of the function are $a = 1, b = 0$, and $c = 3$. This tells us that the vertical

---

[13] In some courses, the real numbers are augmented by the addition of a new element, i, defined as the square root of -1. This results in a larger number system, the complex numbers, which contains the real numbers as a subset. We will restrict our attention to the real number system in this course.

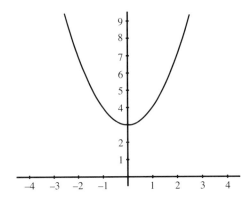

Figure 3.18. Graph of $y = x^2 + 3$.

bisecting line has $x = -b/(2a) = 0$, and that the vertex therefore has $x = 0$ and $y = 0^2 + 3 = 3$. Also, because a is positive, we know that the graph has a $\cup$ shape, and that the vertex is the lowest point. These features, shown in Figure 3.18, imply that the parabola never crosses the x-axis. Thus there are no x-intercepts, and consequently no real number solutions to the equation $x^2 + 3 = 0$.

As discussed above, $\sqrt{13}$ can be approximated to any degree of accuracy using trial-and-error calculation. Of course, almost every calculator includes a button for performing this process. When you push the $\sqrt{}$ button, the calculator will compute as accurate an approximation to the true value as it can. In some cases the result is an approximation, and sometimes it is exactly correct.

As an example of an approximate answer, a calculator will display $\sqrt{7}$ as a long decimal number starting 2.645751. We saw earlier that $\sqrt{7}$ cannot be expressed exactly by a finite decimal, so the finitely many digits shown by the calculator cannot give the exact value. No matter how many digits your calculator shows, the result is only an approximation. Simple examples of exact answers using whole numbers include $\sqrt{1} = 1$ and $\sqrt{4} = 2$. More complicated examples are easily constructed along the following lines. Begin with a (finite) decimal, say 4.17. Squaring this will result in another finite decimal, 17.3889. That means $\sqrt{17.3889} = 4.17$, exactly, and the calculator will find that value.

Usually it will not matter greatly whether the calculator gives an exact square root or an approximation. But there is an easy way to check whether an answer is exact or not: square it (exactly) and see whether you obtain the starting number. For example, a calculator will display $\sqrt{0.25}$ as 0.5. That is exactly correct because $0.5^2 = 0.25$. But here is a more complicated example. On a particular calculator, $\sqrt{64.505313511}$ is displayed as 8.03152. If that is exactly correct, then $8.03152 \times 8.03152 = 64.505313511$. This calculation cannot be performed exactly on most calculators because there are too many digits. But using the same logic we saw earlier (page 176), the final decimal digit of 8.03152×8.03152 will be a 4 in the tenth decimal place. Therefore it cannot exactly equal 64.505313511.

By its very definition, the square root operation provides an exact symbolic solution to a special type of quadratic equation, namely, those that can be expressed in the form

3.3. Quadratic Functions and Equations

$x^2 =$ [a number]. For example, a solution to $x^2 = 13$ is given by $x = \sqrt{13}$. But we can go further. With square roots it is possible to solve *any* quadratic equation.

We illustrate with the following example. Suppose we wish to solve

$$3x^2 - 5x - 8 = 0.$$

Recalling the earlier discussion of graphs, we can interpret the solutions of the equation as the x-intercepts of the curve $y = 3x^2-5x-8$. And now we can use other properties of parabolic graphs. Specifically, we know the curve has a $\cup$ shape, and a center line with $x = 5/6$. Then, by symmetry, if there are x-intercepts, they must be equal distances to the left and right of $5/6$ on the x-axis. These features are shown in Figure 3.19, where the distance from the center line to each x-intercept is labeled w.

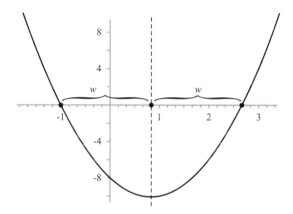

Figure 3.19. Graph of $y = 3x^2 - 5x - 8$. The center line is at $x = 5/6$, and the x-intercepts must be equally spaced to the right and left. If they are w units away from the center line, they must be at $x = 5/6+w$ and $x = 5/6 - w$. We can find the intercepts by finding w.

We do not know a value for w, but whatever it is, $5/6 + w$ must be a root of the equation. Thus, if we replace x by $5/6 + w$, the equation

$$3\left(\frac{5}{6} + w\right)^2 - 5\left(\frac{5}{6} + w\right) - 8 = 0$$

must hold.

At this point, we don't seem to have made much progress. We can find w by solving a new equation, but it seems even more complicated than the original. But an amazing simplification is about to occur. In the last equation expand $\left(\frac{5}{6} + w\right)^2$, then remove the parentheses and combine like terms. The steps are shown below.

$$3\left(\frac{5}{6} + w\right)^2 - 5\left(\frac{5}{6} + w\right) - 8 = 0$$

$$3\left(\frac{25}{36} + \frac{10}{6}w + w^2\right) - 5\left(\frac{5}{6} + w\right) - 8 = 0$$

$$\frac{25}{12} + 5w + 3w^2 - \frac{25}{6} - 5w - 8 = 0$$

$$3w^2 + \frac{25}{12} - \frac{25}{6} - 8 = 0.$$

This still looks messy, but it has a simple algebraic form:

$$3w^2 + (\text{a constant}) = 0.$$

If we combine all the numbers making up that constant, we can see the form more clearly:

$$3w^2 - \frac{121}{12} = 0.$$

The amazing simplification is that there is no w term—the only w in the equation is w^2. And that means we can solve this equation using square roots. Add 121/12 to both sides of the equation, and then divide by 3, and we reach

$$w^2 = \frac{121}{36}.$$

The solutions are now evident,

$$w = \pm\sqrt{\frac{121}{36}}.$$

But remember that we really want the x-intercepts, which are given by $\frac{5}{6} + w$ and $\frac{5}{6} - w$. Thus we find the solutions to the original equation are given by

$$x = \frac{5}{6} \pm \sqrt{\frac{121}{36}}.$$

This entire process is rather involved, so let us review the highlights. We began with an equation in the form

$$ax^2 + bx + c = 0$$

(so the process can be applied to any equation that can be put in this form). Then we recognized that the solutions are x-intercepts, and so are placed symmetrically, a certain distance w to the right and left of the central line, where $x = 5/6$. That means the solutions must be given by $5/6 + w$ and $5/6 - w$. Substituting $5/6 + w$ for x in the original equation, and using algebra, we obtained an equation for w of the form

$$w^2 = \text{a constant}.$$

Then we found w using the square root operation. Finally, we found the solutions to the original equation as $5/6 + w$ and $5/6 - w$.

With all of those steps, it is quite possible an error was made. How can we check if our answers are correct? One approach is to substitute the proposed answers back into the orginal equation

$$3x^2 - 5x - 8 = 0.$$

To that end, use a calculator to find

$$\frac{5}{6} + \sqrt{\frac{121}{36}} = 2.66666667,$$

which we recognize as a decimal approximation of the true value. Substituting, the calculator now shows that

$$3(2.66666667)^2 - 5(2.66666667) - 8 = 0.000000037.$$

This shows that $x = 2.66666667$ comes very close to solving our equation.

3.3. Quadratic Functions and Equations

Usually, this is about the best you can do with a calculator. But in this problem the square root can be found exactly. In fact
$$\frac{121}{36} = \frac{11^2}{6^2}$$
so
$$w = \sqrt{\frac{121}{36}} = \frac{11}{6},$$
and our roots are
$$x = \frac{5}{6} + \frac{11}{6} = \frac{8}{3}$$
and
$$x = \frac{5}{6} - \frac{11}{6} = -1.$$
With a calculator that can manipulate fractions, or working by hand, you can verify that 8/3 really is an exact solution of the original equation. For the other proposed root, -1, verification is even easier, and left to the reader.

The Quadratic Formula. The method illustrated above can be applied to any quadratic equation. However, there are too many algebraic steps to make the method very attractive. Fortunately, what was done above with specific numerical coefficients can be carried out just as well with parameters a, b, and c. That is, start with $ax^2 + bx + c = 0$ and carry out the same steps as before. To start the process, the x-coordinate of the center line will be given by $-b/(2a)$ rather than a specific number. Defining w as before, the x-intercepts must be at
$$x = -\frac{b}{2a} + w \tag{3.16}$$
and
$$x = -\frac{b}{2a} - w. \tag{3.17}$$
This will again lead to a quadratic equation for w of the form
$$w^2 = [\text{constant}],$$
although in this case the constant is represented by a formula involving $a, b,$ and c. Continuing in this fashion eventually leads to the following formulas for the two x-intercepts:
$$x = \frac{-b}{2a} + \frac{\sqrt{b^2 - 4ac}}{2a} = \frac{-b + \sqrt{b^2 - 4ac}}{2a}$$
and
$$x = \frac{-b}{2a} - \frac{\sqrt{b^2 - 4ac}}{2a} = \frac{-b - \sqrt{b^2 - 4ac}}{2a}.$$
These equations are usually expressed as the single equation
$$x = \frac{-b \pm \sqrt{b^2 - 4ac}}{2a},$$
which is called the *quadratic formula*. It can be used to determine the solutions to any quadratic equation immediately. Or, if $b^2 - 4ac$ comes out to be a negative number, the formula does not give any solutions, since there are no square roots of negative numbers. In this case there are no solutions to the quadratic equation.

To use the quadratic formula, a quadratic equation must be expressed in standard form. For example, given the equation $3x - x^2 = 4 + x$, we first rearrange the terms to obtain $x^2 - 2x + 4 = 0$. In this form we can identify $a = 1, b = -2$, and $c = 4$. The quadratic formula then gives

$$x = \frac{-b \pm \sqrt{b^2 - 4ac}}{2a}$$

$$= \frac{2 \pm \sqrt{4 - 4 \cdot 1 \cdot 4}}{2 \cdot 1}$$

$$= \frac{2 \pm \sqrt{4 - 16}}{2}$$

$$= \frac{2 \pm \sqrt{-12}}{2}$$

However, there is no square root of -12. This shows that the original equation has no solutions.

For a more complicated example, let us apply the quadratic formula with the oil consumption model. In that model the position number n represents a year, with $n = 0$ in 2003, and s_n represents the cumulative world oil consumption for years 0 through n, in units of billions of barrels. Our data included a figure of 1,687.9 billion barrels for the known oil reserves as of the end of 2013, and we asked when these reserves would be used up. We then showed that this question could be answered by solving the equation

$$29.5 + 29.88n + 0.38n(n-1)/2 = 2{,}033.3$$

for n (see page 161).

We can see that this is a quadratic equation although it is not in the standard form. The product $n(n-1)$ will produce an n^2 and all other terms will either be constants or constant multiples of n. In fact, the equation can be transformed into the standard form

$$0.19n^2 + 29.69n - 2{,}003.8 = 0,$$

so that $a = 0.19, b = 29.69$, and $c = -2{,}003.8$. Using the quadratic formula, we obtain

$$n = \frac{-b \pm \sqrt{b^2 - 4ac}}{2a}$$

$$= \frac{-29.69 \pm \sqrt{29.69^2 - 4 \cdot 0.19 \cdot (-2{,}003.8)}}{2 \cdot 0.19}$$

$$= \frac{-29.69 \pm \sqrt{29.69^2 + 4 \cdot 0.19 \cdot 2{,}003.8}}{0.38}$$

$$= \frac{-29.69 \pm \sqrt{2{,}404.3841}}{0.38}.$$

To proceed, remember that we are actually finding two answers—one that comes from adding the square root and another that comes from subtracting. Finding the first answer with a calculator produces

$$n = 50.91,$$

to two decimal places. This is consistent with what we found earlier. Without actually computing the second answer we can see that it will produce a negative result. This cannot be meaningful in our model, where n is restricted to positive values. Therefore, only the first answer is of interest in this problem.

The quadratic formula completely clears up the problem of solving quadratic equations. It can be used with numerical values to find a solution to a specific quadratic

3.3. Quadratic Functions and Equations

equation, and the square root operation on a calculator can then be used to obtain a highly accurate numerical approximation to the solution. But the formula is also useful for theoretical purposes. For one thing it gives a definite indication when an equation has no solutions.

It can also be used to invert a quadratic function. To illustrate, we begin with the functional equation for the oil consumption model,

$$s_n = 29.5 + 29.88n + 0.38n(n-1)/2.$$

Again using algebra, we can express this in the standard form

$$0.19n^2 + 29.69n + 29.5 - s_n = 0.$$

Next, apply the quadratic formula, with $a = 0.19$, $b = 29.69$, and $c = 29.5 - s_n$. That produces

$$\begin{aligned} n &= \frac{-b \pm \sqrt{b^2 - 4ac}}{2a} \\ &= \frac{-29.69 \pm \sqrt{29.69^2 - 4 \cdot 0.19 \cdot (29.5 - s_n)}}{2 \cdot 0.19} \\ &= \frac{-29.69 \pm \sqrt{881.4961 - 22.42 + 0.76 s_n}}{0.38} \\ &= \frac{-29.69 \pm \sqrt{859.0761 + 0.76 s_n}}{0.38}. \end{aligned}$$

This is really two equations. For any specified value of s_n we can compute either

$$n = \frac{-29.69 + \sqrt{859.0761 + 0.76 s_n}}{0.38}$$

or

$$n = \frac{-29.69 - \sqrt{859.0761 + 0.76 s_n}}{0.38},$$

provided $859.0761 + 0.76 s_n$ isn't negative.

Each equation expresses n as a function of s_n. However, the second equation, while mathematically correct, is not useful in this model. To see why, we consider an example with a specific value of s_n. Say we wish to know when s_n will equal 123. Substituting this value into each of the equations, we compute two values of n. They are, to three decimal places,

$$n = \frac{-29.69 + \sqrt{859.0761 + 0.76 \cdot 123}}{0.38} = 3.088$$

and

$$n = \frac{-29.69 - \sqrt{859.0761 + 0.76 \cdot 123}}{0.38} = -159.351.$$

The first can be interpreted to say that the cumulative consumption of oil will reach 123 billion barrels after 3.088 years (starting in 2003). The second can be interpreted as referring to a time 159.3 years *before* 2003, but, in the problem context, there is no reason to believe the functional equation will be valid for such a time. More generally, observe that the second equation will always produce a result $n \leq 0$. This refers to a time before 2003, and is not meaningful in the problem context.

Graphical Property Summary for Quadratic Functions. We saw earlier how the parameters a, b, and c can be used to determine four characteristics of the graph of the function $y = ax^2 + bx + c$: the shape as $\cup$ or $\cap$, the y-intercept, the center line, and the vertex. We have also observed the close connection between solutions to a quadratic equation and x-intercepts on the graph of a quadratic function. In particular, using the quadratic formula, we can express the x-intercepts of a quadratic curve in terms of a, b, and c. This rounds out our list of graphical properties, as summarized below.

Quadratic Function Graphical Features: On the graph of $y = ax^2 + bx + c$:

- The shape is $\cup$ for positive a; $\cap$ for negative a.

- The y-intercept is c.

- The vertical center line that divides the graph into symmetric halves crosses the x-axis at $-b/(2a)$.

- The high point for a $\cap$ or the low point for a $\cup$ is the point at which the vertical center line crosses the graph. This is called the *vertex*. The x at this point is $-b/(2a)$; the y can be found by substituting $-b/(2a)$ for x in the function.

- If $b^2 - 4ac$ is negative, the function has no x-intercepts. Otherwise, the quadratic formula gives the x-intercepts as $\dfrac{-b \pm \sqrt{b^2 - 4ac}}{2a}$.

Factored Form. There is one more aspect of solving quadratic equations that should be mentioned. As already discussed, a quadratic function can be written in many different algebraic forms. One form in particular is handy for solving equations. Consider this equation

$$(x - 2)(x - 3) = 0.$$

The solutions to the equation are 2 and 3. Check this by substituting 2 for x in the equation. Do you see why 2 and 3 are the solutions? For another example, consider

$$(3x - 5)(2x + 1) = 0.$$

Can you guess the solutions this time? As in the previous example, one solution for x makes $3x - 5 = 0$, and the other makes $2x + 1 = 0$. Therefore, the solutions are $5/3$ and $-1/2$.

The examples above involve what is called the factored form of a quadratic polynomial. In this context, *factors* has a very specific meaning: items which are multiplied together. When an algebraic form is a multiplication of two or more parts, each part is called a factor. In the examples above, the factors are the parts enclosed in parentheses. In the expression $5xy$ the factors are 5, x, and y. On the other hand, in $9x + 3y$ there are no factors because as written the expression is not a multiplication of two or more parts. But if we rewrite the expression as $3(3x + y)$ then it *is* a multiplication, and the factors are 3 and $(3x + y)$.

You may recall learning how to break a composite number down into its factors, for example, $12 = 2 \cdot 2 \cdot 3$. The numbers on the right side of this equation are called factors of 12 because they are all multiplied together to give 12. In the same way, $(x - 2)$

3.3. Quadratic Functions and Equations

and $(x-3)$ are called factors of the function $(x-2)(x-3)$ because they are multiplied together to give that function.

Using factors to find solutions to an equation only works when the opposite side of the equation is 0. If the equation is

$$(x-4)(x-7) = 1$$

it is no good to make x equal to 4. That produces 0 on the left side of the equation, not 1 as required. It also is no good making $x - 4$ equal to 1. Even if $(x-4)$ does equal 1, there is still the $(x-7)$ to worry about. The point is that 0 is very special. When $(x-4)$ is equal to 0, we can forget about $(x-7)$, because 0 times anything is still 0. It is this special property of 0 that makes the method of factors work, but only for equations with 0 on one side. In fact, the idea works for any number of factors. The equation

$$(x-1)(x-4)(x-2.5)(x+9)(x-17) = 0$$

can be solved just by looking at it. The solutions are 1, 4, 2.5, -9, and 17, because each of these numbers makes one factor equal to 0. However, we are especially interested in problems with just two factors, because those are quadratic equations.

Look again at the first example,

$$(x-2)(x-3) = 0.$$

We can rewrite the expression on the left using algebra, just as we did on page 171. That will produce $x^2 - 5x + 6 = 0$. This is immediately recognizable as a quadratic equation. In general, multiplying two linear factors results in a quadratic. In the example, $x - 2$ and $x - 3$ are both linear factors because they are linear functions of x. Similarly, $(5x - 6)(\frac{x}{8} + 12.4)$ is a product of two linear factors.

It is easy to go from a factored form to the usual descending form. The reverse problem is more difficult. For example, to put $x^2 - x - 20$ into a factored form, you might try something of the form $(x-\square)(x-\diamond)$, where multiplying $\square \cdot \diamond$ has to result in 20. Proceeding by trial and error, take $\square = 2$ and $\diamond = 10$. Then we have $(x-2)(x-10) = x^2 - 12x + 20$. This is similar to the expression we want, but we need a -20 not a 20, and we need a $-x$ not a $-12x$. By trying different combinations, you can eventually find that the factored form for $x^2 - x - 20$ is $(x-5)(x+4)$. But the process is neither direct nor obvious.

We already saw how easy it is to find solutions to an equation if it is given in factored form. But it is also possible to obtain the factored form if we know the solutions. Here is an example. Recall that earlier we found the solutions of $3x^2 - 5x - 8 = 0$ to be -1 and $8/3$ (see page 181). This shows that $(x + 1)$ and $(x - 8/3)$ should be factors of $3x^2 - 5x - 8$. However, $(x+1)(x-8/3)$ has a leading term of x^2 where we need $3x^2$. Therefore, multiplying by 3, we propose

$$3x^2 - 5x - 8 = 3(x+1)(x-8/3).$$

As you can verify by multiplying out the right-hand side, this is a correct factorization.

This concludes our discussion of properties of quadratic functions. We have already seen that these functions arise in applications involving quadratic growth sequences, but they also occur in other applications. We will see some of these applications in the next section.

3.3 Exercises

Reading Comprehension. Answer each of the following with a short essay. Give examples, where appropriate, but also answer the questions in written sentences.

(1) Write short definitions or explanations of the following terms and phrases used in the reading.

   a. quadratic function
   b. quadratic equation
   c. coefficient
   d. x- and y-intercepts
   e. roots
   f. factors
   g. factored form
   h. square root
   i. quadratic formula
   j. vertex

(2) @Explain what it means for two different algebraic expressions to define the same function. How can you tell whether two expressions do or do not define the same function?

(3) Explain what is meant by the statement: $x^2 = -16$ has no solutions. Give both theoretical and graphical reasons in support of the statement.

(4) @Abbie and Bill each has a quadratic function to study as part of a homework assignment. Abbie's is $f(x)$ and Bill's is $g(x)$. Comparing their work, they see that the equations are different. But they also notice that when they start to plot points on a graph, they get the same results. They find $f(1) = 7$ and $g(1) = 7$; $f(2) = 10$ and $g(2) = 10$; $f(3) = 12$ and $g(3) = 12$. Can they conclude that their functions are actually the same function? Or is it more correct to say that $f(x)$ and $g(x)$ will be the same for some x's and different for other x's? Explain.

(5) Describe general features of the graphs of quadratic functions. Tell about the general shape and explain what information about the graph is revealed by the coefficients of the quadratic function.

(6) @Allen and Ellen are studying a quadratic function $y = f(x)$. Among other things, they create a table of x and y values, choosing x values of 2, 2.4, 2.8, 3.2, 3.6, 4.0, 4.4, and 4.8. When they substitute these into the function, they notice a familiar pattern in the values of $f(2)$, $f(2.4)$, $f(2.8)$, $f(3.2)$, etc. What type of pattern do they notice?

(7) The quadratic formula is a well-known mathematical tool. In this section you read an example of how one might discover the formula and why the formula holds true. Write a summary of these ideas as if to a friend who is getting bogged down in the algebraic manipulations and missing the main ideas. You may choose to include a simple example in your work.

3.3. Exercises

(8) @A group of students discusses the implications of using a quadratic growth model for a social media network. The network grew from from 100,000 users to 700,000 users in 5 years. Hasan says "If there were 700,000 users after 5 years, there should be 1,400,000 users after another 5 years." Hongwei says "Since the number of users grew by 600,000 in 5 years, it should grow by an equal amount in the next 5 years. So then there should be 1,300,000 users." Esperanza says "In the first five years, the number of users was multiplied by 7. The same thing should happen in the next 5 years, multiplying by 7 again, for a total of 4,900,000." Natasha says "if we double the number of years we should expect to see 4 times as many users, so after another 5 years there would be 2,800,000 users." Which of these predictions is most consistent with a quadratic growth model, and why?

Mathematical Skills.

(9) @Indicate for each of the graphs shown in Figure 3.20 whether or not it could represent a quadratic function. For the graphs that are *not* quadratics, give a reason for your answer.

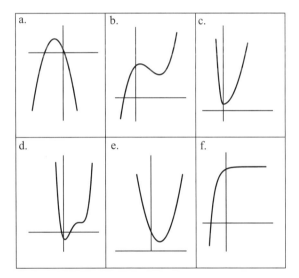

Figure 3.20. Graphs of six functions. Which ones are quadratic?

(10) @Indicate for each of the expressions below whether or not it represents a quadratic function. For the expressions that are *not* quadratics, give a reason for your answer.

  a. @$(x^2 - 5)(x^2 + 12)$
  b. $7(x - 1)x + 10x - 3$
  c. @$\frac{50}{x^2}$
  d. $0x^2 + 1x + 2$

(11) @Put each quadratic function into the standard form $ax^2 + bx + c$.

  a. @$2x - 3x^2 - 5x + 7$

b. $(7x - 2)x + 3$
c. @$(2x - 5)(x + 3)$
d. $(x - 2)(x + 2)$

(12) @Use graphical and numerical methods to solve the following polynomial equations approximately. Your answers should be correct to at least 2 decimal places.

a. @$3x^2 - 10 = 0$
b. $x^2 - x - 1 = 0$
c. @$3x^2 + 1 = 3x + 2$
d. $x^2 + 2x + 5 = 20$

(13) @Use theoretical methods (the quadratic formula or the factored form) to solve these equations exactly:

a. @$(x - 2)(3x - 6) = 0$
b. $3x^2 = 15$
c. @$11x^2 + 3x - 5 = 0$
d. $3x^2 + x - 2 = 0$
e. $(x - 2)(x + 3) = 3$

(14) @Use theoretical methods (such as the quadratic formula) to determine if real number solutions can be found for the following equations. If so, give the exact solutions. If not, say why not.

a. @$10 + 7x - 3x^2 = 5$
b. $x^2 - 5x = 3x - 28$
c. @$x^2 + 25 = 0$
d. $2x^2 - 10x + 8 = 0$

(15) @For each equation below, use the coefficients to find the following graphical features:

- x- and y-intercepts
- orientation ($\cap$ or $\cup$)
- center line
- vertex.

Verify the results by using a graphing calculator or computer graphing application.

a. @$y = 5x^2 - 20x + 15$
b. $y = -2x^2 + 3x - 7$

(16) @Factor the following quadratic expressions. You may use any correct techniques you are comfortable with. Recall that one method of factoring $ax^2 + bx + c$ involves using the exact solutions to $ax^2 + bx + c = 0$ as described on page 184.

a. @$x^2 + 3x + 2$
b. $4 - 3x - x^2$
c. @$x^2 + 8.5x + 4$

3.3. Exercises

   d. $7x^2 + 7x - 84$

   e. @$5x^2 - 34x - 7$

(17) @Give a quadratic equation which has the stated solutions. For example the values $x = 3$ and $x = 7/2$ are solutions to $(x - 3)(x - 7/2) = 0$. They are also solutions to $5(2x - 7)(x - 3) = 0$. You need only provide one quadratic equation for each pair of solutions.

   a. @$x = 17$ and $x = 23$

   b. $x = 1/3$ and $x = -5$

   c. $x = -0.64$ and $x = 1.8$

   d. $x = 2$ is the only solution

Digging Deeper.

(18) @The equation $y = x^2 - 4x$ defines y as a function of x. Use the quadratic formula to invert this function, expressing x in terms of y.

(19) In the Second Differences subsection on page 171 we explored the fact that for equally spaced x values the resulting values of a quadratic function form a quadratic sequence. We will explore that further here.

   a. First, let's verify what was presented. Using the function $f(x) = 3x^2 - 5x + 2$ we defined the sequence $a_0 = f(-2)$, $a_1 = f(-1.5)$, $a_2 = f(-1)$, $a_3 = f(-0.5)$, and so on. For this sequence it was asserted that

   $$a_{n+1} = a_n - 7.75 + 1.5n$$

   because

   $$a_n = f(-2 + 0.5n) = 3(-2 + 0.5n)^2 - 5(-2 + 0.5n) + 2$$

   and

   $$a_{n+1} = f(-2 + 0.5n + 0.5) = 3(-1.5 + 0.5n)^2 - 5(-1.5 + 0.5n) + 2.$$

   Using these definitions of a_{n+1} and a_n, verify that a_{n+1} does equal $a_n - 7.75 + 1.5n$, as stated above.

   b. Would this still work if we used a different arithmetic sequence for x values? That is, if we used x values such as $x_0 + dn$ where x_0 and d are constants. This would mean

   $$a_n = f(x_0 + dn) = 3(x_0 + dn)^2 - 5(x_0 + dn) + 2$$

   and

   $$a_{n+1} = f(x_0 + dn + d) = 3(x_0 + dn + d)^2 - 5(x_0 + dn + d) + 2.$$

   Find $a_{n+1} - a_n$ and, if possible, put it in the form (a constant) + (a constant)n.

   c. What conclusions can be drawn from the previous algebraic work? What next steps could be used to further explore this topic and what questions would they answer?

3.4 Quadratic Models for Revenue and Profit

As we have seen, quadratic functions and equations arise naturally when we use quadratic growth difference equations. They can also appear in models that are formulated without reference to difference equations. If the shape of a data graph suggests a parabola or part of a parabola, an analyst might begin with an equation of the form $y = ax^2 + bx + c$, and select values of a, b, and c to approximate the data as closely as possible. This is similar to what was done in the polar sea ice example (see discussion beginning on page 147).

Now we will focus on another way that quadratic models arise, based on an understanding of the phenomenon or process being modeled. In particular, we will see how linear functions in a model can be combined to arrive at a quadratic function. This is an instance of a structural quadratic function model, similar to structural quadratic growth sequences that arise as sums of arithmetic growth sequences.

To illustrate these ideas, we will consider a simple model for a business selling a single product, including equations for sales volume, revenue, costs, and profits. In this example, electing to adopt linear models for demand and costs leads to quadratic models for revenue and profit.

Linear Demand Model. The context for this example is a student organization selling soft drinks as a fundraising activity. We considered this situation in an earlier example (page 107). There, based on sales survey data, we developed a linear model for the demand on a typical day. Recall that *demand* means the number of items that can be sold. Taking s to be demand, and p to be the price in dollars, the equation we found in the earlier example says

$$s = -\frac{560}{3}(p - 1.20) + 40. \tag{3.18}$$

So, for example, the model predicts that at a price of $0.90 per drink, the demand will be

$$-\frac{560}{3}(0.90 - 1.20) + 40 = 96.$$

As we proceed, we will assume that the demand equals the sales volume.[14] That implies the organization will always have enough supplies to meet the demand, and will sell as many drinks as the market allows.

The analysis to follow will be more convenient if we use decimals instead of fractions, so we round off $560/3 = 186.666\cdots$ to 186.7, amounting to a very small shift in the slope of our linear model. We can justify this by observing that the original linear model was only an estimate to begin with. Moreover, the model only makes sense for p up to about 1.41, beyond which the equation produces negative values for demand. In the range $0 \leq p \leq 1.41$, replacing $560/3$ with 186.7 in (3.18) alters the predicted demand by an amount less than 0.05.

Using the decimal approximation, the equation for s becomes

$$s = -186.7p + 264. \tag{3.19}$$

This equation defines s as a linear function of p because it is in the standard form $s = mp + b$. Recall also that p is the independent variable and s is the dependent

---

[14]That is why we represent demand with the variable s instead of d, the variable used in the earlier example.

3.4. Quadratic Models for Revenue and Profit

variable. This reflects the fact that we can choose to make the price p whatever we wish, without knowing (and therefore independent of) s, while the computed value of demand s depends on the price we have set.

As always, it is important to be mindful of the assumptions that go into our model. As we discussed when it was first developed, this model imperfectly represents the demand data on which it is based. For example, according to the model, charging too high a price will result in a negative demand, an absurdity. We will not now repeat the complete discussion of such considerations. But we emphasize their importance. We will soon see that the demand model is incorporated in our models for cost, revenue, and profit, so all of the conclusions we derive will be subject to the same limitations as the demand model.

One other aspect of our assumptions is worth mentioning here. On any given day, the demand should be a whole number—the students will not be selling fractional parts of a drink. Yet the linear demand equation does not always produce whole number results. Is this a problem? The demand model was described as representing a *typical* day. That means it is a kind of average, and that the demand at a given price will be higher than the model on some days and lower on other days. It is reasonable that averaging demand over different days might not result in a whole number. Moreover, any predictions we make about demand, revenue, and profit, should likewise be understood as average figures, and at best are approximate values. Therefore, let us agree to interpret demand, revenue, cost, and profit as continuous variables, and in particular, to accept non-whole number values for these variables.

Our ultimate goal is to study how the price should be set to obtain the greatest profit. In general, the profits of a venture are determined by two things: revenue (which is a synonym for income) and expenses or costs. We will take up each of these in turn, beginning with revenue.

Quadratic Revenue Model. If we sell s drinks at a price p, that will produce sp dollars of revenue. But s depends on p, as given in (3.19). If the price is set at $0.60, for example, the number of sales will be $s = -186.7 \cdot 0.60 + 264 = 151.98$. Multiplying this number by the price per drink shows the total income is $R = 151.98 \cdot 0.60 = 91.188$. In a similar way, given any choice for p we can calculate the revenue as

$$R = (-186.7p + 264)p, \qquad (3.20)$$

or, after algebraic simplification

$$R = -186.7p^2 + 264p. \qquad (3.21)$$

These equations express the revenue as a quadratic function of price. The first version is in factored form. We will use this later to find the roots of R. The second version is the standard form for a quadratic, and we can read off the coefficients $a = -186.7$, $b = 264$, and $c = 0$.

What we know about quadratic functions reveals quite a bit about how revenue and price are related. For one thing, the equation can be graphed, as shown in Figure 3.21. With a negative a coefficient, we anticipate the $\cap$ shape. In this problem context the high point on the curve has a special significance. It indicates the maximum revenue. That is, finding the vertex in this model tells us what price to charge per drink to obtain the highest possible income, and also predicts what the highest income will be.

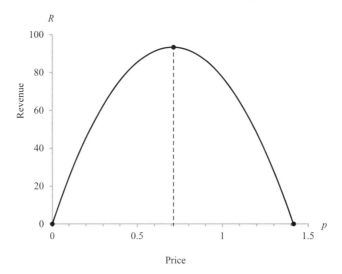

Figure 3.21. Graph of $R = -186.7p^2 + 264p$. Revenue can be zero in two ways. If we charge nothing for our drinks ($p = 0$), then we will collect no income. Also, if we sell no drinks we will collect no income. This happens when the demand equals 0 ($p \approx 1.41$). In the middle of the graph ($p \approx 0.71$) is the vertex, corresponding to the maximum possible revenue ($R \approx 93.33$).

As a first step toward locating the vertex, we note that the center line of the graph crosses the horizontal axis at $p = -b/(2a) = 264/(2 \cdot 186.7) = 0.707016\cdots$. This is consistent with the figure, which shows the axis of symmetry slightly to the right of $p = 0.7$. Substituting $0.707016\cdots$ into (3.21) gives $R = 93.33$ (to two decimal places). This tells us that the vertex is at approximately $(0.707, 93.33)$. Thus, at a price of 0.707 we expect to obtain a revenue of 93.33, and that is as high a revenue as we can obtain. Of course, it is not possible to actually charge a price of 70.7 cents. Instead, we would probably set the price at 70 cents and expect a revenue of about 93.30. Here again we see the effects of using a continuous model. In reality, the possible prices are restricted to a discrete set of values, namely whole numbers of cents. But our equations permit us to treat p as a continuous variable. This simplifies the model and facilitates analysis. And as the vertex calculation shows, it is not difficult to modify the conclusions we reach so as to reflect the true discrete nature of the price variable.

As a final feature of the graph, we find the horizontal intercepts, using the factored form (3.20). Setting $R = 0$ produces

$$(-186.7p + 264)p = 0,$$

which will hold if either $p = 0$ or $-186.7p + 264 = 0$. The second condition holds for $p = 264/186.7 = 1.414$ to three decimal places. Notice that the center line at $p = 0.707$ falls halfway between the horizontal intercepts 0 and 1.414, as it should. This is observable in Figure 3.21.

Linear Cost Model. On first consideration it seems reasonable to set the price so that we obtain the maximum revenue. However, profits depend not only on revenue, but on costs as well. To get the greatest profit, we will need to take cost into account.

3.4. Quadratic Models for Revenue and Profit

How do the costs C vary as a function of price p? We saw a linear cost model in the T-shirt example (page 69). Here we will develop a similar model. To keep matters as simple as possible, let us say that the only expense involved is the cost of buying the sodas. If we buy the soda at a big volume discount store perhaps we can get it for $0.25 per can. We will ignore any costs for driving to the store, ice to keep the soda cool, and so on. The total costs will be modeled as the price for the sodas at $0.25 per can. Of course, the number of cans we buy will depend on how many we expect to sell, and that in turn depends on the price we will charge. Reasoning as before, if the price we charge is set at $0.60 per can, then we need 151.98 cans at a cost of 151.98·0.25 = 37.995. Similarly, for any price p we will sell $-186.7p + 264$ cans, and the costs for these cans will be $C = 0.25(-186.7p + 264)$. This is a linear model. When it is put in standard form

$$C = -46.675p + 66 \qquad (3.22)$$

we can read off the slope -46.675 and the intercept 66, and so we can easily graph this equation.

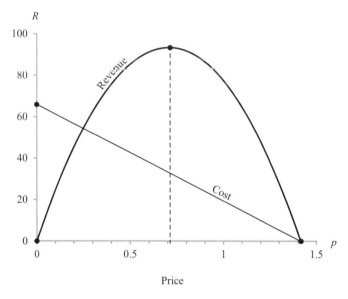

Figure 3.22. Graph for revenue and cost equations. Where the curve and the line intersect, revenue and cost are equal. These are break-even points: the venture shows neither a profit nor a loss.

In Figure 3.22 both the cost function and the revenue function are shown. This graph has some interesting features. For any price, we can find both the costs and the revenue. Anywhere that the cost line is *above* the revenue curve, there will not be enough income to cover expenses. In this case the venture will suffer a loss. On the other hand, where the cost line appears below the revenue curve, the costs will be less than the income. This will produce a profit. The points where the line and the curve cross are called break-even points. They tell us the prices that produce just enough income to meet expenses. Of course, what we are more interested in is the profit, which is the amount of income left after subtracting the expenses. Visually, the profit will be a maximum at the price for which the cost line is farthest below the revenue curve. This is not easy to determine on the graph. But we can observe that the maximum profit will not occur at the vertex of the revenue curve. If we start at that point and increase the price, the costs initially fall faster than the revenue, because just to the right of the center line the cost line slopes down more steeply than the revenue curve. It appears that the maximum profit must occur for a price between the center line and the break-even point where the parabola and the cost line both meet the p-axis. This is the best possible price, from the standpoint of maximizing profits. How can we find it accurately?

One approach is to apply a numerical method, computing both the cost and revenue for a variety of prices to see where the greatest profit occurs. For a more efficient approach, let us develop an equation for the profit and use theoretical methods.

Quadratic Profit Model. For any specified price, profit can be found by subtracting cost from revenue. That gives us the equation $P = R - C$ where P stands for profit. We will use the upper case (capital) letter here because the lower case letter p is being used for the price we charge per drink. We already have (3.21) and (3.22) expressing R and C as functions of p. Therefore

$$P = R - C$$
$$= (-186.7p^2 + 264p) - (-46.675p + 66)$$
$$= -186.7p^2 + 310.675p - 66. \tag{3.23}$$

This gives the profit P as a function of the price p. The graph is shown in Figure 3.23.

More precisely, profit is a quadratic function of price, and the graph again has the characteristic $\cap$ shape. As before, we can read off the coefficients from the equation in standard form, finding $a = -186.7, b = 310.675$, and $c = -66$.

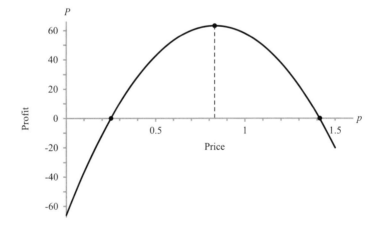

Figure 3.23. Profit as a function of price. The intercepts on the p-axis are break-even points. The vertex is the point of maximum profit.

We can use (3.23) to find both the maximum profit and the break-even points. The center line of the parabola occurs for the price $p = -b/(2a) = 310.675/(2 \cdot 186.7) = 0.832$ to three decimal places. If we charge this price then the highest possible profit will be obtained. Notice that this is about twelve cents higher than the price that produces the maximum revenue.

3.4. Quadratic Models for Revenue and Profit

As mentioned earlier, we can understand this graphically from Figure 3.22. But conceptually, how can we get a higher profit with less than the maximum revenue? Although we are bringing in less than the maximum revenue, more of each dollar that is brought in goes to profit.

We can verify our conclusions numerically. At the point of greatest revenue, the price is 0.707. At the center of the profit curve, the price is 0.832. Using (3.23) we can compute the profit (rounded off to the nearest cent), for each of these prices:

$$P(0.707) = 60.33$$
$$P(0.832) = 63.24$$

If you calculate the profit for a few more prices, you will see that 0.832 really does give the highest possible profit.

There is not much difference in this example between the profit at the point of maximum revenue and the maximum profit. In fact, for convenience in making change we would probably just decide to make the price 80 cents. However, this analysis gives us confidence that a price of 80 cents will result in close to the best profit possible. Moreover, with a profit of about $63 each day of the fundraiser, it appears that selling sodas is a feasible project.

The break-even points are prices that result in a profit of 0, and so are the intercepts of the profit curve on the p-axis. We can find them by solving the equation

$$-186.7p^2 + 310.675p - 66 = 0$$

using the quadratic formula. Alternatively, recalling that the profit is given by $R - C$, and that $R = ps$ while $C = 0.25s$, we can substitute to see that

$$P = ps - 0.25s = (p - 0.25)s.$$

This is a factored form, and it shows immediately that the profit will be 0 if we set $p = 0.25$. That is reasonable. If we sell the sodas for just what we pay for them, there will be no profit. We can also see that, if $s = 0$, there will be no profit. That is also reasonable. There can be no profit if we sell no sodas. These comments show that the two break-even points are for $p = 0.25$ and at the price for which the demand model predicts zero demand.

Reflection. The model development shown above is a bit involved, with equations for several different variables. As always, it is beneficial to reflect on the analysis, asking if the results are reasonable and looking for ideas that might prove useful in the future. In fact, this extended example illustrates several noteworthy ideas.

A Quadratic Model. Here, starting from a linear model for demand, we are led naturally to a quadratic model for revenue. Combined with a linear model for costs, we are led to a quadratic model for profits as well. This shows how quadratic models can arise naturally out of linear models, and also illustrates the more general idea that one model can be built on top of another.

Properties of Parabolas. The example illustrates how a knowledge of graphs of quadratic functions can be useful. From the vertex of the parabola we determine the maximum revenue and profit. From the quadratic formula we can find the break-even points. Even the general shape of the profit and revenue curves is immediately understood in terms of the general results on graphs of parabolas.

Using Algebra. At each step, our understanding of costs and income leads us to combine separate equations: revenue is found by multiplying the price per item by the number of items we can sell; profit is found by subtracting costs from revenue. The algebraic forms that these combinations produced were not immediately useful, so we used rules of algebra to transform them into familiar standard patterns. This illustrates one of the benefits of an algebraic approach. Algebra allows us to change the form of an equation or a function. In some cases one form is easier to work with; in other cases another form is preferred. And often the original form taken by an equation is not the one we want to use. Algebra is the tool that allows us to translate from one form to another.

The Theoretical Approach. Much of the analysis in this example uses a theoretical approach. The ultimate result is an equation that gives the profit as a function of price. This is a powerful result. It allows us to easily study the way profit increases or decreases. We can also generalize the results to determine the effect of changing basic assumptions. Suppose for example that we have to pay 30 cents per can of soda, rather than 25 cents, as assumed. What will the effect be on profits? Would this still be a feasible fundraising activity? In more sophisticated models, the profit might depend on a large number of variables. But the fundamental idea of predicting the profit based on assumed values of the variables is identical to what we did in the example.

Limitations. This example also gives us a chance to remember the limitations of the model. For one thing, we have been acting as if the price per drink is a continuous variable. This is not true. It is not possible to choose a price to be a fraction of a cent, and for convenience we would probably want to set a price that is a multiple of 5 cents. So even after we carefully determine the best price to charge, we will end up rounding that price off. Although we have argued that the entire model has to be understood as a kind of average, it is important to keep in mind that the predictions of the model will probably not be observed exactly.

Contributing to this conclusion is our reliance on the linear demand model. We know that the demand line only approximates the survey data collected, and even the survey data are only an approximation. When concluding that the maximum profit will occur at a price of 0.832, we must remember that this is only approximately true because it is based on other models that are only approximately true. One important way to judge the significance of these errors is to recompute the model based on modified demand or cost equations, and see how much the optimal price and maximum profit change. That is, we can study how *sensitive* our conclusions are to inevitable errors in the model. This is a standard practice in mathematical modeling, and is referred to as a sensitivity analysis.

3.4 Exercises

Reading Comprehension. Answer each of the following with a short essay. Give examples, where appropriate, but also answer the questions in written sentences.

(1) @What is revenue? Profit? A break-even point?

(2) Which is more important for a company, obtaining the highest possible revenue, or the highest possible profit? Why?

3.4. Exercises

(3) @What are some limitations of using a linear demand model?

(4) Explain why demand for a product on any given day should be a whole number. Also explain why considering demand, revenue, cost, and profit to be continuous variables makes sense in the discussion of the student organization selling soft drinks as a fundraising activity.

Problems in Context.

(5) @Range of Prices. In a market analysis, a frozen yogurt chain develops a model for profits as a function of unit price. They find the equation
$$P = -54.63u^2 + 146.4u - 7.18,$$
where P is the average daily profit (in hundreds of dollars) from the stores in the chain, and u is the basic unit price (in dollars) for a cup of yogurt.

  a. @Based on this model, determine what price the chain should set to get the largest possible average daily profits and what profit they will gain.
  b. @Determine the range of prices they can select without losing money. That is, the range of prices for which the profit is 0 or more. Recall that a price for which profit is 0 is also called a break-even point.

(6) Undercutting the Competition. Sally owns a factory currently making backpacks. A model was developed for the way factory profit depends on the price that is charged for the backpacks. According to the model
$$\text{Profit} = -0.3p^2 + 25.2p - 434.3$$
where p is the price charged for each backpack (in dollars) and the profit is given in thousands of dollars. Using the equation answer the questions below.

  a. How much profit would the factory make by selling the backpacks for $30 each?
  b. In order to undercut the competition, the factory owners decide to sell the backpacks at a price that will produce no profit at all. They want to make the price as low as possible, without actually losing any money. Find the price they should set to obtain a profit of zero.
  c. What price should be charged to make the highest possible profit?

(7) @Additional Flat Fee. As a variation on the student fundraising activity described in the reading, suppose the school requires student groups to get a permit to sell goods on campus, charging a flat fee of $25 for each day that goods are sold. Then on a typical day, the students will have a fixed cost of $25 plus whatever they spend to purchase drinks. Develop a cost model for this situation, and use it in place of the cost model in the reading to obtain a model for profit. In this modified model, what price should be charged per drink to obtain the greatest profit, and what is the amount of that profit? Compare these results with the ones found for the original model.

(8) Additional Tax. As a different variation on the student fundraising activity, suppose the students have to pay a 5% sales tax on all the drinks they sell. Assume that the demand and revenue equations are still valid, but incorporate the sales tax into

the cost model. Thus, assuming the students buy and then resell s drinks, the cost equation will include both what they pay for the drinks, as well as the taxes they have to pay on the income from selling the drinks. With this modification work out the equations for cost and profit. In this modified model, what price should be charged per drink to obtain the greatest profit, and what is the amount of that profit? Compare these results with the ones found for the original model.

(9) @Creating a Chromebook Model. At a price of $350, a store sells 200 Chromebook laptop computers in a typical week. A sales survey indicates that the demand would increase by 20 units if the price were $10 lower.

  a. @Assuming that demand s is a linear function of price p, find an equation for demand as a function of price.
  b. @Use your answer from part *a* to find an equation for revenue R as a function of price p.
  c. @What price should the store charge to obtain the greatest revenue? How much will charging this price increase their revenue, compared with the current price?
  d. @Suppose the store pays $175 per Chromebook. Disregarding all other costs, what price should they charge to maximize their profits on Chromebooks?

(10) Selling Smoothies. A fast food restaurant introduces a line of fruit smoothies. Their cost for each smoothie is 47 cents. At a price of $3, the restaurant sells 150 smoothies per day. When they lowered the price to $2.75 their sales increased to 180 smoothies per day. Assuming that demand is a linear function of sales price, develop a model to maximize profits. Include equations for demand, revenue, costs, and profit, and determine the maximum possible profit, as well as the price that should be charged to obtain this profit.

Digging Deeper.

(11) @In the sea ice example in Section 3.2, both arithmetic growth and quadratic growth models were discussed (see page 147). We observed that the data points appeared to follow a parabolic curve more closely than a straight line, and speculated that the quadratic model might therefore be more credible than the linear model. The point of this problem is to consider whether the same reasoning might be used in this section's demand model for selling soft drinks.

Reviewing the development of the linear model, there is a noticeable bend in the graph of the soft drink demand data (Figure 2.22, page 107). Using a trial-and-error approach, the authors found the following quadratic equation to approximate the demand data:
$$s = 92p^2 - 355.12p + 337.69.$$
Figure 3.24 shows the original data as individual dots and the quadratic model as a curve. Visually, this quadratic model does seem to fit the data points more accurately than the linear model. Will it produce more credible results for our revenue and profit analyses? The following steps will help you answer this question.

  a. @Using the development in this section as a guide, incorporate the quadratic demand model into equations for revenue, cost, and profit, all as functions of price.

3.4. Exercises

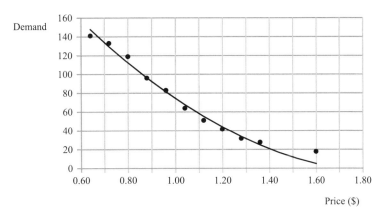

Figure 3.24. Soft drink demand data with a possible quadratic model.

b. @Using your equation for revenue, with graphical and numerical methods, find approximately the maximal revenue and the price for which this revenue is obtained. (We have not developed theoretical methods for solving the type of equation you should have found for revenue.) How do your results compare with those in the reading based on the linear demand model?

c. @Using your equation for profit, with graphical and numerical methods, find approximately the maximal profit and the price for which this profit is obtained. How do your results compare with those in the reading based on the linear demand model?

d. @For the polar ice cap model, predictions based on the linear model and those from the quadratic model were far apart. Is the same true for the revenue and cost models? If predictions for the sales data (based on the linear and quadratic models) are *also* far apart, what key facts about the ice model and the sales model are similar and may influence the variation in predictions between linear and quadratic models? If predictions for the sales data (based on the linear and quadratic models) are *not* far apart, what key facts about the ice model and the sales model are different and may influence the variation in predictions between linear and quadratic models?

4

Geometric Growth

The preceding chapters focused on arithmetic growth and quadratic growth. Each of these can be considered as a family of models. That is, we group all of the arithmetic growth models together because of their common features, and likewise group all of the quadratic growth models together. Now we proceed to a third family, namely, *geometric growth* models. Once again, we will see that the members of this family have several important characteristics in common, including the forms of their difference and functional equations, and properties of their graphs.

Arithmetic sequences are recognized by a simple recursive rule: each term is obtained by adding a constant to the preceding term. Something similar is true for geometric growth sequences, but this time the rule is: each term is obtained by *multiplying* the preceding term by a constant. A second interpretation is that each term of a geometric growth sequence increases or decreases by a constant percentage over the preceding term. This second interpretation accounts for many of the applications of geometric growth.

In organization, this chapter is similar to those that came before. In Section 4.1 the definition and general characteristics of geometric growth sequences will be presented. Models that apply geometric growth sequences are the focus of Section 4.2. The functional equations for geometric growth sequences draw our attention to the family of exponential functions. These are studied in Section 4.3, and Section 4.4 discusses models that apply exponential functions. A final section concerns the mathematical constant e.

4.1 Properties of Geometric Growth Sequences

We begin by formalizing the foregoing description of geometric growth as follows.

> **Geometric Growth Sequence Verbal Definition:** In a geometric growth sequence each term is found by multiplying the preceding term by a constant.

To produce an example of such a sequence, suppose that $a_0 = 10$ and that the constant multiplier is 4. Succeeding terms are then found by repeatedly multiplying by 4. Depicting this in a diagram, we have

$$10 \xrightarrow{\times 4} 40 \xrightarrow{\times 4} 160 \xrightarrow{\times 4} 640 \xrightarrow{\times 4} 2560 \cdots,$$

showing that $10, 40, 160, 640, 2560, \cdots$ is a geometric growth sequence. The rapid increase in the size of the terms is a reflection of the multiplier 4.

Indeed, the multiplier has fundamental importance in geometric growth sequences. We will refer to it so frequently that it is convenient to give it a special name. We call it a *growth factor*. The word *factor* indicates multiplication, so *growth factor* denotes the multiplier that produces the observed amount of growth. Using this terminology, the terms in the preceding example increase rapidly because of the size of the growth factor 4.

In contrast, with a growth factor of 1.1, we obtain the sequence

$$10 \xrightarrow{\times 1.1} 11 \xrightarrow{\times 1.1} 12.1 \xrightarrow{\times 1.1} 13.31 \xrightarrow{\times 1.1} 14.641 \cdots.$$

Observe that these terms are still increasing in size, but much more slowly than in the preceding example.

With a positive initial term, a geometric sequence whose growth factor exceeds 1 always has increasing terms, and the greater the growth factor, the more rapid the increase. If the growth factor is less than 1 (and still positive), the terms decrease in size. For example, if the initial term is again 10 and the growth factor is 0.8, we have

$$10 \xrightarrow{\times 0.8} 8 \xrightarrow{\times 0.8} 6.4 \xrightarrow{\times 0.8} 5.12 \xrightarrow{\times 0.8} 4.096 \cdots.$$

A geometric growth sequence whose terms decrease in this way is sometimes referred to as geometric decay. But it remains correct to refer to it as geometric growth, as dictated by our definition. In this context *geometric growth* can be viewed as synonymous with *geometric change*. In the case of a decreasing sequence, we can also describe the terms as *growing smaller*.

We will generally be concerned with positive growth factors, although no such restriction is included in the definition. In fact, we won't see many instances of growth factors that are negative, 0, or 1. We can certainly work out examples of geometric growth sequences for such values. Thus, with an initial term of 10, taking growth factors of 0 or 1 leads to the sequences $10, 0, 0, 0, \cdots$, and $10, 10, 10, 10, \cdots$, respectively, neither of which is very interesting. Similarly, a growth factor of -0.5 produces the sequence $10, -5, 2.5, -1.25, 0.625, \cdots$. This type of pattern can arise in applications, but we will not emphasize them.[1] Rather, we will concentrate on geometric growth sequences for which the growth factor is positive and not equal to 1.

Difference Equations. There is a standard difference equation for geometric growth sequences, just as there were for arithmetic and quadratic growth sequences. To see the difference equation in a specific example, let us again consider a sequence with growth factor 4. If the terms of the sequence are a_0, a_1, a_2, etc., then the geometric

---

[1]This pattern of alternating signs and decreasing magnitudes arises in the motion of a pendulum, swinging to and fro by successively decreasing amounts, for example.

4.1. Properties of Geometric Growth Sequences

growth pattern becomes

$$a_0 \xrightarrow{\times 4} a_1 \xrightarrow{\times 4} a_2 \xrightarrow{\times 4} a_3 \xrightarrow{\times 4} a_4 \cdots.$$

In particular, a_1 has to equal $4a_0$, a_2 has to equal $4a_1$, and so on. Writing these in the form of equations we obtain

$$a_1 = 4a_0$$
$$a_2 = 4a_1$$
$$a_3 = 4a_2$$
$$a_4 = 4a_3.$$

These equations are equivalent to our verbal understanding: each term is equal to 4 times the preceding term. Thus we see that the sequence satisfies the difference equation

$$a_{n+1} = 4a_n.$$

This example illustrates the following general principle.

> **Geometric Growth Difference Equation:** Every geometric growth sequence follows a difference equation of the form
> $$a_{n+1} = ra_n, \quad (4.1)$$
> where the parameter r represents the constant growth factor.

In any specific application, r will be replaced by a numerical value, as is usual for a parameter. Thus, for the preceding example, $r = 4$, and replacing r by 4 in (4.1) produces the difference equation we derived for that example. Similarly, in the earlier example with a growth factor of 1.1, the difference equation is

$$a_{n+1} = 1.1a_n.$$

As discussed earlier, the examples we consider will usually involve positive growth factors not equal to 1. This can be stated more succinctly in terms of the parameter r: in most cases $r > 0$ and $r \neq 1$.

Recognizing Geometric Growth. In all of the examples so far discussed, we have constructed geometric growth sequences by applying the definition. But what if a sequence arises in some other way? How do we tell whether it is a geometric growth sequence? In preceding chapters we have seen how to recognize arithmetic growth (constant first differences) and quadratic growth (constant second differences). There is an analogous method for recognizing geometric growth sequences.

The key is to calculate growth factors for the given sequence. Here we note that a growth factor can be determined for any two nonzero quantities. To illustrate, pick a pair of numbers with no particular meaning or significance, say 13.5 and 4.59. We call a number r a growth factor from 13.5 to 4.59 if $13.5r = 4.59$. That is, r is a factor by which 13.5 must be multiplied to become 4.59. And using either algebra or a common sense understanding of multiplication and division, we see that $r = 4.59/13.5 = 0.34$. In general, the growth factor from one number to another can be found by dividing the second number by the first.

Using this idea, we can compute the growth factors for successive pairs of terms in a sequence. The growth factor from a_0 to a_1 is $r = a_1/a_0$; from a_1 to a_2 is $r = a_2/a_1$;

and so on. In other words, we compute growth factors by dividing each term of the sequence by the preceding term. Here we see that each growth factor is the *ratio* of successive terms of the sequence. In fact, you can think of r as standing for *ratio*. Growth factors can be computed for any sequence, as long as none of the terms are zero. For a geometric growth sequence, all of the growth factors must be equal. This provides the following alternate verbal definition for geometric growth sequences.

> **Constant Growth Factor Definition of Geometric Growth:**
> In a geometric growth sequence $a_0, a_1, a_2, \cdots$, the growth factors $a_1/a_0, a_2/a_1, a_3/a_2, \cdots$ are constant.

To illustrate, let us look at a few examples. First, consider the sequence

$$3, 6, 9, 12, 15, \cdots.$$

Dividing each term by the preceding term, we find the growth factors

$$6/3 = 2, \quad 9/6 = 1.5, \quad 12/9 = 1.333\cdots, \quad 15/12 = 1.25.$$

These are not all the same, so the given terms do not belong to a geometric growth sequence. For a second example, we have

$$64, 32, 16, 8, 4, \cdots.$$

This time, dividing each term by the preceding term consistently produces the same result, $r = 1/2$. Therefore this is an instance of geometric growth. More precisely, it is an instance of geometric decay. This is indicated both by the evident fact that the terms decrease in value, as well as by the fact that the common growth factor r is less than 1.

Finally, consider

$$5, 55, 555, 5555, 55555, \cdots.$$

Is this an example of geometric growth? Compute the growth factors for successive terms, and see.

When a sequence is to be tested for geometric growth, as in the preceding examples, it is often convenient to list the position numbers and terms in a table, along with an extra column for the growth factors. This extra column is similar to the columns we added for differences between terms when testing for arithmetic and quadratic growth. In the earlier context, we found entries for the difference column by *subtracting* successive terms of the sequence. For geometric growth we instead find growth factors by dividing successive terms. This is illustrated for the preceding two examples in the following tables.

Table 4.1. Two sequences with growth factors between successive terms.

n	a_n	Growth Factor
0	64	
		0.5
1	32	
		0.5
2	16	
		0.5
3	8	
		0.5
4	4	

n	a_n	Growth Factor
0	5	
		11
1	55	
		10.09
2	555	
		10.009
3	5555	
		10.0009
4	55555	

(a) Equal growth factors.   (b) Unequal growth factors.

4.1. Properties of Geometric Growth Sequences

In real data, exactly equal growth factors will rarely be seen. However, if the growth factors are nearly equal, then the data may be closely approximated by a geometric growth model. This is the situation for part (b) of Table 4.1, where the growth factors are all close to 10. We can conclude that the sequence $5, 55, 555, 55555, \cdots$ is approximately geometric. In Section 4.2 we will discuss the use of geometric growth models for such sequences.

Graphs. We turn now to graphs of geometric growth sequences. We look first at two examples of sequences with positive terms and with growth factors greater than 1. For the first sequence, $a_0 = 16$ and $a_{n+1} = 1.5 a_n$. Using the difference equation or the definition of geometric growth, we can compute the first several terms of the sequence, and plot them on a graph. See Figure 4.1. The terms of the sequence appear as dots on the graph; they are connected by straight lines to highlight the increasing trend of the sequence.

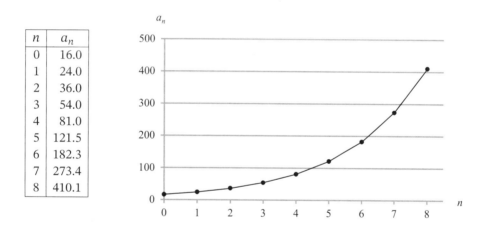

Figure 4.1. Data table and graph for a geometric growth sequence with initial term $a_0 = 16$ and growth factor $r = 1.5$. Each term is 1.5 times the preceding term. Values in the table are rounded to one decimal place.

Observe that the graph has an intercept at a_0 on the vertical axis, and climbs upward from left to right. Moreover, the graph becomes consistently steeper as n increases. This visual impression of increasing steepness corresponds to a numerical pattern in the data table: The increases from one term to the next get consistently larger as we consider terms further along in the sequence.

These are all characteristics of any geometric growth sequence with $a_0 > 0$ and $r > 1$, although the shape of the graph does depend closely on the value of r. This is illustrated by considering a second sequence, with $b_0 = 5$ and $r = 4$. A table and graph are displayed in Figure 4.2.

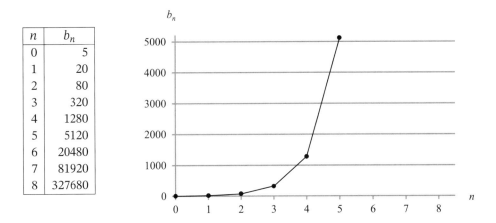

Figure 4.2. Data table and graph for a geometric growth sequence with initial term $b_0 = 5$ and growth factor $r = 4$. These terms grow much more rapidly than in the previous example.

As before, the graph begins at b_0 on the vertical axis, and slopes upward as we move to the right. But notice how much more rapidly the terms increase. The vertical range on the second graph is ten times larger than for the first graph, and even so, it is only possible to plot points up to b_5. The shape of the second graph is flatter at the start and transitions to a nearly vertical line in only a few steps.

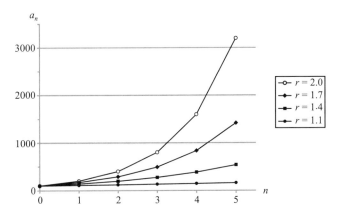

Figure 4.3. Graphs for geometric growth sequences with initial term $a_0 = 100$ and growth factors $r = 1.1, 1.4, 1.7,$ and 2.0, respectively.

We can obtain a better perspective on the graphical significance of r by comparing several different geometric growth sequences in a single graph, as in Figure 4.3. For each sequence the initial term is 100, and the terms have been plotted for n up to 5. To include additional terms would require expanding the vertical range of the graph so much that the lowest sequence would appear to lie on the n-axis. Nevertheless, the graph provides a reasonable overview of the effect of changing r. We can visualize how the graph might look with $r = 1.5$, for example. For $r > 2$ we know that the graph will curve upward even more rapidly than the top curve shown. Similarly, for r between 1

4.1. Properties of Geometric Growth Sequences

and 1.1 we anticipate a graph that is even closer to a horizontal line than the bottom curve shown.

The preceding examples illuminate the nature of graphs for geometric growth sequences with positive initial terms and $r > 1$. These graphs also shed light on what happens with other values of the parameters. In particular, we want to consider what happens if the initial term is negative, or if $0 < r < 1$. Consistent with earlier comments, we will not consider negative values of r.

When a_0 is positive, each succeeding term is also positive, as we repeatedly multiply by a positive growth factor. Thus, in the graphs we have considered, the points all lie above the n-axis. Likewise, if we begin with a negative initial term, the following terms will remain negative, and all the points on the graph will lie below the n-axis. This is shown in Figure 4.4, both numerically and graphically, where two sequences appear. Both sequences have the same growth factor $r = 1.5$. For the first sequence $a_0 = 16$, while for the second $b_0 = -16$. Each term of the b sequence is the negative of the corresponding term of the a sequence, as shown in the table of values. Pictorially, that means that the graphs of the two sequences are mirror images of each other, reflected across the n-axis. These observations hold in general. Thus, our conclusions about graphs with positive initial terms can, with a simple modification, be applied to graphs with negative initial terms.

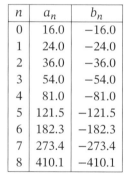
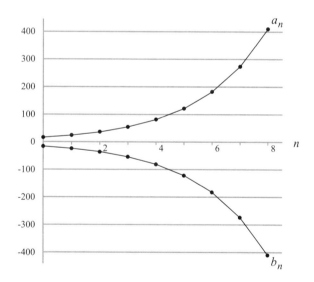

Figure 4.4. The graph for the b_n sequence is the mirror image in the n-axis of the graph for the a_n sequence. Both sequences have the same growth factor, $r = 1.5$, but the initial term is -16 for the b_n sequence and $+16$ for the a_n sequence. Compare with Figure 4.1.

Informally speaking, changing the sign of a_0 flips the graph top to bottom. In a similar way, replacing r by its reciprocal $1/r$ can be seen to flip the graph from left to right. To see how this occurs, consider the sequence $10, 20, 40, 80, \cdots$. This is a geometric growth sequence with initial term $a_0 = 10$, and with growth factor $r = 2$. For ease of reference, we display the first several terms in the following table.

n	0	1	2	3	4	5	6	7	8
a_n	10	20	40	80	160	320	640	1280	2560

As we know, each term of the sequence is twice the preceding term. But that also means that each term is half of the following term. In other words, in the second row of the table, each step to the right has the effect of multiplying by 2, so each step to the left has the effect of dividing by 2, or equivalently, multiplying by 1/2. This shows that writing the terms in the reverse order produces another geometric sequence, this time starting with 2560 and with growth factor $r = 1/2$. Referring to this as the b sequence, we can display it in a table as follows.

n	0	1	2	3	4	5	6	7	8
b_n	2560	1280	640	320	160	80	40	20	10

This illustrates how a geometric growth sequence with $r = 1/2$ can have the same terms as one with $r = 2$, but in the opposite order. To be more correct, we are only working with a finite number of points, and the first point of one sequence is equal to the last point of the other. Moreover, if we disregard the vertical axis, this reversal of points in the table has a clear visual effect when the sequence is graphed, flipping the pattern of dots from left to right. This is shown in Figure 4.5.

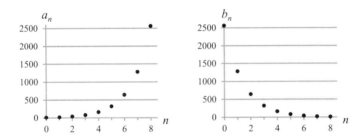

Figure 4.5. Graphs for two geometric growth sequences. On the left, $a_0 = 10$ and $r = 2$. On the right, $b_0 = 2560$ and $r = 1/2$. The pattern of dots in each graph is a horizontal reflection of the other.

All of the foregoing considerations show that the general shape of a geometric growth sequence in the first case examined, namely, $r > 1$ and $a_0 > 0$, can be flipped horizontally, vertically, or both, to obtain the shapes of graphs in the other cases. This is illustrated qualitatively in Figure 4.6. In the upper left corner we see the characteristic shape for $r > 1$ and $a_0 > 0$. It features a curve that begins on the vertical axis, and slopes upward with increasing steepness as it is traced to the right. The graph shown in the figure is meant to suggest the global behavior of the curve: if we extend the n-axis to larger and larger values, the curve continues to climb, without any upper limit. Flipping this curve horizontally produces the general shape for $0 < r < 1$ and a_0 still positive. In such a sequence successive terms decrease, getting closer and closer to 0, without ever reaching 0. This is the shape of geometric decay. When $a_0 < 0$, both of these curves have to be flipped vertically across the n-axis. Note in particular that on any of these curves, either all the a_n values are positive (and so the points lie above the n-axis), or all the a_n values are negative (and so the points lie below the axis).

4.1. Properties of Geometric Growth Sequences

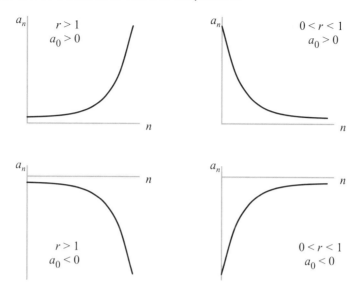

Figure 4.6. Representative graphs for geometric growth sequences. The growth factor r can be either greater than 1, or between 0 and 1. The initial term (and therefore all the following terms) can be either positive or negative. These four graphs show all possible combinations of these alternatives.

Functional Equations. The functional equation for arithmetic growth can be found by observing a simple pattern generated by the difference equation. The same is true of geometric growth. As an illustration of this idea, consider the sequence with difference equation
$$a_{n+1} = 1.5 a_n$$
and with initial term $a_0 = 100$. We compute the first several values, but do not actually perform any of the multiplications:

$$
\begin{aligned}
a_1 &= 1.5 \cdot a_0 = 1.5 \cdot 100 \\
a_2 &= 1.5 \cdot a_1 = 1.5 \cdot 1.5 \cdot 100 \\
a_3 &= 1.5 \cdot a_2 = 1.5 \cdot 1.5 \cdot 1.5 \cdot 100 \\
a_4 &= 1.5 \cdot a_3 = 1.5 \cdot 1.5 \cdot 1.5 \cdot 1.5 \cdot 100.
\end{aligned}
$$

The pattern is clear, and can be used to find any a_n. For example, a_7 is produced by multiplying 7 factors of 1.5 and a final factor of 100. This pattern makes sense, too. If we start with $a_0 = 100$ and multiply by 1.5 for each successive term, then to reach a_7 we will have multiplied by 1.5 seven times, which is just what the pattern says. Of course, it becomes awkward to write out the pattern longhand: $1.5 \cdot 1.5 \cdot 1.5 \cdot 1.5 \cdot 1.5 \cdot 1.5 \cdot 1.5 \cdot 100$. Imagine writing out the pattern for a_{25}! Using exponents simplifies things considerably. Since an exponent indicates repeated multiplication, we can write the pattern for a_7 as $1.5^7 \cdot 100$. Similarly, $a_{25} = 1.5^{25} \cdot 100$. Using this notation in the pattern above leads to

$$
\begin{aligned}
a_1 &= 1.5 \cdot 100 \\
a_2 &= 1.5^2 \cdot 100 \\
a_3 &= 1.5^3 \cdot 100 \\
a_4 &= 1.5^4 \cdot 100.
\end{aligned}
$$

As usual we can summarize the pattern with a single equation:
$$a_n = 1.5^n \cdot 100.$$
This equation fits all the examples above. For $n = 0$ it says
$$a_0 = 1.5^0 \cdot 100,$$
which is also valid because $1.5^0 = 1$. This is a convention many readers will recall from previous studies.[2]

Now we want to generalize from the example to any geometric growth model. For the example we have been using, the functional equation is $a_n = 1.5^n \cdot 100$. Observe that the numbers 1.5 and 100 appear in the original problem as the growth factor and the initial value a_0. This leads to the general equation $a_n = r^n a_0$, although it is more customary to write it in the form $a_n = a_0 r^n$. This is a functional equation for the sequence specified by the difference equation $a_{n+1} = r a_n$. Using the functional equation, we can immediately compute a_n for any n without computing all the preceding a's. Since this gives the functional equation for any geometric growth model, we restate it below for emphasis.

> **Geometric Growth Functional Equation:** If a geometric growth sequence has the difference equation $a_{n+1} = ra_n$ with r a constant, then a functional equation is $a_n = a_0 r^n$ for $n \geq 0$.

Answering Questions with the Functional Equation. Now that we have a functional equation, we can use it to answer find-a_n and find-n questions. Consider again the sequence with $a_0 = 16$ and $r = 1.5$, as shown in Figure 4.1. Although the figure only goes up to $n = 8$, with the functional equation we can easily compute $a_{25} = 16(1.5^{25})$ and with a calculator we find that to be 404,018.69, rounded to two decimal places. This answers the find-a_n question *What is a_{25}?* through function evaluation.

To illustrate a find-n question, we might ask *For what value of n does $a_n = 1000$?* As always, we can apply numerical and graphical approaches. Consulting Figure 4.1 we see that n must be greater than 8, and can make an educated guess that n will be greater than 9. Switching to a numerical approach, we use systematic trial and error. We compute $a_{10} = 16(1.5^{10}) = 922.64$, and $a_{11} = 16(1.5^{11}) = 1383.96$, again rounded off to two decimal places. These results show us that there is no term of the sequence that exactly equals 1000, and that a_{11} is the first term that exceeds 1000.

We can carry our trial-and-error approach further, because most calculators can compute $16(1.5^n)$ for fractional values of n. A little experimentation shows that $16(1.5^{10.19}) < 1000$ and $16(1.5^{10.20}) > 1000$. So the equation $16(1.5^n) = 1000$ appears to have a solution for n somewhere between 10.19 and 10.20. As in earlier chapters, this sort of computation may or may not be meaningful in a particular model, depending on whether n is continuous or discrete, and on whether a functional equation valid when n is a whole number remains valid when n is not a whole number. We will consider these ideas in greater detail in Section 4.2.

Although a numerical approach permits us to find solutions to find-n questions as accurately as we wish, it is not particularly efficient. As we will see in Section 4.3,

---

[2]One of the reasons for adopting the convention $r^0 = 1$ for all $r > 0$ is for consistency in equations like our functional equation.

4.1. Properties of Geometric Growth Sequences

there is a theoretical approach for these problems, based on the concept of logarithms. Until we reach that discussion, numerical and graphical methods will suffice for find-n questions.

Constant Percentage Growth and Decay. To conclude our survey of properties of geometric growth sequences, we discuss an alternative conceptualization. Geometric growth is mathematically equivalent to growth (or decay) by a constant percentage.

As an example, consider the effect of monthly 25% increases in the number of followers of a popular blogger. Suppose there are initially $f_0 = 1600$ followers. Then 25% of that amount is 400, so after a month we would have $f_1 = 1600 + 400 = 2000$. Continuing, 25% of 2000 is 500, so $f_2 = 2000 + 500 = 2500$. In the same fashion we find $f_3 = 2500 + 625 = 3125$. These findings are summarized in table form below.

n	0	1	2	3
f_n	1600	2000	2500	3125

This is a geometric growth sequence, with growth factor 1.25. In verification, let us recompute the terms, starting with 1600, and repeatedly multiplying by 1.25:

$$1600 \xrightarrow{\times 1.25} 2000 \xrightarrow{\times 1.25} 2500 \xrightarrow{\times 1.25} 3125.$$

In general, growth by a fixed percentage is the same as geometric growth, and the growth factor can be found by adding one to the percentage increase, expressed as a decimal.[3] For example, suppose each term of a sequence is 10% more than the preceding term. Expressed as a decimal, 10% becomes 0.10, and the growth factor is $r = 1 + 0.10 = 1.10$. Similarly, a repeated increase of 45% corresponds to a growth factor of $r = 1 + 0.45 = 1.45$.

In a similar way, if each term of a sequence is a fixed percentage *less* than the preceding term, then the sequence is *decaying* geometrically, and the growth factor is one *minus* the percentage, expressed as a decimal. So, if each term is 10% less than the preceding term, that is geometric decay with $r = 1 - 0.10 = 0.90$. Similarly, a repeated decrease of 45% has a growth factor $r = 1 - 0.45 = 0.55$.

As a specific example, suppose a student has a trust fund worth $30,000 and decides to spend 20% each year. The first year, he spends $0.20 \cdot 30{,}000 = 6{,}000$, leaving a balance of 24,000. In the second year, he spends $0.20 \cdot 24{,}000 = 4{,}800$, leaving a balance of $24{,}000 - 4{,}800 = 19{,}200$. Continuing in this fashion, we produce the sequence of annual fund balances,

$$30{,}000, \quad 24{,}000, \quad 19{,}200, \quad 15{,}360, \quad 12{,}288, \quad \cdots .$$

The reader may verify that this is a geometric growth sequence with $r = 0.80$.

Although the examples show that the first few terms of each sequence adhere to a geometric growth pattern, they do not tell us why this is so. One way to understand the situation is to analyze difference equations. Consider the trust fund example again. The sequence of fund balances can be described verbally as follows:

---

[3] When working with percentages, the word *of* translates into multiplication. Thus 10% *of* a number becomes 0.10 *times* the number. Viewed in this way, we can think of 1.45 times x as being 145% of x. So if you already have x (which is the same as 100% of x) and you add 45% more, you then have 145% of x, which is the same as $1.45x$. This is one way to understand why the growth factor is one more than the percentage increase, expressed as a decimal.

Each year's balance is the preceding balance minus 20% of the preceding balance.

Now we reason that if you subtract 20% of something, what remains is 80%. Thus, we can reformulate the statement as

Each year's balance is 80% of the preceding balance.

This translates directly to a difference equation

$$B_{n+1} = 0.80 B_n,$$

which we recognize as geometric growth with $r = 0.80$.

In a similar way, for the blogger follower example, we know:

The number of followers each month is the number from the previous month plus 25% of the number from the previous month.

This translates immediately to the difference equation

$$f_{n+1} = f_n + 0.25 f_n = 1 f_n + 0.25 f_n.$$

Factoring the right-hand side, we obtain

$$f_{n+1} = (1 + 0.25) f_n$$

and hence

$$f_{n+1} = 1.25 f_n.$$

So we again recognize an instance of geometric growth, this time with $r = 1.25$. The analysis of these two examples shows why the growth factor is found by adding the percentage increase to 1 or subtracting the percentage decrease from 1.

We summarize these results as follows.

> **Constant Percentage Growth:** If each term of a sequence grows or decreases by a fixed percentage from the preceding term, the sequence is an instance of geometric growth. If the percentage, expressed as a decimal, is p, then the growth factor is $r = 1 + p$ for an increasing sequence, and $r = 1 - p$ for a decreasing sequence.

As a final comment, just as repeated additions and subtractions of a fixed percentage always produce geometric growth sequences, so too every geometric growth sequence can be interpreted as addition or subtraction by a fixed percentage. For addition, we can rewrite $r = 1 + p$ as $p = r - 1$, to express p as a function of r. Thus, for a geometric growth sequence with a growth factor of $r = 1.85$, we find $p = r - 1 = 0.85$, the decimal equivalent of 85%. Thus each term of the sequence is 85% greater than the preceding term. Similarly, for a geometric sequence with $r = 0.32$, we know the terms are decreasing ($r < 1$), so use the equation $r = 1 - p$. Then $p = 1 - r$ so substituting $r = 0.32$ leads to $p = 0.68$. This shows that successive terms of the sequence decrease by 68%.

For growth factors greater than two, the association of a percentage increase is not as familiar. For example, consider the situation with $r = 3$ and $a_0 = 100$. The first few terms of the geometric growth sequence will be 100, 300, 900, $\cdots$. According to our rule,

4.1. Properties of Geometric Growth Sequences

the percentage should be $p = r - 1 = 2.00$, and that is the decimal equivalent of 200%. But since the terms are tripling at each step, in common usage, many people would describe that as a 300% increase. Technically, it is not the *increase* that is 300%, but rather the multiplier. But the potential for confusion between these two interpretations means that the language of percentage increase should be used with caution when $r \geq 2$. As an alternative, we can describe the sequence $100, 300, 900, \cdots$ as undergoing a *three-fold increase* at every step.

Profile for Geometric Growth. This section has detailed quite a few properties of geometric growth sequences. They can be summarized in a profile, analogous to the profiles for arithmetic and quadratic growth, as follows.

Table 4.2. Profile for geometric growth sequences.

Verbal Description:	Each term is a constant multiple of the preceding term
Alternate Description:	Each term increases or decreases from the preceding term by a constant percentage
Identifying Characteristic:	Constant growth factors; equivalently, constant ratios of consecutive terms
Parameters:	Initial term a_0, growth factor r, usually with $r > 1$ (increasing terms) or $0 < r < 1$ (decreasing terms)
Difference Equation:	$a_{n+1} = r a_n$
Functional Equation:	$a_n = a_0 r^n$
Graph:	In all cases a_0 is the intercept on the vertical axis. For $r > 1$, if $a_0 > 0$ the graph rises to the right, ever more steeply, and with no upper limit; and if $a_0 < 0$, the graph falls to the right, ever more steeply, and with no lower limit. For $r < 1$, if $a_0 > 0$ the graph falls to the right, getting ever closer to the n-axis from above; and if $a_0 < 0$, the graph rises to the right, getting ever closer to the n-axis from below. See the graphs below.

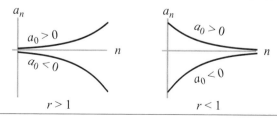

4.1 Exercises

Reading Comprehension.

(1) @For each of the following, write a brief paragraph explaining the meaning of the phrase. It should be clear from your descriptions what makes each item different from the others.

   a. @geometric growth
   b. @geometric decay
   c. fixed percentage of growth
   d. growth factor

(2) @Let's consider the equations that describe geometric growth.

   a. @State the definition of geometric growth in your own words.
   b. @Write down the general form of the geometric growth difference equation, or an example for a specific geometric growth sequence. Explain how this equation corresponds to the definition from part a.
   c. Write down the general form of the geometric growth functional equation, or an example for a specific geometric growth sequence. Explain how this equation corresponds to the definition from part a.

(3) The graphs of geometric growth sequences have a distinctive shape unlike the shapes of arithmetic growth sequence graphs or quadratic growth sequence graphs. Compare and contrast the shapes of the graphs for each of these three types of sequences. What, if anything, do all three types of graph have in common? What distinguishes them from each other?

(4) @The parameters, a_0 and r, of a geometric sequence, $a_n = a_0 r^n$, are evident in the graph of the sequence. The graph of $a_n = 10(2)^n$ is shown below. For each of the following items, describe how the graph would be affected by the stated change.

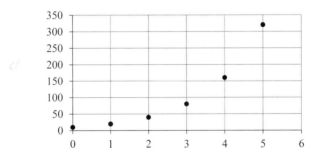

   a. @Change a_0 from 10 to 150.
   b. Change a_0 from 10 to -10.
   c. @Change r from 2 to $\frac{1}{2}$ and a_0 from 10 to 320.
   d. Change r from 2 to 5.
   e. Change r from 2 to 1.3.

4.1. Exercises

(5) Geometric growth can also be thought of as constant percentage growth. Write an explanation that might help a fellow student understand why these two types of growth are the same, how to find the percentage from the growth factor, and how to find the growth factor from the percentage. Examples may be helpful.

(6) A geometric growth sequence has a growth factor of 1.8. What is the percentage of increase? How should it be interpreted?

(7) Is proportional reasoning appropriate for geometric growth sequences? Explain why or why not. It may be helpful to include an example in your answer.

Math Skills.

(8) @For each of the following, determine whether the sequence follows an arithmetic growth pattern, a quadratic growth pattern, a geometric growth pattern or none of these. Give a reason for your answer. If it is arithmetic, quadratic or geometric growth, find the difference and functional equations. In each case, assume the first number in the given list is a_0.

   a. @2, 2.6, 3.38, 4.394, 5.7122, $\cdots$
   b. 5, 6, 9, 14, 21, $\cdots$
   c. $-12, -3, -\frac{3}{4}, -\frac{3}{16}, -\frac{3}{64}, \cdots$
   d. @10, 4, 1, 3, 12, 30, $\cdots$
   e. 10.2, 18, 25.8, 33.6, 41.4, $\cdots$

(9) @Suppose a number sequence has the functional equation $a_n = 2.84 \cdot 16^n$. What is the growth factor for this sequence?

(10) @The following sequences of numbers each follow a geometric growth pattern. Find the growth factor and the percentage of increase (or decrease). It may be helpful to make a table as on page 204.

   a. @3.2, 4.8, 7.2, 10.8, $\cdots$
   b. 1.25, 0.25, 0.05, 0.01, $\cdots$

(11) @Determine if each of the following sequences follows a geometric growth pattern. Justify your answer.

   a. @1.4641, 1.331, 1.21, 1.1, $\cdots$
   b. 1.08, 1.13, 1.18, 1.23, $\cdots$
   c. @1/3, 1/4, 1/5, 1/6, 1/7, $\cdots$
   d. 1/4, 1/8, 1/16, 1/32, 1/64, $\cdots$

(12) @A geometric growth sequence can be defined by specifying either a growth factor or a percentage increase or decrease. In each of the following you are given one of these parameters, and asked to find the other.

   a. @The percentage increase is 72%. What is the growth factor?
   b. The percentage decrease is 18%. What is the growth factor?
   c. The growth factor is 1.08. What is the percentage of increase or decrease, and which is it, increase or decrease?

d. @The growth factor is 0.78. What is the percentage of increase or decrease, and which is it, increase or decrease?

(13) @For each part below determine what kind of sequence is represented and find the difference equation.

a. @The functional equation is $a_n = 4 \cdot 1.3^n$.
b. Each term is 65% greater than the previous term and the initial value is 10.
c. @There is a term $c_3 = 14.2$ and each term is 1.3 less than the preceding term.
d. The functional equation is $d_n = 1 + 5n + 3(n-1)n/2$.
e. The ratio of any two consecutive terms is given by $\frac{e_{n+1}}{e_n} = \frac{5}{8}$ and the initial term is $e_0 = 32{,}768$.

(14) @For each part below determine what kind of sequence is represented and find the functional equation.

a. @A number sequence starts out with $a_0 = 14$ and satisfies the difference equation $a_{n+1} = 0.3a_n$.
b. A number sequence starts out with $b_0 = 14$ and satisfies the difference equation $b_{n+1} = b_n + 9.8$.
c. A number sequence starts out with $c_0 = 14$ and satisfies the difference equation $c_{n+1} = 1.4c_n$.
d. @The initial term is 287 and each term is a 3 percent decrease from the previous term.
e. The initial term is 1,092 and dividing any term by the preceding term results in 0.25.

(15) @Find the next three terms, as well as the difference and functional equations for each of the geometric sequences described below.

a. @This sequence begins with $a_0 = 100$ and has $a_1 = 150$.
b. This sequence begins with $b_0 = 20$ and has $b_1 = 16$.
c. @This sequence begins with $c_0 = -8$ and has $c_1 = -8.8$.

(16) @Match each of the following graphs with the parameters, a_0 and r, of the equation, $a_n = a_0(r)^n$, used to produce the graph.

i. $a_0 = 5$ and $r = 0.3$
ii. $a_0 = 1$ and $r = 3$
iii. $a_0 = -2$ and $r = 1.5$
iv. $a_0 = -7$ and $r = 0.1$

a. @-16

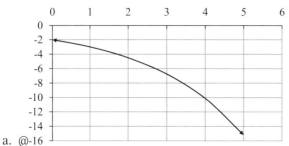

4.1. Exercises

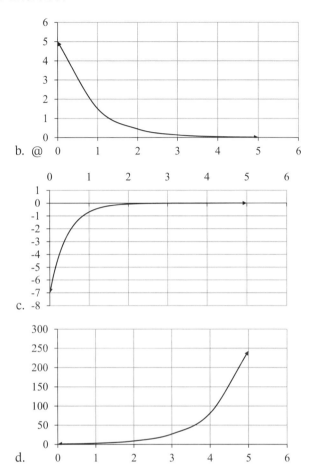

b.

c.

d.

(17) @Given the information about a sequence, find the requested term.

  a. @The functional equation is $a_n = 2(1.05)^n$. Find a_{10}.
  b. @The initial term is $b_0 = 3.2$ and the difference equation is $b_{n+1} = b_n \frac{1}{10}$. Find b_3.
  c. @The initial term is $c_0 = 2{,}600$ and each term increases by 5 percent. Find c_{12}.

(18) A number sequence starts out with $a_0 = 1{,}000$ and has the difference equation $a_{n+1} = 1.08 a_n$.

  a. Use the difference equation to find a_3.
  b. Determine the functional equation and use it to find a_{17}.
  c. Use a numerical approach with either the difference or functional equation to figure out which a_n equals 2,000. Is there a term of the sequence equal to 2,000? If not, based on your calculations, what can be said about when the sequence reaches 2,000?

(19) @A number sequence starts out with $a_0 = -200{,}000$ and has the difference equation $a_{n+1} = 0.15 a_n$.

  a. Use the difference equation to find a_4.

b. Determine the functional equation and use it to find a_{12}.

c. @We know that the sequence never actually reaches 0, but it gets arbitrarily close. Say, for practical reasons, -0.001 is close enough to 0 to be of interest. Use a numerical approach with either the difference or functional equation to figure out which a_n equals -0.001. Is there a term of the sequence equal to -0.001? If not, based on your calculations, what can be said about when the sequence reaches -0.001?

(20) Consider the sequence $a_n = -10(1.5)^n$. Use a numerical approach to find n for which $a_n = -250$. Is there a term of the sequence equal to -250? If not, based on your calculations, what can be said about when the sequence reaches -250?

(21) @Use a numerical approach to figure out when $a_n = 600(0.83)^n$ is closest to 200. Your answer should include both the n and a_n values.

Digging Deeper.

(22) Is it possible for one sequence to be both an arithmetic growth sequence and a geometric growth sequence? Explain.

(23) @In the text we found functional equations based on the values of two consecutive terms of a sequence. It is possible to find a geometric growth functional equation based on any two terms of the sequence. Do that for the following sequences. Notice that there is a hint in part a.

   a. @A geometric growth sequence begins with $a_0 = 100$ and has $a_2 = 25$. Find the difference and functional equations for this sequence. [Hint: If the growth factor is r, then the functional equation gives $a_2 = a_0 r^2$. Substitute known information and solve for r.]

   b. A geometric growth sequence has $b_0 = 81$ and $b_4 = 16$. Find the difference and functional equations for this sequence.

   c. @A geometric growth sequence begins with $c_0 = 100$ and has $c_3 = 1600$. Find the difference and functional equations.

   d. A geometric growth sequence has terms $d_2 = 100$ and has $d_7 = 13.1687$. Find the difference and functional equations.

(24) In Chapter 3 we saw that sums (or running totals) of a sequence can be useful in developing a model. We also saw that the sums of an arithmetic growth sequence always form a quadratic growth sequence. Now we ask, what can be learned about the sums of a geometric growth sequence?

   a. For the geometric growth sequence $1, 2, 2^2, 2^3, \cdots$, calculate the first several terms of the sequence of sums. Can you find a pattern in the sum sequence?

   b. Find a functional equation for the sum sequence of part a.

   c. Repeat the two preceding parts for the geometric growth sequences $1, 3, 3^2, 3^3, \cdots$ and $1, 5, 5^2, 5^3, \cdots$.

   d. Based on the preceding parts, guess a functional equation for the sum sequence of $1, r, r^2, r^3, \cdots$. Does your guess give a correct formula for $r = 4$? For $r = 0.1$?

(25) @In Chapter 2 we studied arithmetic growth sequences, where each term is found by adding a constant to the preceding term. In this chapter we are studying geometric growth sequences, where each term is found by *multiplying* the preceding term by a constant. Now consider combining both of these operations, so that each term of a sequence is found by multiplying the preceding term by a constant, and then adding another constant. In particular, suppose $b_0 = 70$, after which every term is obtained by multiplying the preceding term by $1/2$ and then adding 3. For example, $b_1 = (1/2)70 + 3 = 38$.

  a. @Make a table showing b_0 through b_{10}.

  b. @Find a difference equation for the sequence.

  c. @Create a graph of the sequence.

  d. @On page 209 we found a functional equation for a geometric growth sequence with difference equation $a_{n+1} = 1.5a_n$. Using a similar approach, find a functional equation for the sequence b_n in this problem. [Hint: It is also helpful to refer to the equation found in part *d* of the previous problem.]

  e. @Your graph should show that the terms b_n decrease rapidly at first, but then level off, getting closer and closer to 6. This suggests looking at a new sequence obtained by subtracting 6 from each term of the b_n sequence. Call the new sequence c_n, defined by $c_n = b_n - 6$. Show that c_n is a geometric growth sequence, find its functional equation, and use it to find a functional equation for b_n.

4.2 Applications of Geometric Growth Sequences

In discussing applications of arithmetic growth and quadratic growth in prior chapters, we distinguished between two kinds of examples. In *structural* examples, the decision to use a particular type of model reflects our knowledge of the problem context. For instance, the laws of physics tell us that under water pressure should increase with depth at a constant rate, approximately one atmosphere per ten meters of depth. That leads to an arithmetic growth model. In contrast, in *empirical* or *data based* examples, properties of the data suggest using a particular kind of model. If the first differences are constant or nearly constant, we might choose to use an arithmetic growth model; if the second differences are constant or nearly constant, we might choose to use a quadratic growth model. These choices are not based on knowledge of the structure of the problem, but merely on a pattern in the data.

This distinction will continue to be observed as we consider applications of geometric growth. Sometimes, geometric growth is suggested by the structure of the modeling context. At other times, we observe approximate geometric growth in a set of data. In addition, there are several contexts for which geometric growth has proven effective in the past, and which we recognize as appropriate settings for geometric growth models. In this section we will see examples of all of these situations. In addition, we will further explore the geometric growth assumption, and how it can be extended for continuous variables.

Structural Examples of Geometric Growth.

A Mouse Population Model. In many contexts researchers are interested in the growth of some sort of population, for example, people with a specific disease, or members of an endangered animal species, or a colony of bacteria. As a first approximation, geometric growth is often assumed. These are structural examples, because the geometric growth assumption can be derived from our knowledge of how populations increase or decrease.

As an illustration of this idea, consider a new species of mouse that is introduced into an ecological system. Initially researchers estimate that the mouse population is about 100 in a certain area. A month later a follow-up study finds that the population of mice has grown to 200. How big would you expect the size of the population to be after one more month? At first glance, you might be tempted to say 300. After all, if the population increased by 100 in the first month, maybe the same will occur in the second. That kind of thinking leads to an arithmetic growth model. It amounts to the following assumption, labeled *Assumption A* for future reference.

Assumption A: In any month, the mouse population will increase by 100 mice.

Although an arithmetic growth assumption of this type is useful in many applications, there is good reason to doubt it here. To see why, consider the following even more restricted assumption.

Assumption B: In any month, a population of 100 mice will grow to 200 mice.

In this version, we do not assume that the increase depends only on the amount of time, as in arithmetic growth. Rather, we recognize that the size of the population at the start of the month will have an effect on the number of new members added to the population during the month. We are not willing to say that *any* size population will increase by 100 mice in a month—only that a population of size 100 will increase by 100 in a month.

Using Assumption B, what can we conclude about the future growth of the mouse population? It starts out at 100 mice. After 1 month it grows to 200 mice. Now think of those 200 mice as making up two populations each of size 100. This makes some sense, for as the mouse population grows, the mice will spread out and inhabit a larger territory. Maybe we can divide the territory into two regions, north and south, with 100 mice each. In any case, our assumption says that each of these separate groups of mice will grow from 100 to 200 during a month. Therefore, after the second month, there will be two populations of size 200, for a total of 400 mice. Now repeat this process. Consider the 400 mice as being divided into four separate groups of 100 mice each. In another month, each will grow in size to 200 mice, for a total of 800 mice. To summarize, Assumption B leads us to predict the following growth for the mouse population:

Months	0	1	2	3
Mice	100	200	400	800

This is an example of geometric growth. And because we arrived at our predictions from an understanding of the conditions under which the mouse population would grow, this is a structural example.

4.2. Applications of Geometric Growth Sequences

Here we emphasize an important distinction between what geometric and arithmetic growth models predict. Under geometric growth the population grows by the same *percentage* month after month. In our example there is a 100% increase each month. However, the *amount* of increase, meaning the actual number of additional mice, is *not* constant. We find 100 new mice after the first month, 200 after the second month, 400 after the third month, and so on.

In contrast, under arithmetic growth (Assumption A) the amount of increase would remain constant month after month. Assuming a constant increase of 100 mice, for instance, the arithmetic growth model would lead to the following predicted populations:

Months	0	1	2	3
Mice	100	200	300	400

Comparing the two models numerically, we observe that the population grows much more rapidly according to the geometric growth model. Graphically, we know that an arithmetic growth model is represented by a straight line, whereas a geometric growth model graph curves upward, with increasing steepness. Significantly, geometric growth models are *not* linear; proportional reasoning will generally lead us far astray.

Given such different models, how do we know which to use? Of course, one important step in any model development is to collect data and determine empirically whether the model is accurate. But in the mouse population example, common sense seems to indicate that the arithmetic model is not right. The other assumption, the one that depends on the population size, seems more consistent with what we know about how animals live. It recognizes that the growth of the mouse population in one month will depend on how many mice were alive at the start of the month. The more mice we have to start with, the more females are available to have young, and so the more new mice will be added to the population. Thus, without collecting any data, our knowledge of the structure of the problem context strongly suggests assuming geometric growth, not arithmetic growth.[4]

Let us develop the mouse population example more fully. We consider a sequence $p_0, p_1, p_2, \cdots$, where p_0 is the initial mouse population of 100, p_1 is the population after a month, p_2 the population after a second month, and so on. With n representing the number of months, we thus define p_n to be the population after n months.

For our model, we assume geometric growth. The growth factor is $r = 2$, which we can observe in the first several terms. Alternatively, we can use the fact that, according to Assumption B, the mouse population will grow by 100% each month. Expressed as a decimal, 100% = 1, so the growth factor is given by $r = 1 + p = 1 + 1 = 2$.[5]

Now we can apply our results from the preceding section. We have a geometric growth model with initial term $p_0 = 100$ and growth factor $r = 2$. The difference equation is $p_{n+1} = 2p_n$, and the functional equation is $p_n = 100 \cdot 2^n$. In general terms, the model predicts that the population will grow rapidly and without an upper limit.

---

[4] But geometric growth is not sustainable in the long run. Eventually limited amounts of space and food will restrict the growth of the mouse population. Taking this idea into account leads to a new type of model, *Logistic Growth*, which we will discuss in a later chapter.

[5] Do not confuse the parameter p with sequence terms $p_0, p_1, \cdots$. In one case p stands for percentage. In the other it stands for population. Collisions of notation of this sort sometimes appear in modeling. While it is better to avoid them if possible, you should also be aware that they occur, and recognize them when they do.

As is typical for growth sequence models, we may consider find-p_n or find-n questions. For example, the model predicts the population after 6 months will be $p_6 = 100 \cdot 2^6 = 6{,}400$. When will the population reach one million? That is a find-n question, and translates to the equation

$$100 \cdot 2^n = 1{,}000{,}000.$$

Systematic trial and error shows that $p_{13} = 819{,}200$, which is less than a million, while $p_{14} = 1{,}638{,}400$ which exceeds a million. Therefore, the population will reach one million sometime during the 14th month.

A Tank Model. Knowledge of a problem context sometimes permits us to use a recursive analysis. This is an approach where we ask how one term of a sequence leads to the next. We saw this in connection with network problems in Chapter 3 (see for example the Handshake Problem on page 153). Because it uses knowledge of the problem context, recursive analysis leads to examples that are structural in nature.

Now we will apply recursive analysis to model the way a contaminant can be flushed from a tank of water. In this context, the recursive analysis depends on a knowledge of mixing, and leads directly to a difference equation. But as we shall see, the model we develop here finds application more broadly than in the specific context of the example.

Consider a water tank that holds 10 gallons. Suppose someone accidentally spills a contaminant into the tank. For the sake of discussion, we will imagine that 4 pounds of salt fall into the tank and completely dissolve. In an attempt to purify the water, we drain the tank and refill it with pure water. Unfortunately, the bottom of the tank is a little below the level of the drain valve. Completely opening the valve allows most of the water to flow out, but always leaves 1 gallon at the bottom of the tank. So when we put in fresh water, we only put in 9 gallons, to mix with the one salty gallon that was left in the tank.[6]

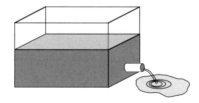

If we repeat the process of draining and refilling several times, the amount of salt in the tank is reduced further and further. In symbols, let s_n be the amount of salt in the tank after draining and refilling n times. Our goal is to develop a model for s_n. In particular, we might wish to answer the following questions. Can we reduce the amount to zero? If not, can we reduce it to a thousandth of a pound? To a millionth? And if so, how many times must the tank be flushed?

To apply a recursive analysis, we have to analyze the effect of one drain and refill step. Let us begin by considering the first step. Initially, there are 4 pounds of salt in

---

[6] Admittedly, this is not very realistic. A ten gallon tank can easily be lifted and poured out, with no need for a drain. However, we have chosen to work with an unrealistic example so that the numbers we encounter will be very simple. The reader can readily imagine a more realistic version of the problem involving a much larger tank. The recursive analysis for all such problems is the same.

4.2. Applications of Geometric Growth Sequences

the water. Draining the tank removes 9 gallons from the tank, representing 9/10 = 90% of the total. The remaining gallon equals 10% of the total, so it will contain 10% of the salt, assuming the liquid is well mixed. Therefore, after one drain and refill step there should remain 10% of 4 pounds of salt, amounting to $0.10 \cdot 4 = 0.4$ pounds.

Now we can apply the identical logic for the second step. This time there are 0.4 pounds of salt in the tank before we drain it. We remove 90% of the liquid, also removing 90% of the salt. What remains will be 10% of 0.4 pounds, equal to $0.10 \cdot 0.4 = 0.04$ pounds.

Now we argue that the same pattern always applies. After n steps, there will be s_n pounds of salt dissolved in the ten gallons in the tank. When we drain the tank, we leave behind one-tenth of the water, and so one-tenth of the salt. That means that there will be $0.1s_n$ pounds of salt. After refilling the tank with pure water, the same $0.1s_n$ pounds of salt remain. Thus, starting with s_n pounds of salt, if we drain and refill the tank once, there will then be $0.1s_n$ pounds of salt. This gives the difference equation

$$s_{n+1} = 0.1s_n,$$

which we recognize as an instance of geometric growth. In this way, a recursive analysis leads us to a geometric growth model for the water tank problem.

Proceeding with the analysis, with a starting term of $s_0 = 4$, we find the functional equation $s_n = 4 \cdot 0.1^n$. According to this model, the amount of salt in the tank will never equal 0, because at each step we are multiplying a positive quantity by 0.1. In fact, for this particular example, there is a simple pattern to the numerical form of the terms:

$$s_0 = 4.0$$
$$s_1 = 0.4$$
$$s_2 = 0.04$$
$$s_3 = 0.004.$$

Each successive multiplication by 0.1 moves the 4 one place to the right, and s_n will have a 4 in the nth decimal place. This demonstrates that the terms never reach 0, but do approach 0 to unlimited accuracy. In particular, the amount of salt will be less than one millionth (0.000001) when $s_n = 0.0000004$, and thus for $n = 7$. That means that flushing the tank 7 times will reduce the amount of salt to less than one millionth of a pound.

Although the problem context of flushing a water tank may seem to be of limited practical significance, there are a great many other problem contexts where similar reasoning applies. In one example, a lake takes the place of the water tank, and the water flowing into and out of the lake each day corresponds to draining and refilling part of the tank. The difference equation for the pollution left in the lake will be of the same form as the one for the tank. In another example, the water tank concept provides a good first approximation to the way the human body removes a drug from the blood stream. In each hour, a certain fraction of the blood is purified, leaving some of the drug behind. If the body purifies about nine-tenths of the blood, then about one-tenth of the drug remains in the blood at the end of the hour. The recursive analysis for the tank model can be adapted to this situation, leading to the difference equation

$$d_{n+1} = 0.1d_n,$$

where d_n is the amount of drug in the body after n hours. For a third example, we are concerned with the way heat escapes from some reservoir (such as an insulated building). The tank model analysis can again be adapted to this context, and again it leads to a geometric growth sequence model.

Continuous and Discrete Variables. In geometric growth models distinctions between continuous and discrete variables arise very naturally, regarding both the position number n as well as the terms of the sequence. We can see this in the two examples already considered.

For the first example, modeling a mouse population, we originally had a growth factor $r = 2.0$, reflecting a 100% increase each month. But suppose instead we found the mouse population increases by 50% in any month. If we start with 100 mice, then a 50% increase will add 50 mice to the population. Now we have 150 mice, and another 50% increase will add 75 mice, for a total of 225 mice. Continuing for another month, the model again predicts a 50% increase. But 50% of 225 will be 112.5. This is clearly nonsense. The population can't have half a mouse. Although the population size should always be a whole number, this model produces fractional values.

This is inevitable in any geometric growth model for which the growth factor is not a whole number. The situation in the example shows why. With a growth factor of $1.5 = 3/2$, the functional equation will be

$$p_n = 100 \left(\frac{3}{2}\right)^n = 100 \left(\frac{3^n}{2^n}\right).$$

This can only produce a whole number when 2^n divides evenly into 100. That occurs for 2^1 and 2^2, but not for any higher power.

Something similar occurs whenever the growth factor is not a whole number, leading to a very broad conclusion: any geometric growth sequence with r other than a whole number must include terms that are not whole numbers. How can we use such a model when we know that the terms of the sequence are supposed to be whole numbers?

One answer is that our model only approximates the real problem context, so we don't necessarily expect the model to produce results that are exactly correct. If the model predicts $p_3 = 112.5$, perhaps there are really going to be 113 or 112 mice. This kind of situation is typical in geometric growth models. It is one of the limitations that must often be accepted when geometric growth is assumed. Geometric growth models seem to fit so well in so many situations, that they are used extensively. The fact that they produce fractional values of the variables is usually no worse than a minor inconvenience. Consequently, we usually elect a continuous interpretation for the variable represented by the terms of our sequence (these represent population figures in our example), even when that is conceptually incorrect.

Turn next to the position number n. For the tank model, s_n is the amount of salt remaining after flushing the tank n times. It makes little sense to assign n a value other than a whole number. For example, what would it mean to ask how much salt remains after flushing the tank 6.27 times? For that example, we should restrict n to be a whole number, and hence a discrete variable.

In contrast, for the population model, the position number n has the additional interpretation as a period of time. And while it doesn't make sense to consider the 4.8th term of a sequence, it makes perfect sense to ask what the population is after 4.8

4.2. Applications of Geometric Growth Sequences

months. At this point, it is important to recall the library fine example (see page 76). There, too, we saw that the position number n could reasonably be interpreted as a continuous variable. But in that model, the functional equation for the sequence was only valid for whole number values of the variable. Returning to the population model with $r = 2$, we know the functional equation,

$$p_n = 100 \cdot 2^n,$$

is valid when n is a whole number. But can the same equation be used when n is not a whole number?

This turns out to be a little tricky. Consider the situation when $n = 3.78$. We understand that as an amount of time, 3.78 months after the starting point for the model. The functional equation becomes

$$p_{3.78} = 100 \cdot 2^{3.78}.$$

But what does that mean? By definition, 2^3 means to multiply together 3 twos ($2 \cdot 2 \cdot 2$). Similarly, 2^4 means to multiply together 4 twos ($2 \cdot 2 \cdot 2 \cdot 2$). But it is not possible to multiply together 3.78 twos. Adding to the mystery, a calculator computes $2^{3.78}$ as 13.737, rounded off to three decimal places. But how is that number determined?

The calculator does not compute fractional powers using proportional reasoning. To see this, note that $2^2 = 4$ and $2^3 = 8$. Proportional reasoning would indicate that an exponent of 2.5, halfway between 2 and 3, should result in an answer of 6, halfway between 4 and 8. However, the calculator evaluates $2^{2.5}$ as 5.657, to three decimal places. The question remains, what is the calculator doing?

One way to understand the answer involves the basic idea of geometric growth. As defined for sequences, that means any two consecutive terms are related by the same growth factor. But there is stronger version of this idea, analogous to the definition of arithmetic growth on page 39. We state it as follows:

> **Strong Geometric Growth Assumption:** Consider a variable that changes over time. Under the strong geometric growth assumption, in equal periods of time the variable changes with equal growth factors.

This is stronger than the earlier definition because it assumes more.

For the mouse population example, with n representing the number of months, the original definition of geometric growth says that the growth factor will be the same for every month. But if we assume that births of new mice occur throughout the month, then we would still expect to find geometric growth with n representing a number of weeks, days, or any familiar unit of time. By the same logic, n could equally well count something like 3.4-hour periods. However time is subdivided, we should get geometric growth, and hence equal growth factors. And that is exactly what the strong geometric growth assumption states. The growth factor for the population over one week will be the same no matter which week is considered. The growth factor will be the same one day as it is for all the other days. The growth factor we observe for some 3.4-hour period, will be the same as the growth factor for any other 3.4-hour period.

Applying this idea in a simple case will illustrate how it extends our functional equation to fractional values of n. For this discussion we will switch from subscripts to parentheses, for example, writing $a(3.27)$ in place of $a_{3.27}$. This is consistent with the convention we adopted earlier when thinking of n as a continuous variable (see page 88).

Recall that in the original mouse population example, we had $a(2) = 400$ and $a(3) = 800$. What should be the value for $a(2.5)$? Instead of attacking this question directly, we focus on the growth factor. From $n = 2$ to $n = 2.5$, there will be some unknown growth factor, x. This would give $a(2.5) = xa(2)$. But the same growth factor must be observed between $n = 2.5$ and $n = 3$, by the strong geometric growth assumption. Therefore

$$a(3) = x \cdot a(2.5) = x \cdot (xa(2)) = x^2 a(2).$$

Now we can isolate x^2 finding

$$x^2 = \frac{a(3)}{a(2)} = \frac{800}{400} = 2.$$

We learned how to solve such equations in Chapter 3. Additionally, the solution we seek must be positive, and so it is given by

$$x = \sqrt{2}.$$

A calculator gives this as 1.414214, to six decimal places. In turn, the value of $a(2.5)$ must be

$$a(2.5) = xa(2) = \sqrt{2} \cdot 400$$

which equals 565.685425, to six decimal places.

This is rather a round-about way to find $a(2.5)$. But it makes sense logically, and moreover, gives the same result as using the functional equation $a(n) = 100 \cdot 2^n$. That is, on a calculator, if we compute $a(2.5) = 100 \cdot 2^{2.5}$, we again get 565.685425, to six decimal places. More generally, the logic above based on the strong geometric growth assumption will always produce the same result as applying the functional equation. This shows that the calculator's programming for fractional exponents is consistent with the strong law of geometric growth. In a similar way, the strong law also leads to the accepted mathematical convention for fractional exponents:

$$r^{m/n} = \left(\sqrt[n]{r}\right)^m.$$

Extending geometric growth functional equations to fractional values of n also has a graphical representation. This is illustrated in Figure 4.7, which shows two graphs for the geometric growth sequence with initial term $b_0 = 5$ and growth factor $r = 4$. On the left, the individual terms of the sequence are plotted as dots. As a visual aid, the dots are connected with straight lines giving the appearance of a curve. (This is the same as Figure 4.2 on page 206.) On the right, the functional equation $b_n = 5 \cdot 4^n$ is graphed for both whole number and fractional values of n, with dots marking the points for whole number values of n. Visually, the second graph connects the points for the sequence more smoothly than the first, providing an appearance that many viewers find more appealing. In some sense, this confirms that the strong geometric growth law is the *right* concept for extending geometric sequences to fractional values of n.

The main point of the preceding discussion is that geometric growth functional equations can be applied with fractional values of n. The results obtained from a calculator or computer are consistent with the strong geometric growth assumption, and create a smoothly curving graph. Whether or not these results are appropriate in any specific model will depend on the model context. But if a continuous interpretation

4.2. Applications of Geometric Growth Sequences

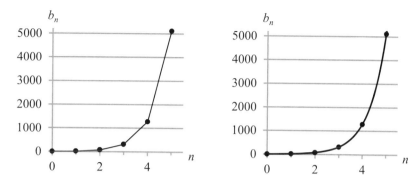

Figure 4.7. Two graphs for the geometric growth sequence with $b_0 = 5$ and growth factor $r = 4$. Both show the terms from b_0 to b_5 as dots. On the left, the dots are connected with straight lines. On the right, the points between the dots are computed using the functional equation $b_n = 5 \cdot 4^n$ for fractional values of n.

of n is reasonable, and if the strong geometric growth assumption is adopted, then extending the functional equation to fractional values of n is justified.

An Empirical Example of Geometric Growth. In the examples we have already discussed, geometric growth is suggested by the structure of the problem context. A geometric growth model may also be adopted because we find data that follow or approximate the geometric growth condition: each term is a constant multiple of the preceding term. Thus, adopting a geometric growth model reflects a pattern in observed data, rather than a logical analysis of the problem context. An illustration is provided by the following example.

A Queen Bee's Family Tree. We consider the problem of counting ancestors for a queen bee.[7] Male bees are produced asexually by the queen, so each male has a mother but no father. Each queen bee, by contrast, has both a father and a mother. Starting with a queen, we can chart the ancestors in a diagram as in Figure 4.8.

In the diagram, each black circle is a male, and each white circle is a female. The circle at the top of the diagram is the queen bee we are interested in. The white circle directly below is the mother of the queen, while the black circle off to the right is the father of the queen. In fact, for any white circle, there is a white circle directly below, indicating the mother, and a black circle off to the side for the father. In contrast, for black circles, there is only one parent, shown by a white circle directly below. We can continue this diagram for as many generations of bees as we wish. Observe that the queen has 2 parents, 3 grandparents, 5 great-grandparents, 8 great-great-grandparents, and so on. It is this pattern of numbers, 2, 3, 5, 8, etc., that we are interested in. To be specific, let a_n be the number of ancestors that the queen has in the nth preceding generation. Then a_1, the number of ancestors in the first preceding generation (that is the parents), is 2, $a_2 = 3$, $a_3 = 5$, and so on. This is the sequence we would like to model.

In the diagram we can count the ancestors in each generation up to $n = 5$. These values are shown in Table 4.3. The table also includes figures for two additional gen-

---
[7]This example is adapted from a discussion starting on page 277 of [30].

228   Chapter 4. Geometric Growth

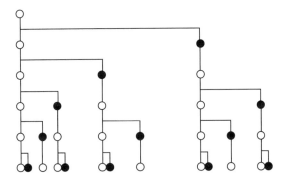

Figure 4.8. Family tree for a queen bee. Males are indicated with black circles, females with white circles. Each bee has one or two parents, lower in the diagram, and connected to the bee by straight lines. The ancestors of the top-most queen bee are aligned horizontally by generation. There are two bees on the level directly below the top queen, three on the level below that, five on the level below that, and so on. Thus, for example, there are 8 bees on the fourth level below the top queen, indicating 8 ancestors in the fourth preceding generation.

Table 4.3. Data table for the a_n sequence, with growth factors for successive terms. The growth factors are not equal, so this is not a geometric growth sequence. However, the growth factors are nearly equal, so this sequence may be approximated with a geometric growth model.

n	a_n	Growth Factor
1	2	
		1.500000
2	3	
		1.666667
3	5	
		1.600000
4	8	
		1.625000
5	13	
		1.615385
6	21	
		1.619048
7	34	

erations beyond what is shown in the diagram. The reader is invited to verify these results by extending the diagram. A column of growth factors has been included in the table to test for geometric growth. These values are shown to six decimal places.

We can see from the table that the bee ancestor sequence is not an instance of geometric growth, because the growth factors are not equal. But they are nearly equal, with values near 1.6. Indeed, if the table is extended to about 20 terms, the growth factors become constant to six decimal places. But even with the information shown in Table 4.3, we can see that the ancestor sequence is approximately geometric. We can produce an approximating geometric growth sequence by choosing an initial term a_1

4.2. Applications of Geometric Growth Sequences

and growth factor r. By trial and error, comparing the model and the actual ancestor sequence for various values of the parameters, a very accurate model was identified with initial term 1.895 and growth factor 1.618. Numerical and graphical results for the model are shown in Figure 4.9.

n	a_n	Model (b_n)	Error
1	2	1.895	0.105
2	3	3.066	−0.066
3	5	4.961	0.039
4	8	8.027	−0.027
5	13	12.987	0.013
6	21	21.014	−0.014
7	34	34.000	0.000
8	55	55.012	−0.012
9	89	89.010	−0.010
10	144	144.018	−0.018

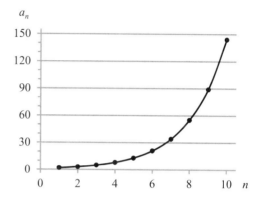

Figure 4.9. Data values and graph comparing the bee ancestor sequence and an approximating model. In the table, for each value of n, the corresponding entry in the a_n column represents the actual number of ancestors in the nth preceding generation, while the *Model* column shows the values of a geometric growth sequence with initial term $a_1 = 1.895$ and growth factor $r = 1.618$. For each n the error is computed as $a_n - b_n$. Entries in the *Model* and *Error* columns are rounded to three decimal places. At the worst, the model is off by 0.105, and for all but two terms the model is off by less than 0.05. On the graph, the a_n values are shown as dots, while the model is represented by a smooth curve. A perfect agreement between the actual a_n values and the model would be indicated by dots perfectly centered on the curve. For this model, the errors are so small that the eye cannot detect them in the graph.

Let us introduce the notation b_n for the terms of the approximating geometric growth sequence. Then $b_1 = 1.895$ and the growth factor is $r = 1.618$. The difference equation, $b_{n+1} = 1.618 b_n$, says that we can produce each successive term by multiplying the preceding term by 1.618. This can be used repeatedly to obtain the values in the *Model* column of the table. We can also find the functional equation as $b_n = b_0 r^n$. However, for this model, we started with b_1, and do not have a value of b_0. We can still find a functional equation, by selecting a value of b_0 that is consistent with the model. That means we want b_0 to satisfy $b_0 r = b_1$, implying that $b_0 = b_1/r = 1.895/1.618 = 1.171$ to three decimal places.[8] Thus, $b_n = b_0 r^n = 1.171 \cdot 1.618^n$.

This problem shows how an observed pattern in a data sequence can lead us to adopt a geometric growth model. For the bee ancestor sequence, the model is quite accurate for the terms we have examined. This gives us confidence that the model

---

[8] Rounding to three decimal places here introduces slight errors. These can be avoided in calculations by using $b_n = (1.895/1.618) \cdot 1.618^n$. Alternatively, $b_n = 1.895 \cdot 1.618^{n-1}$ can be used.

would be a reasonable means for answering questions about the bee ancestor sequence. Is that of practical value? Is there some reason we might need to know how many ancestors a queen bee had in a prior generation? Perhaps not. On the other hand, it does give us some qualitative insight about the nature of the queen bee's family tree: the number of ancestors grows approximately geometrically with a growth factor around 1.6. This might also be useful in some larger study, for example in bee genetics.

Reflecting on the Model. As part of model development we should reflect broadly on the end product. In this regard, four considerations are particularly noteworthy. The first is the distinction between continuous and discrete variables. Here we know that both n and a_n must be whole numbers. For example, n cannot equal 3.5 because it makes no sense to ask how many ancestors were present in the 3.5th preceding generation. Similarly, we cannot have $a_n = 33.9$ because there cannot be a fractional number of ancestors. On the other hand, we know that the terms cannot all be whole numbers in a geometric growth sequence with a growth factor that is not a whole number. Therefore, a certain amount of error in our model is unavoidable.

In such a situation, one approach is to round each approximating term to the nearest whole number. For the data in the table, this step produces perfect agreement between the model and the original sequence. This raises an intriguing question: Is it possible to find a geometric growth model which, when the terms are rounded to the nearest whole number, exactly matches the bee ancestor sequence for every generation? This is our second consideration. More broadly, it can be understood as focusing on the errors produced by a model. For the problem at hand, we want to know whether all the errors remain uniformly small. But in more general terms, we can always look for patterns in the errors, in the hope of obtaining an improved model.

The third consideration is a global review of the entire problem. Have we missed any feature or pattern that might be valuable? In fact, as many readers have probably already noticed, the bee ancestor sequence is a famous progression known as the *Fibonacci* numbers. Those who have studied this subject might very well recognize the a_n terms as a part of the pattern $1, 1, 2, 3, 5, 8, 13, \cdots$. In fact, this sequence appears in some of the exercises in Chapter 1. But even if you did not do those exercises, and never studied the Fibonacci numbers, you might have noticed that the a_n sequence follows a pattern: starting with a_3, each term is the sum of the preceding two terms.

This is a kind of growth we have not yet analyzed, and we do not have a corresponding functional equation. But our results indicate that it is approximately geometric growth. Combining these ideas might launch an investigator into a new area of analysis. Is there a whole family of growth models that are similar to the Fibonacci numbers? Are they all approximately geometric? If so, how can we find the best geometric approximation?

The final consideration is that this example combines aspects of both structural and empirical analysis. The values of a_n shown in Table 4.3 have been treated as data, and in fact they are correct values. But we generated them using our knowledge of the reproduction of bees, as opposed to making measurements or direct observation of bee populations. That is using structural knowledge. Moreover, using a recursive analysis, we can deduce that the Fibonacci pattern must hold for all n. This amounts to recognizing that the ancestors of a queen bee can be divided into two categories, ancestors of the queen's mother and those of the queen's father. Without digressing to

4.2. Applications of Geometric Growth Sequences

explore this result in detail, we mention it here to show that our structural knowledge in this problem does lead directly to a difference equation. However, our analysis did not take advantage of the difference equation. Rather, we shifted to empirical thinking, and found a very accurate approximate functional equation.

Systematically pursuing these ideas is beyond the scope of this book. We mention them as an example of how models can lead to new ideas, and how important the reflection step is. For the reader who is intrigued to know more about this topic, we summarize a few results. The Fibonacci numbers are known to be exactly given by the sum of two different geometric progressions, which we might call the major and minor progressions. The growth factor is greater than one for the major progression and between -1 and 0 for the minor progression. This implies that the terms of the minor progression alternate in sign, while their absolute values diminish rapidly to zero. Indeed, after the first few terms, the minor sequence contributes almost nothing to the a_n values. Thus the Fibonacci numbers are very closely approximated by the major progression, with accuracy that increases steadily as we consider larger and larger values of n. And the errors correspond to the terms of the minor progression. Interestingly, the growth factors for the major and minor progressions can be found exactly by solving a quadratic equation. In fact, the major progression has initial term $M_0 = \frac{1}{\sqrt{5}}$ and growth factor $r = \frac{1+\sqrt{5}}{2}$. This leads to a surprising result: the nth Fibonacci number can be found exactly by rounding

$$M_0 r^n = \frac{1}{\sqrt{5}} \left(\frac{1+\sqrt{5}}{2} \right)^n$$

to the nearest whole number. Who would have predicted so complicated a formula for terms in a progression of whole numbers defined by simple addition?

Other Applications. In the discussion so far we have emphasized two ways that geometric growth models arise, namely, due to the structure of the problem context, or because of patterns observed in data. There is a third reason to adopt a geometric model. Geometric growth has already been found to be effective in a variety of modeling contexts. When you recognize such a context, it is natural to consider a geometric growth model. With this in mind, we conclude the section by giving brief descriptions of several modeling contexts for which geometric growth is known to be applicable.

Compound Interest. The calculation of compound interest is an important application of geometric growth. Here, we are in a slightly different modeling context. Previously, we imagined some natural process which produced data, and we wanted to use models that fit the data. However, for compound interest there is no natural process at work. Rather, a model is adopted by financial institutions, and used as a basis for interest computations. There is no question whether the model is correct or incorrect. We simply have to learn to predict the results the adopted model will produce.

The fundamental concept of compound interest is this: In a specified period of time, the value of money grows by a fixed percentage. This assumes no other activity. For example, say you owe $500 on a credit card, and the bank charges 1% per month on the balance. If you neither make additional purchases nor repay any part of the debt, then the value of what you owe will grow by 1% every month. That is geometric growth.

As discussed in Section 4.1, an increase of 1% corresponds to a growth factor of 1.01. With a starting balance of $500, our knowledge of geometric growth lets us write immediately

$$a_n = 500(1.01)^n$$

for the account balance after n months.

The same basic structure holds for interest charges on loans, including student loans and automobile loans. For these types of loan, financial institutions are required to disclose the interest rate to the borrower. Usually, this will not be given on a per-month basis. Instead, it will be displayed as an annual interest rate, often denoted as APR, for *annual percentage rate*. This is really only a nominal rate. You must divide it by the number of billing cycles per year to find the percentage by which the account balance grows with each cycle. So, if the nominal rate is 8% and the account is billed monthly, each interest charge is actually one-twelfth of 8 percent, which is computed as (8/12)%, or about 0.006667, expressed as a decimal. Then each month the growth factor is 1.006667. Similarly, for a savings account, the bank may credit interest to the account daily. With a nominal rate of 8%, and 365 days per year, each day you would receive (8/365)% in interest.

For these types of accounts, the annual rate is not a true measure of how much interest accrues over a year. It actually indicates how monetary value would increase if an arithmetic growth model were used. To find the true value, use a geometric model to predict the growth for one year. That figure can be used to compute a growth factor for the full year, leading in turn to a percentage of growth. Let's do this for the first example above. We start with a balance of $500. The variable n counts the number of months. So after one year, $n = 12$. The balance owed on the account is then $a_{12} = 500(1.01)^{12} = 500 \times 1.127$, rounded to three decimal places. Relative to the initial value of 500 this is a growth factor of 1.127, so the decimal form of the percentage of increase is 0.127, meaning 12.7 percent. Therefore, although the annual rate is expressed as 12%, what you actually are charged over a year is 12.7%. This actual figure is called the *effective rate*. In the case of a savings account, the actual rate earned is also called the *yield*.

Tank Models. As discussed earlier, the tank model can be adapted to several other application contexts. Here are two examples. Suppose a lake contains a certain amount of pollution. Assume the total number of gallons of water in the lake is known, as is the number of gallons of clean water flowing into the lake per day from springs and rivers. Assume too that the same amounts of water flow in and out each day, so that the amount of water in the lake remains constant. Over the course of the day, the lake is like the water tank. A fraction of the polluted water flows out and is replaced by clean water. As a result, each day the pollution in the lake is reduced by a fixed fraction. That is, the pollution is flushed from the lake in the same way that the salt was flushed from the tank. This leads to a geometric growth model for the amount of pollution remaining in the lake after n days. This example is further explored in the exercises.

Another variation on the tank model concerns the way the human body removes a drug that is present in the blood. It has been found that, as a reasonable first approximation, each hour a certain percentage of the medication is removed. This leads to a geometric growth model for the amount of medicine in the blood. A further variation is to consider the effect of periodic doses of medication. By studying this kind of model

4.2. Applications of Geometric Growth Sequences

it is possible to figure out what dose to administer each time in order to maintain a desired level of medication in the blood stream. This application is also discussed further in the exercises.

Radioactive Decay. Another accepted application of geometric growth concerns atomic radiation. Some elements occur in nature in several different forms, called isotopes, some or all of which are radioactive. As a radioactive isotope gives off radiation, some of it turns into either another isotope, or another element, thereby reducing the amount of the original isotope. For this application, we model how the amount of a specific radioactive isotope decreases over time.[9]

It has been found that each radioactive isotope follows its own particular geometric growth pattern. In each case, some fixed percentage of the material is transformed to something else in each unit of time. That means the amount of the radioactive element left after each unit of time is a fixed percentage of what was there at the start of the unit of time. This is geometric decay.

Empirical study can be performed to find out how fast each element decays. This is commonly expressed by giving the *half-life* of the element. For example, the half-life of the uranium isotope U237 is 6.75 days. This means that an initial amount of the substance will be half gone after 6.75 days. Suppose you have a laboratory sample made up of a mixture of U237 and other material. Say it contains 1 gram of U237 when it is first measured. Then after 6.75 days the sample will contain only 0.5 grams of U237. The total weight of the sample will be essentially the same as it was originally. The 0.5 grams of U237 doesn't disappear, it just changes into something else. But the total amount of U237 within the sample will decrease from 1 gram to 0.5 grams in 6.75 days. We can model this with a geometric growth sequence in which n is a number of 6.75-day periods, and a_n is the amount of U237 after n such periods. Thus $a_n = a_0(1/2)^n$. In Section 4.3 we will see how to convert this to an equation with a more natural measure of time than 6.75-day periods.

Geometric growth models can be used to analyze radioactivity. For example, following an earthquake and tsunami in March 2011, the Fukushima nuclear power plant in Japan suffered catastrophic malfunctions and released significant amounts of radioactive material into the environment. These will continue to pose a threat until the amount of radioactive isotopes is reduced to a safe level. Once an amount of contamination is known, the model can predict how long it will be before the radiation dissipates to a safe level. Notice that this is a find-n question, and involves inverting the functional equation: we know how much is a safe level; we want to know when that level will be reached.

Population Growth. We have already seen an example of a geometric growth population model, involving an imaginary situation with a population of mice. But geometric growth is often applied with real data for real populations. These can be populations of many different kinds, such as domesticated or wild animals, microbes, and plant species. When dealing with human populations, the focus is often on groups of people sharing some common attribute, such as age, economic status, geography, genetic makeup, or exposure to a disease.

---

[9]This does not require the isotope to exist in a pure state. For example, we can model the amount of the radioactive uranium isotope U238 in a mixture containing other matter as well.

As a model for population growth, geometric growth models are a popular first step. However, they are usually only accurate for limited periods of time. As we have seen, these models involve rapid acceleration of growth, and in any biological application, sooner or later something will act to limit the growth. Nevertheless, there are ways to modify the geometric growth model to make it more true to life. One very important example of this is called a logistic model. We will return to that topic in a future chapter. Interestingly, although logistic models can be very useful, in some situations they lead to a particular kind of breakdown of the entire modeling approach. The problem is a pathological behavior referred to as *chaos*. A discussion of chaos in logistic growth models will be presented at the end of the book.

Heat Transfer. As a final example, geometric growth models can be applied to questions concerning the way heat flows from hot to cold objects. Examples include the cooling of a cup of hot coffee, or heat loss from a building with an inactive heating system. The models we will consider here obey what is often called Newton's Law of Cooling. It says that the temperature difference between an object and its surroundings will decay geometrically. The model is structurally justified by a modified tank model. It has also been tested in many experimental settings.

To illustrate how such models are formulated, let us consider a hot apple pie left to cool on a kitchen counter. When the pie was removed from the oven, its temperature was 350°F. The room temperature is 70°F. What we will model is the difference between the temperature of the pie and its surroundings. Let us call this H (it indicates how much *hotter* the pie is than its surroundings). Initially, $H = 350 - 70 = 280$, indicating that the pie is 280° hotter than the room. Suppose also that after ten minutes the pie has cooled to 300°. Then $H = 300 - 70 = 230$ at that time.

According to Newton's Law, the variable H will obey a geometric growth law. In particular, if we make temperature measurements every 10 minutes, and compute H for each measurement, the sequence of H values will form a geometric growth sequence.

Defining our variables, let n count the number of measurements, or equivalently, the number of 10-minute intervals. For example, $n = 3$ refers to the third measurement, which takes place 30 minutes after the pie is removed from the oven. Let H_n tell how much hotter the pie is than the room at each measurement, in units of degrees Fahrenheit.

We know that $H_0 = 280$ and $H_1 = 230$. That means the growth factor between these terms is $r = 230/280 = 23/28$, expressed as an exact fraction.[10] Under the geometric growth assumption, each successive term will be 23/28 times the preceding term. In other words, the sequence satisfies the difference equation $H_{n+1} = (23/28)H_n$. This can be used to generate a table and graph, as in Figure 4.10. We can also formulate the functional equation $H_n = H_0 r^n = 280 \cdot (23/28)^n$.

This model can be used to answer questions about the time and temperature. Thus we might want to know how hot the pie is after an hour, or when the pie will reach a termperature of 100°. Notice that these have to be translated into questions about n and H_n. For the first question, we use the fact that one hour corresponds to six 10-minute intervals. We ask the find-H_n question, *What is H_6?* We can read the value from our

---

[10] A calculator gives $23/28 = 0.8214$ to 4 decimal places. As noted in the discussion of the bee ancestry model, calculations will be more accurate if r is entered as 23/28, rather than the decimal approximation.

4.2. Applications of Geometric Growth Sequences

n	H_n
0	280.00
1	230.00
2	188.93
3	155.19
4	127.48
5	104.71
6	86.02
7	70.66
8	58.04
9	47.67
10	39.16

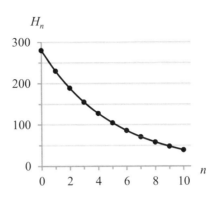

Figure 4.10. Data table and graph for the geometric growth sequence with $H_0 = 280$ and $r = 23/28$. The entries in the H_n column of the table have been rounded to two decimal places.

table, or use the functional equation to find

$$H_6 = 280 \left(\frac{23}{28}\right)^6.$$

Either way, we find $H_6 = 86.02$, to two decimal places. But this is not the temperature of the pie. It says the pie is $86.02°$ hotter than room temperature, $70°$. Therefore we conclude that the pie will be $156.02°$ one hour after it is taken from the oven.

For the second question, we note a temperature of $100°$ is $30°$ hotter than room temperature. So we can restate the question in the form *When does $H_n = 30$?* The table and graph do not extend far enough to answer this question immediately, but we can use a numerical approach. Using either the functional equation or the difference equation, we find $H_{11} = 32.17$ and $H_{12} = 26.42$. Remember that $n = 11$ corresponds to 110 minutes and $n = 12$ to 120 minutes after the pie is removed from the oven. Thus, our preliminary finding is that the temperature will reach $100°$ after cooling for between 110 and 120 minutes. This might be accurate enough. If the question is when to serve the pie, we will probably be content with the answer, *After about two hours*.

But what if we want greater accuracy? In this model, both n and H_n make sense as continuous variables, and the strong geometric growth assumption is reasonable. Thus we can use the functional equation with fractional values of n, switching from subscript to parenthesis notation as usual. With systematic trial and error, we can find that $H(11.35) = 30.03$ and $H(11.36) = 29.97$, to two decimal places, putting n between 11.35 and 11.36. By computing $H(11.355) = 29.998$, we see that n must be between 11.35 and 11.355, indicating that $n = 11.35$ to two decimal places.

In reflecting on this model, one might note the necessity of translating between temperature and time, on the one hand, and variables n and H on the other. How might that aspect be improved? Let us define new variables: t is the time in minutes starting from $t = 0$ when the pie is removed from the oven; T is the temperature of the pie, in degrees F. Then the time after n time intervals will be $10n$, because each interval

lasts 10 minutes. This is expressed in an equation as
$$t = 10n,$$
and solving for n produces
$$n = 0.1t.$$
Similarly, we know that the temperature of the pie is H degrees above the room temperature of 70. This, too, leads to an equation,
$$T = H + 70,$$
and solving for H gives us
$$H = T - 70.$$
Now use these results to replace H and n in the functional equation, as follows.
$$H = 280 \left(\frac{23}{28}\right)^n$$
$$T - 70 = 280 \left(\frac{23}{28}\right)^{0.1t}$$
$$T = 70 + 280 \left(\frac{23}{28}\right)^{0.1t}.$$
In this way we obtain an equation that is expressed in terms of the new variables. To summarize, the geometric growth sequence was most naturally formulated using the variables n and H, but our questions about the model are most naturally expressed using t and T. Our knowledge of geometric growth led us quickly to a functional equation relating H to n, and with algebra we can translate that to an equation relating T to t.

The development in this example can be adapted to many contexts involving heat transfer. The specific equation we found is an instance of the more general form
$$T = C + Ab^{kt},$$
where C is a constant representing the ambient temperature. This form of equation always arises in models based on Newton's Law of Cooling. It shows that the model agrees with our expectations about how things heat and cool. In particular, the temperature of a cooling object will decay toward the ambient temperature, but will not go lower. Similarly, the temperature of a warming object (such as a glass of ice water standing in a room at 70°) will increase toward the ambient temperature, but will not go higher. These conclusions follow from properties of the term Ab^{kt} in the equation for T. Functions of the form Ab^{kt} will be studied in the next section.

4.2 Exercises

Reading Comprehension.

(1) Give brief descriptions of three different problem contexts for which geometric growth models are frequently used. For each example, include the reason geometric growth is a better model than arithmetic or quadratic.

(2) In the mouse population model two different assumptions were stated. They may seem similar at first glance, but they lead to very different models. Compare and contrast the models that would result from these assumptions.

4.2. Exercises

    Assumption A: In any month, the mouse population will increase by 100 mice.

    Assumption B: In any month, a population of 100 mice will grow to 200 mice.

(3) @Let's consider the strong geometric growth assumption discussed in this section.

    a. @State the assumption.

    b. How does the strong assumption differ from other assumptions for geometric growth? For example, *Assumption B* above is a geometric growth assumption but is not a strong geometric growth assumption.

    c. @Under the strong geometric growth assumption, how would we calculate the growth factor for half an interval. For example, if a population grows by a factor of 9 in one year, what factor will it grow by in six months?

    d. Explain briefly how the calculator programming for fractional exponents is related to geometric growth models.

(4) @There are some geometric sequences with only whole number terms, such as $a_n = 2^n$ with terms 1, 2, 4, 8, $\cdots$. There are others which may have several whole number terms, but will eventually have fractional terms. For example, $b_n = 100(1/2)^n$ has terms 100, 50, 25, 12.5, $\cdots$.

    a. How can an equation for a sequence be used (other than by making a list of terms) to determine if the sequence will eventually have fractional terms or not? Your answer should include an application of your method to a few examples of equations for sequences with only whole number terms and a few examples of equations for sequences with fractional terms.

    b. @Populations are counted in whole numbers; it does not make sense to talk about 0.5 mice. And yet, geometric growth models with fractional terms are often used for populations. Why?

(5) @Let's consider compound interest.

    a. @In advertisements, banks often list interest rates for loans and savings accounts in terms of an annual percentage. Write a short paragraph explaining how these annual rates are used to figure out interest payments. Include in your answer an explanation of *effective rate* and *yield*.

    b. @Is the calculation of compound interest, as you described it in part *a*, an example of the strong geometric growth assumption? Justify your answer.

(6) Explain what *half-life* means, and how it is related to geometric growth models for radioactive decay.

(7) The last stage of successfully using a model is to *reflect* on the model and results. Give at least three examples of general questions that should be asked as part of reflection on a model.

Math Skills.

(8) @For each of the following, determine the corresponding growth factor, r, for the given geometric model. The meaning of the terms and n is indicated in each item.

   a. @A population is increasing by 150% per year; a_n represents the population after n years.
   b. A population is decreasing at a rate of 20% per year; b_n represents the population after n years.
   c. @A tank of contaminated water is 95% drained then refilled with pure water repeatedly; c_n represents the amount of contamination in the tank after draining and refilling n times.
   d. A bank account earns 2% interest compounded monthly; d_n is the bank balance after n months, assuming no deposits or withdrawals have been made.
   e. @A debt increases with 5.3% interest compounded monthly; e_n is the loan balance after n months, assuming no payments have been made.

(9) @For each of the following sets of data compute the growth factors for each pair of successive terms and determine if a geometric growth model would be a good approximation for the data. Justify your answer.

   a. @13.4, 30.7, 71.9, 163, 374.8, $\cdots$
   b. 2.7, 4.5, 6.3, 8.1, 9.9, $\cdots$
   c. 5, 3.3, 2.2, 1.5, 1, $\cdots$

(10) Fermium-253 has a half-life of 3 days. Assuming an initial amount of 12 grams, find the following.

   a. How much will remain after 3 days?
   b. How much will remain after 6 days?
   c. How much will remain after 15 days?
   d. When will this sample first contain less than 1 gram of Fermium-253?

(11) @Suppose that a bowl of left-over soup is put into a refrigerator where the ambient temp is 40 degrees. Let H represent how much hotter the soup is than the ambient temperature at any time, and H_n be the value of H after n 20-minute intervals.

   a. @If T is the temperature of the soup, find an equation relating T and H.
   b. @If t is the time in hours starting from when the soup was put in the refrigerator, find an equation relating n and t.
   c. @If the geometric growth model functional equation is $H_n = 60(2/3)^n$, find an equation relating T to t.

(12) @In answering the following questions, consider a model for a population of mice, based on the strong geometric growth assumption. Suppose the growth factor for one year is 16.

   a. @What should the growth factor be for 1/2 of a year? Why?
   b. @What should the growth factor be for 1/4 of a year? Why?

4.2. Exercises

c. Suppose the functional equation for the mouse population model is $p_n = 1{,}000 \cdot 16^n$, where n is the number of years. What does a calculator give you for the population if you set $n = 1/2$? Is this consistent with your answer for part a?

d. Continuing with the model in part c, what does a calculator give you for the population if you set $n = 1/4$? Is this consistent with your answer for part b?

e. @Explain why these questions and answers show that the calculator programming for fractional exponents is consistent with the strong geometric growth assumption.

Problems in Context. Some problems in this section are marked "Trial and Error". For these problems answers may differ from one student to another because there is no single correct answer.

(13) @Lake Replenishment. In this problem you will use a model similar to the one for a water tank in connection with pollution in the Great Lakes.[11] It has been determined that approximately 38% of the water in Lake Erie is replaced each year. That is, in the course of a year, an amount of water flows out of the lake and is replaced by an equal amount that enters the lake; the amount that is replaced in this way is 38% of the total water in the lake. Suppose that we could magically stop all pollution from entering the lake. How long would it be before the pollution in the lake was reduced to 10% of the current level? To 1%? To answer these questions, we start with a difference equation. As in the water tank model, we imagine that 38% of the pollution in the lake drains out with the water that leaves the lake each year. That leaves 62% of the pollution in the lake. The difference equation is then

$$p_{n+1} = 0.62 p_n,$$

where p_n is the total pollution in the lake n years after the start of the model. We also need a starting value, p_0, the total amount of pollution initially in the lake. This is probably thousands of tons, but for ease of reference, we will make up a new unit: eries. We will say that the total amount of pollution in the lake right now is one erie. We want to know when it will be one-tenth as much, or 0.1 eries. We also want to know how long it will take to get down to 1% of the starting amount, or 0.01 eries. To answer these questions, follow this outline:

a. Use the information $p_{n+1} = 0.62 p_n$ and $p_0 = 1$ to compute p_1 through p_5, to get a feel for the model.

b. @Graph the sequence and estimate when the value first reaches 0.1. When it first reaches 0.01.

c. @Find the functional equation for p_n in this model.

d. Use the functional equation and a numerical approach to estimate when the value first reaches 0.1 approximately. When it first reaches 0.01 approximately. Compare these estimates with those you found graphically.

e. @Reflect on this model. What simplifying assumptions were made? What conclusions can be drawn from this model? What is next, that is what other conclusions might be helpful and how could the model be refined to provide those conclusions?

---
[11]This is a simplified version of Example 4.2, pp. 152–154 in [47].

(14) **MRSA.** Methicillin-Resistant *Staphylococcus Aureus* (or MRSA) causes serious and hard to treat infections that can be contracted in hospitals. The table and graph below show how the frequency of MRSA cases increased in US hospitals between 1993 and 2005 [**18**].

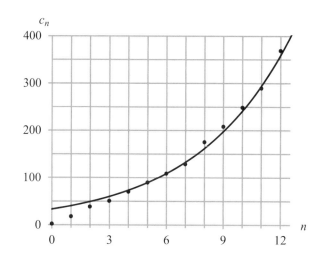

Year	Cases (1000s)
0	1.9
1	17.6
2	38.1
3	50.3
4	69.8
5	89.4
6	108.6
7	128.5
8	175.0
9	207.9
10	248.3
11	289.1
12	368.6

Data and model with $c_0 = 33$ and $r = 1.22$.

In the figure both the data and a model are depicted, with years (n) shown on the horizontal axis, starting with $n = 0$ in 1993. The vertical axis shows the number of cases, in thousands. The data from the table appear as dots in the figure, while the curve represents a geometric growth sequence c_n that approximates the data. The parameters for this sequence are $c_0 = 33$ and $r = 1.22$.

a. Find the difference equation and functional equation for c_n.

b. According to the model, by what percentage did the number of MRSA cases increase each year from 1993 to 2005?

c. Use the model to predict the number of MRSA cases for 2010.

d. Use the model to predict when the number of MRSA infections annually in US hospitals will reach 3 million. This may be done graphically by creating a new graph of the model and extending it as far as necessary. Alternatively, this may be done numerically.

e. Imagine you are part of a task force charged with analyzing the frequency of MSRA in US hospitals. What limitations should be mentioned in discussing the results from this model? Are each of the predictions in the previous two parts of this problem equally reliable? Explain why or why not.

f. **Trial and Error.** Using a graphing application and a trial-and-error approach, try several different values for the parameters c_0 and r. Compare how well each version of the model approximates the data, considering for example the average error, or the maximum error, or how well the model appears to fit the trend of the data points. Recall that the error for each term is found by

4.2. Exercises

subtracting the (model) term from the corresponding data. Of all the alternatives you considered, including the values $c_0 = 33$ and $r = 1.22$, what do you consider to be the best model for the data, and why?

(15) @Surgery Popularity. Radial keratotomy (RK), an early form of surgical vision correction, became popular in the early 1990s. A news account from 1994 reported exponential growth in the number of patients receiving this surgery over the preceding five years [4]. (That is the same as saying that the number of patients has grown geometrically.) The article reported that 30,000 surgeries were performed five years earlier, and that 250,000 were projected for the current year. Use your knowledge of geometric growth to analyze this situation by completing the following steps.

   a. @Define a geometric growth sequence R_n representing the annual number of RK surgeries reported every five years. That is, $R_0 = 30,000$ corresponds to the number of surgeries in 1989, five years prior to the article, $R_1 = 250,000$ for 1994, R_2 for 1999, and so on. Find the difference and functional equations for this sequence, and compute the values of R_n for $n = 2, 3, 4$.
   b. @Draw or use a computer application to create a graph for your sequence.
   c. @Based on your model, how many surgeries were performed each year for 1990 through 1993?
   d. @How many surgeries does your model predict for 2014?
   e. @Imagine you were a venture capitalist in 1994 considering whether to invest in a chain of vision correction clinics. Explain how you might use the model developed in this problem to estimate the potential market for such clinics. Also discuss any limitations you see in the model's applicability.

(16) Population of England and Wales. Geometric growth is often used in population models. In Table 4.4 the population of England and Wales is shown as reported in census data from 1801 to 1911 [51].

Table 4.4. England and Wales population in millions.

Year	1801	1811	1821	1831
Pop.	8.89	10.16	12.00	13.9
Year	1841	1851	1861	1871
Pop.	15.91	17.93	20.07	22.71
Year	1881	1891	1901	1911
Pop.	25.97	29.00	32.53	36.07

   a. Compute a growth factor for each consecutive pair of population values. For example, from 1801 to 1811 the growth factor was $10.16/8.89 = 1.14$. Do the same for the growth from 1811 to 1821, 1821 to 1831, and so on.
   b. Do the growth factors seem pretty constant? Determine a value for the growth factor of a geometric model and explain how you came to your decision.
   c. What percentage growth is represented by the growth factor in your answer to part b?

d. Define a geometric model for the population growth, using your growth factor from part b, and the starting population of 8.89.

e. Analyze the accuracy of your model by following these steps. Plot the true data and the model data on one graph. What do you notice? Create a table which shows the error, that is the difference between each true data value and the corresponding model data value. What is the maximum error? What else do you notice?

f. Discuss any limitations you see in the model's applicability and how the model might be improved.

(17) @Compound Interest. A young couple receives a $15,000 inheritance from the will of a grandparent. They invest the money in bonds that pay 2.3% interest per year.

a. @If the interest is paid monthly, how much will the couple have after 5 years?

b. @What if the interest is paid quarterly?

c. @Which would be a better deal for this couple: an investment that returns 2.3 percent per year, paid quarterly; or one that returns 2.5 percent per year, paid only annually?

d. @The couple wants to save up enough money for a down payment on a condo. They figure they will need $18,000 for the down payment. How long will it take for the original $15,000 to gather enough interest to give them a total of $18,000, assuming they earn 2.3% interest per year, paid quarterly?

(18) Half-life. One of the major concerns about above-ground nuclear testing was that it produced a radioactive element called strontium-90. The fallout from a nuclear test would be deposited on grass, which would be eaten by cows, and the strontium-90 would get into the milk the cows produced. Scientists have determined that the half-life of strontium-90 is 28.9 years.[12] Suppose that in one agricultural area a single above-ground test causes the level of strontium-90 to be 10 times greater than the maximum safe level. Use a geometric growth model to figure out how long it will take before the level of strontium-90 is again at a safe level. For simplicity, let the safe level be the base unit for measurement. Then after the test, the level of strontium-90 is 10 and the safe level you want to reach is 1 or below. Create a geometric growth model with L_n equal to the strontium-90 level after n periods of 28.9 years. Use your model to determine how long it will take for the strontium-90 to get down to a level of 1. [Hint: Use a graphical method to get a rough idea of the answer, then use a numerical method with the functional equation to get a more accurate answer.]

(19) @Smartphone Data Traffic with a Trial-and-Error Approach. The table below shows the average monthly global data traffic for smart phones for each quarter from the first quarter of 2010 ($n = 0$) through the third quarter of 2014 ($n = 18$) [20]. The traffic figures are in units of *petabytes*; each petabyte is one million

---

[12]Various values have been published for the half-life of strontium-90. The 28.9 years figure is from the Table of the Isotopes, pages 11-2 to 11-174 in [11].

4.2. Exercises

gigabytes. Thus, the table tells us that in the first quarter of 2010, 160 petabytes of data were up- or down-loaded each month by smart phones world wide.

n	data traffic	n	data traffic	n	data traffic	n	data traffic
0	160	5	380	10	1005	15	2070
1	200	6	465	11	1230	16	2340
2	240	7	625	12	1460	17	2640
3	300	8	745	13	1670	18	2880
4	325	9	885	14	1820		

a. @Graph the data.

b. @Use a graphing application and a trial-and-error approach to develop a geometric growth model for this data set. Begin by defining a variable to represent the data traffic. Next you must choose values for the initial value and the geometric growth factor. It will be helpful to calculate the growth factors for consecutive terms of the data in order to find a starting value for r. Similarly the data provide a good starting value for the initial term. Try several values for the parameters and choose the model which best fits the data.

c. @Use your model to predict the smart phone data traffic for the year 2020. How reliable do you think your prediction is? Explain.

(20) **Engine Temperature.** When a car is running, the engine temperature remains fairly constant at about 300°F. Once the car is parked, the engine will cool off. A detective wants to use this information to predict how long it has been since a suspect's car was last driven. When the detective got to the suspect's house, at 2 p.m., she checked the temperature of the radiator fluid, and found it to be 150 degrees. At 3 p.m., the temperature was measured again, this time it was 90 degrees. The outside temperature remained constant at about 70 degrees all afternoon. Follow the steps below to create a geometric growth model with n representing time in hours and with H_n representing how much hotter the radiator fluid is than the outside temperature after n hours. Then estimate how long the car had been parked when the detective made the measurement at 2 p.m.

a. Let H_n be the difference between the car temperature and the outside air temperature n hours after 2 p.m. What is H_0? What is H_1?

b. Assume that the terms of H_n follow a geometric growth law. What is the growth factor from H_0 to H_1?

c. What is the difference equation for H_n?

d. What is the functional equation that gives H_n as a function of n?

e. When the car was parked, the temperature of the engine was around 300 degrees. What was H at that time?

f. Assume your functional equation from part d is valid for fractional and negative values of n. Set H_n equal to your answer from part e, and find n using a graphical or numerical approach. [Hint: Expect the answer to be a negative number because we are interested in something that happened *before* 2 p.m.]

g. Based on your model, how long had the car been parked when the detective measured the temperature at 2 p.m.? How reliable is this estimate? Explain.

(21) @Temperature Probe Data [13] with a Trial-and-Error Approach. In an experiment, a temperature probe was taken from a cup of hot coffee; we assume that it was at the same temperature as the coffee. Then the probe was placed in a cold water bath at a temperature of 10°C.

Temperatures from the probe were collected as detailed in the following table, in which times are in seconds, measured from the instant when the probe was put into the cold water, and temperatures are in degrees C.

Time (seconds)	5	10	15	20	25	30
Temperature (°C)	69.39	49.66	35.26	28.15	23.56	20.62

a. @Develop a geometric growth model for this situation by following the steps below. Assume the water in which the probe is immersed remains at 10°C, and use the variable H to define how much hotter the probe is than the water. Let n be the number of 5-second intervals the probe has been in the cold water. Make a new table of n and H_n values. Notice that H_0 is not given. Use a graphing application and a trial-and-error approach to find parameters H_0 and r so that your model sequence is as close an approximation as possible to the n, H_n data.[14]

b. @Provide a graph showing both the curve from your model as well as the (n, H_n) data points.

c. @According to your model, how hot was the probe when it was first put into the cold water?

d. @Using the same original data, create a new revised model by assuming that the cold water is at a temperature of 15°C instead of 10°C. To differentiate the models, let R be how much hotter the probe is than 15°C water. Do the same analysis for this revised model as for the original model. That is, create a table for n and R_n, give the equation you find for the model, provide a graph showing both the curve from your model as well as the original data points.

e. @Compare the models H_n and R_n. Do they have the same initial value? Does one appear to be a better fit? Explain.

f. @The next parts of this problem involve modifying your model so that it is expressed in terms of time and temperature instead of n and H. The first change will be from n, the number of 5-second intervals, to t, time in seconds. Write an equation relating n and t and use it with the equation from part a to write a geometric growth model for H in terms of t.

g. @Create a model for the temperature of the probe, T, rather than how much hotter the probe is than the water, H. To do this, write an equation relating T and H then combine it with the model from the previous part to make a model of T in terms of t. Note: this new model is not an instance of geometric growth. For example, defining T_n to be the probe temperature after n 5-second intervals, we do not observe nearly constant growth factors. This problem

---

[13] Adapted from [25, Exercise 33, p. 362].
[14] Hint: Computing H values for the data, you can find growth factors between consecutive terms. They are between 0.5 and 0.7, so $r = 0.6$ might be a reasonable first guess. Then, with $H_1 = 59.39$, we should have $H_0 = H_1/r = 59.39/0.6$ which is approximately 99. These parameter values can be used as a starting point. You can look for better parameter values by a trial-and-error process.

4.2. Exercises

shows the methods of analyzing geometric growth can sometimes be extended to other sorts of growth.

(22) **Drug Metabolization Part One.** When a drug is introduced into the blood stream, the body has mechanisms to eliminate it. For example, some drugs are removed by the kidneys. How does the amount of the drug in the blood diminish over time? A commonly used model for this process assumes that a fixed percentage of the drug is removed every so many hours. This can be approached as an extension of the water tank model. We imagine that in each period of time, some fraction of the blood is purified, just as in the water tank model, each time we empty and refill the tank we purify a fixed fraction of the water. For example, in the case of aspirin, about half is removed from the blood every half hour. Suppose a patient takes two aspirin tablets, for a total of 650 milligrams (mg) of aspirin. Let a_n be the amount of aspirin in his or her blood after n half-hour periods. Then the difference equation and initial condition for a_n are

$$a_{n+1} = 0.5a_n; \qquad a_0 = 650.$$

a. How much aspirin remains in the blood after 4 hours?

b. Find the functional equation for a_n corresponding to the difference equation above.

c. Find a functional equation that gives the amount of aspirin in the blood after t hours. [Hint: find an equation relating n and t, and use it to replace the n in the functional equation with a function of t.]

(23) **@Drug Metabolization Part Two.** This problem is closely related to the preceding problem. At the Olympics, drug tests are used to check for performance enhancing drugs. These tests have limited sensitivity; there must be some minimum amount of the drug in the blood in order for the test to detect it. Suppose that a test can detect steroids in amounts of 1 mg or more. Suppose also that the body removes about 1/4 of the steroids in the blood every 4 hours. And finally, suppose that an athlete takes a 160 mg dose.

a. @How long will it be before the drug is reduced to an undetectable level in the blood? That is, if the athlete takes the blood test immediately after taking the drug, there will be 160 mg in the blood, and the test will detect that. But the longer the delay between taking the drug and the blood test, the less of the drug will remain in the blood. Eventually there will be less than 1 mg of the drug left in the blood, and the test will not be able to detect that. The question is, how long will that take?

b. @Suppose we can make the blood test 100 times more effective, so that it can detect steroids in the amount of 0.01 mg or more. Then how long after taking the dose would the test be effective?

Digging Deeper.

(24) **Drug Metabolization Part Three.** This problem is also closely related to the two previous problems. As you are aware, it is very common to take medicine on a regular basis. For example, you may take two aspirin every four hours. The model for the way drugs are removed from the body can be modified to include the effect

of taking additional doses. Suppose that 1/4 of a drug is removed from the blood every four hours. An initial dose of 100 mg is taken. Four hours later, one quarter of the drug has been eliminated, leaving 75 mg. At that point another dose of 100 mg is added, giving a total of 175 mg in the blood. After four more hours, one quarter of that amount is removed, and an additional 100 mg is added. This leads to the following difference equation and initial condition:

$$d_{n+1} = 0.75 d_n + 100; \qquad d_1 = 100,$$

where d_n is the amount of drug in the body immediately after taking the nth dose. What will be the long-term effect of taking repeated doses? Will the drug level keep going up until it reaches an unsafe level? Use a numerical method to explore the behavior of d_n in this model. What happens if each dose is 200 mg rather than 100 mg? What if each dose is 50 mg? If the doctor would like to keep the amount of drug in the blood at about 150 mg, how much medicine should be given at each dose?

(25) @Apian Genealogy. In the bee ancestor example, we found that the sequence a_n grows approximately geometrically and constructed a model with the equation $b_n = 1.171(1.618)^n$ for the number of ancestors a queen bee has in the nth previous generation. This leads to further questions.

  a. @About how many ancestors does a queen bee have in the 100th preceding generation? That is, find b_{100}.
  b. @Assuming about 4 bees can stand together in one square inch, how many square miles would b_{100} bees cover, if they were all arranged on a flat surface?
  c. @The preceding computation seems to indicate that according to the model there would be an impossibly large number of bees 100 generations prior to the queen we are considering. And yet, the biology of bee reproduction is correctly portrayed in the model, and bees have certainly been in existence for more than 100 generations. How can this apparent contradiction be resolved?

(26) Fibonacci Numbers. We saw that the Fibonacci numbers are approximately geometric, and that the approximation seems to get more and more accurate as n increases. Complete the outline below to investigate these ideas further. Throughout, the Fibonacci numbers make up the sequence with initial terms $F_0 = 0, F_1 = 1$, and for which each term after F_1 is the sum of the two preceding terms. Thus $F_2 = 0 + 1 = 1, F_3 = 1 + 1 = 2, F_4 = 1 + 2 = 3$, and so on.

  a. Suppose that F_n is eventually almost exactly given by a geometric growth sequence with an unknown growth factor r. Rather than finding the common ratio of terms, we will construct an equation we can solve in terms of r. As a starting point we will choose F_{100}. To simplify the notation, let $F_{100} = c$. Find F_{101} and F_{102} in terms of r and c.
  b. Use the results of part a and the definition of Fibonacci numbers to construct an equation in terms of r and c. In part a the equations involved F_{101} and F_{102}. This new equation involves only r and c.
  c. Did our choice to start with $n = 100$ affect the equation found in part b? To get an idea of the answer to this question, repeat steps a and b beginning with the assumption that $F_{1000} = c$. Show you get the same equation for r.

d. Solve your equation for r to find two possible growth factors for the geometric progression approximating the Fibonacci numbers.

e. Geometric Model. The larger r value found at the preceding step is $r = \frac{1+\sqrt{5}}{2}$. With this value of r and an initial term of $b_0 = \sqrt{0.2}$, define the sequence b_n. Use a numerical method to show that this sequence approximates the Fibonacci sequence very closely for at least 25 terms. Note: this step can be completed most conveniently using a computer spreadsheet to generate the values of b_n, the corresponding Fibonacci numbers, and the errors. Enter r as "(1 + sqrt(5))/2" and b_0 as "sqrt(0.2)" so that these values are approximated to the full accuracy of the spreadsheet program.

f. Continuing the prior step, show that the errors form a sequence that is approximately geometric, with growth factors very close to $r = \frac{1-\sqrt{5}}{2}$, the second possible r value from part d.

g. In the prior step, the sequence of errors $b_n - F_n$ was found to be approximated by a geometric growth sequence. In fact, the errors are closely approximated by the sequence $e_n = \sqrt{0.2} \cdot \frac{1-\sqrt{5}}{2}$. Since we know that $b_n - F_n$ is very close to e_n, it follows that $b_n - e_n$ should be very close to F_n. Verify this numerically. That is, show that $b_n - e_n$ is a much better approximation to F_n than b_n alone. This illustrates how analyzing the errors in a model can sometimes lead to a more accurate model.

4.3 Exponential Functions

The standard functional equation for a geometric growth sequence has the form $a_n = a_0 r^n$. We have seen several examples in the preceding section. In one version of the mouse population example we found

$$p_n = 100 \cdot 2^n,$$

with 100 mice initially and where p_n represents the number of mice after n months. This equation expresses p_n as a function of n, the independent variable, because the value of p_n can be computed for any n by evaluating the expression on the right side of the equation.

Similarly, in the first tank model, we found

$$s_n = 4 \cdot 0.1^n.$$

For this model, there are initially 4 pounds of salt in a tank of water, and s_n represents the number of pounds of salt after draining and refilling the tank n times. Here, too, the equation expresses one variable, s_n, as function of the other, n.

The functions in these examples, and in all geometric growth functional equations, are called *exponential functions,* because the independent variable n appears in the exponent. Although they arise in geometric growth sequences, their applications extend to other types of models as well. In this section we will discuss many of the properties of exponential functions, paying special attention to the following topics:

- Terminology and Notation
- Graphs

- Algebraic Properties
- Solving Equations
- The Number e

Along the way we will introduce the idea of a logarithm.

Terminology and Notation. As you may know, in the expression

$$2^n,$$

n is referred to as an exponent; 2 is called the *base*. In general, in an exponential function, the variable appears in the exponent attached to a constant base. Although the mouse population example defines n as a whole number of months, the functional equation remains valid when n is not a whole number. Thus, the model predicts that the population 1.7 months after the start of the study will be $100(2^{1.7})$. As we saw in Section 4.2, calculator programming is consistent with a strong geometric growth assumption, predicting equal growth factors in equal periods of time. This allows us to think of time as a continuous variable, taking on all possible fractional values. As in previous chapters, we follow the notational convention of changing n to t and writing $p(t)$ instead of p_n. These notations emphasize the idea that the time variable is not restricted to whole numbers and may take on fractional values. Accordingly, the functional equation becomes

$$p(t) = 100 \cdot 2^t$$

which we think of as defining p as an exponential function of the continuous variable t.

Although we have already looked at several examples, before going further we should define exactly what we mean by an exponential function.

> **Definition:** An exponential function $p(t)$ is one that can be expressed in an equation of the form
>
> $$p(t) = \text{(a constant)} \cdot \text{(a positive constant)}^t. \qquad (4.2)$$
>
> We call this the *standard form* for an exponential function. Any equation that can be expressed in the form
>
> $$y = \text{(a constant)} \cdot \text{(a positive constant)}^t$$
>
> is an exponential equation.

The examples we have already considered conform to this definition, and so are definitely exponential functions. But equations that are not in the standard form might still represent exponential functions, if they can be algebraically rearranged into the standard form. This is the meaning of the phrase *can be expressed in* from the definition. Indeed, we have seen something similar with linear functions and with quadratic functions.

As a specific example, we mention the apple pie model (page 234) where we first found a functional equation

$$H_n = 280 \left(\frac{23}{28}\right)^n.$$

4.3. Exponential Functions

Recall that H_n tells how much hotter the pie is than its surroundings after n ten-minute intervals. For more convenient application, we introduced t as the time in minutes, finding $t = 10n$. This can be combined with the functional equation to obtain

$$H = 280 \left(\frac{23}{28}\right)^{0.1t}. \tag{4.3}$$

Note that this is not in the standard form for an exponential equation, because the exponent includes a coefficient of 0.1. Nevertheless, the equation does define H as an exponential function of t, because we can rewrite the equation in the standard form as

$$H = 280 \left[\left(\frac{23}{28}\right)^{0.1}\right]^t,$$

treating everything enclosed in the square brackets as a single constant. Here we have used a rule of exponents, about which we will say more shortly.

Continuing our review of the apple pie model, recall that the pie is cooling in an environment at 70°F. Therefore, if the actual temperature of the pie is T, we know that $H = T - 70$. Substituting this into (4.3) leads to

$$T = 70 + 280 \left(\frac{23}{28}\right)^{0.1t},$$

expressing pie temperature T as a function of time t. But this is *not* an exponential function, because it cannot be expressed in the standard form. Instead, we can describe T as a constant plus an exponential function. Such functions and the models in which they arise will be discussed in Chapter 5.

Parameters. In the standard exponential equation, we can represent the constants as parameters thus:

$$p(t) = Ab^t. \tag{4.4}$$

Often we introduce a variable y, writing

$$y = Ab^t.$$

This shows that each parameter has a specific interpretation. The parameter A is the initial value of y. That is, $y = A$ when $t = 0$. The parameter b is the growth factor for each unit of t. Thus, if y starts at 500 (for $t = 0$) and increases by a growth factor of 1.05 for each increase of 1 in t, we can immediately write the equation

$$y = 500 \cdot 1.05^t.$$

This is consistent with our findings for geometric growth sequences.

The definition of exponential functions requires $b > 0$, because b represents the positive constant in (4.2). This is consistent with the convention we adopted for geometric growth sequences. However, for a continuous independent variable (t in the definition), this restriction is more than just a convention—it is a mathematical necessity. To illustrate the difficulty, with $A = 1$ and $b = -5$ the standard equation defines the function $f(t) = (-5)^t$. How should we compute $y = f(1/2)$? Based on the strong geometric growth assumption, the growth factor from $f(0)$ to $f(1/2)$ is the same as the growth factor from $f(1/2)$ to $f(1)$. We know $f(0) = 1$, $f(1/2) = y$, and $f(1) = -5$, so the growth factor from 1 to y must equal the growth factor from y to -5. This leads to the equation $y^2 = -5$, for which there is no solution in our usual number system. To avoid such difficulties, we only define exponential functions for $b > 0$.

Graphs. In Section 4.1, graphs were shown for geometric growth sequences. For example, look back at Figure 4.1 on page 205. The functional equation for this graph is $a_n = 16(1.5^n)$, and data points are shown for n equal to each of the whole numbers from 0 to 8, with straight lines connecting the data points. Now we want to reexamine graphs of this type using fractional and negative values for the variable. In fact, we want to imagine plotting points on the graph for fractional values of n that are so close together that they appear to form a continuous curve. We will use the conventional variables x for the horizontal axis, and y for the vertical axis. In the functional equation, the n will be changed to x and the a_n will be changed to y, resulting in $y = 16(1.5^x)$. Observe that this matches (4.4), with parameters $A = 16$ and $b = 1.5$, although the independent variable is x rather than t.

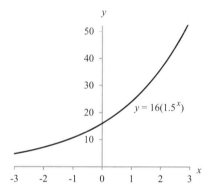

Figure 4.11. $y = 16(1.5^x)$.

A graph for the equation $y = 16(1.5^x)$ is shown in Figure 4.11. If you compare this figure with Figure 4.1, you will see that the two graphs are very much alike. One difference is that in Figure 4.1, it is possible to make out corners between the short straight lines joining the plotted points, whereas in Figure 4.11, the graph is a smooth curve. This is the result of computing $y = 16(1.5^x)$ with fractional values for the variable x, which we saw in Section 4.2 is consistent with the strong geometric growth assumption. However, that earlier discussion only considered $x \geq 0$. But 1.5^x can be computed for all negative values of x as well. The general rule for a negative exponent can be conveyed with an example: we define $1.5^{-2.7}$ to be $1/1.5^{+2.7}$. This too is consistent with (and derivable from) the strong geometric growth assumption. And because 1.5^x can be computed for any value of x, positive, negative, or zero, the graph in Figure 4.11 can be extended as far as wish to the left and to the right of the y-axis.

In the next few paragraphs we will look at general characteristics of the graph of any exponential function $y = Ab^x$. We will also see how to predict from the values of A and b what the general nature of a graph will be. What we find will be compatible with our earlier observations about the graphs of geometric growth sequences.

As with geometric growth sequences, the graphs of exponential functions can be categorized for positive b either greater than 1 or less than 1, and for A either positive or negative. For the case $b > 1$ and $A > 0$, any exponential function has the same general shape shown in Figure 4.11: curving steeply upward as it is traced to the right, and leveling off along the x-axis as it is traced to the left. In fact, as x is taken further and further to the left of 0, the values of $y = Ab^x$ get closer and closer to 0, without

4.3. Exponential Functions

ever reaching it. This is described by saying that the negative side of the x-axis is an *asymptote* of the function, or that y asymptotically approaches 0 as x decreases toward negative infinity.

Just as all exponential function graphs have the same basic shape when $b > 1$ and $A > 0$, so too the other cases each share a common general shape. Here the situation is exactly analogous to what we observed for geometric growth sequence graphs: changing the sign of A flips the graph vertically (that is, across the x-axis), while replacing the base b with $1/b$ flips the graph horizontally (across the y-axis). Thus, by reflecting the general graph for the $b > 1$ and $A > 0$ case in various ways, we can obtain the graphs for all the other cases. This is shown in Figure 4.12.

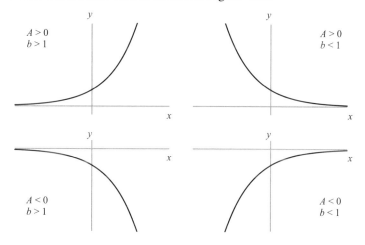

Figure 4.12. Possible shapes for the graph of an exponential function $y = Ab^x$ with $b > 0$ and $b \neq 1$.

Notice that none of the variations cross the x-axis. This is another important feature of all exponential functions: they have no x-intercept. Thus either the entire graph remains above the x-axis, or the entire graph remains below the x-axis. As the figure indicates, the four different variations depend on the values of A and b in the equation $y = Ab^x$. If A is positive, the entire graph must stay above the x-axis. It will approach the x-axis to the left and curve steeply up on the right if $b > 1$, or curve steeply up on the left and approach the x-axis on the right if $b < 1$. When $A < 0$, the entire graph must stay below the x-axis. Then, for $b > 1$ the curve is nearly horizontal on the left and curves steeply down at the right, while for $b < 1$ the situation is reversed. Exponential functions with $A > 0$ are the most common in applications. All of those functions have graphs with the same general shape as one of the first two in Figure 4.12.

The value of A is the y-intercept of the equation $y = Ab^x$. As an example of this, the graph of $y = 100(1.5)^x$ has a y-intercept of 100. In terms of the geometric growth models presented in the previous section, this corresponds to the fact that A is the starting value for the model. Algebraically, we know that the y-intercept can be found by setting $x = 0$. In the example equation, this leads to $y = 100(1.5^0)$. But 1.5^0 is 1 (any positive base raised to the zero power is 1), so $y = 100$. This confirms algebraically that the parameter A is the y-intercept. Usually A is positive. In this case the y-intercept is above the x-axis, and the entire curve must stay above the x-axis. If A is negative, the y-intercept is below the x-axis, as is the entire curve.

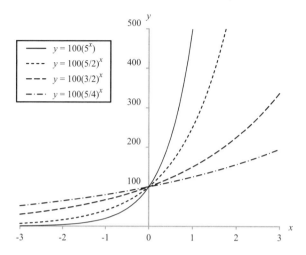

Figure 4.13. Graphs for $b > 1$.

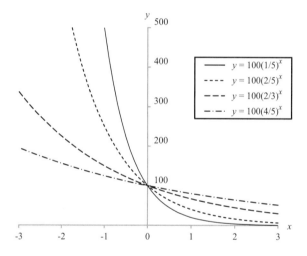

Figure 4.14. Graphs for $b < 1$.

As a final observation about the graphs of exponential functions, the value of b controls how steeply the curve rises or falls. This is what we saw in the discussion of geometric growth sequences, where the base b is the same as the growth factor. Comparisons of graphs for a few different values of b appear in Figure 4.13 and Figure 4.14. In all the curves illustrated, the value of A is 100, so they all have a y-intercept of 100. The b values in Figure 4.13 are $5, 5/2, 3/2$, and $5/4$. Writing these in the form of decimals, $5.0, 2.5, 1.5$, and 1.25, shows that they appear in order, greatest to least. The b values in Figure 4.14 are the reciprocals of those in Figure 4.13. Thus, for example, where $b = 3/2$ in Figure 4.13 the corresponding curve has $b = 2/3$ in Figure 4.14. There are two main lessons to be learned from the graphs. First, in each graph, there is a consistent progression as the b values move further and further from 1. Once you observe that progression, you should be able to make a reasonable guess about the shapes

4.3. Exponential Functions

of the curves $y = 100(6^x)$ (considering Figure 4.13) and $y = (1/6)^x$ (considering Figure 4.14). Second, the graph for one value of b is the mirror image of the corresponding graph with the reciprocal value of b.

This completes the discussion of graphs of exponential functions.

Algebraic Properties. In the equation

$$p(t) = 1{,}000(1.5)^t$$

the symbols on the right side of the equal sign provide a recipe for computing population. For example, if we want to know the population 3 months after the start of the study, the recipe says to raise 1.5 to the third power and then multiply the result by 1,000. There are other possible forms for the recipe. One example is $1{,}500(2.25)^{0.5(t-1)}$. Although this looks quite different from $1{,}000(1.5)^t$, it will produce the same result for any value of t. You can see some evidence of this by using a calculator to compute both $1{,}000(1.5)^t$ and $1{,}500(2.25)^{0.5(t-1)}$ for several values of t. Some representative results appear in Table 4.5.

Table 4.5. Two versions of an exponential function. Decimal figures have been rounded to three decimal places.

t	0	1	1.5	3	10	-2
$1{,}000(1.5)^t$	1,000	1,500	1,837.117	3,375	57,665.039	444.444
$1{,}500(2.25)^{0.5(t-1)}$	1,000	1,500	1,837.117	3,375	57,665.039	444.444

Our discussion of algebraic properties of exponential functions will explain how to recognize when one form of an exponential function can be replaced by an equivalent form. As we shall see, there are situations in which one form is easier to use than another. The relationship between different forms will be important both in solving equations as well as in understanding features of a geometric growth model.

Notationally, the following three formats may be used interchangably:

$$1{,}000(1.5)^t = 1{,}000(1.5^t) = 1{,}000 \cdot 1.5^t.$$

The parentheses in the first two versions and the raised dot in the third all serve a common purpose, to indicate a multiplication between the coefficient 1,000 and the expression 1.5^t. But parentheses can also indicate an order of operations. Thus in the second formula, the operation of raising 1.5 to the t power is to be computed before any other operation. In contrast, for the first formula, the parentheses do not contain any operations, so have no impact on the order.

In general, unless dictated otherwise by parentheses, exponents are applied before the other operations of arithmetic. For example, ab^2 is interpreted to mean $a \cdot bb$. Taking $a = 3$, $3b^2 = 3 \cdot b \cdot b$, and if $b = 5$, $3b^2 = 3 \cdot 5 \cdot 5 = 75$. But suppose we want the multiplication to be applied before the exponent. Then we write $(ab)^2$. This time the multiplication is applied first, because it is within parentheses. That means $(ab)^2 = ab \cdot ab$. Again using 3 for a, $(3b)^2 = 3b \cdot 3b = 9b^2$. With $b = 5$, $(3b)^2 = 15^2 = 225$. Notice that the order convention in the absence of parentheses is assumed in the general exponential function equation $y = Ab^t$. Given $A = 1{,}000$, $b = 1.5$, and $t = 2$, we know we must find y by first computing $1.5^2 = 2.25$ and then multiplying by 1,000, not by first multiplying $1{,}000 \cdot 1.5 = 1{,}500$ and then squaring the result.

In addition to these considerations, the algebra of exponential functions depends on what are usually referred to as rules or laws of exponents. As an example, let us consider what will result from multiplying 2^5 by 2^8. Remember that 2^5 is really $2 \cdot 2 \cdot 2 \cdot 2 \cdot 2$. Similarly, $2^8 = 2\cdot2\cdot2\cdot2\cdot2\cdot2\cdot2\cdot2$. Therefore, $2^5 \cdot 2^8 = 2\cdot2\cdot2\cdot2\cdot2\cdot2\cdot2\cdot2\cdot2\cdot2\cdot2\cdot2\cdot2$, or 2^{13}. To express this in words, when we multiply together a group of five twos with another group of eight twos, the result is the same as multiplying together a group of 13 twos. In symbols, this can be expressed in the form of the equation $2^5 \cdot 2^8 = 2^{5+8}$. The same kind of reasoning can be applied for any base (in place of 2) and for any exponents (in place of 5 and 8). The result, expressed in the equation

$$a^n a^m = a^{n+m}$$

is our first rule of exponents. It can be used with exponential functions as in the following example. In the expression $1{,}500(1.5)^t$ rewrite 1,500 as $1{,}000 \cdot 1.5$. That gives us the alternate version $1{,}000 \cdot 1.5 \cdot 1.5^t$. Now apply the first rule of exponents to show that $1.5 \cdot 1.5^t = 1.5^{t+1}$ (using the fact that $1.5 = 1.5^1$). This leads to $1{,}500(1.5)^t = 1{,}000(1.5)^{t+1}$.

Here is another example of a rule of exponents. Raise 2 to the third power. Now take the result and raise that to the fifth power. What is the final result? In symbols, we wish to consider $(2^3)^5$. Now $2^3 = 2 \cdot 2 \cdot 2$. When we raise that to the fifth power, we multiply together 5 identical copies of the same thing. That gives $(2 \cdot 2 \cdot 2)(2 \cdot 2 \cdot 2)(2 \cdot 2 \cdot 2)(2 \cdot 2 \cdot 2)(2 \cdot 2 \cdot 2)$. This is clearly the same as multiplying together 15 twos. That is, if we multiply together 5 groups of 3 twos each, the result is 15 twos, all multiplied. In symbols, $(2^3)^5 = 2^{3 \cdot 5}$. Again, the same reasoning applies for any base (not just 2) and any exponents (not just 3 and 5), so we have a second rule of exponents:

$$(a^m)^n = a^{m \cdot n}.$$

The first two rules of exponents were explained using examples in which the exponents were whole numbers. But the same rules apply to any exponent, be it a whole number, fraction, or decimal, positive, or negative. You can test this on your calculator. Is $2^{1.6} \cdot 2^{3.2} = 2^{4.8}$? Is $(2^{-1.5})^{2.4} = 2^{-3.6}$? What this indicates is that the calculator has been programmed to compute fractional and negative exponents in such a way that the rules of exponents are followed. You might even say that it is the rules of exponents that tells us how to compute exponents that are not whole numbers. And as we have observed, the programming is also consistent with the strong geometric growth assumption. To be more accurate, making the strong geometric growth assumption is equivalent to adopting the rules of exponents for all sorts of exponents, not just whole numbers. In particular, this allows us to assign meaning to exponential expressions in which the exponent is any number. In contrast, the base to which the exponent is applied must be a positive quantity.

So far we have covered two rules of exponents. Using the same methods, you should be able to come up with a third rule: What is the result of dividing 2^8 by 2^3? What rule of exponents does this suggest? Try to answer this before reading further.

The third rule of exponents, suggested by the preceding example, is

$$\frac{a^n}{a^m} = a^{n-m}.$$

This can be verified most easily for n and m whole numbers with $n > m$. But the rule applies for all n and m. A useful variant for the case of whole numbers n and m with

4.3. Exponential Functions

$n < m$ is

$$\frac{a^n}{a^m} = \frac{1}{a^{m-n}}.$$

For example

$$\frac{a^5}{a^8} = \frac{1}{a^3}.$$

There are some special instances of these rules that are also worth mentioning. We have already noted that any positive base raised to the zero power is 1 (page 251). Combining this fact with the third rule shows that

$$\frac{1}{a^m} = \frac{a^0}{a^m} = a^{0-m} = a^{-m}.$$

This shows that $1/a^m$ and a^{-m} can be used interchangeably. As a related case, in discussing the first rule we observed that $1.5 \cdot 1.5^t = 1.5^{t+1}$. This depends on the rule for adding exponents, and the fact that $1.5^1 = 1.5$. Something similar occurs any time a numerical exponent and a variable exponent appear on the same base. For example, $2.1^3 \cdot 2.1^t = 2.1^{t+3}$ and $2.1^t/2.1 = 2.1^{t-1}$.

The rules of exponents provide ways to modify how exponential functions are written. Let us look at three examples.

Example 1. Suppose that we observe a mouse population that initially is 1,000 and that quadruples in six months. Assuming geometric growth, we can derive the equation

$$p(t) = 1{,}000 \cdot 4^{t/6}$$

for the population after t months. Now using the fact that $4 = 2^2$, we can rewrite the equation in the form

$$p(t) = 1{,}000 \cdot (2^2)^{t/6}$$

and using the second rule of exponents, we derive

$$p(t) = 1{,}000 \cdot 2^{2t/6}.$$

In this case, a rule of exponents allows us to change an exponential equation in which the base is 4 into one in which the base is 2. This provides a glimpse of a much more general phenomenon. Any base can be transformed to any other base. This means that, if we wish, we can express every exponential function using the base 10. We will go into this idea in greater depth a little later.

Simplifying the exponent in the last equation above leads to

$$p(t) = 1{,}000 \cdot 2^{t/3}.$$

We can interpret this to mean that the population starts at 1000 and doubles every 3 months, whereas the original equation says that the population starts at 1000 and quadruples every 6 months. Thus, the two equations not only have different (but equivalent) algebraic forms, they also have different (but equivalent) interpretations.

Example 2. We will start again with the equation

$$p(t) = 1{,}000 \cdot 4^{t/6}.$$

This time, write the exponent as $\frac{1}{6} \cdot t$, and use the second rule of exponents again. We then have

$$p(t) = 1{,}000 \cdot (4^{1/6})^t.$$

According to a calculator, $4^{1/6} = 1.26$ to two decimal places, so our equation can be expressed approximately as

$$P(t) = 1{,}000 \cdot 1.26^t. \tag{4.5}$$

This is in the standard form for an exponential function and shows that the population grows by a factor of 1.26 each month, or that it increases by 26% each month.

However, we should add a word of caution about decimal approximations. Exponential functions are very sensitive to small changes in their parameters. Often, a two decimal place approximation will not give very accurate results. For this specific example, $4^{1/6} = 1.25992105\cdots$, and the correct three decimal approximation would be 1.260. Thus 1.26 is actually accurate to three decimal places, and nearly accurate to four. Its use will produce more accurate results than usual for a two decimal approximation. As a rule of thumb, we recommend using 6 decimal place approximations for the constants that appear in an exponential equation. Following this rule, (4.5) would become

$$P(t) = 1{,}000 \cdot 1.259921^t.$$

We can still say that the population increases by approximately 26% each month, but for computation, the more accurate equation should be used.

Example 3. As a final example, we will convert the equation into an exponential form with the base 10. There is one bit of information that is needed for this process: $10^{0.60206}$ equals 4, to a very close approximation.[15] In the original equation

$$p(t) = 1{,}000 \cdot 4^{t/6}$$

replace the 4 with $(10^{0.60206})$. That gives

$$p(t) = 1{,}000 \cdot (10^{0.60206})^{t/6}.$$

Using the second rule of exponents again, the equation can be re-expressed as

$$p(t) = 1{,}000 \cdot 10^{0.60206 t/6}$$

or, simplifying the exponent,

$$p(t) = 1{,}000 \cdot 10^{0.100343 t}.$$

In this way we have expressed our function as a base-10 exponential function. That might be of interest because we use a base 10 number system, or because most scientific calculators have a built-in function for 10^x. But more importantly, this illustrates an important idea: it is possible to express an exponential function with any (positive) base we desire. In particular, it raises the possibility of adopting some standard base for exponential functions.

Summary of the Examples. The examples showed that the following equations are all equivalent:

$$p(t) = 1{,}000 \cdot 4^{t/6}$$
$$p(t) = 1{,}000 \cdot 2^{t/3}$$
$$p(t) = 1{,}000 \cdot 1.259921^t$$
$$p(t) = 1{,}000 \cdot 10^{0.100343 t},$$

---

[15] As you can verify using a calculator, $10^{0.60206} = 4.00000008\cdots$. This equals 4 when rounded to 6 decimal places, so conforms to the 6 decimal place rule of thumb. Presently we will see how this kind of information is obtained using logarithms.

4.3. Exponential Functions

although the last two are really only approximately equivalent to the first equation. We also saw how rules of exponents can be used to transform one version of the equation into another. These algebraic methods will be particularly important in solving exponential equations. Before proceeding to that topic, several exponent rules we have used are repeated below, for easy reference.

> **Rules of Exponents:** The following equations are valid for any positive base b and for any exponents r and s:
> $$b^0 = 1 \qquad\qquad b^r b^s = b^{r+s}$$
> $$b^1 = b \qquad\qquad (b^r)^s = b^{rs}$$
> $$1/b^s = b^{-s} \qquad\quad b^r/b^s = b^{r-s}$$

Solving Equations. An exponential equation is one for which a variable occurs in an exponent. The simplest form of such an equation has a single exponential expression on one side of the equal sign, and a number on the other. A typical example is

$$10^x = 2. \tag{4.6}$$

As usual, we can approach a problem of this type graphically and numerically. In Figure 4.15 the graph of the equation $y = 10^x$ is shown. To solve the equation $10^x = 2$ graphically, we need to find a point on the curve where the y-coordinate equals 2. A horizontal line is shown in the figure crossing the y-axis at 2. The point we seek is where the horizontal line and the curve meet. Carefully drawing a vertical line in the figure from the intersection point down to the x-axis reveals the x-coordinate to be very near 0.3. That is the value of x we want, where $10^x = 2$.

Now let's switch to a numerical method. Using a calculator, you will find that $10^{0.3}$ is a little less than 2. On the other hand, $10^{0.31}$ is a bit more than 2. So the solution to the equation $10^x = 2$ must be somewhere between 0.30 and 0.31. It particular, the decimal expression for x must start out as 0.30.

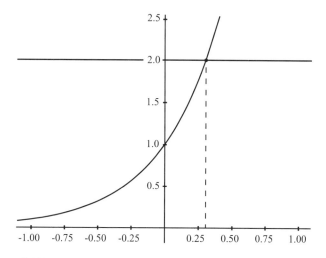

Figure 4.15. Graphical solution of $10^x = 2$. The curve is the graph of $y = 10^x$. The horizontal line is $y = 2$. At the point of intersection, $10^x = 2$.

Proceeding, we can systematically compute $10^{0.301}$, $10^{0.302}$, $10^{0.303}$, and so on, looking for the first result that is greater than 2. In fact, we find it right away. That is $10^{0.301} < 2$ while $10^{0.302} > 2$, showing that the decimal expression for x must begin 0.301.

By using systematic trial and error in this fashion, you can eventually determine the solution to the equation to as many decimal places as the calculator can display. This is the best that can be hoped for. There is no exact solution that can be written down in a finite number of decimal places.

Exponential equations are inevitably encountered in geometric growth models. This is simply another instance of the inversion of a functional equation. For example, one of our first cases of geometric growth modeled a mouse population, deriving the functional equation

$$a_n = 100 \cdot 1.5^n.$$

As we observed then, we can use the equation to answer find-a_n and find-n questions. The first is answered using function evaluation. To illustrate the second, we asked, for what n does $a_n = 1000$? Answering that is equivalent to solving

$$1000 = 100 \cdot 1.5^n$$

for n. This equation does not quite have the same form as (4.6). However, if we divide both sides of the equation by 100, the result is

$$10 = 1.5^n,$$

and exchanging the two sides, we obtain

$$1.5^n = 10.$$

This does have the same form as (4.6), and it can be solved using similar methods. However, those methods are too time-consuming. These equations occur so frequently in applications that we need a streamlined approach.

Logarithmic Functions. The streamlined approach uses what are called *logarithmic* functions. Although there is a different logarithmic function for each base, we will focus on the base-10 logarithm, for now. This is also referred to as a *common* logarithm, and the word *logarithm* is frequently shortened to *log*. On most scientific calculators there is a built-in function for this operation, either as a key labeled *log* or as an item in a menu. The log key automatically calculates accurate approximate solutions to equations of the form $10^x = $ a number. For example, to solve the equation we studied earlier, $10^x = 2$, you must instruct the calculator to apply the log operation to 2. The result should be 0.3010299957. Readers who have such a calculator are encouraged to determine how the log key operates now, and to use the calculator to check the results shown in the examples to follow.

Similar operations can be used to solve an equation with any other number in place of 2. Consider this equation: $10^x = 7$. The solution can be found by applying the log operation to 7. The result should be 0.845098 to the first six decimal places. You can check that this is a good approximate solution to $10^x = 7$ by raising 10 to the 0.845098 power on the calculator.

The log operation is the inverse function for the base-10 exponential function. In the equation $10^x = y$, if x is given and y is to be found, that is *evaluating* the exponential function 10^x. Conversely, if y is given and x is to be found, that is *inverting*

4.3. Exponential Functions

the exponential function. It is also the defining characteristic of the log operation, and we write the solution as $x = \log y$. So, for the equation $10^x = 7$, we express the (exact) solution as $x = \log 7$ and use the calculator to find the numerical approximation 0.845098.

More generally, replacing 7 by the variable r, we define $\log r$ to be the exact value of x that satisfies the equation $10^x = r$. This makes the two equations $\log r = x$ and $10^x = r$ equivalent. Observe that the exponent in the second equation, x, is equal to the logarithm defined by the first equation. For this reason, it is often useful to identify a logarithm as an exponent. Specifically, we can understand $\log r$ as *the exponent that must be applied to* 10 *to obtain r*.

The log operation can be applied to any positive number. It cannot be applied to a negative number, or to 0, because there are no solutions to the equation $10^x = r$ if r is negative or 0. That is easy to verify by reviewing the graphical method used to solve $10^x = 2$ earlier. For this reason, we say that there is no log of a negative number or of 0. The preceding remarks are summarized below for future reference.

> **Definition of Base Ten Logarithm:** For any positive number r, the base ten logarithm of r, denoted $\log r$, is defined to be the solution to the equation $10^x = r$. That is, $\log r = x$ if and only if $10^x = r$. For $r \leq 0$, there is no value defined for $\log r$.
>
> It is also valid to think of $\log r$ as being an exponent. Specifically, it is *the exponent that must be applied to* 10 *to produce a result of r*.

Computation of Logarithms. How can logarithms actually be computed? One method is to use trial and error as in the numerical approach used earlier to solve $10^x = 2$. Alternatively, you could compile a very complete table of values for the function 10^x. The table might start out like this:

x	10^x
0.001	1.0023052
0.002	1.0046158
0.003	1.0069317
$\vdots$	$\vdots$

Now suppose you want to find out when 10^x equals 2. You look down the table until you find an entry in the right-hand column as close to 2 as possible.

x	10^x
$\vdots$	$\vdots$
0.300	1.9952623
0.301	1.9998619
0.302	2.0044720
$\vdots$	$\vdots$

For the data in the table, 0.301 is the exponent that comes closest to giving the desired result of 2. This is essentially a numerical approach using a precomputed table of 10^x

values.[16] For most calculators, logarithms are computed using a combination of table look-up and a numerical approach akin to systematic trial and error.

Solving $b^x = c$. So far, all of this discussion has dealt with a very special case of exponential equations, namely, those with a base of 10. However, the special case can be used to solve any exponential equation through a two-step process. First, we write each of the constants as a power of 10. This is done using logarithms. That will lead to a new equation that is easily solved.

To illustrate, here is how to solve the equation $2^x = 7$. First, express 2 and 7 as powers of 10. You may recall from our earlier discussion that 2 is very closely approximated by $10^{0.30103}$. But even if you had forgotten, you can simply set $2 = 10^r$ and solve for r. The answer is $r = \log 2 = 0.30103$ (correct to 7 decimal places).[17] Similarly, $7 = 10^s$ for $s = \log 7 = 0.845098$, so $7 = 10^{0.845098}$. Now use these results to replace the 2 and the 7 in the original equation:

$$2^x = 7$$
$$(10^{0.30103})^x = 10^{0.845098}$$
$$10^{0.30103x} = 10^{0.845098}.$$

(Notice that we again used the second rule of exponents in the last step.) Now it should be clear that this equation will be satisfied if the exponents are equal, that is, if

$$0.30103x = 0.845098.$$

This leads to the solution $x = 0.845098/0.30103 = 2.807355$. Does this give the answer to the original equation $2^x = 7$? Check using your calculator.

The calculator shows that 2.807355 is close to a solution to the equation $2^x = 7$, but it is not exactly correct. Remember that the two decimal numbers in the fraction are approximations to logarithms. The top number is an approximation to log 7, and the bottom number is an approximation to log 2. If you use more decimal places in this approximation, your answer will be more accurate. For greatest accuracy, express the answer in this form:

$$x = \frac{\log 7}{\log 2},$$

and enter that directly in the calculator.

---

[16] This idea of searching for a solution to an exponential equation in a table is related to the history of the development of logarithms. The Swiss mathematician Burgi, credited as one of the inventors or discoverers of logarithms, compiled tables in essentially this way. His tables were published in 1620. However, compiling tables by this approach requires an incredible amount of calculation, a significant problem prior to the development of computers and calculators! More efficient (and more subtle) methods were developed by the Scot Napier. His tables, first published in 1614, are a part of the first published account of a logarithmic function. Thus, 2014 marked the 400th anniversary of the invention of logarithms. Napier's work was the foundation for tables of logarithms that were essential to scientific computation from his day until the middle of the 20th century. For more information on this subject, see [3].

[17] Usually logs computed on a calculator are not exact values, aside from obvious exceptions such as $\log 100 = 2$ or $\log 0.1 = -1$. However, it is tedious to mention the approximate nature of calculated values at every instance. Accordingly, we will occasionally report logarithms, and results calculated using logarithms, correct to at least 6 decimal places, without specific comment, and will write as if they were exact values. That is, although 7 is only approximately equal to $10^{0.845098}$, we may simply write $7 = 10^{0.845098}$. Answers reported as decimals with six or more decimal places can generally be assumed to be approximations, not exact values.

4.3. Exponential Functions

There is a simple pattern to be observed here. The solution to the equation $2^x = 7$ is $x = \log 7/\log 2$. Can you guess the solution to the equation $3^x = 5$? Following the same pattern, it should be $x = \log 5/\log 3$. Check this on the calculator. In fact, this pattern always holds, and it provides a simple way to solve exponential equations using a calculator with a log button. The pattern is expressed in a general form below.

> **Exponential Equation Solution:** For positive constants b and c, the solution to the equation $b^x = c$ is given by $x = \log c/\log b$.

As an illustration, recall the earlier example where we wished to solve

$$62.5 = 1.5^n.$$

Rewriting this as

$$1.5^n = 62.5$$

and applying the principle above, we immediately find

$$n = \frac{\log 62.5}{\log 1.5} = 10.198576.$$

To check that this is a good approximation to the correct answer, compute $1.5^{10.198576}$ and see how close it is to 62.5.

Here is a slightly more involved example: Find an equation of the form $y = A \cdot 2^{kx}$ given that the graph passes through the points $(0, 100)$ and $(4, 50)$. To find the solution we replace x and y by 0 and 100, respectively, in the equation

$$y = A \cdot 2^{kx}.$$

This results in

$$100 = A \cdot 2^{k \cdot 0} = A \cdot 2^0 = A \cdot 1 = A,$$

and confirms that A is the y-intercept, 100. Thus, our equation becomes

$$y = 100 \cdot 2^{kx}.$$

Now use the second point, replacing x with 4 and y with 50. That gives

$$50 = 100 \cdot 2^{k \cdot 4}.$$

To use the Exponential Equation Solution above, we divide both sides by 100 and exchange the sides of the equation, obtaining

$$2^{4k} = 0.5.$$

This is in the form $b^z = c$, with $b = 2$ and $c = 0.5$, if we think of z as representing $4k$. The solution is $z = \log c/\log b = \log 0.5/\log 2 = -1$. But $z = 4k$ so we have $4k = -1$, which implies that $k = -1/4 = -0.25$. Therefore, our equation must be

$$y = A \cdot 2^{kx} = 100 \cdot 2^{-0.25x}.$$

We can check that this gives the correct values for y when $x = 0$ and when $x = 4$. For example, setting $x = 4$, we find

$$y = 100 \cdot 2^{-0.25 \cdot 4} = 100 \cdot 2^{-1} = 50,$$

as required.

Using a Specified Base for a Given Exponential Function. The preceding discussion shows that a calculator with a log button can be used to solve exponential equations involving any base. As mentioned earlier, this also allows us to express any exponential function with 10 as the base. As an example of this idea, consider again 1.5^x. We can express 1.5 as a power of 10. Specifically, we know $1.5 = 10^r$ for $r = \log 1.5$. Then, since $1.5 = 10^{\log 1.5}$, we have

$$1.5^x = (10^{\log 1.5})^x = 10^{(\log 1.5)x} = 10^{0.176091x}.$$

Multiplying by $a_0 = 1,000$ we find

$$1,000 \cdot 1.5^x = 1,000 \cdot 10^{0.176091x}.$$

This illustrates how an exponential function with base $b = 1.5$ can be reexpressed with 10 as the base. By similar means, we can express *any* exponential function in the form

$$A \cdot 10^{kx},$$

where A and k are numerical constants.

In this regard, there is nothing special about the base 10. We could choose any other number to be our favorite base instead. For example, there are some applications where the most natural base to use is 2. Suppose we decide to express the function 1.5^x using base 2. Arguing as before, we have to express 1.5 as a power of 2, or in other words, to solve $2^r = 1.5$. But we have seen how to solve such equations: $r = \log 1.5 / \log 2 = 0.584963$. This leads to

$$1.5^x = 2^{0.584963x},$$

approximately, or

$$1.5^x = 2^{((\log 1.5)/(\log 2))x},$$

exactly.

Alternatively, using very similar reasoning to what was presented earlier, we could introduce the idea of a base-2 logarithm, defined as follows: $\log_2 x$ is *the exponent you apply to 2 to produce a result of x*. Here, the subscript 2 indicates that we are using base 2. Similarly, we can define a logarithm with respect to any positive base b, indicated as $\log_b x$. When logarithms with respect to several different bases are under discussion, the base ten log is sometimes written $\log_{10}$ for consistency.

Exponential equations can be solved using base-2 logarithms, in perfect analogy with the methods presented earlier for base-10 logarithms. The solution to $2^x = 7$ is simply $\log_2 7$. The solution to $5^x = 7$ is $\log_2 7 / \log_2 5$. The point of these remarks is simply to show that what worked for base 10 would work equally well for any other base.

The Base e. Actually, for most applications, the preferred base is neither 2 nor 10. It is an irrational number that is approximately 2.718281828. Because it is irrational, it cannot be exactly expressed using a finite number of decimal places. So a special symbol has been given to this number, in just the same way that π is used as a symbol for the irrational number $3.14159\cdots$. The symbol is e. On your calculator you should find a key that is marked e^x. That can be used to compute powers of the special base e. In particular, if you apply this operation using 1 as the value of x, the calculator will compute $e^1 = e$ and display a decimal approximation like 2.7182818. Similarly, there is a key marked *ln* that computes the base-e logarithm of any number. This immediately

4.3. Exponential Functions

solves equations of the form $e^x = $ a number. For example, $e^x = 2$ has solution $x = \ln 2 = 0.693147\cdots$. The *ln* label has an *l* for *log* and an *n* for *natural*, and the base-*e* logarithm is often referred to as the *natural log*.[18]

In Section 4.5 more information about *e* will be presented, including some reasons why *e* is the preferred base for so many applications. For now, we focus on some examples showing the use of base-*e* exponential and logarithm functions.

As a first example, we will convert the function $p(t) = 1{,}000(1.5)^t$ to an expression using base *e*. As usual, we begin by expressing the original base, 1.5, as a power of the new base, *e*. We need to find an exponent *r* so that $1.5 = e^r$. By definition, the solution is $\ln 1.5$. The calculator gives this approximately as 0.405465. Then, replacing 1.5 with $e^{0.405465}$, we have

$$p(t) = 1{,}000(e^{0.405465})^t = 1{,}000 e^{0.405465 t}.$$

This is an approximation. We can also leave it in the form

$$p(t) = 1{,}000 e^{(\ln 1.5)t},$$

which holds exactly. In this example, the original equation comes from a geometric growth model, and is naturally expressed in terms of powers of 1.5. Then, when we use the functional equation to describe a continuous model, we convert into a base-*e* exponential. An alternate approach to this kind of modeling leads directly to a base-*e* exponential. We will discuss that in Section 4.4.

Here is a second example of base-*e* calculations. In Section 4.2 we saw that compound interest can be formulated in terms of geometric growth sequences. As a particular instance, suppose that you put $10,000 in a bank account that pays 9 percent per year, compounded monthly. Then each month you will be paid interest in the amount of one twelfth of 9 percent, or $0.09/12 = 0.0075$ expressed as a decimal. That means your account balance will grow by a factor of 1.0075 each month. Letting a_n be the account balance after *n* months, we see that $a_{n+1} = 1.0075 a_n$. This leads to the functional equation $a_n = 10{,}000(1.0075)^n$. Using this equation, we can predict how long it will take to double your money. That is, we would like to find *n* so that $a_n = 20{,}000$. We must solve

$$20{,}000 = 10{,}000(1.0075)^n$$

for *n*. Dividing both sides of the equation by 10,000, we have the equation

$$2 = (1.0075)^n.$$

That can be solved immediately using the methods we discussed earlier, but this time let us use the natural logarithm instead of the base-10 logarithm. The solution is $n = \ln 2 / \ln 1.0075 = 92.766$ (approximately). This is the same solution we would have found using base-10 logarithms. To solve an exponential equation, you can use any base logarithms you please. To emphasize this fact, we repeat the previous boxed comment using the natural logarithm.

> **Exponential Equation Solution:** For positive constants *b* and *c*, the solution to the equation $b^x = c$ is given by $x = \ln c / \ln b$.

---

[18] The base-*e* logarithm can also be denoted $\log_e x$, but the base *e* plays a special role throughout mathematics, so we distinguish its logarithm with a special notation.

To complete the discussion of the problem, we turn again to the solution $n = 92.766$. Remember that this is in units of months. The account will be less than twice the original amount after 92 months, but more than twice the original amount after 93 months. This is a little less than 8 years. In this case, it may not be valid to use a continuous model, because the bank might not pay interest for fractional parts of a month. However, the continuous function analysis gives us a quick way to get the answer of 92.766. It is then necessary to adjust the answer to a whole number of months.

Our final example involves radioactive decay. As we have seen, it is conventional to describe radioactivity in terms of half-life (page 233). For example, the half-life of strontium-90 is 28.9 years. This means that in a sample that contains strontium-90, the amount of strontium-90 will be reduced by half in any 28.9-year period. Using our knowledge of geometric growth, we can immediately write an equation of the form

$$a_n = a_0(0.5)^n \tag{4.7}$$

for the amount of strontium-90 after n 28.9-year periods, where a_0 is the original amount of strontium-90. As in an earlier example, we can introduce the variable t for the time in years corresponding to n 28.9-year periods, and observe that $t = 28.9n$. Dividing by 28.9 leads to $n = t/28.9$. Now substitute this on the right-hand side of (4.7) to find

$$a_0(0.5)^n = a_0(0.5)^{t/28.9}.$$

This expresses the amount of strontium-90 as a function of t. Regarding t as a continuous variable, and therefore using parenthesis notation instead of subscript notation, the function is given by

$$a(t) = a_0(0.5)^{t/28.9}.$$

Now we will change this into an equation that uses e for the base. Following previous examples, observe that $0.5 = e^{\ln 0.5}$. Therefore, $a(t) = a_0(e^{\ln 0.5})^{t/28.9} = a_0 e^{(\ln 0.5/28.9)t}$. This is approximately equal to $a(t) = a_0 e^{-0.023984t}$.

In many areas of application, it has become standard practice to use e as the base for exponential functions. Most scientific calculators include special built-in operations for e^x and ln. In our approach to the subject, we began with geometric growth models, which led in a natural way to exponential functions with various bases. The preceding examples show how to express these exponential functions using the base e. But what is so special about e, and why is it used so much? A complete understanding of this question requires some knowledge of calculus. However, leaving methods of calculus aside, there is still much that can be said about e. We will take up that discussion in Section 4.5. First though, we explore the formulation and use of continuous exponential function models in Section 4.4.

4.3 Exercises

Reading Comprehension.

(1) Explain what is meant by an exponential function. By an exponential equation. How do they differ?

(2) @What can be said about the x- and y-intercepts of the graphs of exponential functions?

4.3. Exercises

(3) Taking $A = 1$ in the standard exponential equation (4.4) leads to $y = b^x$. Describe the graphs for such equations, indicating how the graphs depend on the values of b. Include in your answer an example for which the graph will be rising to the right and an example for which the graph will be falling to the right. Also give a pair of values of b for which the corresponding graphs are left-right mirror images of each other.

(4) Explain why the rules $b^r/b^s = b^{r-s}$ and $1/b^s = b^{-s}$ hold. Include an example of each in your explanation.

(5) @What is meant by *evaluating* a function? What is meant by *inverting* a function?

(6) Explain the concept of the logarithm function. As part of your answer, explain the meaning of log 6. Be sure your answer is consistent with the log definitions given in this chapter.

(7) What is the meaning of $\log_2 7$? How can you approximate the value of $\log_2 7$ using a numerical method?

(8) @Write an explanation of why there is no logarithm of a negative number, nor of 0, using the graph in Figure 4.15.

Math Skills.

(9) @For each of the equations below identify the type of equation as linear, quadratic, exponential, or none of these. State the reason for your identification.

   a. @$y = 10(1.3)^{x/2}$
   b. $a(t) = 5t^2 - 7.2t + 12.8$
   c. @$f(x) = 100 + 2^x$
   d. @$y = \frac{2x-3}{7}$
   e. $a(t) = (\sqrt{5})^t$
   f. $f(x) = 5.2x - 17$

(10) @Using your knowledge of how the parameters effect the graph of an exponential function, match each of the graphs below with one of these functions: 2^x, $-10(1.05)^x$, $\left(\frac{1}{6}\right)^x$, $-3(1.05)^x$, $-2(0.88)^x$. Note that two of the functions do not match any of the graphs.

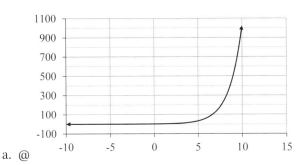

a. @

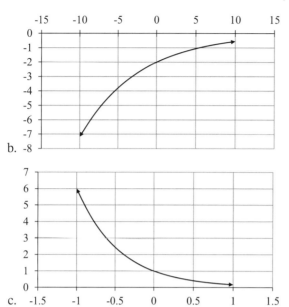

b.

c.

(11) @Let $p(t) = 1{,}000 \cdot 16^{t/4}$ for the parts of this problem.

   a. @Replace 16 by 4^2 and express $p(t)$ as an exponential function with base 4.
   b. @Approximate 16 by $10^{1.204}$ and express $p(t)$ approximately as an exponential function with base 10.
   c. @Approximate 16 by $e^{2.773}$ and express $p(t)$ approximately as an exponential function with base e.
   d. @Writing $16^{t/4}$ as $(16^{1/4})^t$, express $p(t)$ as an exponential function for which the only exponent is simply t.

(12) Let $p(t) = 1{,}000(9)^{1.5t}$ for the parts of this problem.

   a. Express $p(t)$ as an exponential function with base 3. Is your equation exactly or only approximately correct?
   b. Express $p(t)$ as an exponential function with base 10. Is your equation exactly or only approximately correct?
   c. Express $p(t)$ as an exponential function with base e. Is your equation exactly or only approximately correct?
   d. Express $p(t)$ as an exponential function with exponent equal to t. Is your equation exactly or only approximately correct? If stuck, see below for a hint.[19]

(13) @Express each function in the form $f(t) = Ae^{kt}$ in two ways: (1) representing k exactly, and (2) representing k as a six decimal place approximation.

   a. @$f(t) = 25 \cdot 1.04^{3t}$
   b. $f(t) = 150 \cdot 0.86^{t/5}$
   c. @$f(t) = 48.5 \cdot 2^{1.57t}$
   d. $f(t) = 1.012 \cdot (1/2)^{t/4}$

---

[19]Hint: $9^{1.5} = 9^{\frac{3}{2}} = \left(9^{\frac{1}{2}}\right)^3$.

4.3. Exercises

(14) @Express each equation in the form $f(t) = Ab^t$ in two forms, representing b exactly, and as a six decimal place approximation.

   a. @$f(t) = 52e^{2t}$
   b. $f(t) = 1{,}800e^{-1.078t}$
   c. @$f(t) = 85.4 \cdot 1.012^{t/5}$
   d. $f(t) = 510 \cdot 10^{t/15}$

(15) @Follow steps a and b to determine which of the following expressions are equivalent.

   a. Complete the table below.

t	0	1	2	3	4	5
$36(1.2)^t$						
$30(1.2)^{t-1}$						
$30(1.44)^{0.5(t+1)}$						

   b. @The numbers you entered in your table should indicate that two of the rows are the same. Using algebra convert one of the corresponding expressions into the other. You may start with either expression.

(16) @For each of the following, the first 5 decimal places of the solution are given. Use a numeric approach to find the next decimal digit of the solution. Recall from the text if we know a solution is approximately 7.44976 one way to find the next digit is to compute 7.449761, 7.449762, and so on. Alternatively, systematic trial and error can also be used.

   a. @For $1.5^t = 3$, $t = 2.70951$ to 5 places.
   b. For $2^x = 12$, $x = 3.58496$ to 5 places.

(17) @The parts of this problem concern the equation
$$10^t = 13.$$

   a. @Use a numerical method to solve the equation approximately, giving an answer correct to two decimal places.
   b. @Using logarithms, what is the exact solution of the equation?
   c. @Use a calculator to express your answer from part b as a decimal. Does it agree with your answer to part a?

(18) The parts of this problem concern the equation
$$e^t = 4.$$

   a. Use a numerical method to solve the equation approximately, giving an answer correct to two decimal places.
   b. Using logarithms, what is the exact solution of the equation?
   c. Use a calculator to express your answer from part b as a decimal. Does it agree with your answer to part a?

(19) @The parts of this problem concern the equation
$$7^t = 5.$$

a. @Use a numerical method to solve the equation approximately, giving an answer correct to two decimal places.
b. @Using logarithms, what is the exact solution of the equation?
c. @Use a calculator to express your answer from part b as a decimal. Does it agree with your answer to part a?

(20) The parts of this problem concern the equation
$$1.06^t = 2.$$
a. Use a numerical method to solve the equation approximately, giving an answer correct to two decimal places.
b. Using logarithms, what is the exact solution of the equation?
c. Use a calculator to express your answer from part b as a decimal. Does it agree with your answer to part a?

(21) @Solve the equation $3^{1.2t} = 7$ by first defining $1.2t = u$ and solving $3^u = 7$, and then finding t using the value of u. Check your answer by substituting it in the original equation.

(22) @Solve the following equations. Give both an exact answer, and a decimal form correct to six decimal places. [Hint: The method of the previous problem can be applied.]

a. @$12 = 1.04^{3t}$
b. $1.7 = 5^{t/10}$
c. @$0.8^{2t/5} = 10$
d. $1.35^{t/2.3} = 25$

(23) @Solve each of the following equations for t, finding answers that are correct to six decimal places. [Hint: use algebra to change each equation into a form like the ones in the preceding problem.]

a. @$250(1.056^{t/4}) = 800$
b. $1200e^{-0.78t} = 300$
c. @$400e^{0.46t} - 115 = 600$
d. $414\left(\frac{381}{414}\right)^{t/3} = 96$

(24) @For each part below, find a function of the form $f(x) = Ab^x$ that satisfies the conditions.

a. @$f(0) = 60$ and $f(4) = 303.75$.
b. The graph passes through the points $(0, -7)$ and $(10, -43.34)$.
c. @The graph passes through the points $(2, 90)$ and $(5, 2{,}430)$. Note that the y-intercept has not been given so this problem requires a bit more thought.
d. The graph passes through the points $(1, 225)$ and $(3, 506.25)$.
e. @The initial value is 0.78 and the variable grows by a factor of 1.9 over 5 hours. Let x be time in hours.
f. The initial value is 9.8 and the variable decays by a factor of 0.55 over 7 months. Let x be time in months.

4.3. Exercises

(25) @For each part below, find an equation of the form $y = A \cdot 2^{kx}$ that satisfies the conditions.

   a. @When $x = 0$, $y = 1$ and when $x = 6$, $y = 8$.
   b. The graph passes through the points $(0, -15)$ and $(5, -1920)$.
   c. @The graph passes through the points $(4, 48)$ and $(7, 384)$. Note that the y-intercept has not been given so this problem requires a bit more thought.

(26) @For each part below, find an equation of the form $y = Ae^{kx}$ that satisfies the conditions. Give both exact and approximate answers.

   a. @When $x = 0$, $y = 250$ and when $x = 10$, $y = 500$.
   b. The graph passes through the points $(0, -15)$ and $(5, -1920)$.

Digging Deeper.

(27) At the start of this section, it was stated that the equation
$$T = 70 + 280 \left(\frac{23}{28}\right)^{0.1t}$$
cannot be expressed in the standard form for an exponential function. How can we be sure that this is really an impossibility? One approach is to reason as follows. (1) Any exponential function satisfies the Strong Geometric Growth Assumption (page 225). (2) The function T in the above equation does *not* satisfy the Strong Geometric Growth Assumption. Therefore (3) the function T is not an exponential function. To complete this argument, show that (1) and (2) are both true.

(28) @Chapter 4 mentioned the following surprising result: the nth Fibonacci number can be found exactly by rounding $M_0 r^n = \sqrt{1/5}(0.5 + 0.5\sqrt{5})^n$ to the nearest whole number. Use this idea to determine whether 54,321 is a Fibonacci number, and if so, what the corresponding n is. If not, what is the closest Fibonacci number to 54,321?

(29) Show that the base-ten logarithm and the base-e logarithm are related by the equations
$$\log a = \frac{\ln a}{\ln 10}$$
and
$$\ln a = \frac{\log a}{\log e},$$
where a can be any positive quantity. [Hint: $\log a$ is the solution to the equation $10^x = a$. Solve that equation with the Exponential Equation Solution that uses the ln function (page 263). Note that there is also a version of the Exponential Equation Solution that uses the log function (page 261).]

(30) @For positive quantities a and b, define $\log_b(a)$ to be the unique solution x of the equation $b^x = a$. Show that
$$\log_b(a) = \frac{\log a}{\log b}$$
and
$$\log_b(a) = \frac{\ln a}{\ln b}.$$

4.4 Applications of Exponential Functions

For the applications considered in Section 4.2, we developed models based on geometric growth sequences. This is a natural approach when we can conceptualize the problem context in terms of a sequence. An alternative approach is to formulate a model directly in terms of an exponential function. In this approach, we assume that the function of interest has a standard exponential form, such as Ab^x or Ab^{kx} or Ae^{kx}, for appropriate values of the parameters A, b, and/or k. As we have seen before, this assumption may be based on the structure of the modeling context, or some pattern we observe in data. Both of these possibilities will be illustrated with examples in this section.

Structural Geometric Growth: Radioactive Decay. Radioactive decay has been studied comprehensively as part of the larger subject of atomic physics. The idea that the decay process can be accurately modeled by geometric growth emerges as part of the larger theory. While it is beyond the scope of this book to present an account of the theory here, we do wish to observe that there is a structural basis for using geometric growth models in this area.

Our consideration of such models will begin with a particular form for the exponential function equation:

$$A(t) = A_0(1/2)^{t/h}, \tag{4.8}$$

where

- The variable t measures elapsed time from some specified starting point, time 0;
- $A(t)$ is the amount of a radioactive substance at time t;
- A_0 is the initial amount of the substance, i.e., at time 0; and
- The parameter h is the half-life of the radioactive substance, expressed in the same units as t.

Notice that (4.8) is equivalent to the form Ab^{kx} with $b = 1/2$ and $k = 1/h$, because

$$\frac{t}{h} = \frac{1}{h}t.$$

As a reminder, the term *half-life* refers to the length of time required for the amount of the substance to decrease by half. Thus, to repeat an example considered earlier, strontium-90, denoted Sr90, has a half-life of 28.9 years. This means, if we have a sample originally containing 4.8 grams of Sr90, after 28.9 years that amount will be reduced to 2.4 grams, after another 28.9 years the amount will be 1.2 grams, and so on.

To use (4.8), we identify A_0 as 4.8 grams and h as 28.9 years. Thus we have

$$A(t) = 4.8(1/2)^{t/28.9}. \tag{4.9}$$

Using (4.8) in this way provides a direct formulation of an exponential function model for the amount of Sr90 at any time t.

Derivation of the Radioactive Decay Equation. Why does (4.8) take the form it does? To answer this question, we can use a geometric growth sequence in which a_n is the amount of Sr90 after n 28.9-year periods. The starting amount is $a_0 = 4.8$, and a_1 is the amount remaining after one 28.9 year span, a_2 the amount after another such

4.4. Applications of Exponential Functions

span, and so on. As we have already computed, $a_1 = 2.4$ and $a_2 = 1.2$. More generally, we recognize that this is a geometric growth sequence with growth factor $r = 1/2$, so the functional equation is
$$a_n = 4.8(1/2)^n.$$

As part of our structural understanding of radioactive decay, we assume that n can be interpreted as a continuous variable, and that the strong geometric growth assumption applies. But, we recognize that the variable n is not very convenient, because it measures time in 28.9 year increments. It would be better to find an equation that predicts the amount of Sr90 after t years.

More formally, we introduce the variable t, representing the time, in years, starting from the initial observation of a_0. Observe that for a given value of n, the corresponding t is $28.9n$. For example, a_4 is the amount of Sr90 after 4 periods of 28.9 years each, that is, after $t = 28.9 \cdot 4$ years. The general case, with n in place of 4, is expressed in the equation
$$t = 28.9n$$
which we transform into
$$n = t/28.9.$$
Conceptually, this says that the number of 28.9-year periods in a given amount of time t is found by dividing t by 28.9.

Now substituting $t/28.9$ for n we find
$$4.8(1/2)^n = 4.8(1/2)^{t/28.9},$$
expressing the amount of Sr90 as a function of t, rather than n. This is the derivation of (4.9) for our particular example, but it shows more generally why (4.8) is correct.

A More General Continuous Geometric Growth Equation. Indeed, this sort of reasoning applies not only for radioactive decay, but to many other contexs, as well. It is equally valid for any model where we assume a fixed growth factor in a particular amount of time. We can state this more general principle as follows.

> **Continuous Growth Model Equation:** In a continuous model for geometric growth, suppose the variable a is a_0 at time $t = 0$, and suppose that over d units of time there is a growth factor of r. Then
> $$a(t) = a_0 r^{t/d}. \tag{4.10}$$

Applying this to another radioactive decay problem, suppose we have a sample containing 0.1 grams of the uranium isotope U237. How much will remain after 14 days? How long will it take for the amount to be reduced to 0.001 grams?

To answer these questions, we need to know that U237 has a half-life of 6.75 days. That means we will observe a growth factor of $r = 1/2$ every $d = 6.75$ days. Thus, we can write
$$A(t) = 0.1(1/2)^{t/6.75}$$
for the amount of U237 after t days. In particular, after 14 days the amount will be
$$A(14) = 0.1(1/2)^{14/6.75} = 0.024$$
rounded to three decimal places.[20]

---

[20]To compute this on a calculator, both the entire base and the entire exponent should be enclosed in parentheses, like so: $0.1 \times (1/2) \wedge (14/6.75)$. If the parentheses are omitted, many calculators will perform a different calculation than the one desired for this problem.

Turning to the second question, we replace $A(t)$ by 0.001, obtaining

$$0.001 = 0.1(1/2)^{t/6.75}.$$

Now we apply the methods of Section 4.3. Dividing both sides of the equation by 0.1 and exchanging the two sides gives us

$$(1/2)^{t/6.75} = 0.01.$$

This is in the form $b^x = c$, with $b = 1/2 = 0.5$ and $c = 0.01$, and with x representing $t/6.75$. We know the solution is

$$x = \frac{\log c}{\log b} = \frac{\log 0.01}{\log 0.5}.$$

That implies

$$t/6.75 = \frac{\log 0.01}{\log 0.5},$$

and therefore

$$t = 6.75 \frac{\log 0.01}{\log 0.5}.$$

A calculator approximates this as $44.84602928\cdots$. To check this answer, verify that a calculator gives $A(44.8460298) = 0.001$.[21]

As another illustration, consider a population that grows by a factor of $r = 2.8$ in a time period of length $d = 6$ weeks. If we assume the population $p(t)$ follows a geometric growth law, then substituting the values of r and d into (4.10) gives

$$p(t) = p_0(2.8)^{t/6}$$

with the variable t representing the number of weeks from the starting point for the model.

In the preceding examples we are given the exact information necessary to apply (4.8) for a radioactive decay model, or (4.10) for a general geometric growth model. But that is not generally the case in developing a model. Rather, you have to analyze the given information to extract the parts necessary for defining the model. A little analysis of this sort is needed for the following example.

Example: Geometric Growth Model for Water Quality. A scientist is studying how soil runoff after a major rainfall affects the water quality in a local lake. One index of water quality is TDS, Total Dissolved Solids, which measures amounts of chemicals and minerals from such sources as urban runoff, hard water, salinity due to irrigation, and acid rain. The TDS level of a water sample is measured in units of milligrams per liter, or mg/L.

Immediately after a major rainfall, the TDS levels in the lake are elevated due to runoff from the surrounding area. But over time, the effects of the runoff dissipate, and normal TDS levels are restored. The scientist tests geometric growth models for this process. In one investigation, shortly after a major rainfall, a TDS level of 414 mg/L was detected in the lake. Three days later the TDS level was found to be 381 mg/L.

---

[21] Both the value of t and the calculator's result for $A(t)$ are approximations. Even if the calculator displays 0.001, that does not necessarily mean that the value of t is exactly correct. On the other hand, the symbolic expression $6.75 \frac{\log 0.01}{\log 0.5}$ is an exact solution of the equation.

4.4. Applications of Exponential Functions

Assuming geometric growth, find an equation for the TDS level t days after the initial measurement, and use it to predict when the TDS level will be 250 mg/L.[22]

Solution. Let t represent elapsed time, in days, from the initial TDS reading. Let $s(t)$ be the TDS level, in mg/L, at time t. We are given that $s(0) = 414$ and $s(3) = 381$. In particular, the initial value for this problem is 414, defining the constant a_0 for (4.10).

We find a growth factor for our two TDS values as follows. From the initial level of 414 to 381, the growth factor is $r = 381/414$. This occurs in $d = 3$ days. Now, assuming geometric growth, we observe that $s(0) = 414$, and that $s(t)$ grows by a factor of $381/414$ over 3 days. Thus, (4.10) becomes

$$s(t) = A_0 r^{t/d} = 414 \left(\frac{381}{414}\right)^{t/3}. \tag{4.11}$$

In this way we obtain an equation for $s(t)$.

We are also asked when the TDS level will reach 250. Replacing $s(t)$ by 250 produces

$$414 \left(\frac{381}{414}\right)^{t/3} = 250.$$

Dividing by 414,

$$\left(\frac{381}{414}\right)^{t/3} = \frac{250}{414}.$$

This is in the form $b^x = c$, with $b = 381/414$, $x = t/3$, and $c = 250/414$. Therefore, we know the solution will be $x = \ln c / \ln b$, giving us

$$\frac{t}{3} = \frac{\ln(250/414)}{\ln(381/414)}.$$

Therefore

$$t = 3 \frac{\ln(250/414)}{\ln(381/414)}.$$

A calculator gives this as 18.216891, approximately. Substituting this for t in (4.11) produces $s(t) = 250$, verifying that the answer is correct.[23]

Using e as a Base. When we use (4.8) for a radioactive decay model, or (4.10) more generally for a geometric growth model, the parameters have clear meanings in the model context. For radioactive decay, we use the initial value and the half-life; for the more general case we have initial value, an observed growth factor r, and the length of time over which the factor r is observed. These interpretations simplify the task of formulating an equation from given information. But sometimes it is desirable to re-express the equation using a base-e exponential function. Let us review this step for two of the examples just considered.

For the U237 model, we found the equation

$$A(t) = 0.1(1/2)^{t/6.75}.$$

---

[22] This problem is based on a survey of water quality tests published by the South Australian Science Teachers Association [48].

[23] Here and in the following computations, answers expressed to six or more decimal places should be assumed to be approximations, unless noted otherwise. Moreover, calculating with approximations can be assumed to produce approximate results. For the case at hand, computing $s(18.216891)$ may produce a result of 250 on a calculator, but that merely shows that 18.216891 is a very accurate approximation to the exact value of $3 \frac{\ln(250/414)}{\ln(381/414)}$.

To express this using base e, the first step is to express $(1/2)$ as a power of e. That is, we wish to find the exponent k for which $1/2 = e^k$. By the definition, the solution is $k = \ln(1/2) = -0.693147$. Now replace $1/2$ by e^k in the equation for $A(t)$, and we have

$$A(t) = 0.1(e^k)^{t/6.75} = 0.1e^{kt/6.75}.$$

We can replace k either by $\ln(1/2)$ or a decimal approximation. Alternatively, writing the equation in the form

$$A(t) = 0.1e^{(k/6.75)t},$$

we can compute $k/6.75 = -0.102688$, and obtain

$$A(t) = 0.1e^{-0.102688t}.$$

Applying similar reasoning to the population example, we begin with the equation

$$p(t) = p_0(2.8)^{t/6}.$$

Write $2.8 = e^k$ where $k = \ln 2.8 = 1.029619$. Thus we have

$$p(t) = p_0(e^k)^{t/6} = p_0 e^{kt/6} = p_0 e^{(k/6)t}.$$

Then, computing $k/6 = (\ln 2.8)/6 = 0.171603$, the equation becomes

$$p(t) = p_0 e^{0.171603t}.$$

For both of the preceding examples, we began with an instance of (4.10) and then converted it to one of the form $A(t) = A_0 e^{kt}$. But an equation of this form can also be derived directly. To illustrate the method, we provide the following alternate solution of the water quality example.

Direct Formulation of Base e Equation. As in the first solution, we let t represent elapsed time, in days, from the initial TDS reading, and let $s(t)$ be the TDS level, in mg/L, at time t. But now, by assuming geometric growth, we know that $s(t)$ is given by an equation of the form

$$s(t) = A_0 e^{kt}.$$

Although there are quite a few variables in the equation, only two are parameters that must be determined as part of formulating the model. The letter e represents a known numerical constant, and t and s will remain as variables until we apply the model. In contrast, A_0 and k are parameters, and A_0 represents the initial value of the variable s.

As already noted, the initial amount is $s(0) = 414$. Therefore, the equation is

$$s(t) = 414 e^{kt}.$$

To complete the formulation of the model, we need to find k. We know that $s(3) = 381$. But the equation says that $s(3) = 414 e^{k \cdot 3} = 414 e^{3k}$. Combining these facts we have

$$414 e^{3k} = 381,$$

from which we derive

$$e^{3k} = 381/414.$$

This says that $3k$ is the exponent on e that results in $381/414$, so by definition

$$3k = \ln(381/414),$$

hence

$$k = \frac{\ln(381/414)}{3}.$$

4.4. Applications of Exponential Functions

A calculator gives this as
$$k = -0.027689.$$
Therefore, the equation for $s(t)$ is
$$s(t) = 414e^{-0.027689t}. \tag{4.12}$$

Let us use this version of the equation to predict when the TDS level will reach 250. Replacing $s(t)$ by 250 produces
$$250 = 414e^{-0.027689t},$$
thus
$$e^{-0.027689t} = 250/414.$$
Again invoking the definition of natural logarithm, we find
$$-0.027689t = \ln(250/414).$$
This in turn leads to
$$t = -\frac{\ln(250/414)}{0.027689}.$$
A calculator gives this as 18.216803, very nearly the same answer we found before. The slight difference reflects using a six digit approximation for k in (4.12).

All of the preceding examples emphasize adopting a geometric growth model for structural reasons, and then employing various methods of formulating and manipulating equations. Next we look at some examples where data play a more central role in defining a geometric growth model.

Data Based Model Development.

Temperature Probe Data Example. [24] A temperature probe is a type of digital thermometer that can be used to collect temperature data in an electronic form. Such a device could be used in the context of the apple pie example on page 234. In that example, a pie was taken from a hot oven and allowed to cool on a kitchen counter. With a temperature probe inserted into the pie, we could collect and model temperature data for the cooling pie.

As a simple experiment along these lines, the cooling process of the probe itself was investigated. Initially, the probe was left in a cup of hot coffee long enough to reach a temperature nearly the same as the coffee. Then the probe was placed in a cold water bath at a temperature of 4.5°C. Temperatures were recorded at several times, as shown in Table 4.6.

Table 4.6. Temperature probe data for a hot probe placed in cold water at time 0. For example, the first pair of values indicates that 2 seconds after time 0, the probe registered a temperature of 64.8°C.

Time (seconds)	2	5	10	15	20	25	30
Temperature (°C)	64.8	49.0	31.4	22.0	16.5	14.2	12.0

As discussed in Section 4.2, geometric growth is often assumed in such a situation, modeling not the actual temperature but the difference between the temperature of an

---

[24] The experiment and data described in this example are as reported in [**25**, Example 7, p. 360].

object and its surroundings (see page 234). Accordingly, we consider a new variable, H, representing how much hotter the probe is than its surroundings. In this case, the probe is surrounded by cold water at a temperature of 4.5°C, so when the probe temperature is $T = 64.8$, we find $H = T - 4.5 = 60.3$. Computing an H for every T in the table produces the results shown in Table 4.7.

Table 4.7. Adjusted temperature probe data. As before, t is the time, in seconds, starting from when the probe was placed into cold water. The adjusted variable H indicates at each reading how much hotter the probe is than its surroundings, cold water at 4.5°C.

Time (seconds)	2	5	10	15	20	25	30
Observed H (°C)	60.3	44.5	26.9	17.5	12.0	9.7	7.5

Now if we assume in our model that H decays geometrically with time, that implies that H will be an exponential function of t, and therefore expressible in an equation of the form

$$H = Ab^t.$$

To formulate our model, we need to select values for the parameters A and b. If the data points follow a geometric model exactly, we can find A and b using similar methods as in the earlier examples of this section. But that rarely happens with real data. Rather, the data are only approximated by an exponential equation, and exact agreement with all the data points is not possible for any choice of the parameters. Therefore, we would like to choose parameters that make our model as close as possible to the observed data.

As we have observed several times, there are advanced mathematical and statistical methods for choosing the best parameters, but such methods are beyond the scope of this book. As an alternative, we can use a graphical trial-and-error approach to find a reasonably accurate model. For each choice of the parameters, we can compare the model with the data points, both numerically, and graphically. As we try different choices of the parameters, we can see whether the resulting model is better or worse than our previous attempts. Following such an approach, our best results occurred with the equation

$$H = 71 \cdot 0.918^t.$$

A comparison of this model with the data points is provided in Figure 4.16. The values computed with the model are pretty close to the observed data, differing by at most 3.3°. For all but two of the data points, the model is off by 2° or less. This illustrates how a geometric growth model may be formulated, and helps to validate the use of such a model for similar heat flow problems. However, an even better model can be found, as follows.

Our first model includes the assumption that the cold water bath is at a temperature of 4.5°C. In particular, we assume that the water temperature remains constant. This may not be correct. Perhaps the temperature rises a little as a result of inserting the hot temperature probe. Moreover, unless it is perfectly insulated, the water bath will gradually warm up to room temperature. From the given information, we cannot assess either of these aspects. But we can experiment with the effect of modifying the 4.5°C figure.

4.4. Applications of Exponential Functions

t	H	Data	Error
2	59.8	60.3	−0.5
5	46.3	44.5	1.8
10	30.2	26.9	3.3
15	19.7	17.5	2.2
20	12.8	12.0	0.8
25	8.4	9.7	−1.3
30	5.5	7.5	−2.0

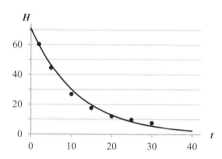

Figure 4.16. Data table and graph for the geometric growth model with equation $H = 71 \cdot 0.918^t$. The entries in the H column of the table have been rounded to one decimal place. Each entry in the *Error* column is found by subtracting an observed data value from the corresponding model value $H(t)$. Thus, a positive error means that the model value is too high; a negative error means the model value is too low.

Before proceeding, let us take one preliminary step. Combining the equations $H = 71 \cdot 0.918^t$ and $H = T - 4.5$, we find

$$T = 71 \cdot 0.918^t + 4.5.$$

In our model, this expresses the temperature of the probe as a function of time. The equation has the form

$$T = Ab^t + C,$$

where C represents the temperature of the water bath. Note that this is not an exponential function, because of the added constant C. As mentioned at the end of Section 4.2, this is the sort of equation that results from Newton's Law of Cooling.[25]

Treating C as a parameter, we can again use a trial-and-error approach to find the best agreement possible between the observed data and a model. Proceeding as before, but this time varying three parameters, A, b, and C, we found our best results with the equation

$$T = 69 \cdot 0.89^t + 10.$$

A graph and table comparing this model with the original data are shown in Figure 4.17.

Although the new figure shows the actual temperatures, and the preceding figure shows instead the values of H, the graphs display the same pattern of data points. Both visually and numerically, the model in Figure 4.17 agrees much more closely with the data values than the earlier model. On this basis, it appears that the second model is a better choice than the first.

Considering the two models further, we can ask how their results differ. For one thing, we can use each model to estimate the temperature of the probe when it was first put into the cold water. This is the temperature at time 0. Using the first model,

---

[25] A function of this form is sometimes called a *shifted* exponential function, because the added constant has the effect of shifting or *translating* the graph of the function vertically by C units. Thus, the general shape of the graph is unchanged, but it levels off at $T = C$ rather than at $T = 0$.

t	T	Data	Error
2	64.7	64.8	−0.1
5	48.5	49.0	−0.5
10	31.5	31.4	0.1
15	22.0	22.0	0.0
20	16.7	16.5	0.2
25	13.7	14.2	−0.5
30	12.1	12.0	0.1

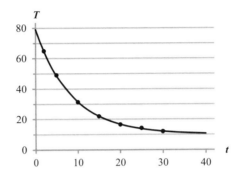

Figure 4.17. Data table and graph for the model with equation $T = 69 \cdot 0.89^t + 10$. The model and data values for this equation agree much more closely than for the preceding case.

we find $T(0) = H(0) + 4.5 = 71 + 4.5 = 75.5$. From the second model we find $T(0) = 69 + 10 = 79$. Similarly, we can compare the long term predictions from each model. In the first model, as time increases, H decreases to 0, and we see that T levels off at 4.5. In the second model, where $T = 69 \cdot 0.89^t + 10$, the exponential part of the function decreases to 0, showing that T levels off at 10. So both initially and in the long term, the second model is about 5° higher than the first model. On the other hand, for times between 2 and 30, the two models are closer to each other, differing by about 3° or less, though in some cases the first model is higher and in others the second is.

Based on the given information, we cannot really tell which model is better. But our comparison does suggest how we might investigate the issue further. If we repeat the experiment, we should try to determine the initial temperature of the probe directly and continue collecting data long enough to see the temperatures level off. We also should be sure to insulate the cold water bath from the surrounding conditions. It may be that the first model gives more accurate predictions of the initial and long term temperatures. In this case, we might attribute the disagreement between the model and the data to measurement error, and conclude that while the second model fits the data more accurately, it would not provide better predictions in the long run. On the other hand, if the second model proves to be the more correct, that would suggest revising the general modeling approach. Rather than forming the variable H and fitting a pure exponential function, perhaps we should consider working directly with T and fitting a model that is an exponential function plus a constant. As we will see in Chapter 5, such functions arise in many interesting contexts.

Although we will not investigate these issues further here, the reader should recognize that this discussion illustrates an important feature of model development and refinement. In the process of developing and testing the first model, we are led to a possible modification that might improve the model. That in turn leads to ideas for further experimentation, and possibly to new families of models. In this way, the modeling process can lead to ever more refined and more accurate models, contributing significantly to the effectiveness of models and modeling.

4.4. Applications of Exponential Functions

As discussed in Section 4.2, geometric growth assumptions have proven valuable in several broad modeling contexts, including compound interest, tank models, radioactive decay, population growth, and, as in the preceding example, heat transfer. When working in one of these contexts, it is reasonable to assume an exponential relationship between two variables, and then to fit a model to the data by carefully selecting the parameters. And we expect reasonably close agreement between the data and the model.

In contrast, we may also choose an exponential function based purely on the shape of the data, even in a context where the data are not all closely arranged along any one curve. Indeed, we saw something similar with the quadratic model for polar ice (see page 147). As a final example, we look at a similar situation using an exponential function to model the relationship between the length and weight of trout in an ecological study.

Trout Length and Weight Example. Data in Table 4.8 were collected as part of a study conducted for the Washington State Department of Ecology. The data consist of weights and lengths measured for rainbow trout taken from the Spokane River. One purpose of the study was to analyze the presence of heavy metals in the fish, and data were collected for several variables, including length, weight, and age of the fish surveyed. Here we will model the relationship between length and weight, using an exponential function.

Table 4.8. Length and weight data for trout. Length (L) is in units of millimeters, weight (W) in units of grams. Data source: Quantitative Environmental Learning Project [**44**].

L	W	L	W	L	W	L	W	L	W	L	W
247	184	309	392	334	406	346	433	368	581	405	715
259	202	317	335	335	472	347	432	368	605	413	754
265	223	318	340	335	410	351	506	385	609	438	840
270	209	324	387	337	363	359	476	387	538	455	975
276	235	324	353	338	460	360	557	390	660	457	855
280	248	326	353	343	390	360	479	392	623	460	895
305	303	332	383	345	438	365	540	395	584	502	1300

Defining variables L and W representing length and weight of each trout, one possible equation based on the data is

$$W = 42e^{0.0068L}.$$

A graph showing the data and the model appears in Figure 4.18.

To further illuminate this example, we answer a series of questions.

How was the equation found?

The adoption of a geometric growth assumption implies that an exponential function will be used. We have seen that such equations can be expressed in several standard forms, including $W = Ab^L$ and $W = Ae^{kL}$. In the preceding example we used the first of these forms, so for this model we have elected to use the second form.

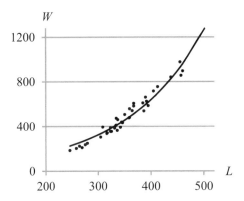

Figure 4.18. Trout length and weight data, modeled by the equation $W = 42e^{0.0068L}$, where L is length, in millimeters, and W is weight, in grams. Dots represent data from Table 4.8. The curve is the graph of the equation.

We plotted the data points using graphing software, and also graphed equations of the form $W = Ae^{kL}$ for several different choices of the parameters A and k. Through a systematic process of trial and error, the curve in the figure was found. To the eye, this appears to show the trend of the data points pretty well. That is, the data points appear to cluster fairly closely around the curve.

How should the model be interpreted?

This example differs from the preceding example in that the data points do not appear to align with a single curve. In fact, in the data we cannot say that W is a function of L, because there are instances where two different fish have the same length, but different weights. Graphically, this means there are places where two data points appear on the same vertical line. Based on this observation, we do not expect any model expressing weight as a function of length to agree with the data exactly. Instead, we think of the model curve as a trend that approximates the variation of weight with length. For the data points in the table, each weight is within about 50 grams of the value predicted by the model equation. Put another way, we can use the model to predict weight, based on length, but we expect the prediction to be off by as much as 50 grams. To put this in perspective, the weights of the trout vary roughly from 200 to 1,000 grams. An error of 50 grams is more significant for a small fish than for a large one. Accordingly, we might wish to consider the error as a percentage of the true weight. This percentage is referred to as relative error. The relative errors in our model range between nearly zero and 26%, and are 15% or less for all but four of the data points.

Why was a geometric growth assumption adopted?

Quoting from the Quantitative Environmental Learning Project web page,

Rainbow trout show an exponential relationship between length and weight. Smaller (younger) fish are relatively long and skinny, whereas larger (older) fish are much more stout. This type of growth sequence is very common among many plants and animals; humans show the same kind of growth behavior. [44]

4.4. Applications of Exponential Functions

This suggests that an exponential model for the trout data is consistent with past findings in similar situations. And certainly the upward curving shape of the data points is similar to the graphs of exponential functions. Even so, the use of an exponential function here is essentially a matter of choice. We have not advanced a structural reason to explain why the relationship should be geometric. This contrasts with contexts such as radioactive decay, heat transfer, and tank models, where geometric growth can be strongly supported with structural properties of the model contexts.

In the polar ice example (see page 147), we observed a similar curving trend in the data points. However, in that context we developed a quadratic model. It is natural to ask whether a quadratic model could also be used with the trout data. Indeed it can. In fact, again using a trial-and-error approach, we can find the equation

$$W = 0.009L^2 - 2.65L + 289.8,$$

which is graphed with the trout data in Figure 4.19. Comparing this with the prior graph, we can see that the quadratic model is very similar to the exponential model, and is actually a better fit for the first several data points.

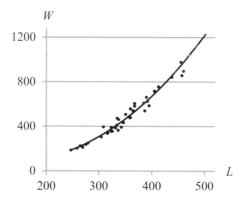

Figure 4.19. Trout length and weight data, modeled by the quadratic equation $W = 0.009L^2 - 2.65L + 289.8$.

It should not surprise us that we obtain a better agreement between the model and the data with a quadratic equation, where we choose three parameter values. In the exponential equation, where there are only two parameters, there is not so much flexibility in adjusting the shape of the curve. But that does not automatically make the quadratic model better. As an alternative, we might try an equation of the form $W = Ae^{kL} + C$, where C is a constant.[26] This produces a curve with an exponential shape, but vertically shifted by the amount C. With this additional parameter it is possible to obtain a model that agrees with the data about as well as the quadratic model does.

Is there any reason to prefer the exponential model? It depends on how the model will be used. If the goal is simply to find an approximating curve, then the quadratic model is as good as the exponential model. But if we hope to gain insights that go beyond the immediate data set, then there is a good reason to prefer the exponential model. Consider what happens when we extend the model graphs to additional values

---

[26]This is another instance of a shifted exponential function, mentioned in an earlier footnote.

of L, between 0 and 200. As shown in Figure 4.20, the quadratic model has the characteristic parabolic shape. To the left of the vertex, it predicts that weights increase as the length of a fish *decreases*, which doesn't make much sense. In contrast, the extended graph for the exponential model shows that higher weights are consistently paired with greater lengths, as we would expect. Based on this comparison, the exponential model seems much more reasonable.

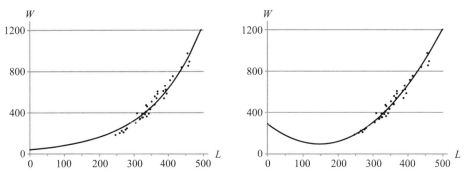

Figure 4.20. Comparison of exponential and quadratic models over an extended range of lengths. The exponential model, at left, shows a consistent increase of weight with length, as would be expected. The quadratic model, at right, shows that as fish length increases, the weight initially *decreases*, reaches a minimum, and then increases. This contradicts expectation, raising strong doubts about the validity of the quadratic model.

This example helps show why it is useful to study general properties of different kinds of models. In our consideration of quadratic and exponential models, we learned about the characteristic shapes of the graphs that arise for each type of model. Now we see that this general information helps us identify an exponential function as a better model for the trout data than a quadratic function would be.

How might the model be used?

As noted at the Quantitative Environmental Learning Project web page [**44**], this model might be used in future data collection efforts. For example, rather than measuring both length and weight for every fish caught, researchers might measure just the lengths, then use the equation to estimate weights. Or, more likely, the weights would be measured, assuming it is easier to get an accurate weight than an accurate length.

Alternatively, a future study could replicate the one considered here, and produce a new version of the length-weight equation. Then the parameters of the two equations could be compared. A significant difference in parameters might indicate some change in the characteristics of the fish population. This might be used to detect that the environmental conditions had improved or degraded.

Such a model might also be part of a larger theoretical development. Imagine performing many similar studies for different populations of fish. If exponential models are found to be generally reliable, the parameters of those models can then be viewed as biometric indicators for different populations. Researchers might then try to discover reasons why two fish populations have different model parameters. For example, they could try to correlate such differences with environmental factors, or investigate whether the differences have a genetic basis.

This completes the discussion of the trout example, and our consideration of exponential function properties and applications. The next section will focus on properties of the constant e. This material is not of great practical significance, but is intended to shed light on some theoretical, conceptual, and historical aspects connected with e. In Chapter 5 we proceed to consider a new category of models, involving sequences combining both arithmetic and geometric growth.

4.4 Exercises

Reading Comprehension.

(1) @Explain how the equations $p(t) = 1{,}000 \cdot 8^{t/6}$, $p(t) = 1{,}000 \cdot 2^{t/2}$, and $p(t) = 1{,}000 \cdot (1.414)^t$ can each be interpreted at first glance. (For example, if t is in units of hours, the first equation indicates that p increases by a factor of 8 every 6 hours.) Also explain how it can be understood from these interpretations that these three formulas all express the same geometric growth model.

(2) Identify all the parameters in the equation
$$a(t) = a_0 r^{t/d},$$
explaining what each parameter represents.

(3) @Based on your understanding of the term *half-life*, what would be the likely meaning of the term *doubling-time*? Would that concept be meaningful in a radioactive decay model? The trout length-weight model? A model for the value of a bank account that collects interest? Explain.

(4) In Table 4.8, find as many instances as possible of one length appearing with two different weights. What is the significance of this?

(5) @Recall that error is the difference between data and a model. Explain what *relative error* means. Would it be useful to find relative error for the temperature probe data example? Why or why not?

(6) Imagine you have been given a set of data and asked to find a model for it. One of the decisions you must make is what type of model to use: arithmetic, quadratic, exponential, or something else. What techniques would you use to determine which type of model to use? What information other than the data itself would be useful in choosing a type of model? It may be helpful to reread the Trout Length and Weight Example.

Math Skills.

(7) @Each part below describes an exponential function. Find an equation for the function. [Hint: Use the Continuous Growth Model Equation (page 271).]

    a. @The variable t is in units of years. Function $a(t)$ equals 150 at $t = 0$ and grows by a factor of 2.3 every 5 years.

    b. The variable t is in units of days. Function $a(t)$ equals 38.7 at $t = 0$ and decays by a factor of 0.45 every 2.5 days.

c. @The variable t is in units of minutes. Function $y = a(t)$ has a y-intercept of 480 and increases by 4% every hour.

d. The variable t is in units of hours. Function $a(t)$ is initially equal to 2,500 and decreases by 20% every 15 minutes.

(8) @Each part below describes an exponential function. Find two equations for the function, one in the form $a(t) = Ab^{t/d}$ and the other in the form $a(t) = Ae^{kt}$.

   a. @The variable t is in units of days. Function $a(t)$ starts at 590 and doubles every 3.5 days.

   b. The variable t is in units of years. Function $a(t)$ is initially equal to 4.87 and has a half-life of 16.38 years.

(9) @Use algebra to express each function in the form $a(t) = a_0 t^{t/d}$, and state the values of $a_0, r,$ and d. Give a description of the function in the manner of the preceding two problems, assuming that t is in units of hours.

   a. @$a(t) = 20 \cdot 1.8^{2.5t}$
   b. $a(t) = 1.69 \cdot 0.95^{0.125t}$
   c. @$a(t) = 450 \cdot 1.5^{t-2}$
   d. $a(t) = 50 \cdot 0.85^{4t+2}$

(10) @In each part two situations are described. With P representing population and t the number of months since January, find an exponential equation for each situation. Then determine if the two equations are equivalent. Explain how you made your determination.

   a. @Situation 1: In January, there is an initial population of 50 which grows by a factor of 8 every 3 months.
   Situation 2: In January, there is an initial population of 50 which doubles every month.

   b. Situation 1: The population grows by a factor of 9 every 2 months.
   Situation 2: The population triples every month.

(11) @Use exact theoretical methods (with logarithms) to answer the following questions.

   a. @Solve this equation for t: $450 \cdot 1.5^{t/3} = 2,000$.
   b. Solve this equation for t: $300 \cdot 0.85^{t/1.8} = 50$.
   c. @Let $a(t) = 20 \cdot 1.8^{2.5t}$. For what t does $a(t) = 100$?
   d. Suppose that t is in units of hours, and function $a(t)$ is initially equal to 8. How long will it be before $a(t) = 2,500$ if $a(t)$ increases by 12% every 5 hours?

Problems in Context.

(12) @Exponential Graph Properties. We know there are 4 possible shapes for an exponential graph and they are given in Figure 4.12 which is repeated here for easy reference.

4.4. Exercises

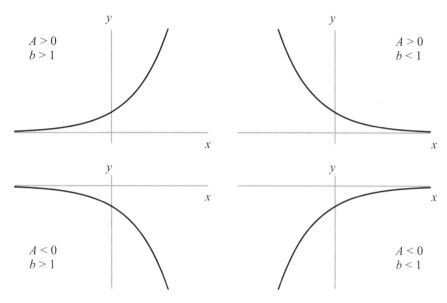

Figure 4.12. Possible shapes for the graph of an exponential function $y = Ab^x$ with $b > 0$ and $b \neq 1$.

Each part below gives some information about a geometric growth model that a student is developing. In each case, tell which of the graphs in Figure 4.12 you would expect most to resemble the graph for the student's model, and why. The graph shape may be repeated and not all shapes need necessarily be used.

a. @The student's model includes the equation $a_n = 4(0.1)^n$.

b. The student's model includes the equation $s_n = -100 \cdot 2^n$.

c. @The student is modeling the size of a spreading population of bacteria.

d. The student is modeling the amount of a radioactive substance.

e. @The student is modeling the height of a deer based on the distance between hoof prints.

f. The student is modeling the temperature of a turkey as it cooks in a 350° oven. The exponential function in the model is $H(t) =$ (turkey temperature)$-$ (oven temperature), and $t = 0$ when the turkey was put into the oven.

(13) Mouse Population Model.

a. A research group is modeling the size of a population of mice. They find the equation $p(t) = 100(2)^{t/3}$ where p is the number of mice in the population and t is time in months. What is the doubling time for this population? That is, how long will it take for the population to reach 200, double the original population? [Hint: Use the Continuous Growth Model Equation (page 271).]

b. Another group has produced a model for the population of a similar type of mouse, but they developed an equation using base e, that is of the form $p(t) = p_0 e^{kt}$ To help the groups compare their models, convert $p(t) = 100(2)^{t/3}$ to an equation with base e. Give both an exact and approximate version of the equation. Identify which is which.

(14) @Radioactive Decay for Iodine-131. The decay of 10 grams of iodine-131 can be modeled by $y = 10e^{-0.086427t}$ where t is time in days. Based on this equation, find the half-life of iodine-131. Then express the equation for y with a base of $1/2$.

(15) Radioactive Decay. In a geometric growth model for a radioactive substance, the variables are t, representing the time in minutes, and $a(t)$, representing the amount of the substance at time t, in units of grams. Suppose $a(0) = 1.60$ and $a(8) = 1.15$. Find an equation for $a(t)$.

(16) @Radioactive Decay. A geometric growth model for a radioactive substance includes the equation
$$a(t) = 14.62(0.85^{t/5}),$$
where $a(t)$ is the amount of the radioactive substance at time t, and t is in units of years. Find the half-life of the substance.

(17) Smart Phone Owners. A geometric growth model for the number of smart phone owners represents time t in years and the number of smart phone owners at time t as $S(t)$, in units of thousands. If the graph of this model passes through the points $(0, 20.8)$ and $(2, 105.6)$, find an equation for $S(t)$.

(18) @Find-n for Methicillin-Resistant *Staphylococcus Aureus*. An exercise in Section 4.2 involved the equation
$$c_n = 33(1.22)^n$$
for the number of MRSA infections in U.S. hospitals in year n. Here, c_n is in units of thousands and $n = 0$ in 1993. According to the equation, in what year will the number of MRSA infections in U.S. hospitals first reach or exceed 5 million?

(19) Find-m for Compound Interest. A savings account offers 3% interest compounded monthly. If $1,000 is deposited into the account and simply left alone, future values of the account balance can be found using the equation
$$B = 1{,}000\left(1 + \frac{0.03}{12}\right)^m.$$
Here B is the account balance in dollars and m is the number of months since the initial deposit. How long will it take for the account balance to grow to $1,500?

(20) @Chilling in a Freezer. A continuous geometric growth model describes the way food is chilled after being placed in a freezer at 0 degrees (Fahrenheit). The model includes the equation:
$$T = 70(0.78^{t/30}),$$
where T is the temperature of the food in degrees Fahrenheit and t indicates how long the food has been in the freezer in minutes. Describe as clearly as possible the initial conditions and change over time of the food temperature.

(21) Removing Impurities. An exponential decay model describes how impurities are removed from a chemical solution in a manufacturing process. The model includes the equation
$$A(t) = 100(0.36)^{t/4},$$
where t is time in hours from the start of the process, and $A(t)$ is the amount of impurities (in grams) after t hours.

4.4. Exercises

a. Explain what each of the parameter values, 100, 0.36, and 4 tells us about the impurities in the chemical solution.

b. How long will it take for the impurities to be reduced to no more than 5 grams?

c. Convert the equation for $A(t)$ to a base e equation. Use the new equation to determine the amount of impurities after 12 hours.

(22) @Inflation. One very simple model of inflation uses geometric growth to describe how the price of some commodity increases over time. For example, suppose a gallon of milk costs $1.45 today, and that the price increases by 2.8% every 6 months.

a. @Find an exponential equation that gives the price of milk m months from now.

b. @Find an exponential equation that gives the price of milk t years from now.

c. @Find an exponential model with base e that gives the price of milk t years from now.

(23) School Enrollment. There were 20,000 students enrolled in a school district in 2010. Over the next several years, enrollment increased by roughly 8% per year. Assuming that enrollment continues to increase by 8% every year, develop a model for the enrollment figures. Use your model to predict the number of students in the school district for the year 2025.

(24) @Chemical Spill. A researcher is monitoring the way pollutants from a chemical spill dissipate over time. In her initial measurement, on August 1, 2014, she found about 20 parts per million of pollution in a soil sample. She continued making measurements once a month, and found that the amount of pollution decreased by 2.8% with each measurement. Assuming that the pollution continues to decrease by 2.8% every month, develop an exponential equation model for the pollution data, and use it to predict the level of pollution (in parts per million) by January 1, 2026.

(25) Investment Earning 8% Annual Interest. Suppose that Adam wants to build some capital. Adam is going to put $97,500 away in an account earning 8% annual interest. Let n be the number of years since Adam made the investment.

a. Find an exponential equation to describe the growth. Be sure to define your variables.

b. How much money will Adam have in this account after 2 years assuming no new deposits or withdrawals are made?

c. Assuming no new deposits or withdrawals, determine how many years it will take for Adam's total balance to first reach at least $170,000?

(26) @Carbon Dating. Carbon has a radioactive isotope carbon-14 with a half-life of 5,730 years. Using factors such as the amount of other carbon isotopes in a sample, it is possible to determine how much carbon-14 was present in the sample when it was alive. After death, the carbon-14 present in the tissue will decay. To get a rough idea of how radio-carbon dating works, complete the following questions.

a. @A sample contains 0.003 micrograms of carbon-14 at time $t = 0$. Find an equation for $C(t)$, the amount of carbon-14 in the sample after t years. (A microgram is a millionth of a gram.)

b. @Suppose at a later time, we find the same sample contains 0.0007 micrograms of carbon-14. How long was that after the initial time?

c. @If a sample is known to have contained 0.003 micrograms of carbon-14 when it was alive and now the sample contains 0.0007 micrograms of carbon-14, about how long ago was the sample alive?

(27) **Investment Earning 1.2% Interest per Quarter.** An investment fund promises a return of 1.2% per quarter. (This is a quarterly rate, not to be confused with an APR as discussed on page 231 and used in problem 19.) If $10,000 are invested with this fund, find an equation for $A(t)$, the amount the investment is worth after t years. Using your equation, determine the value of the investment after 6 years. Also determine when the investment will be worth $25,000.

(28) **@Antibiotic Results.** A medical researcher is studying the effectiveness of a new antibiotic. In one experiment, the antibiotic was applied at time $t = 0$ to a bacteria colony estimated to weigh 156 grams. After 5 hours the colony had been reduced to 140 grams. Assuming geometric growth, find an equation for $B(t)$, the weight of the bacteria colony at time t. Use your equation to determine when the colony will be reduced to 10 grams.

(29) **@Temperature Probe Part 1 with a Trial-and-Error Approach.** The Temperature Probe Data Example beginning on page 275 describes an experiment in which a temperature probe is initially in a cup of hot coffee, and is then placed into a cold water bath. This experiment was later replicated under similar conditions, with a 10° water bath.[27] The following data were collected

t	2	5	10	15	20	25	30
T	80.47	69.39	49.66	35.26	28.15	23.56	20.62

where t represents time, in units of seconds, measured from the instant when the probe was put into the cold water, and T represents the probe temperature in degrees C.

a. @Let H represent how much hotter the probe is than the surrounding cold water. Thus, $H = T - 10$. Compute the H value for each entry in the table. Use a graphing application to graph H versus t. Show both your graph and a table with t and H values in your answer.

b. @As explained in the earlier example, we expect H to be closely modeled by an exponential function of t. Using a trial-and-error approach with a graphing application, find an equation of the form

$$H = Ab^t$$

that provides a good approximation to the data. In reporting your results, include a graph and table as shown in Figure 4.16 on page 277. [Hint: You need to find values for the parameters A and b. Start with $A = 150$ and $b = 0.85$ and then adjust the parameters to improve the accuracy of the model.]

---

[27]This problem and the next are variants on Problems in Context 21 in Section 4.2. In the earlier version of the problem geometric growth sequence methods were used. The new versions emphasize the direct use of an exponential function model, without reference to sequence methods. Note in particular that the data values are at irregularly spaced times, making a sequence approach problematic. This exercise, like the earlier version, is adapted from [**25**, Exercise 33, p. 362].

4.4. Exercises

c. @In your table for part b, you found the error for each data point by subtracting an observed data value for H from the corresponding model value $H(t)$. The *absolute error* E is defined to be the absolute value of the error. This represents how far apart the model and the data values are for each point, without regard to which value is the larger. For your model from part b, what is the largest E value you found? What is the average E value?

d. @According to your model, how hot was the probe when it was first put into the cold water?

e. @Use your model to find an equation for temperature T as a function of time t. Use your equation to determine when T will equal 20°.

(30) **Temperature Probe Part 2 with a Trial-and-Error Approach.** Repeat the preceding problem, but this time assume that the cold water is at a temperature of 15°C instead of 10°C. You should be able to define parameters so the revised model is a better approximation to the data than the original model. Compare the average E value you found for the original and revised models. Also compare how the models differ in what they predict as the temperature of the probe when it was put into the cold water, and how long it will take the probe to reach 20°.

Digging Deeper.

(31) A 2014 article in the Economist [**16**] discussed the growing enrollments in English language schools in countries whose primary language is not English. The following data were obtained from a graph in the article.

Year	2009	2010	2011	2012	2013	2014	2019	2024
Enrollment	2.45	2.76	2.94	3.12	3.36	3.79	5.79	8.94

In the table, enrollment figures are in units of millions, and the final three figures are projections. This problem considers how those projections might have been determined by the authors of the article.

a. Use graphing software to create a graph of the data in the table.

b. Find an exponential function of the form $y = Ae^{kt}$ that is as close as possible to the three projected data points. [Hint: Look at the growth factors for the final three enrollment figures.]

c. Compute the errors between this exponential model and the data for the years 2009 through 2013. Do you think that the projections in the article were made using a similar exponential growth model?

d. For comparison purposes, develop a linear equation of the form $y = mt + b$ for the data for 2009 through 2013. Choose the parameters m and b so that the linear equation produces y values as close as possible to the enrollment figures. Compute the errors between this linear model and the data for 2009 through 2013.

e. Compare the linear model and the earlier exponential model. Considering the accuracy with which each model predicts the data for 2009 through 2013, is one significantly better than the other? Can you think of any structural reason that one might be preferred over the other?

f. The two models predict widely different enrollment figures for 2024. Based only on your analysis in this problem, how credible do you find the projections in the Economist article graph?

4.5 More About e

As mentioned in Section 4.3, exponential functions are often expressed in the form Ae^{kx} where e is the irrational constant $2.7182818284\cdots$. Here, the term *irrational* indicates that the decimal expression of e goes on forever, without a consistent repeating pattern in the digits. It also indicates that e cannot be expressed exactly as a ratio of whole numbers. Consequently, there is no way to give an exact numerical value for e in finite terms. This might seem to make e a strange choice for the standard base of exponential functions.

Curiously, the number e somehow arises naturally in many different contexts and in several different branches of mathematics. And of all the exponential functions, e^x has special properties that make e a good choice for the standard base. Fully understanding these properties requires a knowledge of calculus and would take us far beyond the scope of this text. Indeed, there is a vast literature about e, including an entire book on the subject (see [37]). Our goal in this section is to give some explanation of what distinguishes e from other bases and show a few of its properties.

Euler's Computation of Exponential Functions. One aspect of the significance of e is connected with methods for computing expressions of the form b^x. These were important in the era when e was first identified as a noteworthy constant, long before the invention of modern computing technology. Of course, when b and x are both small whole numbers, we can compute b^x easily using repeated multiplication: e.g., $2^5 = 2 \cdot 2 \cdot 2 \cdot 2 \cdot 2 = 32$. But accurate computation is more difficult when the exponent is not a whole number. For example, without a calculator, how can one find the value of $2^{3.58}$? One approach is based on the strong geometric growth assumption. Thinking of 3.58 as $358 \cdot 0.01$, we first find $a = 2^{0.01}$, then raise *that* to the 358th power. Both of these steps are difficult. The value of a can be obtained by solving the equation $a^{100} = 2$, possibly by systematic trial and error. But computing just one trial, say finding 1.007^{100}, would be no picnic. And even once we get a good estimate for a, which a calculator gives instantly as 1.006956, raising that to the 358th power will still be an arduous task. Before the advent of modern computing, how did mathematicians, scientists, and engineers calculate the values of exponential functions?

One answer is based on a 1748 development by Leonard Euler, one of the most prolific mathematicians in history.[28] Euler's method uses a polynomial to approximate an exponential function. To illustrate the method, consider 2^x for a small value of x. We approximate this using a two-step process:

(1) Multiply x by the constant 0.693. Call the result y.

(2) Compute $1 + \dfrac{y}{1} + \dfrac{y^2}{2 \cdot 1} + \dfrac{y^3}{3 \cdot 2 \cdot 1}$.

---

[28] For more about Euler see [15]. Chapter 2 of that reference includes the ideas sketched here.

4.5. More About e

For example, to compute $2^{0.14}$, approximately, we first find $y = 0.693 \cdot 0.14 = 0.09702$. Then we compute

$$1 + \frac{y}{1} + \frac{y^2}{2 \cdot 1} + \frac{y^3}{3 \cdot 2 \cdot 1} = 1.101878\cdots.$$

We can check this with a calculator, which gives

$$2^{0.14} = 1.101905\cdots.$$

Thus, the estimate is off by less than 0.00003.

Euler knew that the accuracy of this method is progressively worse for larger values of x. For example, the method produces the estimate

$$2^{3.68} \approx 9.566$$

whereas the correct answer is actually $12.817\cdots$. But Euler also knew that the pattern of the computation in step 2 can be extended to as many terms as we like, and that each additional term provides greater accuracy. If we extend the pattern for two more terms, the computation in step 2 becomes

$$1 + \frac{y}{1} + \frac{y^2}{2 \cdot 1} + \frac{y^3}{3 \cdot 2 \cdot 1} + \frac{y^4}{4 \cdot 3 \cdot 2 \cdot 1} + \frac{y^5}{5 \cdot 4 \cdot 3 \cdot 2 \cdot 1}.$$

With this expression, we find

$$2^{3.68} \approx 12.228,$$

which is much closer to the correct answer.

Euler's method does not only apply for powers of 2. It can be adapted to any base, by changing the constant used in step 1. If we want to compute powers of 5 instead of powers of 2, the constant is 1.6094, and the adapted method to compute 5^x is

(1) Multiply x by the constant 1.6094. Call the result y.

(2) Compute $1 + \frac{y}{1} + \frac{y^2}{2 \cdot 1} + \frac{y^3}{3 \cdot 2 \cdot 1} + \cdots$.

This time we have included three dots in the second step to indicate that additional terms in the same pattern should be included to obtain whatever level of accuracy is desired.

As you can see, Euler's two-step method can be applied to calculate any expression b^x, provided we know how to find the appropriate constant for the base b. In fact, having to find the constant accurately is one drawback of the method. For the two examples above, each constant is actually an irrational number, and the decimal values 0.693 and 1.6094 are only approximately correct. In particular, for powers of 2, since 0.693 is only accurate to three decimal places, we should expect that the accuracy of the results will also be limited, no matter how far we extend the pattern in step 2.

Clearly the method would work much better if the multiplier is known exactly. In fact, the best possible case would be for a multiplier exactly equal to 1. In that case, step 1 becomes superfluous, because multiplying x by the constant 1 has no effect. So Euler asked, is there a base, let's call it c, whose step 1 constant is exactly 1? And if so, how can we find this c?

At first glance, these seem like very difficult questions to answer. But what we can say for sure is, the powers of this number c will be easy to compute using Euler's method. To find c^x for any x, we use the pattern in step 2, replacing y by x. That is

$$c^x \approx 1 + \frac{x}{1} + \frac{x^2}{2\cdot 1} + \frac{x^3}{3\cdot 2\cdot 1} + \cdots,$$

where we can continue the pattern on the right for as many terms as we wish, and thus obtain an approximation as accurate as we wish. In particular, applying this with $x = 1$ shows that

$$c \approx 1 + \frac{1}{1} + \frac{1}{2\cdot 1} + \frac{1}{3\cdot 2\cdot 1} + \cdots.$$

In this way we can compute the value of c to any desired number of decimal places.

Of course, this c is really the constant e. In Euler's analysis, it arises naturally as a particularly convenient base for finding accurate approximations of exponential expressions. And as a byproduct, we find that

$$e \approx 1 + \frac{1}{1} + \frac{1}{2\cdot 1} + \frac{1}{3\cdot 2\cdot 1} + \cdots.$$

This has two interpretations. First, we can regard it as a recipe for approximating the irrational constant e to any desired degree of accuracy. In this case, the three dots mean to extend the pattern to however many terms are necessary to obtain the accuracy required. Second, we can imagine extending the pattern forever. In other words, regard the equation as expressing e as a sum of infinitely many terms. In this view, we have an exact equation for e, although it cannot be realized in finitely many steps.

Going further, we can write a similar equation for e^x valid for any x:

$$e^x \approx 1 + \frac{x}{1} + \frac{x^2}{2\cdot 1} + \frac{x^3}{3\cdot 2\cdot 1} + \cdots. \tag{4.13}$$

This has the same two interpretations as the equation for e. And it is this equation for e^x that justifies singling e out as the natural base to use for exponential functions.

After defining e in this way, we can come to a better understanding of Euler's method for other bases. To calculate the value of $2^{0.14}$, for example, one alternative is to convert the expression to one with base e. To do so, we express 2 as a power of e like so:

$$2 = e^{0.693147181\cdots}.$$

The exponent on the right side of the equation is $\ln 2$, as we saw in Section 4.3. Substituting for 2 in our desired calculation, we find

$$2^{0.14} = (e^{0.693147181\cdots})^{0.14} = e^{(0.693147181\cdots)(0.14)}.$$

Or, denoting $(0.693147181\cdots)(0.14)$ by y, we have

$$2^{0.14} = e^y.$$

But we know that powers of e can be computed with (4.13). Thus, we find

$$2^{0.14} = 1 + \frac{y}{1} + \frac{y^2}{2\cdot 1} + \frac{y^3}{3\cdot 2\cdot 1} + \cdots.$$

This is exactly what we did in Euler's two-step process: To find $2^{0.14}$ we multiplied 0.14 by a constant, and then substituted the result for y in the expression $1 + y + y^2/2 + y^3/6$. Originally, we used 0.693 for the constant. Now we know that was only an approximation to the actual value of $0.693147181\cdots$, which equals $\ln 2$. Bringing this discussion full circle, we can now see that the constant required in Euler's two-step method for powers of a base b is none other than $\ln b$. We summarize this observation in the following reformulation of the two-step method.

4.5. More About e

> **Two-Step Approximation of** b^x: For any $b > 0$ and any x, we can compute an accurate approximation of b^x by this two-step process:
>
> (1) Multiply x by the constant $\ln b$. Call the result y.
>
> (2) Compute $1 + \dfrac{y}{1} + \dfrac{y^2}{2 \cdot 1} + \dfrac{y^3}{3 \cdot 2 \cdot 1} + \cdots$,
>
> where we extend the pattern indicated by the three dots for as many terms as necessary for the desired level of accuracy.

It should be emphasized here that Euler's two-step method is not suggested for use in our modern world. A calculator with an exponent key provides immediate and highly accurate approximations of powers such as $2^{3.68}$. In Euler's day the two-step method was of practical interest because it involves only multiplication and addition, operations that can be carried out by hand. And Euler also knew methods to estimate $\ln b$. But what is more significant for our purposes is how Euler's two-step method leads in a natural way to the identification of e as a particularly useful base for exponential functions.

Other Properties of e.

Slope Crossing the y-Axis. As indicated earlier, there are many other special properties of e. We will mention two more here. One has to do with the graphs of curves of the form $y = b^x$ for various choices of the base $b > 0$. Our findings from Section 4.3 tell us these graphs all cross the y-axis at the point $(0, 1)$, and all have similar shapes.[29] The curves differ, however, in that each crosses the y-axis with a different slope. Thus, if you looked at a magnified graph that just showed points very near the y-axis, the curves would look like straight lines. Some are very steep, like 10^x. Others are very flat, like 1.01^x. In between, for some base, the curve crosses the y-axis at a 45-degree angle. What is this base? It is e.

Comparing a^b and b^a. Another special property has to do with comparisons of the following sort: which is greater, $0.6^{1.4}$ or $1.4^{0.6}$? Using a calculator this question can easily be answered. But what we have in mind here is whether there are general properties of exponential functions that allow such questions to be answered without using a calculator. One way to pursue this goal is to look more broadly at the effects of interchanging the base and exponent in an expression such as $0.6^{1.4}$.

In particular, we ask how the curves $y = 0.6^x$ and $y = x^{0.6}$ compare. The first curve is an exponential function with the base 0.6. The second is the power function with the power 0.6. If you graph these two curves, you will see that for some values of x the power function is greater, and for some values of x the exponential function is greater. This makes it difficult to formulate a rule of thumb to predict whether or not $0.6^{1.4}$ is greater than $1.4^{0.6}$. It all depends on the location of $x = 1.4$ relative to the crossing points of the two curves.

A similar situation occurs if we change 0.6 into another number, say 2.5. Because 2.5^x is greater than $x^{2.5}$ for some values of x and less for other values of x, it is difficult

---

[29]This situation is similar to what is shown in Figure 4.13, page 252, except that all the curves in that figure have a y-intercept at 100 instead of at 1.

to predict whether or not $2.5^3 < 3^{2.5}$, without actually computing both 2.5^3 and $3^{2.5}$. And this is almost always true when comparing $y = b^x$ and $y = x^b$ for any base b.

There is just one exception. For just one number b, the exponential function b^x is never less than the power function x^b for positive x's. That one number is e. That is, for all positive x,

$$e^x \geq x^e.$$

For any other base b, there are some positive x values for which $b^x > x^b$, and there are other x values for which $b^x < x^b$. This is another of the many properties that distinguish e as a special choice for the base of an exponential function.

4.5 Exercises

Reading Comprehension.

(1) @The constant e is an irrational number. What does that mean?

(2) On a calculator, the value of e can be found by computing e^x with $x = 1$. On one popular calculator, the result appears as 2.718281828. Should that be understood as the start of the repeating decimal 2.7 1828 1828 1828 $\cdots$? Explain.

(3) @One special property of e is related to the graph of $y = e^x$. What is special about this graph compared to any other curve of the form $y = b^x$, with $b \neq e$?

(4) As discussed in the reading, Euler developed an accurate method for approximating the value of b^x for a given b and a given x using polynomials. Why was that important in Euler's time?

(5) @Why is Euler's method more convenient for approximating e^x than it is for b^x with a base $b \neq e$?

(6) @Which is greater, $2.5^{3.1}$ or $3.1^{2.5}$? Which is greater $1.5^{2.2}$ or $2.2^{1.5}$? Which is greater, $e^{2.9}$ or 2.9^e? What special property of e makes the last question much easier to answer than either of the first two? Explain.

Math Skills.

(7) @Use a polynomial approximation to estimate $2^{0.126}$ as on page 291. Compare with the answer the calculator gives for $2^{0.126}$.

(8) Use a polynomial approximation to estimate $5^{-0.178}$ using the adapted two-step method on page 291. Compare with the answer the calculator gives for $5^{-0.178}$.

(9) @Use a polynomial approximation to estimate $1.06^{2.3}$ using the two-step method shown in a box on page 293. Compare with the answer the calculator gives for $1.06^{2.3}$.

Digging Deeper.

(10) @Instant Interest. Banks typically compound interest for loans or for savings accounts on a monthly or daily basis, but interest can be computed on the basis of any compounding period whatsoever (see the discussion of *Compound Interest* on

4.5. Exercises

page 231). The idea of *continuously* compounded interest is that the compounding period is effectively zero. That means there would be some amount of interest accrued over a minute, or a second, or any fraction of a second. It turns out that instantly compounded interest can be computed using a simple formula with a base e exponential function. With an initial balance of B and an annual interest rate of 4%, the balance including interest after t years would be $Be^{0.04t}$, where t is a continuous variable. Similarly, at an annual interest rate of 2.6%, the formula would be $Be^{0.026t}$. The general formula, with an annual interest rate r expressed as a decimal is Be^{rt}. In this exercise you will explore the rationale for the formula.

For all of the parts below, assume that $100 are deposited in a savings account with an annual interest rate of 4%, and that $B(t)$ is the balance after t years.

a. @If interest is compounded quarterly, what will the balance be after t years?
b. @If interest is compounded monthly, what will the balance be after t years?
c. @If interest is compounded daily, what will the balance be after t years?
d. @Suppose interest is compounded m times per year. What will the balance be after t years?
e. @Using the answer to the preceding question, find an equation for $B(t)$ when interest is paid every hour. [Hint: what is m if interest is compounded hourly?]
f. @Using graphical and numerical methods, compare the function $B(t) = 100e^{0.04t}$ with the $B(t)$ functions found above for quarterly, monthly, daily, and hourly compounding. What do you observe?

5

Mixed Growth Models

In this chapter we will consider what we shall refer to as mixed growth models: models that combine arithmetic and geometric growth.[1] These models involve sequences in which each term is obtained from the preceding term by both multiplying by a constant and adding a constant. The constants may be the same or different. Mixed growth arises in many contexts. One example concerns paying off a debt with regular monthly payments, as in the case of a loan to purchase an automobile. Each month the loan balance increases due to interest charged by the lender, and decreases by the amount of the monthly payment. The interest charge is a fixed percentage of the previous month's balance. This is a form of geometric growth, and its effect can be determined by multiplying the balance by a constant growth factor. Meanwhile the loan payment is a constant amount that is subtracted from the balance. That is a form of arithmetic growth. To calculate the balances after successive payments, both of these aspects must be combined. That gives us a mixed growth sequence.

Following a similar organization as in prior chapters, we focus on the mathematical properties of mixed growth sequences in Section 5.1, followed by a discussion of applications in Section 5.2. Because the functions that arise in mixed growth models are so similar to functions we have already studied, additional sections to consider the properties and applications of these functions will not be needed.

5.1 Properties of Mixed Growth Sequences

The foregoing description of mixed growth can be formalized as follows.

> **Mixed Growth Sequence Verbal Definition:** In a mixed growth sequence each term is found by multiplying the preceding term by a constant and then adding a constant. The constants may be equal or unequal.

---

[1] This terminology, *mixed growth,* is not in common use in mathematics and its applications. We have adopted it for this book to remind students that it combines arithmetic and geometric growth characteristics. In contrast, the terms *arithmetic growth* and *geometric growth* are widely used and quite standard.

As an example we consider the effect of taking repeated doses of a drug. Suppose a patient is taking an antibiotic and we know that in any four-hour period half of the drug in the system will be consumed or eliminated. What happens if the patient takes an initial dose of 160 mg, followed by repeated 120 mg doses every four hours?

In the four hours following the initial dose, the amount of drug in the system decreases from 160 to 80, and then jumps up to $80 + 120 = 200$ with the first repeated dose. In the next four hours the amount of drug decreases from 200 to 100 and then jumps up to $100 + 120 = 220$ with the next repeated dose. We can continue in this fashion to compute the amount of drug in the system immediately after each repeated dose. This produces the sequence

$$160, 200, 220, 230, 235, \cdots,$$

and we recognize that each term is obtained by multiplying the preceding term by 0.5 and then adding 120. Depicting this process in a diagram,

$$160 \xrightarrow{\times 0.5 + 120} 200 \xrightarrow{\times 0.5 + 120} 220 \xrightarrow{\times 0.5 + 120} 230 \xrightarrow{\times 0.5 + 120} 235 \cdots$$

highlights the recursive pattern: multiply by one constant and then add another. That is the key concept of mixed growth.

Difference Equations. The sequence in the antibiotic example conforms to the recursive rule *each term is equal to half the preceding term plus* 120. This translates directly into a difference equation

$$a_{n+1} = 0.5a_n + 120.$$

This is a standard form for mixed growth difference equations. As in earlier chapters, we can understand this as one instance of a general mixed growth difference equation

$$a_{n+1} = ra_n + d, \qquad (5.1)$$

where the parameter r represents the constant multiplier or growth factor, as in geometric growth, and the parameter d is the added constant, as in arithmetic growth. In any specific application, r and d will be replaced by numerical values, as is usual for parameters. Thus, in the preceding example, $r = 0.5$ and $d = 120$.

As in geometric growth, the mixed growth examples we consider will usually involve r values for which $r > 0$. In the case that $r = 1$, (5.1) becomes $a_{n+1} = a_n + d$, which is the standard arithmetic growth difference equation. While this is technically an instance of mixed growth, we usually consider it only as arithmetic growth. Accordingly we usually assume $r \neq 1$ for mixed growth sequences.

Similarly, we usually assume that d is either positive or negative, corresponding to a constant amount either added or subtracted. Otherwise, if $d = 0$, (5.1) becomes $a_{n+1} = ra_n$, which is the standard geometric growth difference equation. While this, too, is technically an instance of mixed growth, we generally consider it only as geometric growth.

These observations about (5.1) and the parameters r and d are summarized below.

5.1. Properties of Mixed Growth Sequences

> **Mixed Growth Sequence Difference Equation:** A mixed growth difference equation is one that can be expressed in the form
> $$a_{n+1} = ra_n + d,$$
> where r and d are constants. Usually, we assume $r > 0, r \neq 1$, and $d \neq 0$.

Note that a mixed growth difference equation might not be stated in the standard form. For example, consider

$$a_{n+1} = \frac{a_n}{4} - 20,$$

which is not in the standard form. Using algebra, the equation can be restated as

$$a_{n+1} = \frac{1}{4}a_n + (-20),$$

which *is* in the standard form, with $r = 1/4$ and $d = -20$. So, as in geometric growth, multiplication by a constant can be understood more broadly to mean multiplication *or division* by a constant. And as in arithmetic growth, addition of a constant can be understood to mean addition *or subtraction* of a constant. Other variations are also possible. You should be able to verify that

$$a_{n+1} = a_n + 3(a_n - 5) + 11$$

is a mixed growth difference equation, even though it is not given in the standard form.

Recognizing Mixed Growth. If we begin with a mixed growth difference equation and an initial value, it is easy to produce terms of the corresponding sequence. For example, if $a_{n+1} = 1.5a_n - 24$ we obtain each term by multiplying the preceding term by 1.5 and then subtracting 24. With $a_0 = 80$, this leads to $a_1 = 80 \cdot 1.5 - 24 = 96$, $a_2 = 96 \cdot 1.5 - 24 = 120$, $a_3 = 120 \cdot 1.5 - 24 = 156$, and so on.

But if we had simply been given the terms

$$80, 96, 120, 156, 210, 291, \cdots$$

how could we tell that the sequence is an instance of mixed growth? There is a test, similar to the ones we have seen in prior chapters. The first step is to compute the first differences for the sequence. Next we compute the growth factors for the differences. If those are all the same, the original sequence is a mixed growth sequence. This is illustrated in Table 5.1 for the preceding example.

Table 5.1. Testing for mixed growth. The growth factors for the first differences are all equal.

n	a_n	1st Diffs	Growth Factors
0	80		
		16	
1	96		1.5
		24	
2	120		1.5
		36	
3	156		1.5
		54	
4	210		1.5
		81	
5	291		

A succinct description of the test is: for a mixed growth sequence, the first differences constitute a geometric growth sequence. For this reason, this test is a natural extension of the tests for arithmetic and quadratic growth. A first step in classifying a sequence of unknown type is to compute the differences. If those are constant, we have an arithmetic growth sequence. Otherwise we can look at the pattern of the differences. Do they exhibit arithmetic growth? Geometric growth? We can test the first possibility by looking at the differences of the differences, i.e., the second differences of the original sequence. If they are constant, the original sequence grows quadratically. Similarly, to test the second possibility, we look at successive ratios of the differences. If they are constant, the original sequence is an instance of mixed growth. This leads to an alternate succinct description: for a mixed growth sequence, the growth factors of the first differences are constant.

Returning to the example, notice that the constant growth factor in the table is equal to the parameter $r = 1.5$ for our sequence. This is true in general. Thus, not only does the test indicate the presence of mixed growth, it also reveals the parameter r. And once we know r, we can use the difference equation and two consecutive terms of the sequence to find d, as follows.

With $r = 1.5$, we know the difference equation for the sequence must have the form

$$a_{n+1} = 1.5a_n + d,$$

and in particular,

$$a_1 = 1.5a_0 + d.$$

Substituting 80 for a_0 and 96 for a_1 leads to

$$96 = 1.5 \cdot 80 + d.$$

Solving this equation shows that $d = -24$. Therefore the difference equation for the sequence is

$$a_{n+1} = 1.5a_n - 24.$$

Note that while it is usually convenient to use this method with the initial terms a_0 and a_1, it is equally valid with any two consecutive terms of the sequence. For example, using the terms 120 and 156 leads to the equation

$$156 = 1.5 \cdot 120 + d,$$

which again shows that $d = -24$.

In a similar fashion we can test any sequence for mixed growth, and find the parameters if the test is positive. To illustrate, let's consider a new sequence,

$$5, 55, 555, 5555, 55555, 555555, \cdots.$$

Is this a mixed growth sequence? To find the answer we construct Table 5.2.

From the table we can tell that a_n is a mixed growth sequence with parameter $r = 10$. Therefore, the difference equation is

$$a_{n+1} = 10a_n + d.$$

To find d we can apply the difference equation with $n = 0$, using the fact that $a_1 = 55$ and $a_0 = 5$. Thus we have

$$55 = 10 \cdot 5 + d,$$

5.1. Properties of Mixed Growth Sequences

Table 5.2. Another positive test for mixed growth. The growth factors of the first differences all equal 10. This shows that the original sequence is an instance of mixed growth with parameter $r = 10$.

n	a_n	1st Diffs	Growth Factors
0	5		
		50	
1	55		10
		500	
2	555		10
		5000	
3	5555		10
		50000	
4	55555		10
		500000	
5	555555		

showing that $d = 5$. And now that we know both parameters, the difference equation is given by

$$a_{n+1} = 10a_n + 5.$$

For a different sort of example, consider the sequence $3, 4, 6, 12, 36, 156, \cdots$. Testing this for mixed growth, we construct Table 5.3. This time the entries in the growth factor column are not all the same, so the given sequence is not an instance of mixed growth.

Table 5.3. This is not a mixed growth sequence. Although the growth factors for the first differences follow a simple pattern, they are not all the same.

n	a_n	1st Diffs	Growth Factors
0	3		
		1	
1	4		2
		2	
2	6		3
		6	
3	12		4
		24	
4	36		5
		120	
5	156		

A word of caution for the mixed growth test: it can lead to the misidentification of geometric growth sequences. For example, suppose we apply the test to the sequence $2500, 2000, 1600, 1280, 1024, \cdots$. As shown in Table 5.4 the test does indicate mixed growth, with $r = 0.8$. However, proceeding as in the earlier examples, we can again find the parameter d. The result in this example is $d = 0$, and the difference equation is $a_{n+1} = 0.8a_n$. Therefore the sequence is an instance of geometric growth.

In fact, every geometric growth sequence is also a mixed growth sequence in a trivial way, with $d = 0$. Consequently, for every geometric growth sequence, the first differences will have constant growth factors. In practice, it is preferred to identify a geometric growth sequence as such, rather than to classify it as a mixed growth sequence. Therefore, as part of testing for mixed growth, the possibility of geometric growth should also be considered. This point is included in the following summary statement.

Table 5.4. Geometric growth sequences test positive for mixed growth. Here, the growth factors for the first differences are all equal, but the original sequence is an instance of geometric growth.

n	a_n	1st Diffs	Growth Factors
0	2500		
		−500	
1	2000		0.8
		−400	
2	1600		0.8
		−320	
3	1280		0.8
		−256	
4	1024		

> **Mixed Growth Sequence Test:** Identify a mixed growth sequence as follows.
>
> (1) Compute the first differences.
>
> (2) Compute the growth factors or ratios of successive differences.
>
> (3) If the growth factors are constant, the original sequence is mixed growth, and possibly also geometric growth.
>
> (4) Either test the original sequence directly for geometric growth, or calculate the mixed growth parameter d. If $d = 0$ the sequence is a geometric growth sequence. Otherwise, it is a mixed growth sequence but not a geometric growth sequence.

Functional Equations. As we have seen in earlier chapters, one approach to seeking a functional equation is to work through several applications of the difference equation and look for a pattern. As an example, we examine a sequence that starts with $a_0 = 50$ and for which each term is obtained following the rule: multiply by $\frac{3}{4}$ and then add 100. The standard form of the difference equation for this rule is

$$a_{n+1} = \frac{3}{4} a_n + 100.$$

However, for this example we will instead use the equivalent form

$$a_{n+1} = a_n \cdot \frac{3}{4} + 100$$

which will permit the pattern we seek to emerge more easily.

Using the difference equation and the initial value of 50 we find

$$a_1 = 50 \cdot \frac{3}{4} + 100.$$

5.1. Properties of Mixed Growth Sequences

For the next term, we compute

$$\begin{aligned}a_2 &= a_1 \cdot \tfrac{3}{4} + 100 \\ &= \left(50 \cdot \tfrac{3}{4} + 100\right)\tfrac{3}{4} + 100 \\ &= 50 \cdot \left(\tfrac{3}{4}\right)^2 + 100 \cdot \tfrac{3}{4} + 100 \\ &= 50 \cdot \left(\tfrac{3}{4}\right)^2 + 100\left(\tfrac{3}{4} + 1\right).\end{aligned}$$

Next, we have

$$\begin{aligned}a_3 &= a_2 \cdot \tfrac{3}{4} + 100 \\ &= \left[50 \cdot \left(\tfrac{3}{4}\right)^2 + 100 \cdot \tfrac{3}{4} + 100\right]\tfrac{3}{4} + 100 \\ &= 50 \cdot \left(\tfrac{3}{4}\right)^3 + 100 \cdot \left(\tfrac{3}{4}\right)^2 + 100 \cdot \tfrac{3}{4} + 100 \\ &= 50 \cdot \left(\tfrac{3}{4}\right)^3 + 100\left[\left(\tfrac{3}{4}\right)^2 + \tfrac{3}{4} + 1\right].\end{aligned}$$

One more:

$$\begin{aligned}a_4 &= a_3 \cdot \tfrac{3}{4} + 100 \\ &= \left[50 \cdot \left(\tfrac{3}{4}\right)^3 + 100 \cdot \left(\tfrac{3}{4}\right)^2 + 100 \cdot \tfrac{3}{4} + 100\right]\tfrac{3}{4} + 100 \\ &= 50 \cdot \left(\tfrac{3}{4}\right)^4 + 100 \cdot \left(\tfrac{3}{4}\right)^3 + 100 \cdot \left(\tfrac{3}{4}\right)^2 + 100 \cdot \tfrac{3}{4} + 100 \\ &= 50 \cdot \left(\tfrac{3}{4}\right)^4 + 100\left[\left(\tfrac{3}{4}\right)^3 + \left(\tfrac{3}{4}\right)^2 + \tfrac{3}{4} + 1\right].\end{aligned}$$

There is a pattern to these results. To make it clearer, the results we have already found are repeated below:

$$\begin{aligned}a_1 &= 50 \cdot \tfrac{3}{4} + 100 \\ a_2 &= 50 \cdot \left(\tfrac{3}{4}\right)^2 + 100\left[\tfrac{3}{4} + 1\right] \\ a_3 &= 50 \cdot \left(\tfrac{3}{4}\right)^3 + 100\left[\left(\tfrac{3}{4}\right)^2 + \tfrac{3}{4} + 1\right] \\ a_4 &= 50 \cdot \left(\tfrac{3}{4}\right)^4 + 100\left[\left(\tfrac{3}{4}\right)^3 + \left(\tfrac{3}{4}\right)^2 + \tfrac{3}{4} + 1\right].\end{aligned}$$

According to this pattern, we expect a_{10} to be

$$a_{10} = 50 \cdot \left(\tfrac{3}{4}\right)^{10} + 100\left[\left(\tfrac{3}{4}\right)^9 + \left(\tfrac{3}{4}\right)^8 + \cdots + \left(\tfrac{3}{4}\right)^2 + \tfrac{3}{4} + 1\right].$$

Notice that this pattern allows us to write an immediate expression for a_{10} without first computing all the preceding terms. If we use the variable n in place of 10, we obtain a functional equation of sorts:

$$a_n = 50 \cdot \left(\tfrac{3}{4}\right)^n + 100\left[\left(\tfrac{3}{4}\right)^{n-1} + \left(\tfrac{3}{4}\right)^{n-2} + \cdots + \tfrac{3}{4} + 1\right]. \tag{5.2}$$

Why is this called a functional equation *of sorts*? It does tell us how to compute a_n as soon as we specify a value of n, and in that sense it expresses a_n as a function of n. But the part of the equation that requires us to add up all the powers of 3/4 is not really in the spirit of a functional equation. For one thing, it includes a pattern indicated with three dots, and therefore is not as concise as we expect for a functional equation. And while the pattern is clear, a functional equation should be explicitly expressed without reference to a pattern. More significantly, the number of computations required by the pattern gets larger for each successive n. To use the pattern with $n = 1000$ requires us to sum 1000 terms. That is not how a functional equation should operate.

A Geometric Sum Shortcut. Fortunately, there is a shortcut that can be applied here, similar to the one we used with quadratic growth sequences (page 130). The new shortcut, which is often referred to as the geometric series formula, says

$$\left(\frac{3}{4}\right)^{n-1} + \left(\frac{3}{4}\right)^{n-2} + \cdots + \left(\frac{3}{4}\right)^2 + \frac{3}{4} + 1 = \frac{1-\left(\frac{3}{4}\right)^n}{1-\frac{3}{4}}.$$

This can be substituted in (5.2), producing the result

$$a_n = 50 \cdot \left(\frac{3}{4}\right)^n + 100 \frac{1-\left(\frac{3}{4}\right)^n}{1-\frac{3}{4}}. \tag{5.3}$$

This is a true functional equation for the sequence a_n. With it, we can easily compute a_n for any value of n. For example, with $n = 15$ we have

$$a_{15} = 50 \cdot \left(\frac{3}{4}\right)^{15} + 100 \frac{1-\left(\frac{3}{4}\right)^{15}}{1-\frac{3}{4}}.$$

This can be entered in a calculator to find

$$a_{15} = 395.32$$

to two decimal places.

Shortly we will see a simpler version of this equation. But first we observe that the preceding analysis can be applied in any mixed model. If the difference equation is $a_{n+1} = ra_n + d$, then in place of (5.2) we would have

$$a_n = a_0 r^n + d(r^{n-1} + r^{n-2} + \cdots + r + 1).$$

The sum in the parentheses, $r^{n-1} + r^{n-2} + \cdots + r + 1$, is called a geometric sum or geometric series.[2] The same shortcut we used before now leads to the equation

$$a_n = a_0 r^n + d\left(\frac{1-r^n}{1-r}\right), \tag{5.4}$$

which is the general version of (5.3). An equivalent form is given by

$$a_n = a_0 r^n + d\left(\frac{r^n - 1}{r - 1}\right).$$

---

[2]In normal conversation, the word *series* has the same meaning as *sequence*. We understand *a series of unfortunate events* to refer to an ordered collection: the first event, the second event, and so on. In mathematical language, however, *series* refers to a sum, in which a list of terms is added. This usage is deeply ingrained, as illustrated by the following anecdote. Antoni Zygmund, a Polish mathematician who fled Europe to escape the Nazis, taught at the University of Chicago. According to legend, he once encountered a group of students who were listening to a radio broadcast of the World Series. When they explained what the World Series is, Zygmund replied "I think it should be called the 'World Sequence.'" [35, p. 139]

5.1. Properties of Mixed Growth Sequences

Both forms are correct. Either one may be used as the functional equation for a mixed model, though we usually prefer the first when $r < 1$ and the second when $r > 1$.

An Alternate Functional Equation. In the example considered above, we found the equation

$$a_n = 50\left(\frac{3}{4}\right)^n + 100\frac{1-\left(\frac{3}{4}\right)^n}{1-\frac{3}{4}}.$$

This can be algebraically rewritten as follows. First, replace $1-\frac{3}{4}$ with $\frac{1}{4}$, obtaining

$$a_n = 50\left(\frac{3}{4}\right)^n + 100\frac{1-\left(\frac{3}{4}\right)^n}{\frac{1}{4}}.$$

Next, rearrange the fraction to combine the factors of 100 and $\frac{1}{4}$:

$$a_n = 50\left(\frac{3}{4}\right)^n + \frac{100}{\frac{1}{4}}\left[1-\left(\frac{3}{4}\right)^n\right]$$
$$= 50\left(\frac{3}{4}\right)^n + 400\left[1-\left(\frac{3}{4}\right)^n\right].$$

Now distributing the factor of 400 leads to

$$a_n = 50\left(\frac{3}{4}\right)^n + 400 - 400\left(\frac{3}{4}\right)^n.$$

Finally, combining the two terms involving $\left(\frac{3}{4}\right)^n$ yields

$$a_n = 400 - 350\left(\frac{3}{4}\right)^n.$$

This version of the equation is much more compact than (5.3), and so more convenient to apply. Moreover, it shows that a_n is a constant plus an exponential function. This fact has important consequences for the graphs and applications of mixed growth sequences, as we shall see presently.

If the same algebraic simplifications are applied to (5.4), we obtain

$$a_n = \frac{d}{1-r} + \left(a_0 - \frac{d}{1-r}\right)r^n.$$

For reasons that will become clear, we introduce a new parameter, $E = \frac{d}{1-r}$, so that the preceding equation becomes

$$a_n = E + (a_0 - E)r^n. \tag{5.5}$$

This can be considered an alternate standard form for a mixed growth sequence functional equation. To illustrate its use, suppose we modify the preceding example retaining r as $\frac{3}{4}$, but changing a_0 to 120 and d to 75. We compute

$$E = \frac{d}{1-r} = \frac{75}{1/4} = 300$$

and

$$a_0 - E = 120 - 300 = -180.$$

Substituting these into (5.5) thus gives us the simplified functional equation

$$a_n = 300 - 180 \left(\frac{3}{4}\right)^n.$$

We have now seen three different versions of a mixed growth functional equation. The alternate form is algebraically simpler than the other two, and so may be more convenient to use for computation. On the other hand, the first two forms show more directly the dependence of the functional equation on the parameters r, d and a_0. Each version of the functional equation is useful in certain contexts, and it is worthwhile to be familiar with all of them. They are repeated below for emphasis.

> **Mixed Growth Sequence Functional Equation:** If a mixed growth sequence has an initial value of a_0 and obeys the difference equation $a_{n+1} = ra_n + d$, then a functional equation can be expressed in any of the following ways:
>
> $$a_n = a_0 r^n + d\left(\frac{1-r^n}{1-r}\right), \tag{5.6}$$
>
> $$a_n = a_0 r^n + d\left(\frac{r^n - 1}{r - 1}\right), \tag{5.7}$$
>
> or
>
> $$a_n = E + (a_0 - E)r^n, \tag{5.8}$$
>
> where $E = \frac{d}{1-r}$.

Validating the Shortcut. The geometric series formula is

$$r^{n-1} + r^{n-2} + \cdots + r^2 + r + 1 = \frac{r^n - 1}{r - 1}. \tag{5.9}$$

This is true for any positive integer n and any real number $r \neq 1$. To see why it works, rewrite it by multiplying both sides of the equation by $r - 1$. That leads to

$$(r - 1)(r^{n-1} + r^{n-2} + \cdots + r^2 + r + 1) = r^n - 1. \tag{5.10}$$

This form can be justified algebraically by multiplying out the two parenthetical expressions on the left. While a true algebraic proof requires us to leave n as a variable, looking at an example with a specific value of n will highlight the key idea. With $n = 3$, we wish to verify that

$$(r - 1)(r^2 + r + 1) = r^3 - 1.$$

But using the distributive law we see that

$$(r - 1)(r^2 + r + 1) = r^3 - r^2 + r^2 - r + r - 1 = r^3 - 1,$$

as desired. Notice that most of the terms cancel, leaving only the product of the first terms in each parenthesis, $(r \cdot r^2)$ and the product of the last terms, $(-1 \cdot 1)$. Carrying out the same steps with $n = 4$ and $n = 5$ produces the same pattern of cancellations. Applying the same reasoning with n as a variable verifies (5.10), and therefore (5.9).

5.1. Properties of Mixed Growth Sequences

Graphs. Mixed growth sequence graphs have the same general shape as graphs of geometric growth sequences. As we have seen, the functional equation for a mixed growth sequence can be expressed as $a_n = E + (a_0 - E)r^n$. We recognize this as a constant added to an exponential function. The graph of the exponential part takes one of the four shapes shown in Figure 4.12 (page 251). The effect of the added constant E is to shift the graph vertically.

As an illustration, consider again the example with functional equation

$$a_n = 300 - 180\left(\frac{3}{4}\right)^n.$$

Here the constant $E = 300$ is added to the exponential function $-180\left(\frac{3}{4}\right)^n$. This exponential part is expressed in the standard form Ab^n with $A = -180 < 0$ and $b = \frac{3}{4} < 1$. For such a function, we know that the graph will lie entirely below the n-axis, and approach that axis from below as the curve is traced to the right. Adding the constant 300 shifts the entire graph upward by 300 units. Both graphs are shown in Figure 5.1. Notice that the shifted graph will approach a horizontal line 300 units above the n-axis. This shows graphically that $a_n < 300$ for all n, and will be very nearly equal to 300 for large n. We describe this situation by saying that a_n approaches an *equilibrium value* of 300.

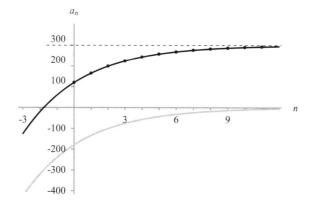

Figure 5.1. The graph for $a_n = 300 - 180\left(\frac{3}{4}\right)^n$. The dots are the points of the sequence. The graph of the exponential part of the functional equation is shown in gray. The graph of a_n lies on the same curve shifted up 300 units, shown in black. Shifting the n-axis up 300 units produces the dashed line, which is approached from below as the black curve is traced from left to right.

The graph in this example reveals characteristics that are found in all mixed growth graphs. In particular, these graphs are shifted exponential curves. Specific details depend on the values of the mixed growth parameters r and d, as well as a_0. In the following remarks we discuss graphical properties that depend on the parameters for mixed growth sequences with $r > 0$.

For an exponential function Ar^n with $0 < r < 1$, the graph will approach the horizontal axis as it is traced from left to right. This shows that in a mixed model with

$0 < r < 1$ the sequence will always approach an *equilibrium value*. It is given by the constant $E = \frac{d}{1-r}$ in (5.8). The letter E was chosen to stand for *equilibrium*.

The terms of the sequence approach E from above when $a_0 > E$, and from below when $a_0 < E$. If $a_0 = E$, the coefficient of the exponential part of the function will be zero. In this case the terms of the sequence are all equal to the equilibrium value E. Indeed, this is what it means to be in equilibrium: the terms of the sequence remain constant. This idea will be discussed more fully later in this section.

We can deduce more from the equation $E = \frac{d}{1-r}$. The denominator of this fraction is positive when $r < 1$, in which case the sign of the fraction is determined by the sign of d. When $d > 0$ the equilibrium value will be positive; when $d < 0$ it will be negative. For example, if a mixed growth sequence obeys the difference equation $a_{n+1} = 0.36a_n + 80$, we can immediately conclude that the terms will approach a positive equilibrium. For $a_{n+1} = 0.36a_n - 80$, on the other hand, the terms will approach a negative equilibrium.

All of these conclusions apply for mixed growth sequences with $r < 1$. When $r > 1$ the exponential part of the functional equation curves steeply upward or downward as it is traced to the right, provided $a_0 \neq E$. In these sequences the terms do not approach an equilibrium, and yet the E value is still significant. For one thing, it is still an equilibrium value. Thus, taking $a_0 = E$ will still produce a constant sequence. In addition, E determines whether a particular a_0 produces a sequence that curves steeply up or down to the right. From (5.8), we see that the coefficient for the exponential part is $A = a_0 - E$. Thus, when a_0 is above E, A is positive so the graph will curve steeply up to the right; when a_0 is below E, A is negative so the graph will curve steeply down to the right.

These ideas are illustrated by mixed growth sequences with $r = 1.5$ and $d = -10$. We find $E = d/(1-r) = -10/-0.5 = 20$. If we take $a_0 = 20$, then $a_1 = 1.5 \cdot 20 - 10 = 20$, and all the following terms are likewise equal to 20. This is the constant sequence case. If a_0 is greater than 20, for example, with $a_0 = 20.1$, then the sequence will increase more and more rapidly as the terms proceed. This is shown by the functional equation, $a_n = E + (a_0 - E)r^n = 20 + 0.1 \cdot 1.5^n$. In contrast, with a_0 less than 20, for example with $a_0 = 19.9$, the functional equation $a_n = E + (a_0 - E)r^n = 20 - 0.1 \cdot 1.5^n$, showing that successive terms of the sequence decrease by ever greater amounts. All three cases are shown in Figure 5.2.

As a final observation, note the distinction between the parameter d and the equilibrium value $E = d/(1-r)$. The first is the added constant in the difference equation, the second is the added constant in the functional equation. To highlight this distinction, let us compare two difference equations, $a_{n+1} = 0.8a_n$ and $a_{n+1} = 0.8a_n + 50$. The first is a geometric growth difference equation, and its functional equation is of the form $A \cdot 0.8^n$. The second is a mixed model, and its functional equation is of the form $A \cdot 0.8^n + 250$. For the mixed growth sequence, there is an extra added constant in both the difference and functional equations, but they are not the same constant. For the difference equation the added constant is $d = 50$. For the functional equation the added constant is $E = 250$. And these two constants are related by the equation $E = d/(1-r)$.

Using the Functional Equation. As we have seen in past chapters, functional equations can be used to answer find-a_n questions and find-n questions. For mixed

5.1. Properties of Mixed Growth Sequences

n	a_n	a_n	a_n
0	20	20.10	19.90
1	20	20.15	19.85
2	20	20.23	19.78
3	20	20.34	19.66
4	20	20.51	19.49
5	20	20.76	19.24
6	20	21.14	18.86
7	20	21.71	18.29
8	20	22.56	17.44
9	20	23.84	16.16
10	20	25.77	14.23
11	20	28.65	11.35
12	20	32.97	7.03
13	20	39.46	0.54
14	20	49.19	−9.19
15	20	63.79	−23.79
16	20	85.68	−45.68

Figure 5.2. Data table and graph for three sequences obeying the same difference equation, $a_{n+1} = 1.5a_n - 10$. Each version of the sequence has a different initial value a_0. In the first a_n column, the initial value is $a_0 = 20$, the equilibrium value. This produces a constant sequence with every term equal to 20. The second sequence has $a_0 = 20.1$, just slightly greater than the equilibrium value. The terms increase very slowly at first, but eventually are seen to increase rapidly. The final sequence has an initial value of $a_0 = 19.9$, just slightly less than the equilibrium value. For this sequence the terms decrease very slowly at first, but eventually are seen to decrease rapidly.

growth sequences, the functional equation can also be used to find the equilibrium value, E, the significance of which was noted earlier. To illustrate these three uses of the functional equation, let us again consider the example for which

$$a_n = 300 - 180\left(\frac{3}{4}\right)^n.$$

What is the twentieth term of the sequence? This is a find-a_n question, and we answer it by evaluating the functional equation with $n = 20$. That leads to

$$a_{20} = 300 - 180\left(\frac{3}{4}\right)^{20}.$$

A calculator gives the value as 299.43, rounded to two decimal places.

When does a_n first reach a value of 295? This is a find-n question. We substitute 295 for a_n in the functional equation, obtaining

$$295 = 300 - 180\left(\frac{3}{4}\right)^n.$$

Now we can solve this equation for n. An effective method is to isolate the part of the equation that involves n and then use methods for exponential equations. Proceeding, we subtract 300 from both sides of the equation,

$$-5 = -180 \left(\frac{3}{4}\right)^n,$$

and then divide both sides by -180,

$$\frac{5}{180} = \left(\frac{3}{4}\right)^n.$$

This equation has the form $b^x = c$ (page 263), and so has the solution

$$n = \frac{\ln(5/180)}{\ln(3/4)}.$$

Using a calculator, we find

$$n = 12.4565,$$

to four decimal places. To check the accuracy of this answer, we substitute it into the functional equation and find

$$300 - 180 \left(\frac{3}{4}\right)^{12.4565} = 294.999964.$$

In the context of a specific model, fractional values of n may not be meaningful. But even if this is the case, we can still conclude that a_{13} is the first term that exceeds 295.

Finally, from the form of the functional equation,

$$a_n = 300 - 180 \left(\frac{3}{4}\right)^n$$

with $r = 3/4 < 1$, we know that the terms of this sequence will approach an equilibrium. What is the equilibrium value? We can identify it as the constant term of the function, namely 300.

Equilibrium Values and Fixed Points. As the preceding example shows, the equilibrium value is easily identified from the functional equation. But an equilibrium can also be found directly from a difference equation. The method is related to an important concept, namely, a *fixed point* of a difference equation.

The basic idea is this: if you begin a sequence with an initial term equal to an equilibrium value, the same value will be repeated over and over when you apply the difference equation. For example, with the difference equation

$$a_{n+1} = 0.8 a_n + 50,$$

the equilibrium value is

$$E = \frac{d}{1-r} = \frac{50}{0.2} = 250.$$

Starting the sequence with $a_0 = 250$, the difference equation gives

$$a_1 = 0.8 \cdot 250 + 50 = 250.$$

This provides a connection between the equilibrium value and the recursive rule of the difference equation: multiply by 0.8 and then add 50. Usually, these steps result in a net change. If we start with 100, for example, multiplying by 0.8 and then adding 50 results in 130. But with a starting value of 250, the process does *not* produce a net

5.1. Properties of Mixed Growth Sequences

change. We say that the value 250 is left fixed by the recursive process and call it a *fixed point* of the difference equation.

Why is 250 left unchanged? Notice that the recursive process has two parts. First, we multiply by 0.8. When we start with a positive quantity, this first step makes it smaller. Next we add 50, and that makes the result larger. For the equilibrium value, these two effects are in balance. The amount of the decrease from the first step exactly matches the amount of the increase from the second. Because the two effects exactly compensate for each other, the number 250 is in a state of equilibrium relative to the difference equation.

In general, an equilibrium value for a difference equation must be a fixed point. We can use this fact to find equilibrium values. We want to find a number E that is unchanged by the recursive process of the difference equation. For the example we have been discussing, that means we want $0.8E + 50$ to equal E. This leads to the equation

$$E = 0.8E + 50,$$

which can be solved to find $E = 250$.

Let us apply the same method to another example. If the difference equation is $a_{n+1} = 0.6a_n + 28$, replace both a_{n+1} and a_n with E. We obtain

$$E = 0.6E + 28.$$

Subtracting $0.6E$ from both sides leads to

$$0.4E = 28,$$

so

$$E = 28/0.4 = 70.$$

You can verify that this value is unchanged by the recursive rule *multiply by 0.6 and add* 28, and so is a fixed point.

Using this same approach with parameters r and d in place of the constants 0.6 and 28, we can rederive the equation $E = d/(1-r)$. So this new approach has not given us a new conclusion. But it is worth understanding for three reasons. First, the new approach is a simpler way to derive the formula for E, compared to algebraically transforming (5.6) into (5.8). Second, the new approach applies to many more kinds of sequences than the original approach. For example, we will see what are called logistic growth sequences in the next chapter. For these sequences, functional equations almost never exist, making it impossible to apply our original method. However, the new method does apply. Third, the new method broadens our understanding of an equilibrium to include the fixed point concept. In many cases this turns out to be a key to understanding how models behave.

The idea of a fixed point is an important one and enters into the analysis of difference equation models in a surprisingly significant way. Interestingly, in a mixed model (and in many other kinds of models) it is not possible to reach a fixed point exactly, unless the model starts there. Instead, the fixed point serves as a kind of ideal state to which the sequence values are attracted. So when we say that the model levels off, that is only approximately true. It is a very good approximation. For all practical purposes, the model can be assumed to reach a state of equilibrium. But the sequence values cannot ever reach the fixed point with perfect, mathematical, equality. For mixed models we can see this from the functional equation, which is an exponential function plus a constant. Because the exponential function is never equal to 0, the functional equation cannot ever reach the constant. The exception is that if the initial value of the sequence equals the fixed point, then the exponential part of the functional equation has a coefficient of zero. In this case, the functional equation is just equal to a constant and the terms of the sequence remain unchanged forever.

Profile for Mixed Growth. The following profile for mixed growth sequences summarizes many of the key ideas presented in this section.

Table 5.5. Profile for mixed growth sequences.

Verbal Description:	Each term is found by multiplying the preceding term by a constant and then adding a constant
Identifying Characteristic:	Growth factors for the first differences are constant; equivalently, first differences form a geometric growth sequence
Parameters:	Initial term a_0, growth factor r, added constant d. Usually $r > 0, r \neq 1$, and $d \neq 0$.
Difference Equation:	$a_{n+1} = r a_n + d$
Functional Equations:	$a_n = a_0 r^n + d \frac{r^n - 1}{r - 1} = a_0 r^n + d \frac{1 - r^n}{1 - r}$ $a_n = E + (a_0 - E) r^n$ where $E = d/(1 - r)$
Graph:	Exponential curve shifted vertically by E units. When $r < 1$ the curve approaches a horizontal line at height E as the graph is traced to the right. It approaches from below if $a_0 < E$ and from above if $a_0 > E$. These two cases are illustrated below. When $r > 1$ the graph curves steeply upward to the right if $a_0 > E$ and curves steeply downward to the right if $a_0 < E$.

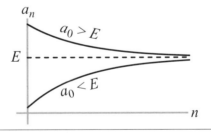

5.1 Exercises

Reading Comprehension.

(1) Explain what is meant by a *mixed model*. Why is it called *mixed*? Include at least one example in your answer.

5.1. Exercises

(2) @Recall the four types of exponential graphs discussed in Figure 4.12 (page 251). Mixed growth sequence graphs have the same general shape as graphs of geometric growth sequences, but differ in one important aspect. Explain that difference. Also describe the similarities and differences in the functional equations of mixed growth sequences and geometric growth sequences. How do the differences in the functional equations correspond to differences in the graphs?

(3) In this section we learned about a parameter E associated with mixed growth sequences.

  a. Explain what the parameter E is and how it is found.

  b. Describe at least three aspects of a mixed growth sequence that are related to the parameter E.

(4) @In this section the idea of a *fixed point* was introduced. Write an explanation of fixed points. Your answer should include what a fixed point is, how it relates to an equilibrium point, how a fixed point can be found, and what it reveals about a difference equation model.

Math Skills.

(5) @The following are mixed growth sequences. For each one, (i) determine the parameters d and r, (ii) write the difference and functional equations, and (iii) provide the graph. The graph may be accurately hand drawn or produced with a graphing application.

  a. @1, 3.3, 6.29, 10.177, 15.2301, $\cdots$

  b. 10, 8, 6.2, 4.58, 3.122, $\cdots$

  c. 5, 6, 7.7, 10.59, 15.503, $\cdots$

  d. 100, 57, 39.8, 32.92, 30.168, $\cdots$

(6) @Decide whether each of the following sequences is an instance of arithmetic growth, geometric growth, quadratic growth, mixed growth, or none of those. For each sequence that is arithmetic, quadratic, geometric, or mixed growth, find the difference equation and functional equation. Assume the first term is a_0.

  a. @8, 16, 20, 22, 23, 23.5, $\cdots$

  b. 81, 54, 36, 24, 16, 32/3, $\cdots$

  c. @12.45, 14.85, 17.25, 19.65, $\cdots$

  d. 1,000, 900, 790, 669, 535.9, $\cdots$

  e. @100, 80, 64, 52, 44, 40, $\cdots$

  f. 1.3, 2.6, 3.9, 5.2, 6.5, $\cdots$

(7) @State for each of the following graphs whether it most resembles arithmetic growth, geometric growth, quadratic growth, or mixed growth. Explain your choice. State any parameter values that you can determine from the graph.

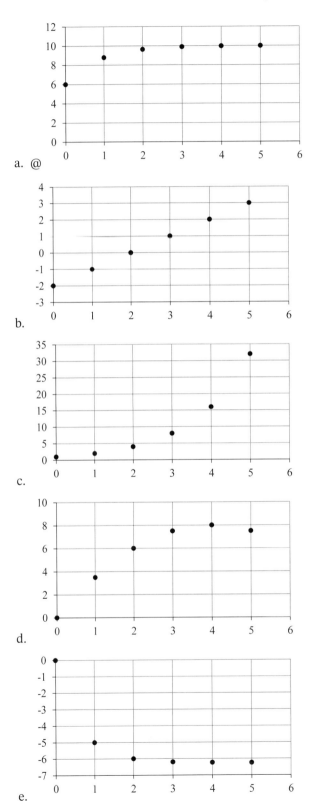

a.

b.

c.

d.

e.

5.1. Exercises

(8) In a mixed growth growth sequence, the recursive pattern is to multiply by 1.02 and then subtract 50 each time. If the starting value is 1,200, find the following.

   a. The first 5 terms of the sequence.
   b. A difference equation for the sequence.
   c. A functional equation for the sequence.

(9) @In a sequence, each new number is found by first adding 18 to the previous number, and then dividing the result by 3. The starting value is 90.

   a. @Find the first 5 terms of the sequence.
   b. @Is this a mixed growth sequence? If so, find the difference and functional equations. If not, explain why not.

(10) @Determine the fixed point for each of the difference equations below. Choose a value for a_0 as indicated and sketch a possible graph for a sequence generated using the difference equation. Does the sequence level off at the fixed point? Is this consistent with the definition of E? Explain.

   a. @$a_{n+1} = 0.75a_n + 10$. Choose a value for a_0 less than the fixed point.
   b. $a_{n+1} = 2a_n + 5$. Choose a value for a_0 greater than the fixed point.
   c. $a_{n+1} = 1.08a_n + 30$. Choose a value for a_0 less than the fixed point.
   d. $a_{n+1} = 0.94a_n + 200$. Choose a value for a_0 greater than the fixed point.

(11) @A mixed growth sequence has the difference equation $a_{n+1} = 1.6a_n - 95$ and starts with $a_0 = 500$.

   a. @Find a functional equation.
   b. @Find a_{20}.
   c. @When does the sequence first reach or exceed 10,000?
   d. @Does this sequence level off? Why or why not?

(12) A mixed growth sequence has the difference equation $a_{n+1} = 0.75a_n - 10$ and starts with $a_0 = 50$.

   a. Find a functional equation.
   b. Find a_{20}.
   c. When does the sequence first reach or fall below 0?
   d. Does this sequence level off? Why or why not?

(13) @A mixed growth sequence has the difference equation $a_{n+1} = 0.4a_n + 200$ and starts with $a_0 = 125$.

   a. @Find a functional equation.
   b. @Find a_{20}.
   c. @When does the sequence first reach or exceed 300?
   d. @Does this sequence level off? Why or why not?

(14) In the reading we saw an example with the difference equation
$$a_{n+1} = a_n + 3(a_n - 5) + 11.$$
Verify that this is a mixed growth difference equation, even though it is not given in a standard form, by converting it with algebra to the form $a_{n+1} = ra_n + d$.

(15) @Each part below shows a functional equation from a mixed growth sequence. Use algebra to rewrite the given functional equation in the form of a constant added to an exponential function, that is, in the form $a_n = E + (a_0 - E)r^n$. Note that $a_0 - E$ may be expressed as a single number.

   a. @$a_n = 100 \cdot 1.05^n - 50 \left(\frac{1.05^n - 1}{0.05} \right)$

   b. $a_n = 500 \cdot \left(\frac{4}{5} \right)^n + 350 \left(\frac{1 - \left(\frac{4}{5} \right)^n}{\frac{1}{5}} \right)$

(16) Consider a sequence with the following difference equation $a_{n+1} = 2.2a_n + 3$.

   a. Using $a_0 = -2.5$ find the first 5 terms of the sequence.
   b. Using $a_0 = -2.4$ find the first 5 terms of the sequence.
   c. Explain why the behavior of the sequence is so very different for the two initial values.

(17) @Consider the difference equation $a_{n+1} = 0.75a_n + 3$ and initial value $a_0 = 20$.

   a. @Write the functional equation in the form $a_n = E + (a_0 - E)r^n$ where $a_0 - E$ may be expressed as a single number.
   b. @Provide a graph of the sequence. Include a horizontal line at $y = E$ on your graph.
   c. @Is $r < 1$ or $r > 1$? Is the graph approaching an equilibrium value as the position number, n, increases?
   d. @Is $a_0 < E$ or $a_0 > E$? Is the graph above or below the equilibrium value?

(18) A mixed growth sequence has the difference equation $a_{n+1} = 1.3a_n + 12$ and starts with $a_0 = 40$.

   a. Find the functional equation in the form $a_n = E + Ar^n$ where $A = a_0 - E$.
   b. Provide a graph of this sequence.
   c. What can be said about this sequence in terms of equilibrium values and fixed points? Review the discussion of equilibrium values and fixed points, then make your answer as thorough as possible.

(19) @A mixed growth sequence has the difference equation $a_{n+1} = 1.95a_n - 50$ and starts with $a_0 = 40$. Answer parts a through e by algebraic or theoretical methods before examining a graph of the sequence. Then answer part f.

   a. @Find the functional equation in the form $a_n = E + Ar^n$ where $A = a_0 - E$.
   b. @Determine when a_n first reaches or falls below 0.
   c. @What is the equilibrium value?
   d. @Does the sequence level off? How can the parameters be used to determine this?
   e. @Are successive terms of the sequence increasing or decreasing? How can the parameters be used to tell?
   f. @Provide a graph of this sequence. Does the graph confirm your answers to the preceding items?

5.1. Exercises

(20) A mixed growth sequence has the difference equation $a_{n+1} = 0.3a_n + 72.8$ and starts with $a_0 = 600$. Answer parts a through e by algebraic or theoretical methods before examining a graph of the sequence. Then answer part f.

   a. Find the functional equation in the form $a_n = E + Ar^n$ where $A = a_0 - E$.
   b. Use the functional equation to determine when a_n first reaches or falls below 105.
   c. What is the equilibrium value?
   d. Does the sequence level off? How can the parameters be used to determine this?
   e. Are successive terms of the sequence increasing or decreasing? How can the parameters be used to tell?
   f. Provide a graph of this sequence. Does the graph confirm your answers to the preceding items?

(21) @Consider the graph below. This is the graph of a mixed growth sequence, with a smooth line passing though all the sequence points.

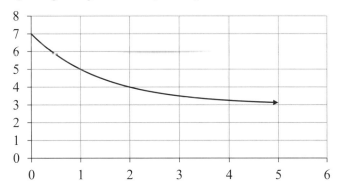

   a. @Use the graph to determine E and a_0.
   b. @What does the shape of the graph tell you about r? Explain.
   c. @The graph is above the horizontal line $y = E$. What does this tell you about $a_0 - E$? Is this consistent with the values you found for E and a_0 above?

(22) Consider the graph below. This is the graph of a mixed growth sequence, with a smooth curve connecting the sequence points.

a. Use the graph to determine a_0.

b. What does the shape of the graph tell you about r? Explain.

c. Could E be less than a_0 for this sequence? Explain.

(23) In the text we used a geometric sum shortcut: $r^{n-1} + r^{n-2} + \cdots + r^2 + r + 1 = \frac{r^n - 1}{r - 1}$ for $r \neq 1$. This was verified for $n = 3$ on page 306. Carry out the same steps with $n = 4$ and $n = 5$ to get a better sense of why the shortcut works.

Digging Deeper.

(24) Using the difference equation approach,[3] investigate whether fixed points can be found for arithmetic growth, quadratic growth, and geometric growth difference equations. Give at least one example and a summary of your findings for each type of model.

(25) @In this section we saw that $5, 55, 555, 5{,}555, \cdots$ is a mixed growth sequence with $r = 10$ and $d = 5$. Let us denote the terms of the sequence in the form a_n, so that $a_0 = 5$, $a_1 = 55$, and so on.

a. @Find a functional equation for this sequence in the form of (5.7). Without using a calculator, verify that this equation gives the correct values for a_4, a_5, and a_{10}.

b. @Find a functional equation for this sequence in the form of (5.8). Without using a calculator, verify that this equation gives the correct values for a_4, a_5 and a_{10}. [Hint: $5/9 = 0.5555 \cdots$.]

c. @Based on your results for the preceding questions, find a functional equation for the sequence $3, 33, 333, 3{,}333, \cdots$.

(26) Consider a sequence that starts with $a_0 = 2$ and that conforms to the recursive rule: add 5 to the preceding term and then multiply the result by 11. This is similar to the definition of a mixed growth sequence, but the operations of multiplying and adding are done in the opposite order. The point of this exercise is to show that adding a constant first and them multiplying by a constant always results in a mixed growth sequence.

a. For the sequence a_n defined in the above statement, with $a_0 = 2$, find the next 5 terms.

b. Find a difference equation for this sequence.

c. Use algebra to express your difference equation in the standard form for a mixed growth sequence, and identify the values of r and d.

d. Complete this sentence: For given constants v and w, a sequence satisfying the difference equation $a_{n+1} = (a_n + v)w$, is a mixed growth sequence with parameters $r = \underline{\qquad}$ and $d = \underline{\qquad}$.

(27) @Consider the mixed growth sequence b_n with $b_0 = 200$ and parameters $r = 0.75$ and $d = 150$. Let s_n be the sequence of sums $s_0 = b_0$, $s_1 = b_0 + b_1$, $s_2 = b_0 + b_1 + b_2, \cdots$.

---

[3] See *Equilibrium Values and Fixed Points* starting on page 310.

a. @Show that the functional equation for b_n can be expressed in the form $b_n = 600 - 400(0.75)^n$.

b. @Show that the sequence of sums satisfies the difference equation $s_{n+1} = s_n + 600 - 400(0.75)^{n+1}$.

c. @Use the difference equation for s_n to find a functional equation for s_n. [Hint: review the derivation of a functional equation for mixed growth sequences, and apply a similar method.]

(28) A mixed growth difference equation is defined by $a_{n+1} = 1.3a_n + 2.4$ and $a_0 = 5$. Define a new sequence by $s_{n+1} = s_n + a_n$ and $s_0 = 0$. Determine a pattern for s_n and use it to find a functional equation for s_n.

5.2 Applications of Mixed Growth Sequences

Like the other kinds of models we have considered, mixed growth models may be developed either because of the structure of the system being modeled, or because graphical or numerical properties of the data appear to be compatible with mixed growth. These will be illustrated with several extended examples in this section, starting with structural examples. Generally, calculated decimal values have been computed to full accuracy, then rounded to the number of places shown. Exceptions to this rule include certain exact decimal expressions such as 0.75 for 3/4.

Structural Example 1: Drug Dosage Model. In Section 5.1 we saw an example of mixed growth involving repeated doses of a drug. Here we consider this type of model in greater detail. Actually, we will look at a family of models, all with similar assumptions and with parameters whose values depend on the context for a particular model. In all the models, it is assumed that drugs are removed from the body according to a geometric growth law: in any time period of a fixed length, the amount of drug decreases by a fixed percentage. We also assume that doses of the drug are administered at a regular interval, such as every four hours. The amounts of these repeated doses are all the same, except for the initial dose which may be of a different amount. Under these assumptions, we consider the amount of drug present in the body immediately after each dose. These constitute the terms of a mixed growth sequence.

As an example, suppose that for one particular drug the amount present in the body decreases by 1/4 every four hours, and that there is an initial dose of 100 mg followed by repeated 100 mg doses administered every four hours. Immediately after the initial dose, there are 100 mg of the drug present in the system. Over the next four hours that amount decreases by 1/4, leaving 3/4 of the original amount, or 75 mg. Then the next dose adds 100 mg more. At that point the amount of drug in the system is

$$\frac{3}{4} \cdot 100 + 100 = 175.$$

Now repeat the process. After four hours, 3/4 of the 175 mg remain, and then another 100 mg are added. Immediately after taking this dose the amount is

$$\frac{3}{4} \cdot 175 + 100 = 231.25.$$

Continuing in the same fashion, we recognize a recursive pattern in which each term is multiplied by 3/4 and then increased by 100. This shows that the terms satisfy the difference equation

$$a_{n+1} = \frac{3}{4}a_n + 100,$$

where a_n is the amount of drug in the system immediately after the nth repeated dose.

Both the recursive rule and the difference equation tell us that we are dealing with a mixed growth sequence, with parameters $d = 100$ and $r = 3/4$, and with initial term $a_0 = 100$. Each of these constants have specific interpretations in the context of our model: a_0 is the amount of the initial dose, d is the amount of each repeated dose and r represents the decay factor between doses. Put another way, r tells us what fraction or percentage of the drug is retained from one dose to the next. For our specific example, with $r = 3/4$, we know that three quarters of the initial dose will remain just before the first repeated dose is taken. Likewise, whatever amount of drug is present immediately after any repeated dose, three quarters of that amount will remain just before the next repeated dose. As we consider variations of the repeated drug dosage model, these interpretations of d and r will always apply.

Recognizing that our sequence is an instance of mixed growth, we apply our findings from Section 5.1. Because the parameter $r = 3/4$ is less than one, we know that the terms approach an equilibrium value E. We can find E using the fixed point method, by solving the equation

$$E = \frac{3}{4}E + 100.$$

The result, $E = 400$, is also confirmed using the equation

$$E = \frac{d}{1-r}$$

with $d = 100$ and $r = 3/4$.

Next notice that the initial term, a_0, is less than E. We conclude that the terms will increase from the initial value of 100 and level off as they approach 400, the equilibrium value.

Applying numerical and graphical methods produces Figure 5.3 and Table 5.6. Both the data and the graph confirm our conclusions: The amount of drug in the body increases rapidly at first, then levels off and seems to remain nearly steady at about 400 mg.

Table 5.6. Data from the drug dosage model with $d = 100$ and $r = 3/4$.

n	0	1	2	3	4
a_n	100.00	175.00	231.25	273.44	305.08
n	5	6	7	8	9
a_n	328.81	346.61	359.95	369.97	377.47
n	10	11	12	13	14
a_n	383.11	387.33	390.50	392.87	394.65
n	15	16	17	18	19
a_n	395.99	396.99	397.74	398.31	398.73

5.2. Applications of Mixed Growth Sequences

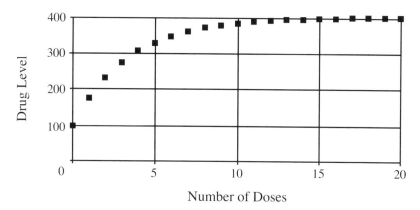

Figure 5.3. Graph for the drug dosage model with $d = 100$ and $r = 3/4$.

We can also formulate the functional equation for this model. In the simplified form of (5.8), we find

$$a_n = 400 - 300 \left(\frac{3}{4}\right)^n. \qquad (5.11)$$

As ever, the functional equation can be used to compute a_n directly for a specific value of n. Thus, we can determine that after 48 hours, the amount of drug in the system will be

$$a_{12} = 400 - 300 \left(\frac{3}{4}\right)^{12} = 390.50,$$

to two decimal places. We can also use the functional equation to answer find-n questions. As a specific instance, if we wish to know when the amount of drug will reach 399 mg, we solve the equation

$$399 = 400 - 300 \left(\frac{3}{4}\right)^n$$

for n. Repeating the method shown in Section 5.1, we isolate the exponential term, obtaining first

$$300 \left(\frac{3}{4}\right)^n = 400 - 399 = 1,$$

then

$$\left(\frac{3}{4}\right)^n = \frac{1}{300}.$$

This shows that

$$n = \frac{\ln(1/300)}{\ln(3/4)}$$

or, in approximate decimal form, $n = 19.83$. Here, we recall that n only makes sense for integer values, because we only consider a whole number of repeated doses. Accordingly, we interpret the result $n = 19.83$ to mean that the amount of drug will not rise above 399 miligrams until at least the twentieth repeated dose, which is administered 80 hours after the initial dose. To confirm this conclusion, we can compute a_{20} directly from the functional equation, finding

$$a_{20} = 400 - 300 \left(\frac{3}{4}\right)^{20} = 399.05.$$

Interestingly, for this model, these sorts of functional equation computations are not as important as in earlier examples. We know that the terms of the sequence are approaching 400, and by the fifteenth repeated dose, the terms have increased above 395. It is unlikely to be important, for therapeutic reasons, to predict precisely what a_{20} or a_{50} will be. We know that all the terms beyond a_{15} will be between 395 and 400, and that they will continue to get closer to 400 as n increases. Without a need to compute a_n precisely for arbitrary values of n, the functional equation has limited utility.

Fractional Values of n. In the current context, as in many of the examples in earlier chapters, the variable n can be assigned a meaning for fractional values. We can consider n to represent time following the initial dose, in four hour increments. Then, for example, $n = 3.5$ corresponds to a time half way between the third and fourth repeated doses, or equivalently, two hours after the third repeated dose. The amount of drug in the system at that time would be denoted $a(3.5)$ instead of $a_{3.5}$, following our usual convention of using function notation for values of n that are not integers. It is also true that the functional equation (that is, (5.11)) can be evaluated with $n = 3.5$. However, the result will *not* be $a(3.5)$. In other words, substituting $n = 3.5$ in the functional equation does not produce the amount of drug in the system two hours after the third repeated dose.

We can see this through a combination of logic and graphical analysis. By logic, we know that $a(3.5)$ should be less than $a(3)$, reflecting the fact that the body continuously consumes or eliminates the drug between repeated doses. In particular, during the entire period between the third and fourth repeated doses, the amount of drug in the body is decreasing. This tells us that $a(3.5)$ will be less than $a(3)$. In contrast, we know that the graph of the functional equation is a shifted exponential, and is curving steadily upward. Substituting $n = 3.5$ into the functional equation therefore produces a result that is more than $a(3)$. Combining both conclusions, we see that the functional equation cannot produce correct values of $a(n)$ when n is not a whole number.

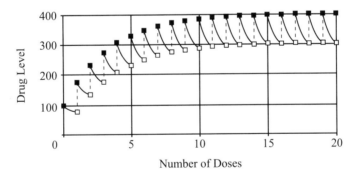

Figure 5.4. Graph for the drug dosage model with fractional values of n. The vertical dashed lines represent the abrupt increases that occur each time a repeated dose of the drug is administered. Each is immediately followed by a geometric decay curve showing how the drug is consumed or eliminated in the interval before the next dose.

This conclusion can be illustrated graphically. The graph in Figure 5.3 shows only the individual points for a_n with whole number values of n. But we can imagine the

5.2. Applications of Mixed Growth Sequences

graph for $a(n)$ with n a continuous variable. Between two of the dots in Figure 5.3 there must be an exponential curve going downward to the right. This is illustrated in Figure 5.4. Thus the graph of $a(n)$ is made up of a collection of downward sloping arcs, with abrupt vertical shifts indicated by the dashed vertical lines. This is clearly not the graph of a shifted exponential function, showing that the functional equation does not produce accurate results for fractional values of n.

Adjusting the Dosages. So far we have focused on a specific example. With parameters $r = 3/4$, $a_0 = 100$, and $d = 100$, we found an equilibrium level of 400 mg. Perhaps we would prefer the equilibrium to be higher, or lower, for a particular patient. As long as we are considering one specific drug, and intend to repeat doses every four hours, the value of the parameter r will remain fixed. But the attending doctors can easily vary the size of the initial dose, a_0, or the repeated dose, d. By modifying these parameters it is possible to change the equilibrium level.

Using numerical and graphical methods we can analyze the effect of changing the parameters. As a first step, let us compare the effects of several different values of the initial dose. We keep $r = 3/4$ and $d = 100$, and consider initial doses of 40, 160, 280, 400, 520, and 640 mg. Applying the difference equation, we compute a_n for several values of n, presenting the results in Table 5.7 and Figure 5.5.

Table 5.7. Comparing drug dosage models with different initial terms a_0. Each line of the table shows a_0 through a_8 for a model with $r = 3/4$ and $d = 100$.

a_0	a_1	a_2	a_3	a_4	a_5	a_6	a_7	a_8
40.00	130.00	197.50	248.13	286.09	314.57	335.93	351.95	363.96
160.00	220.00	265.00	298.75	324.06	343.05	357.29	367.96	375.97
280.00	310.00	332.50	349.38	362.03	371.52	378.64	383.98	387.99
400.00	400.00	400.00	400.00	400.00	400.00	400.00	400.00	400.00
520.00	490.00	467.50	450.63	437.97	428.48	421.36	416.02	412.01
640.00	580.00	535.00	501.25	475.94	456.95	442.71	432.04	424.03

Both the table and the graph give the impression that in all of these models, the terms a_n approach an equilibrium of 400. This suggests that the initial value has no effect on the equilibrium value E and only influences how quickly the equilibrium is reached. In fact, if we administer an initial dose of 400 mg, equilibrium is achieved immediately.

The hypothesis that initial value has no effect on equilibrium value is supported by examining the equation

$$E = \frac{d}{1-r}.$$

This shows that once values of r and d are specified, the equilibrium value is determined. Because a_0 does not appear in the equation, changing the value of a_0 has no effect on E.

The preceding comments illustrate that an algebraic expression provides a highly efficient means for understanding how parameter values influence a model. Looking at several examples numerically and graphically, as we did first, produces concrete and

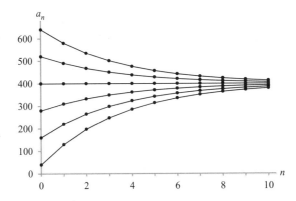

Figure 5.5. Graph for drug dosage models with different initial values. Terms of each sequence are shown as dots, joined by straight lines for visual emphasis. Each initial value and the succeeding ten terms are shown.

readily understood evidence. Although this approach is more concrete than analyzing the equation, it is also much more time consuming. More significantly, the approach is inconclusive. There is always the possibility of observing a pattern in the results that does not extend to cases that have not been considered. The algebraic approach is both more general and more immediate.

As another example of this approach, we can use the equation for E to see how changing the amount of the repeated doses affects the model. Because we are keeping r fixed at 3/4, the equation becomes

$$E = \frac{d}{1 - \frac{3}{4}} = \frac{d}{\frac{1}{4}} = 4d.$$

This establishes a simple rule for the specific models we have been examining, where $r = 3/4$: the equilibrium level will be four times the amount of the repeated doses. This is consistent with the examples we have considered so far, with a repeated dose of 100 mg and an equilibrium value of 400 mg. And it provides guidance for defining dosage amounts. If we wish the patient to reach an equilibrium level of 600 mg, then we should prescribe repeated doses of 150 mg. To reach this level as rapidly as possible, an initial dose of 600 mg should be administered. Alternatively, if the drug produces fewer side effects when it is introduced gradually, we can set a much lower initial dose, such as 75 mg. As shown in Figure 5.6, the numerical and graphical results confirm that with an initial dose of 75 mg and repeated doses of 150 mg, the drug levels in the system immediately after each repeated dose gradually increase to the desired equilibrium level of 600 mg.

Broadening our analysis to include drugs with different rates of absorption or elimination, and to other intervals between doses, we can consider the effects of varying the parameter r. For our initial examples we assumed that 1/4 of the drug is removed from the system between doses, and found $r = 1 - 1/4 = 3/4$. That in turn led to the equation $E = 4d$. We can also express this equation in the form $\frac{1}{4}E = d$. Similarly, if we assume that 1/3 of the drug is removed between doses, then $r = 1 - 1/3 = 2/3$. The fixed point

5.2. Applications of Mixed Growth Sequences

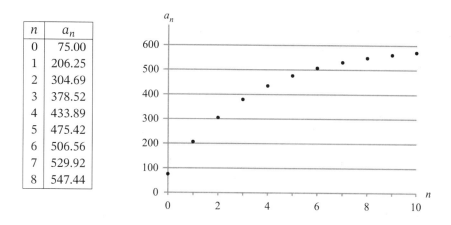

Figure 5.6. Data table and graph for a drug dosage mixed model with $r = 3/4$, $a_0 = 75$, and $d = 150$. As predicted, the amount of drug in the system after each repeated dose increases gradually from the initial amount, approaching an equilibrium value of 600 mg.

equation,

$$E = \frac{d}{1-r}$$

now becomes

$$E = \frac{d}{1-\frac{2}{3}} = \frac{d}{\frac{1}{3}} = 3d.$$

Again we can rewrite this equation in the form $\frac{1}{3}E = d$. The pattern that emerges here holds for any fraction. Thus, if for some drug, 3/7 of the drug disappears from the system between doses, then the equilibrium level and the repeated dosage amount are related by the equation $\frac{3}{7}E = d$. Or, working with decimals, if 0.45 of the drug disappears between doses, then $0.45E = d$. This is also easily expressed in terms of percentages: the body removes 45% of the drug between doses, and the repeated dose amount is 45% of the equilibrium level. Expressed this way, it is immediately clear why these results occur. At equilibrium the increase due to the repeated dose must exactly equal the amount of the drug eliminated or absorbed between doses.

As our analysis of the drug model illustrates, mixed models are rather more involved than the simplest models we studied in the introductory chapters. Here we begin to see how models can provide significant insights about a problem context. In the particular case of the drug model, the idea of an equilibrium level has been a central concern. Next we turn to a mixed model that does not approach an equilibrium, and for which our methods of analysis are a bit different.

Structural Example 2: Repeated Loan Payments. Have you ever wondered how installment payments are worked out for a purchase on credit? Say you buy a $2,000 stereo and spread the payments over two years. How does the store or finance company figure out how much you should pay each month? They consider that you have been loaned $2,000, and every month they charge interest on the unpaid balance of the loan.[4] Part of each payment you make is to pay off the interest. The rest of your payment is used to repay part of the loan. The monthly payment is calculated so that this process results in a zero balance after a prespecified number of payments.

For a given loan amount, interest rate, and monthly payment, the process of paying off the loan can be modeled using a mixed growth sequence. The methods of Section 5.1 apply, predicting how many payments are needed to fully repay the loan. We will look at several examples in detail, observing how the number of payments depends on the model parameters. Ultimately this will allow us to determine what monthly payment is needed to reach a zero balance in a specified number of months, though the discussion leading up to that conclusion will extend over several pages.

As a first step, let's consider a specific example, purchasing a $2,000 stereo on January first. We will assume that the interest charged is one percent per month. On February first, there is a payment due. You have incurred one percent interest on the $2,000 loan, amounting to $20. But your payment will be more than just the interest, otherwise the loan will never get paid off. So say that the monthly payment is $50. This amount was not chosen with any rationale, other than being greater than $20, and we do not expect it to be the correct amount to repay the loan in two years. Indeed, ignoring interest for a moment, 24 payments of $50 would amount to $1,200, which is less than the purchase price. However, our goal in this example is merely to understand the loan repayment process. Refining the model to reach a zero balance after two years will come later.

Let us work out the process in detail. For the first $50 payment the interest is $20. The remaining $30 pays off part of the loan. That leaves a balance of $2,000 − $30 = $1,970. Next, on March first, another payment is due. This time you only incur interest on $1,970. That amounts to one percent of $1,970, or $19.70. Your $50 payment includes the $19.70 of interest, with the remaining $30.30 reducing the loan balance. After the second payment, the amount you still owe is $1,970 − $30.30 = $1,939.70. Now you can see that it will be pretty tedious to work this out all the way to the point where the loan is paid off. Fortunately, we can use difference equations to model the process.

To formulate a difference equation, we focus on how much is still owed after each payment. In symbols, let b_n be the balance in dollars after n payments. Then $b_0 = 2,000$, because after no payments you owe the full amount. In the discussion above, we computed $b_1 = 1,970$ and $b_2 = 1,939.30$. Now we need to see a pattern in those calculations showing how the balance owed changes with each payment. Reviewing the steps followed before, observe that we computed one percent of the old balance, subtracted the result from 50, and then subtracted *that* result from the old balance. In words, we can write this as follows

$$\text{next balance} = \text{current balance} - [50 - 0.01 \cdot (\text{current balance})].$$

---

[4] For installment loans, the interest is usually designated in terms of an annual percentage rate, referred to as *APR*. This is converted to a monthly rate for the purposes of computation. Thus, if the APR is 8%, the monthly interest rate will be 8/12% = 2/3%. However, an interest charge of 2/3% per month will accumulate to something more than 8% over a year, due to the effect of compounding. For this reason, the APR is referred to as a *nominal* rate. In contrast, the disclosures for most credit cards with revolving balances specify the monthly interest rate directly. In either case, the computations are based on a monthly rate, and that is how the interest rates will be defined in the examples of this section.

5.2. Applications of Mixed Growth Sequences

That translates to this difference equation:
$$b_{n+1} = b_n - (50 - 0.01b_n).$$
Using a little algebra, we can simplify this slightly.
$$\begin{aligned} b_{n+1} &= b_n - (50 - 0.01b_n) \\ &= b_n - 50 + 0.01b_n \\ &= 1.01b_n - 50. \end{aligned}$$
This is another mixed growth difference equation. It has the form
$$b_{n+1} = rb_n + d$$
with $r = 1.01$ and $d = -50$. Notice that the growth factor r is exactly what we would have found in a geometric growth sequence with a constant 1% increase between consecutive terms. The value of d represents a constant subtraction with each successive term. So we can see that the difference equation translates directly as a combination of geometric and arithmetic growth. Each month the balance grows by 1% and decreases by $50, defining the parameters r and d of the mixed growth difference equation.

As was the case for the repeated drug dosage models, the initial term b_0 and the parameters r and d each have an interpretation in the loan payment context: b_0 is the original amount of the debt, d is the negative of the monthly payment, and r is the growth factor due to interest. Because the interest is defined in terms of constant percentage increase, we can express r as one plus the monthly interest rate, expressing the percentage as a decimal. That is, with an interest rate of 2% we would find $r = 1 + 0.02 = 1.02$.

Proceeding with our example, with a payment amount of $50, we can use the difference equation to find the first few b_n's. The results are shown in a table and graph in Figure 5.7. As expected, the successive balances are declining, but we would have to extend the table and graph to much greater values of n to find where the balance becomes zero.

Let us find the functional equation for his sequence. As a preliminary step, compute the equilibrium
$$E = \frac{d}{1-r} = \frac{-50}{1-1.01} = 5{,}000.$$
This leads us to the functional equation
$$\begin{aligned} b_n &= E + (b_0 - E)r^n \\ &= 5{,}000 + (2{,}000 - 5{,}000)1.01^n \\ &= 5{,}000 - 3{,}000 \cdot 1.01^n. \end{aligned}$$

A balance of $0 will occur when
$$0 = 5{,}000 - 3{,}000 \cdot 1.01^n,$$
or equivalently, when
$$3{,}000 \cdot 1.01^n = 5{,}000.$$
Dividing both sides by 3,000,
$$1.01^n = 5/3,$$
so
$$n = \frac{\ln 5/3}{\ln 1.01} = 51.34.$$

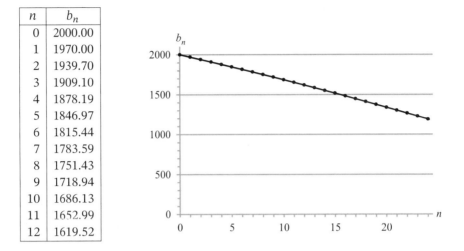

Figure 5.7. Data table and graph for installment payment model, with initial balance of $2,000, one percent per month interest charges, and monthly payments of $50. The table shows the initial balance and the balances after the first twelve payments. The graph shows the first 24 payments.

This shows that after 51 payments there will still remain a balance, though not enough to require another full payment. In fact, we can compute $b_{51} = 16.77$ to the nearest penny. Making one additional payment of $50 would lead to

$$b_{52} = -33.07,$$

indicating an overpayment of $33.07. Thus, the 52nd and final payment would be reduced by $33.07 to $16.93, completing the repayment of the loan.

Reflecting on the Model. We developed our difference equation based on structural knowledge of the problem context. In contrast to many of the models we have considered, in this case our structural knowledge is exactly correct. In essence, the rules governing interest charges and payments are part of a mathematical system that has been adopted by our financial institutions, and our analysis is conducted within that system. It is possible to predict the interest charges and payments each month, exactly. Nevertheless, our difference equation is *not* perfectly correct, because it does not take into account the way financial figures are rounded to two decimal places.

To demonstrate this effect, let us recompute the first few balances, assuming that the lender always rounds the outstanding balance up to the next cent. According to the difference equation, we would have

$$b_1 = 1.01 \cdot 2000.00 - 50 = 1970.00$$
$$b_2 = 1.01 \cdot 1970.00 - 50 = 1939.70$$
$$b_3 = 1.01 \cdot 1939.70 - 50 = 1909.0970.$$

However, the lender would round the third balance to $1909.10. Thus, the value of b_3 predicted by the difference equation does not exactly match the lender's figure. Notice

5.2. Applications of Mixed Growth Sequences

that whenever rounding occurs, it has the effect of increasing the balance. Also each successive b_n includes interest on the preceding value. For both of these reasons, the effects of rounding will grow and accumulate.

Usually, the errors that arise in this way are not significant. For the specific example we have been considering, the model predicts that the loan would be retired after 52 payments, with a final payment of $16.93. Including the effects of rounding shows that the final payment would actually be $17.23. So the prediction of the model is incorrect by 30 cents. Although this is not so great an error as to invalidate the model, it is something to be aware of. More importantly, this example illustrates again the process of reflecting on a model.

Changing the Monthly Payment. The preceding analysis shows that the entire loan will be paid off in a little more than four years if the monthly payment is $50. We return to the original question: What should the payment be to reach a zero balance after two years? Evidently, we should increase the amount of each payment to decrease the number of payments. More generally, we proceed to consider what will happen if we set either a higher or lower monthly payment. That means altering only the parameter d. The initial balance of $2,000 will not change. Likewise, if we keep the interest charge fixed at 1% per month, the parameter r will remain constant at 1.01. Thus, we are interested in geometric growth sequences with $b_0 = 2,000$, $r = 1.01$, and various values of d.

Based on a general understanding of loan payments, we anticipate three possibilities. First, as in the example we have already considered, the loan balances may decrease every month, eventually resulting in a zero balance. Installment loan plans are designed to bring about this result. Second, if the payment is too low, the loan balances can actually increase. This occurs when each loan payment is insufficient to cover the preceding month's interest. Third, the balance can be held constant by paying just the interest each month.

For our specific example, what monthly payment will hold the monthly balances constant? The interest charge is 1% per month. With an initial balance of $2,000, the first month's interest will be $20. Paying just that amount will leave the balance unchanged, and the same computations will apply month after month. Thus, the balances will remain constant if the monthly payment is $20, which corresponds to $d = -20$.

This is the dividing line between the other two cases. If the payment is more than $20, meaning $d < -20$, then the balances will decrease. If the payment is less than $20, so $d > -20$, then the balances will increase.

These conclusions are consistent with the findings of Section 5.1. Recall that the equilibrium amount is given by $E = d/(1-r)$. In our general results for $r > 1$, we found that the terms of the sequence will remain constant when $b_0 = E$, increase when $b_0 > E$, and decrease when $b_0 < E$. For our example $r = 1.01$ so

$$E = \frac{d}{1-1.01} = \frac{d}{-0.01} = -100d.$$

Thus, the condition $b_0 < E$ becomes

$$2,000 < -100d,$$

or

$$-20 > d,$$

where the inequality symbol is reversed because both sides were divided by a negative quantity. Therefore, the condition $b_0 < E$ is equivalent to $d < -20$. Similar logic shows that the conditions $b_0 = E$ and $b_0 > E$ are equivalent to $d = -20$ and $d > -20$. Thus can we verify our earlier conclusions: the balances will decrease when $d < -20$, increase when $d > -20$, and remain constant when $d = -20$.

The reader might wonder whether E has an interpretation, similar to interpretations we have seen for $b_0, d,$ and r. In the case of a mixed model with $r < 1$, we know that the sequence terms approach E as an equilibrium. However, that is not the case when $r > 1$, as in the repeated loan payment context. In a purely mathematical sense, we can identify E as the added constant in the functional equation, and observe that the sequence terms are moving consistently away from E, increasing if the initial term exceeds E and decreasing if the initial term is less than E. But beyond these mathematical aspects, there does not appear to be an interpretation of E that is specific to the repeated loan payment context.

This is actually quite common in the development of complex mathematical models. While some of the parameters and constants have specific contextual meanings, there are frequently other quantities that are useful mathematically but have no ready interpretation in the problem context. Sometimes such a quantity is so useful in so many different contexts that it acquires a meaning on that account, and is referred to using special terminology. A specific example occurs in geometric decay models. We have seen functional equations in the form $a_n = a_0 r^n$, and the initial value and the parameter r have natural interpretations. However, the same equation can also be expressed in the form $a_n = a_0 e^{-kn}$ for a particular constant k. While k has no obvious intrinsic meaning, it is useful mathematically, just as E is in the repeated loan payment model. The constant k arises so often that it has been given a name, the *decay constant* for the model. Repeated use attaches a meaning to the idea of decay constant, reflecting the properties of the models in which it appears. Thus an experienced modeler might know immediately what a decay constant of 0.1 or 1 or 5 signifies.

Increasing the Payment. The preceding analysis shows that the loan balances will decrease as long as the monthly payment is greater than $20. For the specific case of a $50 monthly payment, we found that it takes over four years to completely pay off the loan.

With a higher payment, we expect the loan to be paid off sooner. But how much sooner? If we triple the monthly payment to $150, will the loan be paid off three times as fast? If so, the number of payments should be reduced from 52 to approximately 17. Is that what happens?

To answer this question we repeat the steps from the original example, this time with $d = -150$, again obtaining a table and a graph (Figure 5.8). This time we see that b_n becomes negative for the first time after the fifteenth payment. In practice, we would reduce the fifteenth payment by $92.60 to $57.40 so that after the payment, instead of a balance of -92.60, the remaining balance would be zero.

The same conclusion can be found using a functional equation. With a monthly payment of $150 the equilibrium amount becomes

$$E = \frac{d}{1-r} = \frac{-150}{-0.01} = 15{,}000.$$

5.2. Applications of Mixed Growth Sequences

n	b_n
0	2000.00
1	1870.00
2	1738.70
3	1606.09
4	1472.15
5	1336.87
6	1200.24
7	1062.24
8	922.86
9	782.09
10	639.91
11	496.31
12	351.27
13	204.79
14	56.84
15	−92.60

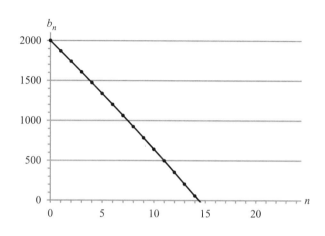

Figure 5.8. Data table and graph for a modified installment payment model, with monthly payments of $150. The graph shows an intercept on the n-axis between 14 and 15. This means that if the fifteenth payment is made in full, the balance would be reduced below zero. In other words, just before the fifteenth payment, the balance is less than the monthly payment of $150. In practice, a final payment of less than $150 would be made to reduce the balance to zero.

Thus the functional equation is
$$b_n = 15{,}000 - 13{,}000 \cdot 1.01^n$$
so we can determine when the balance reaches zero by solving
$$0 = 15{,}000 - 13{,}000 \cdot 1.01^n.$$
Using the same method as before, this leads to
$$n = \frac{\ln(15/13)}{\ln(1.01)} = 14.38.$$
This confirms what we found before.

Thus we see that by tripling the amount of the monthly payment, the number of payments is reduced from 52 to 15. That is actually a little more than three times faster, but an estimate of 17 payments is pretty close.

Further Reflection. This example shows that using proportional reasoning is not exactly correct but does provide a reasonably accurate method to determine how long it will take to pay off a loan. Examining the figures for the two variations of this model, we see that the graphs are nearly straight lines. It is not surprising, therefore, that proportional reasoning is approximately valid. But loan payments extended over longer periods do not have graphs that are nearly linear. As an example, Figure 5.9 depicts the situation for a thirty year mortgage. This time the nonlinearity of the b_n model

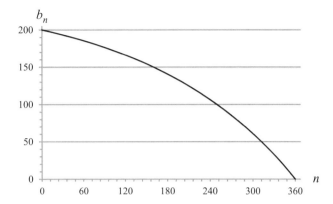

Figure 5.9. Graph for a thirty year mortgage. For this model, the loan amount is $200,000 and the interest rate is 0.48% per month, or nominally 5.76% per year. The monthly payment is $1,168.42. Therefore, the loan payment model has parameters $b_0 = 200{,}000$, $r = 1.0048$ and $d = -1{,}168.42$. The units for the axis labels are thousands of dollars for b_n and months for n.

is readily observed. Because this model is not nearly linear, proportional reasoning cannot be expected to provide reasonably accurate predictions. As a case in point, for this model, tripling the monthly payment would reduce the number of payments from 360 to 67. In contrast, if we applied proportional reasoning, we would expect the higher payment amount to reduce the number of payments to one third of 360, or 120. That is too high by nearly a factor of two.

Paying Off the Loan in Two Years. In the mortgage example, the payment is defined so that the loan will be paid off in exactly 360 payments. How is that determined? To consider a similar question, we return to the two earlier examples with a loan $2,000 and an interest rate of 1% per month. What should the monthly payment be if we want to pay off the loan in 24 months?

In the first example we found that a monthly payment of $50 would require 52 months to pay off the loan. That is too long. In the second example we found a payment of $150 leads to paying off the loan after 15 months. That is too fast. Presumably, for some payment amount between 50 and 150, the balance will just reach zero with the 24th payment. Systematic trial and error is one method to find the correct payment amount. However, repeating the full analysis of the preceding examples, with a table and a graph for each new payment amount, will be too time consuming. A more efficient option is provided by working exclusively with the functional equation.

To illustrate, let us consider as a third example, a payment amount of $100. That means $d = -100$, and
$$E = \frac{-100}{-0.01} = 10{,}000.$$
Thus, evaluated at $n = 24$, the functional equation is
$$b_{24} = 10{,}000 - 8{,}000 \cdot 1.01^{24}.$$
With a calculator we find $b_{24} = -157.88$. Finding that b_{24} is negative shows that a payment of $100 per month is still too much.

5.2. Applications of Mixed Growth Sequences

To try again with $d = -90$, we could repeat all of the manipulations above. However, anticipating that we will be repeating the same manipulations several more times, we can use algebra to streamline the process. The idea is to leave d as an unknown quantity, and do the manipulations one time. That is, with

$$E = \frac{d}{-0.01} = -100d,$$

the functional equation is

$$\begin{aligned} b_n &= E + (b_0 - E)r^n \\ &= -100d + (2{,}000 + 100d)1.01^n \\ &= (2{,}000 + 100d)1.01^n - 100d. \end{aligned}$$

Taking $n = 24$ now produces

$$b_{24} = (2{,}000 + 100d)1.01^{24} - 100d.$$

Now we can easily substitute various values for d and compute b_{24} for each. This provides an efficient means to find a value of d for which $b_{24} = 0$. A few trials are shown in the table below.

d	-100.00	90.00	95.00	-94.00	-94.50	-94.20	-94.10
b_{24}	-157.88	111.86	-23.01	3.96	-9.52	-1.43	1.27

The results show that a monthly payment of $94.10 is too low, leaving a positive balance after 24 payments, but that a monthly payment of $94.20 is too high, leaving a negative balance after 24 payments. With a bit more effort we could determine the correct payment down to the penny.

However, using algebra again leads to even greater efficiency. As is often the case, we can replace systematic trial and error by solving an equation algebraically. Returning to the equation

$$b_{24} = (2{,}000 + 100d)1.01^{24} - 100d,$$

we use algebra to combine the two terms containing d. The first step is to distribute 1.01^{24} into the parenthetical quantity, producing

$$b_{24} = 2{,}000 \cdot 1.01^{24} + 100d \cdot 1.01^{24} - 100d.$$

Then we factor $100d$ from the last two terms to obtain

$$b_{24} = 2{,}000 \cdot 1.01^{24} + 100d(1.01^{24} - 1). \tag{5.12}$$

Now we can set b_{24} equal to 0, and solve

$$2{,}000 \cdot 1.01^{24} + 100d(1.01^{24} - 1) = 0$$

for d. The result is

$$d = \frac{-2{,}000 \cdot 1.01^{24}}{100(1.01^{24} - 1)}.$$

The result is -94.15, indicating that the monthly payment should be $94.15. To check that this is correct, we substitute $d = -94.15$ into (5.12), and find $b_{24} = -0.08$. That is as close as we can get to reaching a balance of zero after 24 equal payments.

A Monthly Payment Formula. In the foregoing discussion we used functional equations and our understanding of mixed models to determine a monthly payment in one specific example. Having done this analysis once, we need never do it again. In the equation

$$d = \frac{-2{,}000 \cdot 1.01^{24}}{100(1.01^{24} - 1)}$$

it is possible to identify how each numerical constant arises from the original problem context. We use 24 because we wish to make 24 payments; 1.01 is the parameter r for an interest rate of one percent per month; 2,000 is the original amount borrowed; and the factor of 100 is equal to $1/(r-1)$. The equation computes the value of d, which is the negative of the monthly payment. Using the same pattern you can make calculations for any monthly payment problem. If the interest is 1.25 percent per month instead of 1 percent, it should be clear what to change in the equation. Likewise, for the mortgage example considered earlier, we know:

- There will be 360 payments
- The interest rate is 0.48% per month, so $r = 1.0048$
- The amount borrowed is 200,000.

Therefore we can calculate the monthly payment using the equation

$$d = \frac{-200{,}000 \cdot 1.0048^{360}}{208.33(1.0048^{360} - 1)}.$$

(The factor of 208.33 is the approximate value of $1/(1.0048 - 1)$.)

Having to keep track of exactly what number goes where can be a little confusing. It is for precisely this situation that we use parameters. In fact, we can replace all of the specific constants in our equation with parameters, as follows

$$d = -\frac{b_0 \cdot r^N}{\frac{1}{r-1}(r^N - 1)}.$$

The negative of this d is the monthly payment necessary to repay an initial debt of b_0 in N equal payments with an interest growth factor of r each month. We can make the parameters a little more memorable as follows. First, suppose that the interest charge is I each month, so a 1% charge corresponds to $I = 0.01$, a 1.4% monthly interest charge corresponds to $I = 0.014$, and so on. Then our growth factor is $r = 1 + I$ and $r - 1 = I$. Also, we write B for the original amount of the debt, and P for the monthly payment. Note that $P = -d$. Now writing the preceding equation using the new parameters leads to

$$P = \frac{BI(1+I)^N}{(1+I)^N - 1}. \tag{5.13}$$

As an example of using this equation, let us compute the monthly payment for a five year automobile loan of $10,000 at a quarter percent per month. The parameters are

$B = 10{,}000,$ the amount borrowed, in dollars

$I = 0.0025,$ the interest charged each month, expressed as a decimal, and

$N = 60,$ the number of payments.

5.2. Applications of Mixed Growth Sequences

Substituting in (5.13) gives

$$P = \frac{10{,}000 \cdot 0.0025 \cdot 1.0025^{60}}{1.0025^{60} - 1}.$$

This allows us to compute the payment $P = 179.69$.

Almost nobody uses this formula in this way. People who need to compute payment amounts typically use calculators or web pages that are preprogrammed to make the calculation for them. All they have to do is enter the parameters B, I, and N into an appropriate application. So we are not suggesting that this equation is important to know or understand for the common person. Rather, we have derived this equation to illustrate several important points about models and modeling. First, it is worthwhile to see how the methods of mixed models can be used to obtain an equation like (5.13). The equation is complicated enough to appear completely mysterious, if simply viewed in its final form. But as we have just seen, using a mixed growth model to analyze installment loan plans leads to the equation through a series of understandable steps. Second, we emphasize the fact that we have solved an inverse problem here. A functional equation allows us to find, for any particular payment plan, what the loan balance will be after a specified number of months. But what we are really interested in is defining the payment plan so that after a specified number of months the balance is zero. The inverse nature of these alternatives may be easier to see in the specific case of our example, where the loan was to be paid off in 24 months. With the direct use of the functional equation, once we specify the payment amount $-d$, we can find b_{24}. The inverse problem is to specify $b_{24} = 0$ and find the corresponding value of d. Third, as is often the case with inverse problems, systematic trial and error can be used to investigate the problem numerically. For this purpose the functional equation from our model is extremely useful. Fourth, while trial and error will find an answer for us, it can be time consuming and inefficient. Instead of trying many different values of the parameter d, it is better to leave that as an unknown and then apply an algebraic method. Then we find the desired value in one step by solving an equation, instead of systematically producing closer and closer approximations to the answer. And finally, the preceding analysis shows again how useful it is to include parameters in a family of models. This approach provides a kind of leverage, extending an analysis from a single context to a broad range of similar contexts. Instead of working out the monthly payment for one specific problem, we find an equation giving the monthly payment for any combination of loan amount, interest rate, and number of payments.

We have now looked in depth at two instances of structural mixed models. In both cases, repeated drug doses and repeated loan payments, a mixed growth difference equation arises structurally from the problem context. Mixed growth also arises structurally when we consider sums or running totals for a geometric growth sequence. We turn to that topic next.

Structural Example 3: Geometric Sum Models. It is often of interest to add up the terms of a number sequence. For example, suppose we study an epidemic, using a discrete variable c_n to model the number of *new* cases of the disease each day. As a simplification we count only the patients who become ill during the period covered by the model, and assume that each person who becomes infected remains in the infected patient population for the duration of the model. Then initially there are no infected people, and the c_0 new cases on day zero comprise the total number infected by the end of that day. Similarly there will be $c_0 + c_1$ infected people at the end of day 1, $c_0 + c_1 + c_2$ infected people at the end of day 2, and so on. The pattern here should be apparent: the total number of infected people at the end of day n equals the sum $c_0 + c_1 + \cdots + c_n$.

We have considered this situation before. In general, summing the terms of a sequence in this way gives rise to a new sequence referred to as the *sums* or *running totals* of the original sequence. In Chapter 3, we saw that summing the terms of an arithmetic growth sequence always produces a quadratic growth sequence. Now we will find that summing the terms of geometric growth sequence always produces a mixed growth sequence. As a specific example, we consider a model for the growth of vision correction surgery in the twentieth century.[5]

A Surgical Procedure Model. Radial keratotomy (RK), an early form of surgical vision correction, became popular in the early 1990s. A news story from 1994 [4] reported exponential growth in the number of patients receiving this treatment each year, with 30,000 procedures performed in 1989 and a projected 250,000 for 1994. Assuming geometric growth means that the number of procedures each year will grow by a constant factor r. In units of thousands, the initial value of 30 would grow to $30r$ in 1990, $30r^2$ in 1991, and so forth. Therefore, for 1994 we would find $30r^5$, and that is supposed to be 250. Thus we find the equation

$$250 = 30r^5.$$

This leads to

$$r = (25/3)^{1/5}$$

which a calculator gives as 1.528. Therefore, assuming geometric growth means that the number of procedures had been growing by a factor of approximately 1.528 each year.

Now a 1994 market analyst might have used this approximate growth factor to develop a model of future growth as follows. Take year zero to be 1994, and let p_n be the projected number of RK procedures, in units of thousands, in year n. Then $p_0 = 250$. Assuming an annual growth factor of $r = 1.528$, we formulate the difference equation

$$p_{n+1} = 1.528 p_n.$$

Next let us compute several terms of the sequence, and also tabulate the running totals, as in Table 5.8.

The sequence of sums $s_0, s_1, s_2, \cdots$ is a mixed growth sequence. We can test that using the method discussed in Section 5.1. We compute the first differences of the sum sequence, and then the ratios of successive entries in the differences column. The results are shown in Table 5.9. Because the ratios are all equal, we conclude that the terms $s_0, s_1, s_2, \cdots$ belong to a mixed growth sequence.

We can also see that something similar will happen any time we compute the sums of a geometric growth sequence. Notice that the differences in Table 5.9 are consecutive terms of the original sequence p_n. For example, with

$$s_5 = p_0 + p_1 + p_2 + p_3 + p_4 + p_5$$

and

$$s_6 = p_0 + p_1 + p_2 + p_3 + p_4 + p_5 + p_6,$$

---

[5]The model is based on the same data as an exercise in Section 4.2.

5.2. Applications of Mixed Growth Sequences

Table 5.8. Values of terms p_n and s_n for $0 \leq n \leq 6$. Each s_n value equals the sum $p_0 + p_1 + p_2 + \cdots + p_n$.

n	p_n	s_n
0	250.00	250.00
1	382.00	632.00
2	583.70	1215.70
3	891.89	2107.58
4	1362.80	3470.39
5	2082.36	5552.75
6	3181.85	8734.61

Table 5.9. Verifying that the sums $s_0, s_1, s_2, \cdots$ constitute a mixed growth sequence.

n	s_n	1st Diffs	Factors
0	250.00		
		382.00	1.528
1	632.00		
		583.70	1.528
2	1215.70		
		891.89	1.528
3	2107.58		
		1362.80	1.528
4	3470.39		
		2082.36	1.528
5	5552.75		
		3181.85	
6	8734.61		

the difference between s_5 and s_6 is p_6. In the same way, the difference between any two consecutive terms of the sum sequence will be one term of the original sequence. We know that one way to recognize a mixed growth sequence is that the first differences make up a geometric growth sequence. If we are looking at the cumulative sums of a geometric growth sequence, then the differences of those sums reproduce the original geometric growth sequence. Therefore the sum sequence is an instance of mixed growth.

Returning to the RK procedure example, now that we know the sums form a mixed growth sequence, we can find the difference and functional equations. The difference equation has the form
$$s_{n+1} = rs_n + d.$$
Moreover, we know that the parameter r is the constant ratio of 1.528 in Table 5.9, so the difference equation is
$$s_{n+1} = 1.528 s_n + d.$$
Applying this with $n = 0$ leads to
$$s_1 = 1.528 s_0 + d.$$
But we can read $s_1 = 632$ and $s_0 = 250$ from Table 5.9. Thus we have
$$632 = 1.528 \cdot 250 + d,$$
and that implies
$$d = 250.$$

Notice that for this example the mixed growth parameters in the sum sequence depend in a simple way on the original geometric growth sequence. Specifically, the mixed growth parameter d equals the initial term p_0 of the geometric growth sequence, and the mixed growth parameter r equals the r parameter of the geometric growth sequence. These observations are valid whenever we consider the sums of a geometric growth sequence. The reader is invited to explain why.

Proceeding to the functional equation, with $s_0 = 250$, $r = 1.528$, and $d = 250$, we know that
$$s_n = E + A \cdot 1.528^n$$

where
$$E = \frac{d}{1-r} = \frac{250}{-0.528} = -473.48$$
and
$$A = s_0 - E = 250 + 473.48 = 723.48.$$
Therefore,
$$s_n = 723.48 \cdot 1.528^n - 473.48.$$

We can check this result by comparing it with the entries in Table 5.8. For example, with $n = 4$ the equation becomes
$$s_4 = 723.48 \cdot 1.528^4 - 473.48 = 3{,}470.37,$$
to two decimal places. This closely agrees with the value of s_4 in the table. The 0.02 discrepancy can be attributed to using two decimal place approximations for the values A and E.

We can also use the functional equation to obtain values of s_n beyond those in the table. Let us find the total number of RK procedures for years 0 through 10. Setting $n = 10$ we find
$$s_{10} = 723.48 \cdot 1.528^{10} - 473.48 = 49{,}721.48.$$
This means a total of 49,721,480 procedures, because s_n and p_n are in units of thousands.

Projections of this sort might be useful for forecasting the potential costs or revenues associated with RK procedures over an extended period of time. On the other hand, the pool of potential RK patients is limited to a particular subset of the population, based on age, affluence, medical suitability, and other factors. With a total U.S. population on the order of 300 million, it may be unrealistic to project nearly 50 million people will have RK procedures. In fact, we would not expect the geometric growth observed in the early years following the introduction of RK to be sustainable indefinitely. But how long might it be sustained? As much as 10 years? 20 years? The model for s_n provides one tool for answering such questions, by showing how the cumulative number of procedures would grow if the geometric growth in p_n continues.[6]

The preceding development illustrates how we can find a functional equation for the sums of a geometric growth sequence. These sums will always be a mixed growth sequence. Using the methods of Section 5.1 we can find the parameters, and hence the functional equation. In turn, the functional equation provides an efficient tool for accumulating geometric growth results over an extended period.

This entire development could be used by our imagined RK market analysts to consider variations of the model. As originally formulated, our model has a growth factor of $r = 1.528$. That implies an increase of nearly 53% in the number of RK procedures performed each year. But how confident can we be that this growth will occur? By repeating the analysis with higher and lower growth rates, we can derive a range of values for the number of RK procedures expected over the next ten years. This is a more realistic approach than to assume one particular growth rate, and it provides some insight about how changes in the assumed rate will affect our projections.

---

[6] A different limitation of the geometric growth model for p_n is that it makes no allowance for technological improvements. As it turned out, RK was supplanted by laser vision correction after only a few years.

5.2. Applications of Mixed Growth Sequences

Sum sequences can be of interest in many different contexts. We have taken an extended look at their use for projecting growth in the delivery of one type of vision correction procedure. At the start of this discussion we mentioned more briefly an example with the spread of a disease. If we model the number of new patients each day as a sequence $p_0, p_1, p_2, \cdots$, then the sequence of sums tracks the total number of patients with the disease. This might be useful for predicting the amount of medication that will be needed. As another example, suppose a sequence $h_0, h_1, h_2, \cdots$ tabulates the number of new houses built each year in a particular city. The sum sequence would then track the cumulative total of new houses built over a range of years. Such a model would be helpful to urban planners if the city has a limited number of lots available for construction. If the number of new houses built each year grows geometrically, then the cumulative number built follows a mixed growth sequence. All of these examples have in common the rate of utilization of a fixed resource: potential RK patients, or doses of medicine, or residential lots. As a final example in this section, we present an actual model that was used in analyzing global demand for copper.

A Copper Demand Model. Based on data published by the U.S. Geological Survey,[7] Figure 5.10 shows the annual global production of copper from 1900 through 2011. Although the data graph is jagged and irregular, the general shape is approximated fairly well by an exponential function, the graph of which appears as a smooth black curve in the figure. This exponential function is given by

$$y = 0.7 \cdot 1.029^x, \tag{5.14}$$

where x is in years, starting with $x = 0$ in 1900. The constant 1.029 is the annual growth factor for the exponential function, corresponding to a constant increase of 2.9% per year.

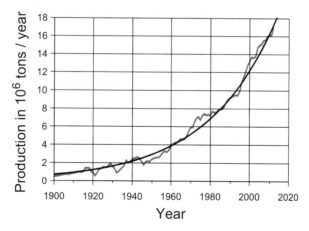

Figure 5.10. Annual global copper production. The jagged gray curve is based on data published by the U.S. Geological Survey. The smooth black curve represents an exponential function that approximates the data.

Assuming that copper production is a fairly accurate indicator of the demand for copper, we can formulate a model for future global demand for copper, as follows. We

---

[7]The graph of the USGS data is from [24]. We have added the approximating exponential curve.

will initiate the model in 2015, taking that as year $n = 0$. Our goal is to define a sequence $c_0, c_1, c_2, \cdots$, with c_n representing the demand for copper in year n. As in the graph, we will express c_n in units of millions of tons. And we assume that future demand will grow at about the same rate as in the past, that is, by 2.9% per year. Thus we construct a geometric growth sequence with parameter $r = 1.029$.

The initial value, c_0, is the demand in 2015. Based on (5.14), with $x = 115$ in 2015, we estimate $c_0 = 18.744$. Then, using our knowledge of geometric growth sequences we find the functional equation

$$c_n = 18.744 \cdot 1.029^n.$$

Going a step further, let us also consider the sum sequence defined by

$$s_n = c_0 + c_1 + c_2 + \cdots + c_n.$$

For example, s_3 is the cumulative demand for copper from 2015 through 2018. From the earlier analysis in the RK example, we know that summing the copper demand sequence in this way defines a mixed growth sequence, with $r = 1.029$ and $d = c_0 = 18.744$. That is, the sum sequence follows the difference equation

$$s_{n+1} = 1.029 s_n + 18.744.$$

Proceeding to the functional equation, we compute

$$E = d/(1-r) = 18.774/(1-1.029) = -646.34$$

and

$$A = s_0 - E = 18.744 + 646.34 = 665.08.$$

Thus

$$s_n = Ar^n + E = 665.08 \cdot 1.029^n - 646.34.$$

As a sample application, we find

$$s_{25} = 665.08 \cdot 1.029^{25} - 646.34 = 712.78.$$

This projects that the global demand for copper will total nearly 713 million tons over the period from 2015 through 2040. This is comparable to an analysis published in 2013 addressing the possibility of a shortage of copper over the succeeding quarter century.[8]

Structural Example 4: Heat Transfer. In Chapter 4 we considered models for the temperature of a warm object that is cooling down (or a cool object that is heating up) to the temperature of its surroundings, referred to as the ambient temperature. Initially, we used a geometric growth model to describe the difference between these two temperatures. For example, suppose a pie is removed from a hot oven and allowed to cool in a room at 70°F. If we model pie temperatures measured every ten minutes with a sequence $T_0, T_1, T_2, \cdots$, the differences between pie temperatures and the ambient temperature would be $T_0 - 70, T_1 - 70, T_2 - 70, \cdots$. Our model assumes that these differences decrease geometrically.

---

[8] See [8]. The author, Patrice Christmann, was a deputy director of France's Bureau de Recherches Géologiques et Minières, described as France's leading public institution in Earth science applications.

5.2. Applications of Mixed Growth Sequences

In symbols, with T_n the pie temperature at the nth measurement, we define $H_n = T_n - 70$, representing how much hotter the pie is than its surroundings at the nth measurement. We also assume that H_n decreases geometrically, and thus according to a geometric growth difference equation

$$H_{n+1} = rH_n.$$

To relate this situation to mixed growth models, let us substitute $T_n - 70$ for H_n and $T_{n+1} - 70$ for H_{n+1}. That produces

$$T_{n+1} - 70 = r(T_n - 70).$$

Adding 70 to both sides and distributing the multiplication by r leads to

$$T_{n+1} = rT_n - 70r + 70.$$

This is a mixed growth difference equation, with growth factor r and added constant $d = 70 - 70r = 70(1 - r)$. We therefore recognize the sequence of temperatures as an instance of mixed growth.

This development shows, more generally, that assuming *geometric* growth or decay in the temperature *difference* between an object and its surroundings is equivalent to assuming *mixed* growth in the *temperature* of the object. Thus we can formulate a heating or cooling model working directly with temperatures T rather than the temperature differences H.

With our knowledge of mixed growth sequences we can go further. Notice that the difference equation for T derived above has the form

$$T_{n+1} = rT_n + C(1 - r),$$

where C is the ambient temperature. Assuming $r < 1$, we know the mixed growth sequence approaches an equilibrium value $E = d/(1 - r)$. So, with $d = C(1 - r)$ we find $E = C$, verifying that the equilibrium value is the ambient temperature as expected. Accordingly, we can rewrite the difference equation as

$$T_{n+1} = rT_n + E(1 - r).$$

We also know that the functional equation for temperature can be written in the form

$$T_n = Ar^n + E, \tag{5.15}$$

for some constant A. If we know the ambient temperature E and the initial temperature T_0, then the constant A is given by $A = T_0 - E$. But the form of the functional equation can be used even when we do not have prior knowledge of the parameters r, E, and T_0. This would be the case if we have only a sequence of temperature measurements. Indeed, this is exactly the situation we discussed in the temperature probe data example beginning on page 275. See Figure 5.11.

To recap this example, we worked directly with the time t, in seconds, rather than n, the number of 5-second intervals. However, that change does not alter the form of the functional equation. Thus, we were looking for an equation of the form

$$T = Ar^t + E. \tag{5.16}$$

We used a trial-and-error approach to find coefficients A, r, and E providing the closest possible agreement between (5.16) and the data values. The best equation we found was

$$T = 69 \cdot 0.89^t + 10,$$

t	T	Data	Error
2	64.7	64.8	−0.1
5	48.5	49.0	−0.5
10	31.5	31.4	0.1
15	22.0	22.0	0.0
20	16.7	16.5	0.2
25	13.7	14.2	−0.5
30	12.1	12.0	0.1

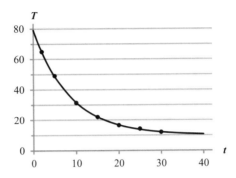

Figure 5.11. Data table and graph for the temperature probe data example. The curve in the graph corresponds to the equation $T = 69 \cdot 0.89^t + 10$. Agreement between this equation and the data values is within a half a degree.

corresponding to the coefficients $A = 69$, $r = 0.89$, and $E = 10$. The entries in the error column in Figure 5.11 show that the resulting equation approximates the data values to within half a degree.

This example demonstrates the interaction between structural and empirical methods. Our structural analysis demonstrated that the temperatures of a cooling object (in this instance, a temperature probe) should conform to a mixed growth model, and in particular, satisfy a mixed growth functional equation. Working directly with observed temperatures, we found values for the parameters in the functional equation that provide for close agreement with the data. But our structural model also supplies interpretations for the parameters. In particular, the constant E gives the ambient temperature as 10°, and the equation $T_0 - E = A$ implies that the initial temperature is $T_0 = E + A = 10° + 69° = 79°$.

Empirical Examples of Mixed Growth. Now we turn to two examples in which a mixed growth model is chosen purely because of the appearance of the data. The first concerns the viewer statistics for a popular YouTube video, as presented in Figure 5.12.

Cumulative Internet Views of a Popular Video. The data for this example were estimated from a graph posted on the YouTube Trends Blog [**1**]. The posted graph showed how the number of viewers of a popular Justin Bieber video grew over time. Our estimated data represent the cumulative number of viewers for the video every ninety days, between July 1, 2010, and September 18, 2012. These values appear in the data column in the table in Figure 5.12, and are represented as dots on the graph.

Looking just at the dots in the graph, it would appear that a mixed growth model might provide a reasonable approximation. To test whether the data values make up a mixed growth sequence, we compute the differences of the terms, and then the ratios of the differences. See Table 5.10.

Although the ratios of the first differences are not constant, they generally fall into a range between 0.7 and 1.1. Nor is there any apparent trend or order to the way the values fluctuate within that range, which might suggest another model. Accordingly, we proceed to consider a mixed growth sequence. Rounding the first data value to the

5.2. Applications of Mixed Growth Sequences

n	Data	v_n	|Error|
0	221.64	222.00	0.4
1	338.98	332.81	6.2
2	420.47	426.56	6.1
3	501.96	505.87	3.9
4	573.66	572.97	0.7
5	625.81	629.73	3.9
6	681.22	677.75	3.5
7	720.34	718.38	2.0
8	749.67	752.75	3.1
9	782.64	781.82	0.8

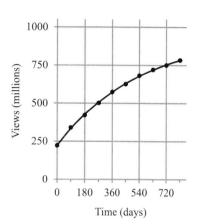

Figure 5.12. Modeling views of Justin Bieber's YouTube video "Baby." The entries in the Data column were estimated from a graph posted on the internet [1]. Each represents, in units of millions, the number of times the video had been viewed as of a particular date, starting with July 1, 2010, for $n = 0$, and occurring every 90 days. The v_n values are terms of a mixed growth sequence that approximates the data. In the graph, the dots show the data values and the curve represents the mixed growth model.

Table 5.10. Testing whether the cumulative view data comprise a mixed growth sequence.

n	Data	1st Diffs	Ratios
0	221.64		
		117.34	
1	338.98		0.69
		81.49	
2	420.47		1.00
		81.49	
3	501.96		0.88
		71.70	
4	573.66		0.73
		52.15	
5	625.81		1.06
		55.41	
6	681.22		0.71
		39.12	
7	720.34		0.75
		29.33	
8	749.67		1.12
		32.97	
9	782.64		

nearest integer, we define $v_0 = 222$. For r we take 0.85, about the center of the observed range of ratios. To determine d, we define v_1 to equal 339, the second data value rounded to the nearest integer. Then, using the mixed growth difference equation, $v_1 = rv_0 + d$, and using the defined values of $v_0, v_1,$ and r we see

$$d = v_1 - rv_0 = 339 - 0.85 \cdot 222 = 150.3.$$

Thus, as a first approximation, we have the mixed growth difference equation

$$v_{n+1} = 0.85v_n + 150.3; \quad v_0 = 222.$$

This produces a sequence that is reasonably close to the observed data.

Next we refined the model by trial and error. Always seeking to reduce the discrepancies between the data values and the terms v_n, we varied the parameters several times. Ultimately we settled on the sequence defined by

$$v_{n+1} = 0.846v_n + 145; \quad v_0 = 222.$$

This is the version of the model that appears in Figure 5.12. Visually, the graph of the model is quite close to all of the data points. Examining the results in the table, we see that the discrepancies between our data values and the corresponding sequence terms are at most 6.2, and are less than 4 for most of the points. At the worst, the errors are less than 2% of the observed data, and the majority are less than 1%. Visually, at least, this is quite good agreement between the model and the data: the curve in the graph appears to pass nearly through the center of each dot.

Once we have settled on the specific parameter values, we can find the functional equation. Since $r < 1$, we know that in the long term, our mixed growth sequence should approach an equilibrium value of

$$E = \frac{d}{1-r} = \frac{145}{1-0.846} = 941.6,$$

approximately. The functional equation will thus be given by

$$v_n = E + (v_0 - E)r^n = 941.6 - 719.6 \cdot 0.846^n.$$

In the usual way, we can use this equation to make predictions. For example, we can find when the number of views should reach 900 million by solving

$$941.6 - 719.6 \cdot 0.846^n = 900,$$

for n. The answer is

$$n = \frac{\ln(41.6/719.6)}{\ln(0.846)} = 17.05$$

to two decimal places. That means n is approximately 17, so the number of days is a bit more than $90 \cdot 17 = 1530$.

However, this methodology of selecting a model based solely on the appearance of the data is not as reliable as developing a model structurally. As an illustration, we know that in the long term, our mixed growth sequence should approach an equilibrium value of $E = 941.6$. approximately. Thus, the model predicts that the number of new viewers of the "Baby" video will dwindle to zero, and the total number of viewers will top out at about 941.6 million. As it happens, that prediction is incorrect. As of June 14, 2015, the cumulative number of views was over 1,177 million.

Although our mixed growth model is a reasonably close fit for the data between July 2010 and September 2012, sometime during 2013 the rate of new viewers accelerated in a way that the model did not predict. A model that has a strong structural analysis at its foundation can be expected to provide reasonably accurate predictions, at least while the structure of the problem context remains unchanged. In contrast, we had no structural reason to expect the viewer data to exhibit mixed growth. Accordingly, we should not have much confidence that the pattern of growth we see in the data will persist into the future.

5.2. Applications of Mixed Growth Sequences

We leave this example with two general observations. First, the method of tuning a model to fit a set of data is commonly seen in applications, and can be very effective. But second, predictions from such models should be viewed with caution.

We conclude with a final example of fitting a mixed growth model to a set of data.

Atmospheric Carbon Dioxide. In the preceding example we developed a model based on a mixed growth sequence. An alternative approach is to bypass sequences and work directly with functions. As we saw in Chapter 4, a geometric growth model can be developed by fitting an exponential function of the form $y = Ae^{kx}$ to a set of data. That is, numerical values for A and k can be selected so that the equation provides as close an approximation to the data as possible. In a similar way, a mixed growth model can be developed by fitting a shifted exponential function to a set of data. Moreover, while the general form of a shifted exponential function can be given as $y = Ar^x + B$, we can also express the exponential part of the function, Ar^x, using base e. Thus, we can assume an equation of the form $y = Ae^{kx} + B$, and try to select values of A, k, and B in close agreement with the data. Among some model developers it is standard practice to use base e. Accordingly, to illustrate the practice, we follow it in this final example.

The data set consists of annual average atmospheric CO_2 concentrations for years 1959 through 2014, published by the National Oceanic and Atmospheric Administration (NOAA) [**40**]. When graphed, the data points appear to slope upward with a shape reminiscent of an exponential curve. Accordingly, we initially attempted to model the data with an exponential function, using an equation of the form $y = Ae^{kx}$. Statistical analysis software was used to select parameters A and k providing the best possible agreement between the data and the model. The result is shown in Figure 5.13.

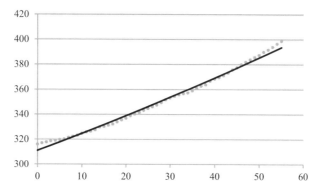

Figure 5.13. Modeling annual average CO_2 concentrations with an exponential function. The horizontal axis represents time in units of years, starting with 1959 as year zero. The gray dots represent the data, in units of parts per million. The black curve is the graph of the equation $y = 311.0257e^{0.004283x}$, which is statistically the best possible exponential function for the given data. The average absolute error is 1.81 rounded to two decimal places. This is the average discrepancy between data values and the corresponding values obtained from the equation, ignoring the signs of the differences.

Although the curve is fairly close to all the data points, it is visually apparent that this model does not correctly capture the trend of the data. Notice that there is a clear pattern to the differences between the dots and the curve, with the curve too low on the ends and too high in the middle. Also, the pattern of dots and the curve seem to be diverging at the ends of the graph. These observations suggest that a better fit should be possible. Moreover, the fact that the curve is defined by the best possible exponential function tells us that we will need to consider a different kind of function to get a better fit.

It is natural therefore to consider alternative equations. In particular, one likely option is a shifted exponential, for which we know the graph will resemble an exponential curve. Accordingly, we adopt an equation of the form $y = Ae^{kx} + B$. Using a trial-and-error approach, we found the data were very well approximated using the equation $y = 66e^{0.015x} + 248$. This is shown graphically in Figure 5.14.

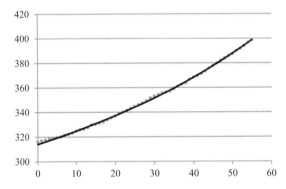

Figure 5.14. Modeling annual average CO_2 concentrations with a shifted exponential function. The axes and data portrayed are the same as in the preceding figure. This time the black curve is the graph of $y = 66e^{0.015x} + 248$, which was found by trial and error. For this model, the average absolute error is 0.57 rounded to two decimal places.

Comparing the two figures, the errors associated with the second model appear to be consistently smaller than those for the first model. And numerically, the average absolute error for the second model is about 0.57, compared to 1.81 for the first model. More significantly, the second curve appears to follow the trend of the data much more closely than the first curve. For this reason, we should be more confident of predictions based on the second model. However, the cautions discussed in the YouTube video example still apply.

It is not surprising that a shifted exponential function is better able to follow the contour of a given curve than an unmodified exponential function. The shifted exponential has three parameters as compared to the two parameters of the pure exponential, suggesting that the shape of the shifted exponential is more easily fine tuned. In any case, when we use a shifted exponential, that is actually a mixed growth model. However, as this example illustrates, we can formulate such a model with almost no regard to the properties of mixed growth sequences. Rather, it is only the form of the mixed growth functional equation that has been adopted. On the other hand, having found a reasonably close fit between a shifted exponential and the data, we might next consider whether there are any structural aspects of the problem context that would lead to a mixed growth model. This might lead in turn to a structural basis for the shifted exponential model, and hence increase our confidence in the model.

One nice characteristic of the modeling approach we have followed above is that we immediately obtain a functional equation giving

$$CO_2 \text{ Concentration} = 66e^{0.015n} + 248,$$

for year n. We can use this to predict future levels of atmospheric CO_2 by computing the right side of the equation for a particular value of n. Alternatively, we can predict at what time in the future a particular level will be observed, by solving the functional equation for n. As an example, a level of 500 will be reached when

$$66e^{0.015n} + 248 = 500.$$

Solving this equation for n using the methods of Section 4.4[9] leads to

$$n = \frac{\ln(252/66)}{0.015} = 89.3,$$

to one decimal place. That corresponds to year 89, or sometime in 2048.

Bearing in mind the cautions already given, we emphasize that year 89 is far in the future, and the prediction might well be inaccurate. As in the YouTube example, we cannot be too confident that the close agreement between the data and the model for past years will persist for the years that come after. Climate scientists use many methods to develop models, including both structural and empirical considerations. The analysis we have presented here is an extreme simplification, intended only to give a flavor of the development and use of such models.

5.2 Exercises

Reading Comprehension.

(1) @In some examples we had a structural foundation for a mixed growth model. In other examples the mixed growth model had no structural basis but was convenient. In at least one case it was mentioned that finding a mixed growth model fit the data well might cause us to consider if some structural aspect of the scenario would lead to a mixed growth model. What is meant by "structural basis" and how is this different from an empirical basis? Does having a structural basis increase or decrease our confidence in predictions made based on the model? Justify your answer.

(2) @In this section we saw a model for the effects of repeated doses of a certain medicine. To be specific, we defined a mixed growth sequence in which the term a_n represents the amount of medicine in the body immediately after taking the nth dose. Is the functional equation for this model valid for fractional values of n? That is will the equation give an accurate prediction of the level of drug in the body for a value such as $n = 2.5$ half way between the second and third dose? Why or why not?

---

[9]See page 275 for an example.

(3) Two students are analyzing a loan payment problem, where the payments are made weekly and b_n is the balance immediately after the nth payment. The first student finds the functional equation
$$b_n = 2{,}500(1.002)^n - 20\left(\frac{1.002^n - 1}{0.002}\right),$$
while the second student finds the equivalent equation
$$b_n = 10{,}000 - 7{,}500(1.002)^n.$$
Use your understanding of this type of model to determine the amount borrowed, the interest rate being charged, and the weekly payment being made. Which of the two functional equations provides this information more directly? Briefly explain your answers as if to a classmate.

(4) @In this exercise we consider loan repayment.
   a. @Matt borrowed $1,500 at 3% monthly interest and would like to repay the loan with payments of $45 per month. What will happen to his balance if Matt pays $45 per month?
   b. The preceding question is one case of a general feature of repeated loan payment models. For every such model there is a threshold for the payment amount. If the monthly payment is above the threshold the loan balances decrease consistently and the loan will eventually be fully paid. Otherwise the balances increase or remain constant and the loan will never be repaid. How can this threshold be found in any repeated loan payment model? Explain the process as if to a classmate who does not yet understand this. Include at least one example in your answer.

(5) @Starting with a geometric growth sequence such as $a_n = a_0 r^n$ we can consider the sums: $s_{n+1} = s_n + a_{n+1}$, $s_0 = a_0$. Explain why the sequence s_n will always be a mixed growth sequence. It may be helpful to review the definition of a mixed growth sequence or the Mixed Growth Sequence Test (page 302).

(6) In the Radial Keratotomy (RK) Surgery example repeated analysis of the model with more than one growth rate was discussed (see page 338). Give an example of another scenario where it would be useful to repeat the analysis of a model with more than one growth rate. Explain why the technique is useful for both RK surgery and your example.

(7) As we have studied various types of sequences, we have summarized important properties and parameters in profiles like the one on page 312. In a similar way, we can summarize models for specific application contexts. For example, a profile is shown below for the repeated loan payment models discussed in this section. Using this example as a guide, create profiles for each of the following contexts: Drug Dosage Models, Geometric Sum Models, and Heat Transer Models. Feel free to change the format as long as all pertinent information is included.

5.2. Exercises

	Profile: Mixed Growth Sequences for Repeated Loan Payments
Verbal Description:	The initial term of the sequence is the amount of money borrowed. Each term gives the balance owed immediately after a payment.
Meanings of Variables:	b_n is the balance immediately after the nth payment; when payments are made monthly, b_n is also the balance after n months.
Parameters:	Initial term, $b_0 > 0$ equals the original debt amount. The negative of the monthly payment amount is d. The payment is assumed to exceed the interest accrued before the first payment. This implies that $d < (1-r)b_0$. Growth factor, r, equals 1 plus the monthly interest rate, expressed as a decimal, so $r > 1$. When the annual interest rate is given the monthly rate can be found by dividing by 12. For example, an annual rate of 2.5% corresponds to a monthly rate of 0.2083%.
Difference Equation:	$b_{n+1} = rb_n + d$
Fixed Point:	$E = d/(1-r)$. This is algebraically useful as in formulating the functional equation, but has little or no direct significance in the problem context.
Functional Equation:	$b_n = E + (b_0 - E)r^n$
Graph: 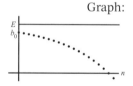	The graph is a shifted exponential. To pay off the loan we must have $b_0 < E$ and $r > 1$. Then the graph is decreasing away from the equilibrium. We are usually interested in finding where the graph crosses the horizontal axis. That indicates the time when the loan is repaid.

Chapter 5. Mixed Growth Models

Math Skills.

(8) @For each of the following difference equations, find the corresponding functional equation of the form $a_n = E + Ar^n$ where $A = a_0 - E$.

   a. @$a_{n+1} = 0.85a_n + 100$, $a_0 = 100$
   b. $b_{n+1} = 1.015b_n - 200$, $b_0 = 1{,}500$
   c. @$c_{n+1} = 0.7c_n + 150$, $c_0 = 600$
   d. $d_{n+1} - 75 = 0.85(d_n - 75)$, $d_0 = 100$
   e. @$e_{n+1} - 80 = 1.015(e_n - 80)$, $e_0 = 50$

(9) @Given the following parameters, find the equilibrium value, E, and determine if the values of the sequence increase or decrease.

   a. @$a_0 = -10$, $r = 1.2$, $d = -0.5$
   b. $b_0 = 350$, $r = 0.96$, $d = 200$
   c. @$c_0 = -45$, $r = 1.04$, $d = 37$

(10) @Recall that the difference equation for a mixed growth sequence can be written $a_{n+1} = ra_n + d$. In each part of this problem, use the given equilibrium value, E, and common ratio, r, to determine the parameter d.

   a. @$E = 275$, $r = 0.2$
   b. $E = 100$, $r = 0.65$
   c. $E = 590$, $r = 1.78$

(11) @Each of the following gives the initial term, d, and r for a mixed growth sequence. In each case write the functional equation for the sequence and use it to find the smallest n for which the corresponding term of the sequence is 0 or less.

   a. @$a_0 = 350$, $d = -50$, $r = 1.005$
   b. $b_{n+1} = 1.01b_n - 150$, $b_0 = 10{,}000$
   c. @$c_0 = 213{,}950$; $d = -2{,}100$; $r = 1 + 0.045/12 = 1.00375$

(12) @A student is exploring a difference equation model. She obtains the following pattern of results.

$$c_0 = 0.4 + 25$$
$$c_1 = 0.4(1.6) + 25$$
$$c_2 = 0.4(1.6)^2 + 25$$
$$c_3 = 0.4(1.6)^3 + 25$$
$$c_4 = 0.4(1.6)^4 + 25$$

Based on this pattern answer the following questions.

   a. @What is c_8? What is c_n?
   b. What do you get if you add up c_0 through c_4?
   c. What do you get if you add up c_0 through c_8?
   d. @What do you get if you add up c_0 through c_n? Express your answer as a formula involving the variable n.

5.2. Exercises

Problems in Context.

(13) @Drug Elimination. A doctor is studying the way a medication is eliminated from the body. She finds that every four hours, one fifth of the drug is eliminated. She starts the patient off with 1,000 units of the medication, followed by 250 units every four hours. Let a_0 be the amount of drug in the body immediately after the initial dose. For $n \geq 1$ let a_n be the amount in the body just after the nth 250 unit dose. So for example, a_3 is the amount of drug in the body right after the 3rd 250 unit dose is taken.

   a. @Find the numerical values of a_1, a_2, and a_3.
   b. @What is the difference equation for a_n?
   c. @Because $r < 1$ we know the model predicts that the amount of drug in the body will approach an equilibrium value. What is that value?
   d. @How many doses are necessary before we expect the sequence values to stay within 10 units of the equilibrium?
   e. @In the previous part of this exercise you found a value for n, the number of 4-hour intervals, after which the sequence values stay within 10 units of the equilibrium value. Does this mean the level of drug in the patient's blood remains within 10 units of the equilibrium value? Explain.

(14) Changing Drug Dosage. A drug company is testing a pain reliever. The researchers know that in six hours the body removes about 4/5 of the drug. In one experiment, each patient is given an initial dose of 200 units of the drug, followed by repeated doses of 100 units every six hours. That leads to the difference equation $p_{n+1} = (1/5)p_n + 100$, where $p_0 = 200$ is the amount of pain reliever in the body immediately after taking the initial dose, and for $n \geq 1$, p_n is the amount of pain reliever in the body immediately after taking the nth dose of 100 units.

   a. Find a functional equation for p_n.
   b. Use your functional equation for p_n to compute p_6.
   c. According to the difference equation, if $p_n = 125$, what will p_{n+1} equal? What does this show?
   d. A doctor would like the amount of drug to level off at 150 units for a patient. This can be achieved by changing the amount of drug that the patient takes every six hours from 100 units to something else. Find the amount of the new repeated dosage. Demonstrate that your answer gives the correct results.

(15) @Drug Equilibrium Values. A certain drug is eliminated from the body at a rate of 25% every three hours. The level of the drug in the body can be modeled by $a_{n+1} = 0.75a_n + d$ where d is the amount of drug given every three hours, the repeated dose.

   a. @If a repeated dose of 100 units is given every three hours, what will the equilibrium value be? That is, the amount of drug in the body will level off to how many units?
   b. @If a dose of 30 units is given every three hours, what will the equilibrium value be? That is, the amount of drug in the body will level off to how many units?

c. @The doctor would like the amount of drug to level off at 280 units. What dose should be given every three hours to achieve this?

d. @There is a simple pattern relating the equilibrium level to the repeated dose in this problem. What is the pattern and how could that be used to answer the preceding part quickly?

(16) **Drug Model Fixed Point.** The general difference equation for a drug dosing model may be given by $a_{n+1} = ra_n + d$, where d is the repeated dose, and where the fraction of the drug remaining in the body between doses is given by r. Solve the equation $x = rx + d$ for x to find the fixed point of the model in terms of r and d.

(17) **@Pollution Model.** A pollution study resulted in a model for the amount of polychlorinated biphenyls (PCBs) in a large lake. In the study, p_n is the amount of PCBs (in parts per million) at the start of month n where the study is considered to have started at the beginning of month 0. The model used the difference equation

$$p_{n+1} = 0.82 p_n + 0.005, \quad p_0 = 0.02.$$

a. @Compare and contrast this model for pollution in a lake with the models you have studied for repeated drug doses. What assumptions are we making by adopting the pollution model? Why might these assumptions be reasonable?

b. @Find a functional equation for the model given above.

c. @According to the model, what prediction can be made for the future levels of PCB pollution in the lake?

(18) **Dynamic Chemical Bath.** In a factory, water is constantly flowing through a large vat, where it is mixed with a solid chemical. In studying the way the chemical dissolves, a chemist develops a mixed growth model. At the start of the experiment 100 pounds of the chemical in a solid form are added to the vat. Each hour, about one tenth of the chemical dissolves, leaving 9/10 still lying at the bottom of the vat, and another 20 pounds of the solid chemical are added to the vat. Develop a mixed model for this situation. Let s_n be the amount of solid chemical at the bottom of the vat after n hours, starting with $s_0 = 100$. Write a brief report on your model. Include at least the following.

a. Compare and contrast drug level in the body to the situation described above. Why does using the same type of model make sense? What assumptions are we making by adopting this model?

b. The values of the first few terms: s_1, s_2, and s_3.

c. A difference equation for s_n.

d. A functional equation for s_n.

e. A discussion of the long range predictions that can be made based on the model.

(19) **@Loan Repayment.** Mariel borrows $5,000 from her uncle. They agree to a monthly interest rate of 1.25% and a monthly payment of $200. This situation can be analyzed using the difference equation $b_{n+1} = (1.0125)b_n - 200$.

a. @Find a functional equation for this model.

b. @What will the balance be after 2 years?

5.2. Exercises

    c. @How long will it take to pay off the entire loan?

    d. @The last payment will be different from the usual $200. How much should it be?

    e. @As soon as Mariel has repaid the loan her uncle surprises her. He has saved all of her payments, and after subtracting the original $5,000 loaned, he now wants to give the excess money back to her. How much money is he giving Mariel?

(20) **Credit Card Repayment.** Suppose you have a balance of $12,000 on your credit card and the finance charge is 2% per month on the unpaid balance. You decide to make payments of $400 per month until the balance is paid off, without charging any new purchases to the card in the meantime. Develop a mixed model for this situation, and use it to determine how long it will take you to pay off everything you owe. Also note how much the last payment should be so that you reduce the balance to exactly zero. As always, be sure to define your variables and give the difference and functional equations.

(21) @**Annual Pay Raise.** A large company gives its workers a raise every year. The raise is computed as 6 percent of the previous year's salary, plus $500. Suppose a new employee starts out with a salary of $25,000. Let s_n be the salary in year n, where the first year of employment is considered to be year 0. Develop a difference equation for s_n. Is this a mixed growth model? Why or why not?

(22) **Mortgage Offer Comparison.** A couple buys a condo for $180,000. They use their savings to make a down payment of $30,000 and take out a 30-year mortgage for the rest.

    a. One lender specifies the interest as 4 percent per year (which really means $4/12 = 1/3$ percent per month). Use the monthly payment formula ((5.13), page 334) to determine what the monthly mortgage payments will be.

    b. Another lender will give the couple a mortgage at 3.75 percent per year. However, the couple will have to pay a fee of $1,800 to get that loan. How much lower will the payments be at the lower interest rate?

    c. How long will it take the savings in the mortgage payments to equal the extra expense of $1,800 required to get the lower interest mortgage?

(23) @**Monthly Mortgage Payment Comparison.** Roberto buys a home for $300,000. He makes a down payment of $65,000 and takes out a 30-year mortgage for the remainder. His annual interest rate is 3.75% and his monthly payments are $1,088.32. Just after purchasing the home Roberto decides to let a friend live with him. The friend pays rent enabling Roberto to double his monthly mortgage payments. How long will it take Roberto to pay off his loan?

(24) **Savings Plan.** Jeff is planning to help his grandson, Spencer, pay for college. When Spencer begins middle school (7th grade), Jeff puts $1,000 into an account earning half a percent per month and makes monthly contributions of $200.

    a. Develop a model for b_n, the balance immediately after the nth $200 deposit. Use the model to determine what the balance will be when Spencer begins 12th grade, after 60 payments of $200 have been made.

354   Chapter 5. Mixed Growth Models

b. Spencer earns a full scholarship to college and ends up not needing any financial assistance from Jeff. But Jeff decides to keep making monthly contributions to the account and to give the balance to Spencer later. How long will it take for the balance to reach $25,000?

(25) @Savings Model. Suppose you begin a savings plan in which you deposit $100 per month in an account that pays 1.5% percent interest per month. On the first day of each month the bank adds interest to the account, and you deposit $100. Develop a model for the amount in the account, a_n, after n deposits. Is the initial deposit represented by a_0 or a_1 in your model? How much will be in the account after 5 years? 10 years? 15 years? How long will it take to accumulate $100,000?

(26) Sums of Cases of Methicillin-Resistant *Staphylococcus Aureus*.

An exercise in Section 4.2 involved the equation

$$c_n = 33(1.22)^n$$

as part of a geometric growth model. Here, c_n is the number of MRSA infections in U.S. hospitals each year, in units of thousands. The position number n is in units of years, starting with $n = 0$ in 1993. Extending that earlier model, let S_n be the sum $c_0 + c_1 + \cdots + c_n$. Note that $S_0 = c_0$ and for $n > 0$, $S_n = S_{n-1} + c_n$.

a. Compute the first five terms of the sequence, S_n.

b. Explain what S_n represents in terms of MRSA infections and years since 1993.

c. What assumptions are implicit in this model of MRSA infections?

d. Find difference and functional equations for S_n.

e. According to the model, how many patients contracted MRSA infections during the time span from 1993 through 2015?

(27) @Radial Keratotomy (RK). Beginning on page 336, a model was developed for the popularity of RK vision correction surgeries. The model assumed that the number of procedures performed each year would grow at an annual rate of 52.8%. Of course it is unrealistic to imagine that the growth rate can be predicted exactly. It is more reasonable to consider a range of possible growth rates. In this problem we will repeat the analysis using a 25% growth rate then combine the results to give a range of predictions. Recall that the number of procedures for the initial year of the model is 250,000.

a. @Let $\hat{p}_n$ (read as p_n hat) be the number of procedures (in thousands) performed in year n assuming a 25% annual increase.[10] This model will have the same starting point as the model in the text: $\hat{p}_0 = 250$ represents the number of procedures performed in 1994. Write a difference equation for $\hat{p}_n$.

b. @Compute the values of $\hat{p}_n$ through $n = 6$ then tabulate the running totals. Let $\hat{s}_n$ be the sequence of running totals.

c. @Verify that the terms of $\hat{s}_n$ constitute a mixed growth sequence.

d. @Find the difference and functional equations for $\hat{s}_n$.

---

[10]The point here is to compare two closely related models for the same variable. We attach a *hat* to the p_n notation when we are referring to the modified model, reserving the plain p_n notation for the original version of the model.

5.2. Exercises

e. @Use $\hat{s}_{10}$ to predict the total number of procedures performed in years 0 through 10.

f. @Considering both the value just derived for $\hat{s}_{10}$ as well as the value of s_{10} originally derived in the text, write a summary market outlook that might have been appropriate for a convention of vision correction professionals meeting in 1994.

(28) **Computer Virus.** A computer virus is released by hackers and spreads over the internet. The number of computers that get infected with the virus each day is given in the table below.

Day (n)	1	2	3	4
Number of computers becoming infected on day n	128	160	200	250

Develop a model for the total number of infected computers after n days. For example, according to the table, there would be $128+160 = 288$ infected computers after 2 days, and $288+200 = 488$ after 3 days. Your model should give a difference equation and functional equation for the number of computers infected after n days. As always, define your variables, and give difference and functional equations. Also, give a justification for the particular kind of model you use.

(29) **@Cooling Pie.** Starting on page 340 a pie-cooling example was analyzed. In this problem we consider a similar situation with slightly different parameters. The pie is checked every 10 minutes and the difference between its temperature and the ambient temperature of 65°F is found to be decreasing geometrically with a constant growth factor of $r = 2/3$. Find a model for the temperature of the pie, T_n, after n 10-minute intervals. The initial temperature of the pie is 350°F.

(30) **Cooling Soup.** A bowl of leftover soup has cooled to 74°F on the counter before being put into the refrigerator. The refrigerator is 40°F and the difference between the temperature of the soup and that of the refrigerator is decreasing by 20% every five minutes. Give a model for the temperature of the soup, T_n, after n five-minute intervals.

(31) **@Temperature Probe.** An exercise in Chapter 4 concerned data collected using a temperature probe.[11] Initially the temperature probe was placed in a cup of hot coffee long enough to reach essentially the same temperature as the coffee. Then the probe was placed in a cold water bath. Temperatures from the probe were collected as detailed in the following table, in which times are in seconds, measured from the instant when the probe was put into the cold water, and temperatures are in degrees Celsius.

Time (seconds)	5	10	15	20	25	30
Temperature (°C)	69.39	49.66	35.26	28.15	23.56	20.62

As discussed in the Heat Transfer example starting on page 340, we expect the temperatures to follow a mixed growth model. Following the same approach as in the example, develop a mixed growth sequence that models the data in the table above as closely as possible. Provide a graph showing both the data in the table and

---
[11] Adapted from [**25**, Exercise 33, p. 362].

a curve representing your model, and describe briefly how you found your model. Based on your model, what was the temperature of the probe when it was placed in the cold water? What was the temperature of the water?

Digging Deeper.

(32) A doctor is testing a new drug. She knows from the research on the drug that the body eliminates 40% of the drug every three hours.

   a. For one patient an initial dose of 1,000 mg is ordered, with repeated doses every three hours. What should the amount of each repeated dose be in order for the equilibrium level of drug in the system to be equal to 1,000 mg?
   b. It is more convenient for the patient to take the drug every six hours. How much of the drug will be eliminated in six hours?
   c. With repeated doses every six hours, what should the amount of the repeated dose be to again produce an equilibrium value of 1,000 mg?
   d. Compare your answers for part a and part c. The time between doses for part c is twice the time between doses for part a. Are the doses twice as large? Why or why not?
   e. In both models, the equilibrium level is 1,000 mg. Are the models equally effective at keeping the level in the body close to 1,000 mg between doses? Explain.

(33) @In this problem we will investigate changing the time between doses for a given drug. We'll model drug elimination according to a strong geometric growth assumption: in equal periods of time, the amount of drug in the body decreases by an equal percentage. For this drug, the amount in the body decreases by 36% in any three-hour period.

   a. @If no additional doses were given, what percentage of the drug would remain after six hours?
   b. @What percentage of the drug would remain after one hour? Is this more or less than 64% and does that make sense?
   c. @How could your answer from part b be used to find the amount of drug remaining after n hours, assuming no additional doses were given?
   d. @Using the results of the preceding parts of the problem, develop a sequence modeling the amount of the drug in the body after each dose, assuming an initial dose of 500 units and repeated 200 unit doses administered every six hours.
   e. @Develop a sequence modeling the amount of the drug in the body after each dose, assuming an initial dose of 500 units and repeated 200 unit doses administered every four hours.

(34) Emily buys a home in Georgia for $183,000. She makes a down payment of $36,600 and takes out a 20-year mortgage for the rest. Emily borrows $146,400 at three percent annual interest ($3/12 = 0.25$ percent per month). Her monthly payments are $811.93. A friend advises that by paying $405.97 (half the monthly payment) every two weeks Emily will repay her loan about two years sooner. To be exact she would have to make 467 payments which is one less than 18 years of payments.

5.2. Exercises

a. Emily is intrigued and asks you to help her analyze the alternative repayment plan. First, check the veracity of the statement. Notice that in the new analysis n would no longer be the number of months, but rather the number of two-week periods. Similarly the interest accrued in each period will be different from that of the monthly payments. Determine if the friend's claim is true.

b. Now let's think about why there is a difference in repayment time. At first glance it seems that the friend was suggesting paying half the amount twice as often, which results in the same annual payments. That is the same amount being paid each year. Is that the case? What other factors have been changed?

c. Emily does not plan to stay in her home until it is paid off. She plans to sell the house after five years and is wondering what difference, if any, the more frequent payments will make in this shorter time frame. Consider how best to compare the payment plans over only five years. Explain what values should be compared. For example, one might compare total payments made, remaining loan balance or something else. Then do the necessary calculations and answer Emily's inquiry.

(35) @A geometric growth sequence with initial term p_0 and constant growth factor r has functional equation

$$p_n = p_0 r^n.$$

Use this to explain each of the following statements about the sum sequence, defined by $s_n = p_0 + p_1 + \cdots + p_n$.

a. @The mixed growth parameter r for the sum sequence is equal to the geometric growth parameter r for the original sequence.

b. @The mixed growth parameter d for the sum sequence is equal to the initial term p_0 for the original sequence.

c. @The sum sequence has functional equation

$$s_n = \frac{p_0}{1-r}(1 - r^{n+1}).$$

(36) If a mixed growth sequence has functional equation

$$p_n = Ar^n + B,$$

show that the sequence of sums has a functional equation in the form

$$s_n = C + Dn + Er^n$$

for suitable constants C, D, and E. Here again, s_n is defined to be $p_0 + p_1 + \cdots + p_n$. Your answer should include each of the new parameters, C, D, and E in terms of r and the old A and B.

6
Logistic Growth

In Chapter 4 we considered population models based on geometric growth sequences. Typically in such a model the terms p_0, p_1, p_2, etc., represent population sizes at regularly spaced time intervals, for example, every day, every month, or every year. The geometric growth assumption is that each successive term increases or decreases by an equal percentage, or equivalently, by an equal growth factor r, over the preceding term. Under this assumption, the functional equation is $p_n = p_0 r^n$. And when $r > 1$ this function grows exponentially. Thus the model predicts that the population will grow at an ever increasing pace, without any upper limit.

In practical terms we know that this type of model must be unrealistic in the long term. Certainly, for biological populations, exponential growth cannot continue indefinitely. Competition for resources must eventually restrict population growth. But even outside of biology, it is difficult to imagine a context where unrestricted growth is a practical possibility.

In this chapter we introduce more realistic models that represent *limited population growth*. They assume not a *constant* growth factor, but rather a growth factor that gets smaller as the population grows larger. As a first specific instance of this idea, we assume that the growth factor decreases linearly as a population increases. That leads to a family of models we refer to as *logistic growth*.[1]

Logistic growth sequences will be defined and analyzed in Section 6.1. As in past chapters we will discuss common properties of these models, including difference equations and graphs. As we will see, in many cases logistic growth sequences agree with our expectations about how a population should grow, subject to limited resources. On the other hand, logistic sequences differ in important ways from the sequences we have seen in earlier chapters. For one thing, there is no logistic growth functional

---

[1] Unfortunately this terminology is not consistent in the literature. Different authors have attached *logistic growth* to different families of models. Some, like us, reserve this name for models where the growth factor r decreases linearly as the population grows. Others apply the name logistic growth to *any* model where r decreases as the population grows, whether or not that decrease is linear. And still others only use the name logistic growth when r decreases nonlinearly in accord with one specific type of equation. This last model will be considered in Section 6.3, where it will be referred to as *refined logistic growth*.

equation. For another, in some cases a logistic growth sequence can exhibit quite unpredictable growth, at times seeming to follow no pattern at all. These aspects of logistic growth will be taken up in detail in Section 6.2. Then in Section 6.3, an attempt to understand why these disorderly aspects occur in logistic growth models leads to a modified set of assumptions, defining what we will call *refined logistic growth*. The refined version of the model retains the attractive features of logistic growth, while avoiding the disorderly characteristics that sometimes arise in the original version of the model.

6.1 Properties of Logistic Growth Sequences

In a logistic growth sequence, successive terms are related by a growth factor r that decreases linearly as the population increases. We will illustrate these ideas in the context of a specific example. This will be a conceptual model, based on simple assumptions, and not intended as an accurate representation of an actual population.

Imagine that we are going to introduce a new species of game fish, such as trout, to a large lake without a natural trout population. Now we argue that initially the trout population should grow geometrically, at least approximately. Say that we stock the lake with 1,000 trout. We know that some fraction of the trout will be fertile females, probably about 50%. Depending on her weight, each female will produce some number of eggs, perhaps as many as 1,000. But after the eggs hatch, most of the offspring will not survive to adulthood. For concreteness, we may suppose that only 10 out of 100 survive. Also, some of the original 1,000 fish will probably die over the year, say one in 10. Now all of these figures can be combined to predict the number of trout in the following year. The exact numbers are not important. What is important is that all of the numbers are reasonably interpreted as fixed percentages, as long as the environmental factors are roughly constant. If we start next year with a different population, we can reasonably expect the same percentage of fertile females, the same number of offspring per female, the same fraction of offspring to survive, and the same fraction of the existing population to die. That is why a geometric growth assumption is plausible. Without trying to determine the specific percentages at each stage of the process, we recognize that the overall increase in the population will be a fixed percentage. If the net result is a 10-percent increase one year, it is reasonable to predict about a 10-percent increase in the following year, too.

These ideas lead to the geometric growth difference equation:

$$p_{n+1} = rp_n,$$

where the parameter r is a kind of net fertility factor. For our population of trout, if the lake can support many thousands of fish, and if there are few natural predators, we expect the population to grow rapidly. Suppose that in one year the population will increase by a factor of 5. That means that after all the eggs are hatched, and taking into account the adults and hatchlings that die during a year, the net result will be that the population will be 5 times larger after a year. In the difference equation, this corresponds to a parameter value of $r = 5$, leading to

$$p_{n+1} = 5p_n.$$

Now as the years go by, this rate of growth cannot continue. Eventually, there will be greater competition for food. Fewer of the hatchlings will survive. Indeed, many of

6.1. Properties of Logistic Growth Sequences

the baby fish will be eaten by the adults. So in that case, the increase will be much less than a factor of 5. Maybe the population will only grow by 50%. Then r will equal 1.5 and the difference equation will be

$$p_{n+1} = 1.5 p_n.$$

As the population continues to grow, it can approach a kind of balance, where the competition for resources will effectively prevent the population from growing at all. At that point, each year's population size will be the same as the previous year's. That means that $r = 1$ and the difference equation is

$$p_{n+1} = p_n.$$

What if the population somehow got over the balance point? This could happen if we overstock the lake. Or, if the conditions in the lake change, the balance point might be reduced to a smaller population size. Then, for a while, there might be more fish than the lake could sustain on a continuing basis. In any case, with too many fish in the lake, we expect the population to actually shrink. The r in that case would be less than 1, say for example, 0.8. This would correspond to a difference equation:

$$p_{n+1} = 0.8 p_n$$

What these ideas suggest is that r should not be expected to be constant. It should change with the population size. If r is as large as 5 when the population size is small, we expect it to be much smaller when the population size is larger.

In Figure 6.1 a qualitative graph illustrates the kind of relationship we have been discussing. The growth factor decreases as population increases, producing a curve that slopes downward from left to right.

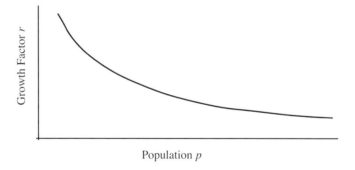

Figure 6.1. Growth factor r varies with population.

A Linear Growth Factor Model. How can we incorporate these ideas into our model? In the difference equation we need to replace the constant r with a function of the population size. Algebraically, the simplest approach is to use a straight-line model for the variation of r with population p. Then we can easily find an equation for r as a function of p. As illustrated in Figure 6.1, the true relationship between r and p is probably not actually a straight line. In order to obtain a good approximation with a linear model, we should probably avoid the extreme ends of the graph, where the population is very small or very large. Instead, we want to draw a line that approximates the correct situation for a wide range of populations in the center of the graph.

In a genuine modeling situation, we would collect data on real fish populations. In this discussion, since we are simply exploring the conceptual ideas of this type of model, we will continue to invent the numbers in the model. We expect that at some point the fish population will be in balance with the environment. Suppose that occurs with a population of 50,000 fish. If there are 50,000 fish this year, then there will be 50,000 next year, too. In fact, once the population reaches 50,000, it will remain the same year after year. From that point on, with a size of 50,000, the population is described by the difference equation $p_{n+1} = p_n$, so the growth factor is $r = 1$ when $p = 50,000$. Therefore, $(50,000, 1)$ is one point on a graph of r as a function of p. See Figure 6.2, which will be referred to as a p–r graph.

For a population lower than 50,000, we know the environment will provide an excess of resources, so we expect the population to grow. In general, the smaller the population, and so the more abundant the resources, the more rapidly we expect growth to occur. A small enough population could even double in a single year. Perhaps this will happen if the population is as small as 10,000. Then, with a population of 10,000, the growth factor would be $r = 2$. In this case $(10,000, 2)$ is also on the p–r graph.

Because we are assuming that the p–r graph is linear, it is completely determined by two points. Thus, we can use the two data points $(10,000, 2)$ and $(50,000, 1)$ to create the straight-line graph of Figure 6.2 and derive an equation for the line.

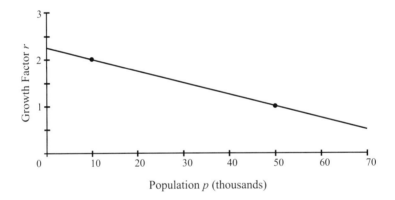

Figure 6.2. Growth factor r based on two points.

We have the points $(10,000, 2)$ and $(50,000, 1)$. This gives us a slope of $-1/40,000 = -0.000025$. Applying the point-slope form for a linear equation (see page 95) leads to

$$r - 2 = -0.000025(p - 10,000)$$

or after simplifying,

$$r = 2.25 - 0.000025p.$$

This is a simple linear model for the relationship between the growth factor r and the population size p. It says, for instance, that when the population is 20,000, the growth factor should be $r = 2.25 - 0.000025 \cdot 20,000 = 1.75$. That is, if the population is 20,000 one year, we expect it to grow to $1.75 \cdot 20,000$ by the next year. On the other hand, if the population is $p = 70,000$, then the growth factor should be $r = 2.25 - 0.000025 \cdot 70,000 = 0.5$. This means that if the population somehow got to be 70,000, in the following year it is expected to shrink to $0.5 \cdot 70,000$. Using this linear equation makes

6.1. Properties of Logistic Growth Sequences

the growth factor r depend on the size of the population. For smaller populations there is more rapid growth. For larger populations there is slower growth, or even a decline in the size of the population.

Now we can incorporate this model for the growth factor into our difference equation for the population growth. In place of r in

$$p_{n+1} = rp_n$$

we will use the growth factor that will apply with a population of size p_n, namely $r = 2.25 - 0.000025 p_n$. That gives us the equation

$$p_{n+1} = (2.25 - 0.000025 p_n) p_n \tag{6.1}$$

This form of difference equation defines logistic growth. More generally, a logistic growth model follows a difference equation of the form

$$p_{n+1} = r \cdot p_n,$$

where r is given as a linear function of p_n.

What does (6.1) predict for our population of fish? Using a numerical method, we can compute p_n for several years, starting with $p_0 = 1{,}000$. Rounded to whole number answers,[2] the first few computations are as follows:

$$p_1 = (2.25 - 0.000025 \cdot 1{,}000)1{,}000 = 2{,}225$$
$$p_2 = (2.25 - 0.000025 \cdot 2{,}225)2{,}225 = 4{,}882$$
$$p_3 = (2.25 - 0.000025 \cdot 4{,}882)4{,}882 = 10{,}390$$

Continuing this process for several more repetitions produces the results shown in Figure 6.3.

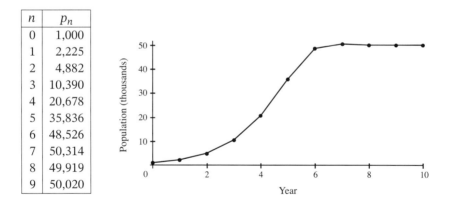

n	p_n
0	1,000
1	2,225
2	4,882
3	10,390
4	20,678
5	35,836
6	48,526
7	50,314
8	49,919
9	50,020

Figure 6.3. Fish population model: p_n represents the population size in year n.

The table and the graph show the kind of behavior that we expect to see. Initially, the population growth looks very much like a geometric model. During this stage, the environment is not limiting growth very much. However, as the population continues

---

[2] Here and throughout the chapter, each p_n is computed with full accuracy, but the printed results are rounded. Using a rounded value of each term to compute the next term would produce slightly different results than appear in the table.

to grow, the rate of growth tapers off. Each year the percentage of increase is less than for the preceding year. Eventually, the population starts to level off near 50,000. If you examine the data carefully, you will see that the population actually goes a little above 50,000. Then it jumps back and forth. One year it is a little less than 50,000, the next it is a little more, then a little less, then a little more. The population size continues to fluctuate back and forth, all the while edging closer to the 50,000 figure. Remember that we picked 50,000 as the size at which the population would remain fixed from year to year.

This introductory discussion shows how a simple model can be modified to obtain more realistic behavior. Earlier models with geometric growth assumed that the growth factor would remain constant no matter what the population size. In this presentation we built up a new model that includes variation in the growth factor. Admittedly, using a linear model for this variation imposes its own limitations. But the resulting model is much more realistic than simple geometric growth. In a true application, we might next try to measure the way the growth factor actually varies as the population grows, and fit the best linear model possible to the data. That would give us a logistic growth model that could be tested for accuracy in the real world.

As a matter of fact, logistic models often appear in applications, and not just for growth of biological populations. Another example of limited growth is provided by sales of new consumer electronics devices, such as cell phones. Initially, sales increase rapidly, but new sales decline as the market becomes saturated. That is another context for which a logistic model might be adopted. At the end of this section we will see an example of fitting a logistic growth sequence to actual data. But now we turn to an investigation of the properties of logistic growth sequences.

General Features of Logistic Sequences. In past chapters, we have introduced difference equation models of several different types: arithmetic growth, quadratic growth, geometric growth, and mixed growth. In each case, we were able to describe a general difference equation by using parameters. In any actual application the parameters would be replaced by numbers, but the general form with the parameters provided a pattern for all the difference equations for each type. We were also able to find functional equations in each case, again expressed in terms of parameters. And finally, we found general characteristics of graphs of each model type.

Now we would like to proceed in the same way for logistic growth. Unfortunately, because these models are inherently more complex than the earlier examples, we will not be as successful in finding results that apply to all logistic models. In particular, we will not be able to derive a functional equation. However, we can formulate a general difference equation, and make some observations about the graphs. To begin, we state a general description of logistic growth, based on the discussion in the preceding section.

> **Logistic Growth Sequence Verbal Definition:** Logistic growth, like geometric growth, defines a relationship between successive terms of a sequence a_n. In a logistic model, the change from one a_n to the next involves a growth factor that depends (linearly) on the size of a_n. This contrasts with geometric growth, for which the change from one a_n to the next involves a *constant* growth factor.

6.1. Properties of Logistic Growth Sequences

Next we will develop general forms for the difference equations that occur in logistic growth models.

Logistic Growth Difference Equations. The difference equation from the example with the fish population took the form

$$p_{n+1} = (2.25 - 0.000025 p_n) p_n \qquad (6.2)$$

As in previous cases, there are several different ways to express this difference equation, and each one gives us different insights. As it stands, we might introduce the parameters r_0 in place of 2.25 and m in place of 0.000025. This gives the general form

$$p_{n+1} = (r_0 - m p_n) p_n.$$

We use r_0 to remind us of an initial growth factor. In the example, this value is 2.25, and for very small populations, the logistic sequence will grow like a geometric sequence with growth factor 2.25. However, as the population grows, the growth factor varies according to a straight-line model. The slope of that straight line is $-m$. For the cases we will consider, the parameter m is positive and the slope, $-m$, is negative.

An equivalent version of (6.2) is

$$p_{n+1} = 2.25 p_n - 0.000025 (p_n)^2.$$

The parameter form of this version is

$$p_{n+1} = r_0 p_n - m(p_n)^2$$

This version shows that the logistic growth difference equation can be thought of as the result of adding one term to a geometric growth difference equation. Where the difference equation for geometric growth had only a constant times p_n, the logistic model has an additional term made up of a constant times p_n *squared*. This is reminiscent of the way polynomials are built up from linear functions. The simplest polynomials are linear, then the next simplest, with squared variables, are the quadratics. Seen in this light, the logistic growth model is a natural extension of geometric growth, and the next simplest form that can be expressed using polynomials. In addition, this form of the difference equation shows how the model starts out like geometric growth. Because the coefficient of $(p_n)^2$ is very small, that term will not have much impact on the final result for p_{n+1} unless p_n is fairly large. With $p_n = 1,000$, for example, we have $p_{n+1} = 2.25(1,000) - 0.000025(1,000)^2$. That works out to $2,250 - 25$, and subtracting the 25 does not have much effect. We would have gotten pretty much the same result by just computing $p_{n+1} = 2.25 p_n$. This illustrates how, for small values of p_n, the formula for p_{n+1} is closely approximated by just $2.25 p_n$, which is the kind of expression that produces geometric growth.

Here is one final reformulation of the difference equation from the example. As you can verify using algebra, the original difference equation (6.2) is equivalent to

$$p_{n+1} = 0.000025(90,000 - p_n) p_n.$$

In this form, we can easily see some of the limitations of the logistic model. Clearly, if $p_n = 0$, the next value p_{n+1} will also equal 0. That is entirely reasonable. However, if p_n ever reaches 90,000, we can see that the next year's population will be zero. That is not such a reasonable effect. We can imagine that the lake cannot support 90,000 fish, but we would probably not expect the entire population to die off in a single year. Yet that is what the model says: if the population is 90,000 this year, it will be zero the next.

Here is an even more questionable aspect of the model. If somehow a population above 90,000 were reached, the difference equation predicts a negative population for the following year. That is just nonsense. These unreasonable aspects are consequences of using a linear model for the way the growth factor varies with the population size. Of course, on a linear *p-r* graph (as illustrated in Figure 6.2), the line will cross the horizontal axis somewhere, and so predict a negative growth factor r for some values of p. That is not reasonable, and would not occur if we used a graph like Figure 6.1.

In spite of this questionable behavior for large population sizes, the logistic model is often very useful. We simply have to keep in mind that the model is reasonable for a limited range of populations. In the fish population example, the limited range runs from 0 to 90,000. In the more general setting, we use the parameter L in place of 90,000, to remind us that it is the limiting case of the model. We simply cannot use the model for populations above L. The coefficient 0.000025 doesn't have such an apparent meaning in this context, so let us continue to use the parameter m for that value, as before. Then we have

$$p_{n+1} = m(L - p_n)p_n$$

as the general form for this version of the logistic growth difference equation. Comparing this equation with the earlier versions, observe that $mL = r_0$.

For easy reference, the three forms of the logistic growth difference equation are shown below:

> **Logistic Growth Sequence Difference Equation:** Logistic growth can always be described by a difference equation in each of the following forms:
>
> $$p_{n+1} = (r_0 - mp_n)p_n \qquad (6.3)$$
> $$p_{n+1} = r_0 p_n - mp_n^2 \qquad (6.4)$$
> $$p_{n+1} = m(L - p_n)p_n \qquad (6.5)$$
>
> where $mL = r_0$ and $L = r_0/m$.

As in previous chapters, we have formulated these general difference equations using parameters to stand for numerical constants that appear in any specific example. In the previous chapters, we found ways to describe characteristics of a model based on the parameter values. Here is a very simple example from geometric growth. The general geometric growth difference equation is $a_{n+1} = ra_n$. Depending on whether the parameter r is greater than or less than 1, the a_n's will either grow larger or decrease over time. So, by looking at the numerical value of r in a specific example, we can draw conclusions about the behavior of the model.

In a similar way, we can formulate characteristics of logistic growth models in terms of the parameters in their difference equations. That is what we will do in the next few paragraphs.

Keeping Within the Proper Range. Earlier, we saw that a logistic growth model only makes sense when the population stays within an appropriate range. In terms of the parameters in (6.5), the model is invalid if p_n gets larger than L, for in that case a negative growth factor occurs. In this section we will see that for some values of the parameters, we can be sure that p_n will always remain in the valid range.

6.1. Properties of Logistic Growth Sequences

To see how this works, let us consider our numerical example again. The difference equation for this example is

$$p_{n+1} = 0.000025(90{,}000 - p_n)p_n.$$

For this discussion, we will focus on how any term of the sequence depends on the preceding term, rather than how the terms vary with the position number n. To emphasize this distinction we will use a different notation. Let x denote the population for some (unspecified) year and write y for the population the following year. That means that in the difference equation above, x will take the place of p_n while y replaces p_{n+1}. Then we have

$$y = 0.000025(90{,}000 - x)x = 2.25x - 0.000025x^2.$$

In this context we can see that y is a quadratic function of x, and so, using what we know about the graphs of quadratics, we can establish several conclusions. First, the general shape of the graph is a downward opening parabola. The parabola crosses the x-axis at the roots 0 and 90,000, and so the highest point must occur just half way between these roots, at $x = 45{,}000$. There, $y = 0.000025(90{,}000 - 45{,}000)45{,}000 = 50{,}625$. These features are illustrated in Figure 6.4.

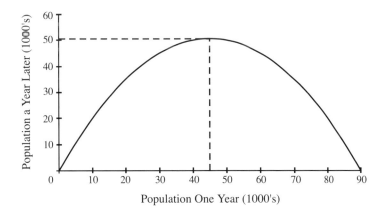

Figure 6.4. y stands for p_{n+1}, x stands for p_n.

One feature of the graph has special significance: the position of the high point with $y = 50{,}625$. This shows that no matter what we take for x, we will never see a value of y above 50,625. In particular, if x is anywhere between 0 and 90,000, then y will also be between 0 and 90,000. Remembering what x and y represent, this gives us the following conclusion: If the population in any year is between 0 and 90,000, the population in the following year will remain between 0 and 90,000. This shows that the logistic model will never lead to a value at 90,000 or above if we start with an initial population within the acceptable range. In fact we can see a little more. With a starting population anywhere between 0 and 90,000, the population after one year will be between 0 and 50,625, and will remain in that range forever after.

A similar analysis can be carried out using parameters instead of numbers. We think of x as representing any term of the sequence, and y as the next term. If the difference equation is

$$p_{n+1} = m(L - p_n)p_n \tag{6.6}$$

the equation for x and y becomes

$$y = m(L - x)x = -mx^2 + mLx.$$

Using the same steps as before, we can derive Figure 6.5 as follows. First, observe that y is a quadratic function of x, and the graph will be a downward opening parabola. Looking at the factored version of the equation, we see that $x = 0$ and $x = L$ both result in $y = 0$, and so correspond to the x-intercepts of the curve. The high point on the parabola must occur for x halfway between 0 and L, that is, at $L/2$. But at $x = L/2$ we compute $y = m(L/2)(L/2) = mL^2/4$. Therefore, the highest point on the curve is $(L/2, mL^2/4)$.

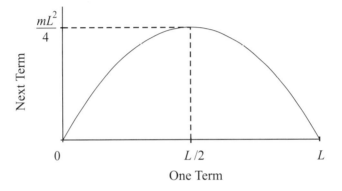

Figure 6.5. Using parameters L and m.

We would like this highest value to be less than or equal to L, so that the population will never grow too large (meaning so large that our linear model for the growth factor will become negative). In symbols, we should insist that

$$mL^2/4 \leq L.$$

Because we assume $L > 0$, algebra can be used to simplify this condition to the form

$$mL \leq 4.$$

So, when we use the logistic model in (6.6) we should be sure to check whether $mL \leq 4$. As long as that is true, we need not worry about the unrealistic behavior that occurs when the population gets too large. On the other hand, if $mL > 4$, we need to be on the alert for unreasonable behavior.

In our original example, with $p_{n+1} = 0.000025(90{,}000 - p_n)p_n$, observe that $mL = 2.25 \leq 4$, as desired. Now let us consider a modified example, with difference equation $p_{n+1} = 0.000085(90{,}000 - p_n)p_n$. For this example we calculate that mL is *not* less than or equal to 4. In fact, $mL = 0.000085 \cdot 90{,}000 = 7.65$.

Starting with $p_0 = 1{,}000$, our modified difference equation leads to the following results (computed using rounded values of each successive term):

$$\begin{aligned}
p_1 &= 0.000085(90{,}000 - 1{,}000)1{,}000 &&= 7{,}565 \\
p_2 &= 0.000085(90{,}000 - 7{,}565)7{,}565 &&= 53{,}008 \\
p_3 &= 0.000085(90{,}000 - 53{,}008)53{,}008 &&= 166{,}674 \\
p_4 &= 0.000085(90{,}000 - 166{,}674)166{,}674 &&= -1{,}086{,}263.
\end{aligned}$$

6.1. Properties of Logistic Growth Sequences

In this example, the population grows so rapidly that it exceeds the range in which the model is valid. Even though $p_3 = 53{,}008$ is a valid population size for the model, it leads to $p_4 = 166{,}674$ which is too large. Our straight-line model for the growth factor is invalid for such a large population, and results in a negative number for p_5. This shows that the model with difference equation $p_{n+1} = 0.000085(90{,}000 - p_n)p_n$ can lead to unrealistic results. This can happen any time we use a model with $mL > 4$. To emphasize this fact, it is presented below in a box.

Logistic Condition for $0 \le p_n \le L$: For a logistic growth model with difference equation

$$p_{n+1} = m(L - p_n)p_n$$

and $0 \le p_0 \le L$,

- if $mL \le 4$, every p_n will remain within the range from 0 to L;
- if $mL > 4$, the model may lead to unrealistic results, including p_n values that are greater than L, or negative.

Where Does the Model Level Off? One of the attractive features of the logistic model in the fish example is the way the population levels off. If wildlife managers want to introduce a species of fish into a lake, that is just what they would like to see. At first the new fish population grows, but it eventually levels off and approaches a steady size. That is a sign that the fish population is in harmony with its environment.

If the population does level off at a steady level, it is easy to predict just what the steady level will be. After all, if the population is steady, it must remain the same year after year. That means that p_{n+1} must equal p_n. Such a population is called a *fixed population*. This is the same concept as the fixed points introduced in the discussion of mixed models. As before, we will find the fixed points or fixed populations by solving the equation $p_{n+1} = p_n$, and once again the answer can be expressed in terms of parameters of the model, although the form of this expression is different for logistic growth than it was for mixed models.

To illustrate, let us find the fixed populations for the logistic growth example considered above. We seek an unknown population size, x, that is fixed by the difference equation. That is, if p_n ever reaches x, then p_{n+1} will also be x. From the difference equation

$$p_{n+1} = 0.000025(90{,}000 - p_n)p_n$$

we can therefore obtain

$$x = 0.000025(90{,}000 - x)x.$$

This is a quadratic equation, and can be solved using factoring or the quadratic formula. But here is a simpler approach. One root of the equation is $x = 0$, as the reader is invited to verify. To find a nonzero root, we can divide both sides of the equation by x, obtaining

$$1 = 0.000025(90{,}000 - x).$$

Notice that the expression on the right represents the growth factor r for a population of size x. This shows that the nonzero fixed population corresponds to a growth factor of 1, which is consistent with our understanding of growth factors. Be that as it may, our equation is now linear, and so can be solved using algebra. The result is

$$x = 90{,}000 - 1/0.000025 = 50{,}000.$$

Thus, the fixed populations for this example are 0 and 50,000.

Recall that a fixed population of 50,000 was built into this example as one of the defining assumptions. Nevertheless, there is a value in rederiving this result using algebra: now we can repeat the process using parameters. So starting with the difference equation in the form

$$p_{n+1} = m(L - p_n)p_n$$

we can see that the growth factor is $m(L - p_n)$. As before, it is clear that 0 is a fixed population. To find the other fixed population, we follow the same steps we used before. We are looking for a population size x that makes the growth factor 1. So write $m(L - x) = 1$ and solve for x. This leads to the equation

$$x = L - \frac{1}{m}.$$

When $p_n = L - 1/m$, we know that $p_{n+1} = 1 \cdot p_n$, which means that the population won't change the next year. To summarize, the general logistic model always has fixed populations of size 0 and $L - 1/m$. The equivalent forms $L - 1/m = (mL - 1)/m = (b - 1)/m$ are all sometimes useful.

For a given logistic model, the fixed population $L - 1/m$ can be positive, negative, or zero. These possibilities lead to different conclusions about the properties of the model. In exploring this idea, it is useful to formulate each case algebraically. The positive case is expressed by the inequality $L - 1/m > 0$, which can be algebraically rearranged to $mL > 1$. Thus $L - 1/m$ is positive exactly when $mL > 1$. In a similar way, $L - 1/m$ is negative when $mL < 1$, and zero when $mL = 1$.

Returning to the main thread of this discussion, we are interested in the possibility that a logistic growth sequence will approach an equilibrium, as in the example of the fish population. This does not occur in all logistic growth models, but when it does, the equilibrium must be one of the fixed populations, 0 or $L - 1/m$. But which one? The answer depends on whether $L - 1/m$ is positive, negative, or zero, as we now proceed to show.

We consider first the case that $L - 1/m$ is negative, which occurs when $L - 1/m < 0$, or equivalently, when $mL < 1$. In this case, our previous results show that the population must remain between 0 and L: since mL is less than 1, it is certainly less than 4. And as the population remains between 0 and L, it cannot approach $L - 1/m$, a negative value. So if the population does level off, it has to be at the other fixed population, 0.

Analyzing this case further, we can also deduce that the population must steadily decline. We first observe that with $mL < 1$, the growth factor will always be less than 1. Look back at Figure 6.2. Remember that we are using a linear model for the way that the growth factor changes as the population changes. What is more, that straight line has a negative slope. So the highest possible growth factor occurs on the left side of the diagram, that is, where $p = 0$. We can find this algebraically using the growth factor formula, $m(L - p)$, and setting $p = 0$. The result is mL, and that is the maximum value for the growth factor. But we also know $mL < 1$. So in this case, the growth factor always stays less than 1. But that means the population always gets smaller from one year to the next because it is being multiplied by a factor less than 1.

6.1. Properties of Logistic Growth Sequences

Combining all of these findings, in the case that $L - 1/m$ is negative, the following conclusions holds:

- The population gets smaller each year.
- The population stays between 0 and L.
- The only possible fixed populations are 0 and something that is negative.

There is only one possible conclusion: the population is going to steadily decrease toward 0. With a few minor modifications, these same conclusions can also be reached when $L - 1/m$ equals 0. To summarize, if the fixed population $L - 1/m$ is negative or zero, which occurs when $mL \leq 1$, we know what will happen. The population will decrease steadily, leveling off at 0. Eventually, all the fish die.

This is an important possibility, and, if it occurs in a real model, would obviously be cause for concern. At the very least, it would be a signal that further study should be undertaken. And our results show that this case definitely occurs when the fixed population $L - 1/m$ is negative or zero.

Next let us consider the case where the population does not die out, but levels off at a steady population (greater than 0). This can only happen if $mL > 1$. In this case, for sufficiently small populations, the growth factor would exceed 1, and the population would increase. Indeed, as argued earlier, from a small initial population the logistic growth sequence will start out approximating geometric growth with a growth factor about equal to $r_0 = mL > 1$. That means the population cannot decrease toward an equilibrium of 0. So, if the population does level off, it would have to approach the fixed population of $L - 1/m$.

This concludes the discussion of what happens when the fixed population $L - 1/m$ is positive, negative or zero. For future reference, a summary is given in the box below.

> **Logistic Growth Steady Levels:** For a logistic growth sequence with difference equation $p_{n+1} = m(L - p_n)p_n$ and with $0 \leq p_0 \leq L$, there are two possibilities:
>
> - If $mL \leq 1$, then p_n will steadily decrease, leveling off at 0.
> - If $mL > 1$, then p_n may level off to a steady positive value. If so, the steady size of the population will be $L - 1/m$.

When Does the Model Level Off? The preceding results tell us how to predict the equilibrium value for a logistic growth sequence that levels off. But that does not always occur. This can be illustrated by modifying the earlier fish population example. For future reference, we will refer to the modified model as the *second fish population example*.

Recall that in the original example we developed a linear equation expressing the growth factor r as a function of the population p. In particular, we assumed a growth factor of 2 for a population of 10,000, meaning that a population of 10,000 would double in a year. We also proposed a growth factor of 1 for a population of 50,000, meaning that the population would remain steady year after year, once it reached 50,000. From these assumptions we found the equation

$$r - 2 = -0.000025(p - 10{,}000).$$

Now we will modify this slightly, by changing the growth factor from 2 to 2.8 for the population of size 10,000. That results in a different equation for r and p :

$$r - 2.8 = -0.000045(p - 10{,}000).$$

Notice that changing from 2 to 2.8 produces two changes in the linear equation. The number subtracted from r on the left changed in the expected manner. However, the coefficient -0.000025 on the right changed to -0.000045. It is less obvious why this change occurred. To really understand this, you should retrace the steps that led to the linear model for r in the first place. Exercise (20) deals with this issue.

Proceeding as before, we isolate r

$$r = -0.000045(p - 10{,}000) + 2.8 = 3.25 - 0.000045p$$

and so obtain the difference equation

$$p_{n+1} = (3.25 - 0.000045 p_n) p_n.$$

This can also be put into the form

$$p_{n+1} = 0.000045(72{,}222 - p_n) p_n.$$

If we again start at $p_0 = 1{,}000$, how will this model work out? Because we simply increased the assumed growth factor for a population of 10,000, you might expect that the population will grow a little faster at first, but eventually level off at the same size as before. But that is not what occurs.

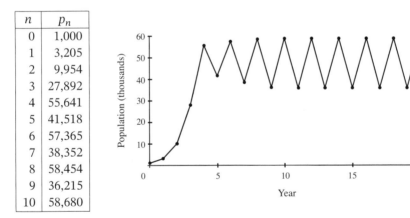

n	p_n
0	1,000
1	3,205
2	9,954
3	27,892
4	55,641
5	41,518
6	57,365
7	38,352
8	58,454
9	36,215
10	58,680

Figure 6.6. Numerical values and graph for the second fish population example. Increasing the assumed growth factor for a population of 10,000 from 2 to 2.8 leads to the difference equation $p_{n+1} = (3.25 - 0.000045 p_n) p_n$. In this model the population does not approach a steady equilibrium. Rather, it settles into a regular pattern of oscillation between two values, approximately equal to 59,000 and 36,000.

In Figure 6.6, the results of the model are shown for the first 10 years in a table and for the first 20 years in a graph. As the table and graph show, instead of leveling off at a steady value, the population jumps back and forth between two different levels, one

6.1. Properties of Logistic Growth Sequences

around 59,000, and the other around 36,000. According to this model, in even numbered years, the population will swell to nearly 59,000 fish. But in the odd-numbered years it will be reduced to around 36,000 fish.

This *oscillation* is quite different behavior than what we found earlier. It offers a glimpse of the greater levels of complexity that can arise in logistic growth sequences, and a phenomenon referred to as *chaos*. In the next section we will explore this situation in greater detail. Then you will see why chaos has important implications for models in biology (and other subjects). But for now we want to concentrate on order (not chaos).

To recap the discussion this far, we started out observing that the first model behaved just as we hoped—the population started out resembling geometric growth, but then leveled off. The revised version of the example showed that logistic models cannot be counted on to act that way all the time. In fact, with a very minor change in the numerical values of the parameters, we obtained quite a different picture of the future growth of the fish population. As in the preceding section, the parameters of the logistic model can be used to predict which of these cases will occur. The mathematical methods that lead up to this prediction are beyond the scope of this course. They involve applying methods of calculus to study characteristics of polynomial equations. However, the prediction itself is not difficult to understand, and is very similar to the kind of prediction we found earlier. It is stated as follows[3]:

> **Logistic Growth Positive Steady Level:** Suppose a logistic growth model has the difference equation
> $$p_{n+1} = m(L - p_n)p_n$$
> and that p_0 is between 0 and L. Then, if $1 < mL < 3$, the population will eventually level off at a positive steady value of $L - 1/m$.

We already knew from previous remarks that the steady level, if one occurs, must be either 0 or $L - 1/m$. There are two new items of information given in this box: (1) when $mL < 3$, a leveling off will always occur, avoiding more complicated behavior such as the zigzag pattern of Figure 6.6; and (2) when mL is both greater than 1 and less than 3, the population will level off at a positive size, and will not die out.

In the original example, we had $m = 0.000025$ and $L = 90,000$. In that case $mL = 2.25 < 3$, so we predict that the model will level off eventually for any initial value p_0 between 0 and 90,000. The level population is supposed to be $90,000 - 1/0.000025 = 50,000$, and that is what was observed. On the other hand, the revised model had $m = 0.000045$ and $L = 72,222$. For that case we have $mL = 3.25 > 3$, so we cannot be sure if the population will level off. And for the example we tried, with $p_0 = 1,000$ the population did not level off.

When we formulate a logistic growth model, if the parameters m and L satisfy $1 < mL < 3$, we can be sure that the population will stay within the range that makes sense for the model, and, in fact, will level off to a positive fixed population size, or equilibrium. This leveling-off process can occur in three ways, as illustrated in Figure 6.7. In some cases the population will simply grow steadily until it reaches the constant level. A second possibility is that we start at a population above the constant level, and then decrease steadily to the constant level. The third possibility is that eventually the

---

[3]See [**31**], pages 41–46.

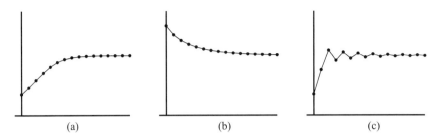

Figure 6.7. Logistic growth sequences approaching a nonzero equilibrium. In (a) the terms consistently increase toward the equilibrium. In (b) they decrease toward the equilibrium. In (c) they oscillate above and below the equilibrium.

population wavers up and down, alternating between a little above and a little below the constant level, but always getting closer to the constant level. That is what we saw in the original example. After an initial period of rapid growth, the population steadied down toward a constant level while wavering up and down. Using the same kinds of methods that lead to the preceding boxed comment, it can be shown how to predict what will happen just by looking at the value of mL, but we will not delve into the details here. We simply observe that this is one more example of how the parameters of the logistic model can convey general information about the behavior of the model.

As mentioned earlier, there is no simple functional equation for logistic models. At first, this may seem to limit our ability to use these models. In past experiences, the functional equation was one of our most important tools. However, we have seen now that a great deal of information can be determined about the general trends in a logistic model. If $mL < 4$, we can predict confidently that the model will remain within the valid range for the growth factor model. If $mL \leq 1$, we know that the model will decrease steadily to 0, meaning that the entire population will die out. Otherwise, if $1 < mL < 3$, we can conclude that the population will level off. We can even say what it will level off to: $L - 1/m$. Given these results, it is not so critical to have a functional equation. In the next section we will explore what happens when the model cannot be relied on to settle down to a steady level. As already remarked, that will involve the concept of chaos. We conclude this section with some examples illustrating the construction of logistic growth models.

Constructing A Logistic Growth Model. A logistic growth model can be constructed in two steps. First, we develop a linear equation defining the growth factor r as a function of the population size p. Second, we substitute that function for r in the general difference equation $p_{n+1} = rp_n$.

One simple way to carry out the first step is to assume or determine values for r at two different population sizes. Here is an example of this process.

Modeling Mold Growth. A laboratory is developing a procedure to grow a certain kind of mold that will be used to make a new antibiotic. The mold is grown in a vat with a nutrient solution made up of sugars, water, and other ingredients. In one experiment with the nutrient solution, the laboratory makes two tests, in each case observing how a known amount of the mold grows over a 24-hour period. For each of the two test amounts, the growth factor for the 24-hour period is recorded. In the first test, with

6.1. Properties of Logistic Growth Sequences

an initial population of 100 a growth factor of 2.0 was observed. In the second test, the initial population was 500 and the growth factor was found to be 1.2. Using this information, we wish to develop a logistic growth model for mold growing in this nutrient solution.[4]

Proceeding with our first step, we will find a linear equation for the growth factor r as a function of the population p. The given information defines two points of the form (p, r). This information can be displayed as a table and graphed as shown in Figure 6.8.

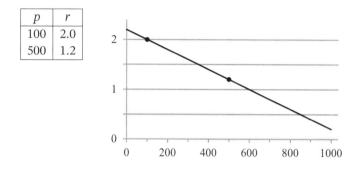

Figure 6.8. A linear equation for growth factor r as a function of population p is determined by two data points.

Applying methods of Chapter 2, we find an equation for the line as follows. The slope of the line is given by

$$m = \frac{1.2 - 2}{500 - 100} = -\frac{0.8}{400} = -0.002.$$

We apply the point-slope equation $y - y_0 = m(x - x_0)$ replacing m by -0.002 and x_0 and y_0 by the coordinates of one data point, $(100, 2)$. That produces

$$y - 2 = -0.002(x - 100)$$

but because our variables are p and r, rather than x and y, we should actually write

$$r - 2 = -0.002(p - 100).$$

Adding 2 to both sides of this equation we find

$$r = 2 - 0.002(p - 100) = 2 - 0.002p + 0.2.$$

As a last simplification, combine the two constant terms on the right to obtain

$$r = 2.2 - 0.002p.$$

Now we are ready for the second step of the model development. In the generic logistic difference equation

$$p_{n+1} = r p_n$$

---

[4]This description of the model context is slightly unrealistic. In an actual laboratory situation, population size would probably be estimated from a small sample, possibly by counting individual organisms within a microscopic area or volume, or by determining the weight of the organisms in the sample. In either case, what is actually measured is more correctly described as a population *density*, in units of weight or number of organisms per square inch or per cubic inch, rather than an absolute count of organisms in the entire vat of nutrient solution. We have chosen to formulate the model as if p_n is an actual count of the population size for the sake of simplicity.

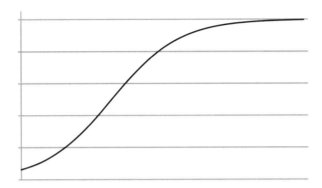

Figure 6.9. A logistic growth model might provide a good fit to data that have a graph similar to this one.

replace r with $2.2 - 0.002p_n$, which our linear equation indicates is the appropriate growth factor when the population is p_n. The result is

$$p_{n+1} = (2.2 - 0.002p_n)p_n. \tag{6.7}$$

To analyze the sequences defined by this difference equation, let us identify the values of the parameters m and L. This can be done by algebraically reformulating the equation in the form of the standard difference equation

$$p_{n+1} = m(L - p_n)p_n.$$

Observe that

$$2.2 - 0.002p_n = 2.2 \cdot \frac{0.002}{0.002} - 0.002p_n = 0.002 \cdot \frac{2.2}{0.002} - 0.002p_n.$$

Now we can factor out 0.002 to obtain

$$0.002\left(\frac{2.2}{0.002} - p_n\right) = 0.002(1{,}100 - p_n).$$

This shows that (6.7) can be rewritten

$$p_{n+1} = 0.002(1{,}100 - p_n)p_n,$$

revealing that $m = 0.002$ and $L = 1{,}100$ for this model.

With these values identified, we can draw several conclusions. First, we compute $mL = 0.002 \cdot 1{,}100 = 2.2$, and observe that this value is between 0 and 4. According to our earlier findings, starting from any initial value p_0 with $0 < p_0 < 1{,}100$, the logistic growth sequence constructed from our difference equation (6.7) must remain within the range from 0 to 1,100. But we can go further. The computed value of mL is more specifically between 1 and 3. This shows that the terms of the sequence must actually approach an equilibrium value. Finally, we compute the fixed population to be $L - 1/m = 1{,}100 - 1/0.002 = 600$. So, in summary, any logistic growth sequence we construct in this model will approach a steady population of 600. These conclusions can be tested by choosing a variety of initial values p_0, and for each one computing the succeeding terms of a logistic growth sequence using our difference equation. Studying the results numerically and graphically will show that the conclusions we have reached are correct.

6.1. Properties of Logistic Growth Sequences

Fitting a Logistic Growth Model to Data. In the preceding example we constructed a model based on hypothetical values for p and r. Now we will consider fitting a logistic growth model to actual growth data. This may be appropriate when the data appear to grow exponentially at first, but then level off approaching an equilibrium, especially when this behavior is suggested by our knowledge of the modeling context. Logistic growth may also be suggested when a graph of the data has a shape similar to Figure 6.9. However, even when these conditions are observed, a logistic growth model might not be very accurate. We will consider two examples. A logistic growth model provides a reasonably accurate approximation to the data in the first example, but not in the second.

Modeling Pumpkin Growth. For our first example we use data published in an early treatise on mathematical models for biological growth.[5] The data and a graph are shown in Figure 6.10. The shape of the graph suggests the idea of logistic growth.

Days	Grams
5	267
6	443
7	658
8	961
9	1498
10	2200
11	2920
12	3366
13	3758
14	4092
15	4488
16	4720
17	4864
18	4980
19	5114
20	5176
21	5242
22	5298
23	5352
24	5360
25	5366

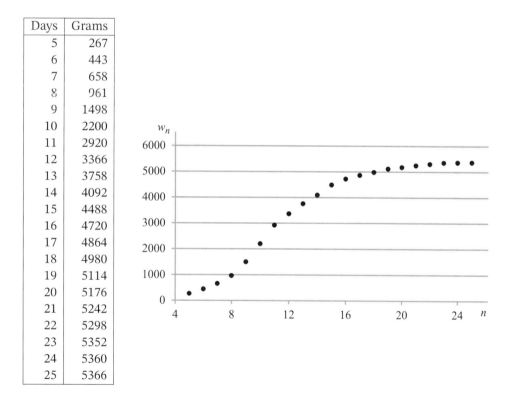

Figure 6.10. The data show daily observations of the weight of one fruit growing on a *Cucurbito pepa* vine. This is a member of the squash family, possibly a pumpkin. The first column of the table represents a number n of days, and the second represents the weight w in grams.

To develop a logistic growth model based on the data, we follow the same approach as in the preceding problem. The first step is to formulate an equation for the growth

---
[5]The data were included in [45], with an attribution to [2].

factor r as a function of weight w. To be more precise, because the data values are listed at one-day intervals, r is intended to predict the growth factor for one day. Thus, if we find that for a weight of 2200 grams the growth factor is $r = 1.33$, that means one day later the weight should be $1.33 \cdot 2200 = 2926$.

We can use the given w values to compute a set of corresponding r values. For example, the growth factor between the initial weight of 267 and the following value of 443 is given by $r = 443/267 = 1.66$ (rounded to two decimal places). This corresponds to one derived data point $(w, r) = (267, 1.66)$. Proceeding similarly we can compute an r value for each of the first 20 w values. The results are shown in a table and a graph in Figure 6.11.

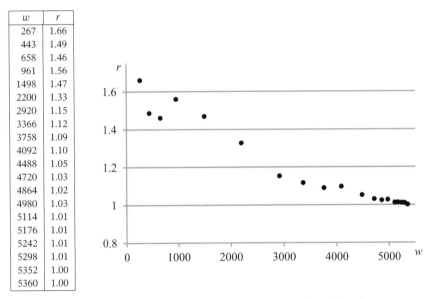

Figure 6.11. Weight w and growth factor r values based on the pumpkin data. Each r is computed as a ratio a/b where b is the corresponding w value and a is the following w value. The r values in the table have been rounded to two decimal places.

Now we wish to define a linear function whose graph is as close as possible to the data points in the figure. Applying a trial-and-error approach as discussed in Chapter 2, we can obtain the equation $r = -0.00011w + 1.6$. The graph of this equation is shown on the left in Figure 6.12.

The equation for r leads to the difference equation

$$w_{n+1} = (-0.00011w_n + 1.6)w_n.$$

Knowing the parameters $m = 0.00011$ and $b = 1.6$, we can easily compute $mL = b = 1.6$, so the model sequence w_n will definitely approach an equilibrium. The fixed population is conveniently computed as $L - 1/m = (b-1)/m = 0.6/0.00011$, which is 5455 to the nearest integer. Starting with an initial value of $w_5 = 267$ produces the graph shown on the right in Figure 6.12.

The logistic growth sequence provides a very close approximation to the first half of the data points, but does not fit very well for most of the remaining points. As a next

6.1. Properties of Logistic Growth Sequences

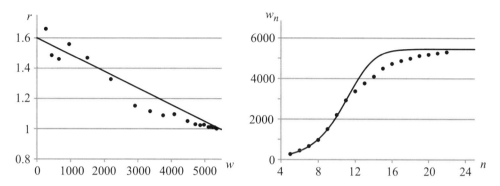

Figure 6.12. Linear model for r as a function of w and logistic model for w_n. The graph on the left shows the line $r = -0.00011w + 1.6$ superimposed on the derived data points (w, r). The graph on the right shows the corresponding logistic growth model with initial value 267, together with the original pumpkin data points.

step, we can vary the parameters m, L, and the initial value w_5, in order to obtain a closer agreement between the data and the model. Again working by trial and error, the following alternative parameter values can be found: $m = 0.000076$, $L = 18550$, and $w_5 = 450$. These lead to the results shown in Figure 6.13. The figure also shows the corresponding linear model for r as a function of w.

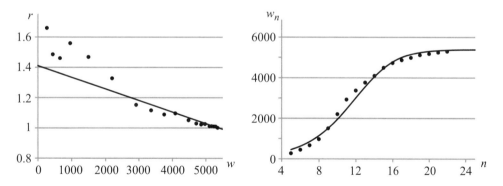

Figure 6.13. Refined linear model for r as a function of w and logistic model for w_n. The linear graph, on the left, does not agree with the derived r data as well as in the first version of the model, but the logistic graph on the right is a better fit to the original weight data.

The figures provide a visual comparison of the two versions of our model. Overall, the second logistic model is a more accurate approximation to the data than the first. But in the second model the linear equation for r does not fit the derived data very well. It is difficult to judge in the abstract which version is better. In an actual modeling context, the criteria for choosing a *best* model typically depend on the way the model will be used.

Properties of logistic growth tell us that an equilibrium will be approached in both models. For the first model the equilibrium level is 5455 and in the refined model it is $L - 1/m = 5392$. Note that these are quite similar values, so that the two models provide nearly equal predictions of the weight of the pumpkin at maturity. As it happens, the original publication included one final data point, giving a weight of 5400 grams after 47 days. This falls between the predictions of the two versions of the model, and shows that each of the predicted values is reasonably accurate. This is a partial validation of the use of a logistic growth model, given the approximate nature of our models.

The trial-and-error approach we used here is not very sophisticated. As we have commented in prior chapters, there are statistical methods for finding parameters that provide the best possible agreement between data and a model. Such methods would be likely to produce a logistic model that approximates the pumpkin data better than either of our models. However, it is not our goal with this example to find the *best* model. Rather, we are content to demonstrate one possible approach to formulating and refining a logistic model for a given data set. In addition, this example shows that even very simple trial-and-error methods can sometimes lead to models that agree very well with observed data.

Modeling Cumulative iPod Sales. The second example concerns the cumulative sales of iPods, starting soon after their introduction in 2001. The data and a graph are shown in Figure 6.14.

Year	Cumulative Sales
0	0.2
1	1.1
2	5.9
3	28.2
4	68.0
5	119.4
6	174.2
7	228.3
8	279.0
9	322.0
10	357.0
11	383.4

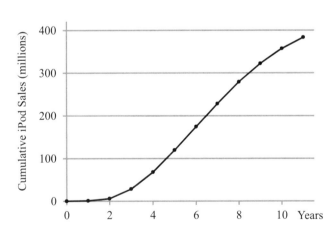

Figure 6.14. Cumulative sales of iPods, 2002 to 2013. The entries in the table were estimated from a graph posted on the internet [13]. Years are counted from 0 in 2002; cumulative sales are in units of millions. On the graph, the data points are connected with straight line segments to emphasize the shape.

6.1. Properties of Logistic Growth Sequences

The shape of the graph suggests logistic growth: although the right side of the curve has not leveled off very much, it appears to be bending down. Moreover, eventually iPod sales should drop off, as the market becomes saturated and newer technologies are developed. Thus, we expect the total cumulative sales to reach a maximum and then remain essentially constant. For both of these reasons, we might expect a logistic growth model to be appropriate for this data set.

However, that expectation turns out to be incorrect. Whereas it was possible to obtain good agreement between a logistic model and the data in the preceding example, that does not occur for this example. One reason is that the growth factors derived from the data are not very well approximated by a straight line. See Figure 6.15.

The data points in the figure suggest some other shape curve than a straight line, in marked contrast to what we saw in the pumpkin example. Because a linear model for r will not be very accurate in this example, we should likewise not expect a logistic growth model to be very accurate. In fact, we can try various linear models for r, but none will produce an accurate approximation of the data. Some specific cases are discussed in the exercises. On the other hand, as we will find at the end of the chapter, using a nonlinear function for r does lead to a reasonably accurate model.

s	r
0.2	5.500
1.1	5.364
5.9	4.780
28.2	2.411
68.0	1.756
119.4	1.459
174.2	1.311
228.3	1.222
279.0	1.154
322.0	1.109
357.0	1.074

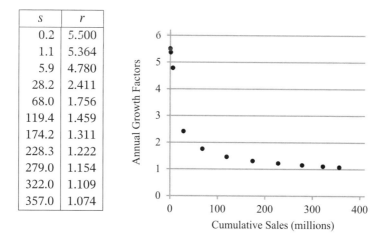

Figure 6.15. Cumulative sales s and growth factors r for data in Figure 6.14. For each s, r is the corresponding annual growth factor, shown to three decimal places.

Profile for Logistic Growth. The following profile for logistic growth sequences summarizes many of the key ideas presented in this section.

Table 6.1. Profile for logistic growth sequences.

Verbal Description:	To find p_{n+1} multiply p_n by $r = f(p_n)$ where $f(p)$ is a linear function that decreases as p increases.
Parameters:	m and r_0, coefficients for an equation $r = -mp + r_0$. $L = r_0/m$, upper limit. Model is valid for $0 \leq p_n \leq L$. p_0, initial term. $0 < p_0 < L$ assumed.
Difference Equations:	$p_{n+1} = (r_0 - mp_n)p_n$ $p_{n+1} = r_0 p_n - mp_n^2$ $p_{n+1} = m(L - p_n)p_n$.
Functional Equation:	none.
Properties:	Fixed populations at $p = 0$ and $p = L - 1/m$. For $0 < mL \leq 1$: p_n decreases, levels off at 0. For $1 < mL < 3$: $0 < p_n < L$, p_n levels off at $L - 1/m$. For $3 \leq mL \leq 4$: $0 < p_n < L$, p_n doesn't level off. For $mL > 4$: p_n values may exceed L and become negative.
Graph:	For $1 < mL < 3$, and with a small value of p_0, the graph rises rapidly at first, but then levels off. For $3 \leq mL < 4$ the graph remains between $p = 0$ and $p = L$ without leveling off. Oscillation or more complictated graphs are possible. Typical examples of these two cases are shown below.

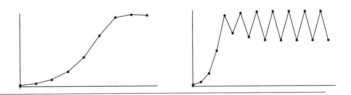

6.1 Exercises

Reading Comprehension.

(1) Compare and contrast arithmetic growth, geometric growth, and logistic growth. Refer to the concept of growth factor where appropriate.

(2) Write a short essay explaining what logistic growth is and how it is related to geometric growth.

6.1. Exercises

(3) @The choice of the letter L for one logistic growth parameter is supposed to remind us of the *limiting case* of the model. What does that mean? Why is this an important consideration in discussing logistic growth models?

(4) What is meant by *fixed populations*? Why are they important in discussing logistic growth? What calculations or parameters are used to determine fixed populations?

(5) Describe the possible long-term patterns that are observable in logistic growth models. Is it possible to predict for a particular logistic growth sequence what the long term pattern will be? Explain.

(6) In your own words, write a summary covering all the key points of the discussion headed *Keeping Within the Proper Range* (pages 366–369). In writing your summary use your understanding of quadratic functions to explain the pertinent results.

(7) @Explain the example connected with the table and graph shown in Figure 6.6. What is the significance of this example?

(8) @In the logistic growth models we have considered, p_n typically represents the size of some population after n periods, and r represents a growth factor. In this context we have studied several types of graphs. In one type, the points of the graph were of the form (p_n, p_{n+1}), in a second they were of the form (p, r), and in a third, (n, p_n). Each of the three graphs below has the shape we would expect for one of these types of graph. Identify the type of each graph, indicating what each axis represents, and give a brief explanation of how the graph is useful in studying logistic growth.

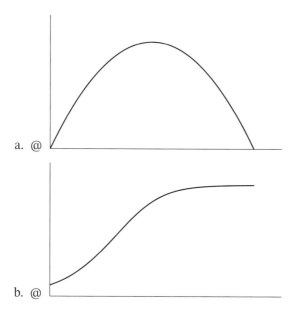

a. @

b. @

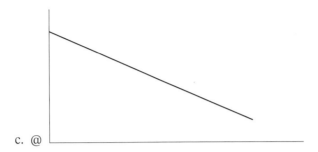

c. @

Mathematical Skills.

(9) @Consider each of the following difference equations. Determine what type of model (arithmetic, quadratic, geometric, mixed growth, logistic, or other) each represents. Briefly explain your choice. Not all types of models are necessarily represented.

   a. $a_{n+1} = 3a_n$
   b. @$b_{n+1} = (3 - b_n)b_n$
   c. $c_{n+1} = c_n + 3$
   d. @$d_{n+1} = 0.1d_n - 4$
   e. $e_{n+1} = e_n + e_{n-1}$
   f. @$f_{n+1} = 2.5f_n - 0.03f_n^2$
   g. @$g_{n+1} = g_n + 5 + 0.6n$
   h. $h_{n+1} = 0.07(150 - h_n)h_n$

(10) @For a certain model, when the population is 1,000, the growth factor is 3, and there is a fixed population size of 25,000. Find a linear equation relating population p and growth factor r.

(11) For a certain model, the growth factor is 5 when the population is 1,000 and the growth factor is 2 when the population is 25,000. Find a linear equation relating population p and growth factor r.

(12) @Find a linear equation giving the growth factor r in terms of the population p (in thousands).

   a. @When the population is 5 thousand the growth factor is 0.015 and when the population is 2 thousand the growth factor is 0.024.
   b. When the population is 10 thousand the growth factor is 0.0014 and when the population is 5 thousand the growth factor is 0.0049.
   c. @The growth factor for a population of 7,000 is 0.02 and for a population of 3,000 is 0.06.
   d. The growth factor for a population of 2 million is 0.1 and for a population of 2.5 million is 0.075.

(13) @For all of the parts of this problem, use the difference equation

$$a_{n+1} = 0.025(100 - a_n)a_n$$

   a. @Assuming $a_0 = 10$, find a_1, a_2, a_3, and a_4.

6.1. Exercises

b. @Predict approximate values a_n for large n, such as a_{50} or a_{100}.

c. @Find a fixed point for this model. That is, find a number for a_n that leads to exactly the same number for a_{n+1}.

(14) A logistic growth model has the difference equation

$$p_{n+1} = 0.0026(1{,}500 - p_n)p_n$$

What are the values of m, L, and r_0 for this sequence? For what values of p_0 is it certain that the entire sequence remains within the range between 0 and 1,500?

(15) @A logistic growth model has the difference equation

$$p_{n+1} = 0.0035(220 - p_n)p_n$$

What are the values of m, L, and r_0 for this sequence? For what values of p_0 is it certain that the model will predict that the population will eventually die out?

(16) A logistic growth model has the difference equation

$$p_{n+1} = 0.000045(17{,}500 - p_n)p_n$$

What are the fixed populations for this model? Can you be sure that the model levels off eventually? If so, to what value?

(17) @A logistic growth model has the difference equation

$$p_{n+1} = 0.00016(17{,}500 - p_n)p_n$$

What are the fixed populations for this model? Can you be sure that the model levels off eventually? If so, to what value?

(18) A logistic growth model has the difference equation

$$p_{n+1} = 0.0002(17{,}500 - p_n)p_n$$

What are the fixed populations for this model? Can you be sure that the model levels off eventually? If so, to what value?

(19) @For each of the models below, find the fixed populations. Then determine if we can be sure the model will level off eventually, and if so, to what value.

a. @$p_{n+1} = 0.005(40 - p_n)p_n$
b. $p_{n+1} = 0.0023(1{,}000 - p_n)p_n$
c. @$p_{n+1} = 0.035(100 - p_n)p_n$
d. $p_{n+1} = 0.035(200 - p_n)p_n$

(20) On page 372 the equation

$$r - 2.8 = -0.000045(p - 10{,}000)$$

is given. Show the steps that lead to this equation, assuming that $r = 2.8$ when $p = 10{,}000$ and $r = 1$ when $p = 50{,}000$.

Problems in Context. In any problem that asks for a graph, the use of graphing software is recommended. Similarly, to compute several terms of a sequence, the use of a spreadsheet or comparable technology is recommended.

(21) A scientist is studying a certain type of plant virus and how it spreads through fields of wheat. In one experiment, there were initially 100 infected plants, and in one week that number had doubled to 200 infected plants (a growth factor of 2). In another experiment the scientist started with 600 infected plants, and after a week that number had grown to 900 infected plants (a growth factor of 1.5). Based on this information develop a linear equation in which the variables are r, the growth factor for infected plants over a week, and p, the number of infected plants at the start of that week. Then use this equation to develop a logistic growth model for the number of infected plants week by week as an infection spreads.

(22) @Mold Example with Various Initial Populations. In the mold example (beginning on page 374), we found the logistic growth difference equation $p_{n+1} = (2.2 - 0.002p_n)p_n$. Then we applied known properties of logistic growth to draw several conclusions about sequences governed by this difference equation. For each value of p_0 shown below, use the difference equation to compute the values of p_1 through p_{10}, and graph the results. Do your results confirm the stated conclusions?

    a. $p_0 = 50$
    b. $p_0 = 200$
    c. $p_0 = 500$
    d. $p_0 = 800$
    e. $p_0 = 1,090$

(23) Growth Factors for Bacteria. A logistic growth model is being developed to describe how bacteria grow in un-refrigerated hamburger. For a one pound sample of hamburger, it is found that the bacteria increase each hour with a growth factor r given by the equation $r = 0.0004(6,000 - p)$ where p is the number of bacteria at the start of the hour.

    a. At the start of the experiment, there are 1,000 bacteria in the hamburger. What is the growth factor r at that point, and how many bacteria will there be after an hour?
    b. Later in the same experiment it is found that there are 3,200 bacteria. What is the growth factor r at that point, and how many bacteria will there be after another hour?
    c. What is the difference equation for this model?

(24) @Long Term Behavior of Weeds. A science lab is investigating several different models for the way weeds grow in a lake. In each model a_n is the number of square feet covered by weeds after n weeks. In each part below, a difference equation is given for one of the models. For each one, tell what the long range predictions of the model will be (if any), and whether you think these predictions make sense.

    a. @$a_{n+1} = 0.004(1300 - a_n)a_n$
    b. @$a_{n+1} = 0.002(1300 - a_n)a_n$
    c. @$a_{n+1} = 0.0007(1300 - a_n)a_n$

6.1. Exercises

(25) **Long Term Behavior of Algae.** At the start of the summer, algae starts to grow in a pond. An ecologist finds that at first the algae spreads rapidly, but over time the algae spreads more and more slowly. An equation for this situation is given by $A_{n+1} = 0.04(60 - A_n)A_n$ where A_n is the number of acres covered by algae at the end of the nth week of the study.

  a. What kind of long term prediction does this model lead to? How can you tell?

  b. Assuming that at the start of the study the algae covers 10 acres, show that your prediction from the preceding part is correct.

  c. If the starting amount of algae is different, say 5 acres, or 20 acres, how does that affect the long term prediction. Explain.

(26) **@Developing a Bacteria Model.** In this problem, you will use methods similar to those of the mold example to construct a logistic model.

Bacteria are grown in a sealed environment in a laboratory. The size of the population of bacteria is measured daily. Early in the experiment, the size grows very rapidly. For example, one day the population was 500, and the next it had doubled to 1,000. (This means the growth factor for that day was 2.) Later, when the population was much larger, the rate of growth slowed down. When the population reached a size of 10,000 it was only growing by about 10 percent each day. (That is a growth factor of 1.1.)

  a. @Find a linear equation for growth factor r as a function of population p. [Hint: The information above gives you two data points: when p is 500 r is 2, and when p is 10,000 r is 1.1.]

  b. @Use your linear equation to formulate a logistic difference equation for the bacteria population.

  c. @Explore the future growth of the population according to the model. Use numerical and graphical techniques, and assume a starting population of 500.

  d. @From your numerical and graphical work, does the population seem to level off? At what level? Can you derive that result using properties listed in the logistic growth profile?

  e. @Try a variety of different starting populations. Do they all lead to a steady population eventually? Can you derive that result using properties listed in the logistic growth profile?

(27) **Developing an Information Dissemination Model.** There are about 4,000 students at Anonymous University. Early one morning the student body president tells the vice president that a famous actor (whose identity is supposed to be kept secret) has agreed to be the commencement speaker. In this exercise you will develop a logistic model for the way that information is spread to the entire student body.

Let p_n stand for the total number of students who have heard the news after n hours. We will assume that each student tells another student every hour. Initially, that would cause p_n to double every hour. However, as the number of students who have heard the news increases, the situation changes. When a student tells the news to another student, he or she might have already heard the news. In fact, once p_n reaches 4,000, it cannot possibly rise any higher, since there are only 4,000 students at the university. So assuming that p_n doubles initially (when $p_0 = 2$,

r also equals 2) and that p_n remains constant when it reaches 4,000, develop a linear equation for r and p, as in the previous problem. Then formulate a logistic difference equation for this model, and use it to study how long it takes the news to spread over the whole university.

(28) @Developing a Legume Vine Length Model. *Lupinus albus* is a type of legume (bean) found widely in the Mediterranean region, and cultivated in some places as a food crop. The table below shows daily measurements of the length of one particular lupinus vine, as reported in [**50**, p. 160].

Table 6.2. Measured length of a *lupinus albus* vine, recorded daily. Lengths are in units of millimeters.

day	length	day	length	day	length
4	10.5	10	77.9	16	149.7
5	16.3	11	93.7	17	155.6
6	23.3	12	107.4	18	158.1
7	32.5	13	120.1	19	160.6
8	42.2	14	132.3	20	161.4
9	58.7	15	140.6	21	161.6

Using the same methods as in the pumpkin example, develop a logistic growth sequence to approximate the data in the table. In your answer, let p_n be the model value of the length after n days, and include the values of the parameters m, L, and p_4. Also include graphs and tables like those in the pumpkin example.

(29) iPod Example. Follow the outline below to analyze the iPod example discussed on page 380.

a. Compute the annual growth factors r for the cumulative sales figures s in the table in Figure 6.14. Graph r as a function of s. Verify that the graph and table in Figure 6.15 are correct.

b. As one attempt to fit a straight line to the growth factor graph, consider the equation $r = 3.5 - 0.0074s$. Add the line defined by this equation to your $r - s$ graph. Does it seem to be a reasonable line for the data points?

c. The logistic growth difference equation $s_{n+1} = (3.5 - 0.0074s_n)s_n$ is based on the linear equation in the preceding part. Use this difference equation and an initial value of $s_0 = 4$ to compute model values of s_1 through s_{11}. Compare the results with the original data, both numerically and graphically. Does the model approximate the original data very accurately? Try again with a few other values of s_0. What do you observe?

d. In the equation $r = 3.5 - 0.0074s$, the intercept on the r-axis is 3.5. This is also the value of mL in the logistic growth model. But we know that $mL = 3.5$ implies oscillation or more complicated variation in the associated logistic growth model. Therefore, to approximate the given data, we should try to use a linear equation with an r-intercept that is less than 3. The equation $r = 1.95 - 0.0024s$ is one possibility. Create a graph showing this line and the (s, r) data points. Do you agree that this equation is a good fit for the (s, r) points, if the first four data points are disregarded?

6.1. Exercises

e. The logistic growth difference equation $s_{n+1} = (1.95 - 0.0024s_n)s_n$ is based on the linear equation in the preceding part. Use this difference equation and an initial value of $s_0 = 4$ to compute model values of s_1 through s_{11}. Compare the results with the original data, both numerically and graphically. Does the model approximate the original data very accurately? Try again with a few other values of s_0. What do you observe?

f. The preceding parts have guided you to consider two different equations for r as a linear function of s. By trial and error, consider other variations. Examine both the accuracy of your r equation compared to the (s, r) points, and also the accuracy of the associated logistic growth sequences (using several choices of s_0) as compared to the observed values for s_n. Your answer should include the parameters for several trials. For any of your alternatives, does the graph of the model seem to capture accurately the shape of the graph of the data points? Why or why not?

Digging Deeper.

(30) @One reason that the logistic growth model does not provide a very accurate approximation to the iPod data is that the $r - s$ graph is not very linear. Accordingly, we might try to use a different form of equation for r as a function of s. One possibility is to replace $r = ms + b$ with something of the form

$$r(s) = \frac{1}{ms + b}.$$

Thus, r is not a linear function of s, it is the reciprocal of a linear function. By trial and error, the authors were led to the following equation of this type:

$$r(s) = \frac{1}{0.00122s + 0.51}.$$

Using this equation to create the difference equation $s_{n+1} = r(s_n)s_n$, and taking $s_0 = 4.4$, compute and graph s_1 through s_{11}. Compare with the original iPod data both numerically and graphically. How does the accuracy of this model compare with the accuracy of the logistic growth models in problem 29 of the Problems in Context section?

(31) @The preceding problem introduces an alternative form of logistic growth, where the growth factor is assumed to be the reciprocal of a linear function. That is, we assume $p_{n+1} = r(p_n)p_n$ where

$$r(p) = \frac{1}{mp + b}. \tag{6.8}$$

Put another way, this assumes that the reciprocal of r is a linear function:

$$\frac{1}{r(p)} = mp + b.$$

Let us apply this to derive a new difference equation for the mold example (page 374).

a. @Recall that we used the assumptions $r = 2$ when $p = 100$ and $r = 1.2$ when $p = 500$. These assumptions can be restated as $1/r = 1/2$ when $p = 100$ and $1/r = 1/1.2$ when $p = 500$. Use these assumptions to find a linear equation for $1/r$ as a function of p, thus identifying the parameters m and p in equation (6.8).

b. @Using the parameters found in the preceding step, formulate a difference equation for p_n.

c. @Investigate the behavior of this model using several different values of p_0, considering both numerical and graphical results. How do these results compare with the original logistic growth model developed for the mold example, as found in problem 22 of the Problems in Context section?

d. @Determine where the new model levels off by finding a fixed point of the difference equation.

(32) Repeat the preceding problem for the second fish population model (page 372). Recall that we assumed $r = 2.8$ when $p = 10,000$ and $r = 1$ when $p = 50,000$ for that model. Using these same assumptions, and the methods of the preceding problem, express r as a function of p in the form of equation (6.8). Then derive the corresponding difference equation and investigate the behavior of the model. How do your results compare with the results in Figure 6.6?

6.2 Chaos in Logistic Growth Sequences

Now we come to the topic of chaos, a subject that is the focus of research today in important areas of mathematics, engineering, and the physical and social sciences. Chaos in mathematics is closely related to the subject of fractals, widely publicized in the form of stunningly intricate and fascinating color images. James Gleick's popular book on the subject [29] was a best-seller. Chaos has even been referenced in popular culture, including in the film *Jurassic Park,* and an episode of The Simpsons.

But what exactly is chaos? And why is it of so much interest? Answering these questions in a limited way is one of the goals for this section. Following a brief general account of the main ideas of chaos, we will look in detail at chaos as it can arise in logistic models. Although this is only the tip of the chaos iceberg, it should give you an idea of several important aspects of chaos. Gleick's book discusses chaos in many other contexts, and describes the historical development of the subject, without going into much mathematical detail.

The Modeling Methodology. A key theme of this text is the application of mathematics to real world problems through the methods of mathematical modeling. We hope that the reader will gain a realistic understanding of how and why mathematics is useful, and that this understanding will extend beyond the specific details of the various types of models we have considered. In particular, we hope readers will witness the power and applicability of the modeling methodology.

It is appropriate to review this methodology here. We will highlight common aspects of the models we have studied up to this point. Note that these are all nonchaotic models. As we will see, the modeling methodology breaks down for chaotic models.

One key idea we have seen is the distinction between an actual phenomenon or problem context and a mathematical model. The model is a mathematical framework with functions, equations, graphs, and so on. We can apply mathematical procedures (such as algebra) to derive definite conclusions about the model. But we should always remember that these are not directly conclusions about the original problem context.

In almost all cases, models are based on simplifying assumptions. We have seen how such assumptions lead to models of several different types, for example arithmetic

6.2. Chaos in Logistic Growth Sequences

growth, geometric growth, etc. We have also seen the use of parameters, and how parameter values can be adjusted to improve the agreement between a model and observed data values. But even when our model agrees very closely with the data, errors remain possible. If our simplifying assumptions are only approximately true, if the observed data can only be measured approximately, then our model can only be expected to approximate the actual context or situation.

Given the approximate nature of a model, we are sometimes more interested in qualitative descriptions of whatever we are modeling than we are in highly accurate predictions of the future. Will the amount of a drug in the blood stream level off? Will a population of fish die out? The models give us qualitative answers to these questions.

The ability to formulate qualitative predictions is closely tied to the appearance of simple numerical and visual patterns in our models. In fact, all of our models have been developed by assuming simple numerical patterns hold in our data, at least approxmately. This is an important feature of our nonchaotic models. We will see that the sequences in chaotic models do not have simple numerical patterns.

There is another important feature of the models we have examined: moderately changing the values of the parameters does not significantly change the qualitative behavior. So, for a geometric growth model, if we have an incorrect value of r, we can still be pretty sure that the long-term behavior will be accelerating growth to ever larger values (for $r > 1$) or a decrease to 0 (for $r < 1$). Similarly, for certain mixed models and logistic growth models, we have found conditions under which the number sequence levels off. For these models, errors in the parameter values need not invalidate the qualitative prediction of leveling off. As long as the errors are within an acceptable range, we know that the number sequence will level off at a predictable value.

Conclusions that remain valid over a range of parameter values are sometimes described as *robust*. For a mixed model, the qualitative conclusion that the model will level off is robust because it doesn't change when the parameter values are slightly modified. This increases our confidence that the same qualitative conclusion will hold for the actual problem context. Even though the mixed model assumptions are only approximately true and the values of the parameters are only approximately correct, because leveling off will occur in a range of model formulations, we expect that it will also occur in the real context. Thus, robustness is an important aspect of using models to make qualitative predictions.

Two Important Aspects of Chaos. The preceding remarks lead us to two key aspects of chaotic models. First, they do not have simple numerical or visual patterns that can be used to predict future behavior. A chaotic model can shift among many kinds of behavior, seemingly at random. This often includes wild unexpected swings from one extreme to another. The term *chaos* reflects an apparent absence of order or pattern.

Second, it is a characteristic of chaos that the results are *not* robust: even miniscule changes in the parameters can lead to vastly different future behaviors. To emphasize this point, we compare two fictitious examples, first one that is not chaotic, then one that is chaotic.

A Nonchaotic Example. Imagine that you are a doctor interested in the way medicine is absorbed by the body. Your favorite applied mathematician describes a model based on a geometric growth assumption: a fixed percentage of the medicine is absorbed in a fixed amount of time. You are told:

> "If the drug is absorbed at a rate of 10 percent each hour, and if 50 mg are taken every four hours, then over time the amount of drug in the blood will level off at about 145 mg."

But you point out that the studies are not conclusive. For some patients the absorption rate is as high as 12 percent an hour. For others it is as low as 8 percent an hour. The analyst responds:

> "In that case, the model still indicates that the amount of drug in the blood will become steady, somewhere between 125 and 176 mg."

That is a reliable model. It allows qualitative predictions even when there are errors in the assumptions. And a small change in the assumptions results in a small change in what the model predicts.

A Chaotic Example. In contrast, suppose you are trying to forecast the weather. Your favorite mathematician describes a model that takes into account temperature, pressure, humidity, and wind data from weather stations all over the globe. It uses equations that are known to be highly reliable descriptions of how the weather evolves. When you ask about the weather for a special event, you are told:

> "Based on the model, two weeks from tomorrow it will be sunny and warm in New York for your campaign speech."

But then one of the mathematician's assistants comes in, and apologizes that the temperature reading from Tokyo was entered in the model incorrectly. It was accidently entered as 67.8° rather than 68.7°. So the model had to be recomputed. Your consultant reviews the new results, then confidently purrs:

> "Ah, now we have the correct answer—expect a blizzard for your speech."

That is typical of a chaotic model. Even very slight changes in the parameters or starting values for the model lead to vastly different qualitative predictions. That is chaos.

The discovery that certain kinds of equations lead to chaotic models is fairly recent (in historical terms). Although Henri Poincaré raised this as a theoretical possibility in the early 1900s, the significance for real models was not appreciated until the advent of modern computing technology some half a century later. Prior to that time it was expected that models with simple equations could lead to accurate long-range predictions if the parameters could be determined accurately enough. Now scientists and mathematicians have shown that for some specific models chaos is unavoidable. For these models, even the slightest changes in the parameters and starting values would lead to wildly altered predictions. Since real measurements almost always include some error, the parameters in our models cannot be expected to be exactly correct. In the presence of chaos, that means the predictions produced by our models simply cannot be trusted. A weather forecast calling for either fair weather or snow, with equal validity, is no forecast at all.

So far, this discussion has been pretty abstract. Shortly we will look at a particular case of chaos, as it appears in the logistic model. There you will see in detail how the

6.2. Chaos in Logistic Growth Sequences

ability to make predictions breaks down, and how very slight changes in the starting value can lead to vast differences in the future evolution of the model. However, before proceeding to that topic, one more point should be made about the difference equations we have studied.

Linear and Nonlinear Difference Equations. There is an algebraic distinction between the difference equations for chaotic and nonchaotic models. It can be seen in the difference equations we have studied in this book, which is summarized in Table 6.3.

Table 6.3. Summary of difference equation models we have studied.

Type	Description	Sample	General Form
Arithmetic	Add a constant	$a_{n+1} = a_n + 0.3$	$a_{n+1} = a_n + d$
Quadratic	Add an amount that grows linearly with n	$a_{n+1} = a_n + 0.3 + 0.02n$	$a_{n+1} = a_n + d + en$
Geometric	Multiply by a constant growth factor	$a_{n+1} = 2a_n$	$a_{n+1} = ra_n$
Mixed	Multiply by a constant growth factor and then add a constant	$a_{n+1} = 2a_n + 0.3$	$a_{n+1} = ra_n + d$
Logistic	Multiply by a growth factor that depends linearly on the current term of the sequence	$a_{n+1} = 0.001(1{,}000 - a_n)a_n$	$a_{n+1} = m(L - a_n)a_n$

We can think of each difference equation in the table as a kind of function. In every case, the equation allows us to compute a_{n+1} as soon as we know the value of a_n. That means that each equation gives a_{n+1} as a function of a_n, or, in one case, as a function of both a_n and n.

Up to this point, when the terms *functional equation* were used in connection with a difference equation, we were interested in giving a_n as a function of n. This was meant to emphasize the distinction between two different approaches to compute a particular term, such as a_{25}. One method is *recursive*: compute a_1, then a_2, then a_3, and continue in this way until a_{25} is reached. The other method is direct: specify that $n = 25$, and compute a_{25} immediately. This is possible if there is a functional equation, that is, an equation that gives a_n as a function of n.

But the idea of *function* has a broader meaning. Any time one quantity can be computed directly from the value of another quantity, there is a function at work. So, it makes sense in a difference equation to say that a_{n+1} is a function of a_n: as soon as the value of a_n is known, we can immediately compute a_{n+1}.

A general formulation for all the difference equations in the table is
$$a_{n+1} = f(a_n),$$
where the right-hand side indicates a function of a_n. In the case of quadratic growth, we have a slightly different form:
$$a_{n+1} = f(a_n, n)$$
indicating that we need to know both a_n and n before we can compute a_{n+1}.[6]

---

[6]To make this idea more concrete, suppose we are told that one of the numbers in a data sequence is 16, and asked to predict the next one. For every case except quadratic growth, all we need to know is the parameters in the model, and then we can compute the next number after 16. But for quadratic growth, we need to know where 16 shows up in the pattern. If it is the fifth number, we will get one result; if it is the tenth number, we will get another. So for quadratic growth, we need to know both a_n and n.

Having reached this perspective, we can classify the different kinds of difference equations by classifying the functions they involve. For arithmetic growth, the function has the form $f(a_n) = a_n + d$. For geometric growth, it is $f(a_n) = ra_n$. And the mixed-model function is $f(a_n) = ra_n + d$. In each of these cases the function f is *linear*, because it has the form

$$f(x) = \text{(a constant)} \cdot x + \text{(a constant)}.$$

Put another way, for arithmetic growth, geometric growth, and mixed models each term in the sequence depends linearly on the preceding term. For quadratic growth, although both the preceding term and n enter into the calculation, the function is again linear. As a result, we refer to the difference equations in all of these cases as linear difference equations.

In contrast, for the case of logistic growth, the dependence of each term on the preceding term involves a quadratic function: $f(a_n) = m(L - a_n)a_n$. This is a nonlinear equation because, when expanded, it includes the term $m \cdot a_n^2$. It is only nonlinear difference equations that give rise to chaos.

Do not confuse the concept of a linear difference equation with a linear functional equation. The linearity of the difference equation is not concerned with how some term in the sequence depends on n, but rather on how it depends on the preceding term. This distinction is illustrated graphically in Figure 6.16 and Figure 6.17 for the difference equation

$$a_{n+1} = 0.5a_n + 20 \tag{6.9}$$

with a starting value of $a_0 = 10$.

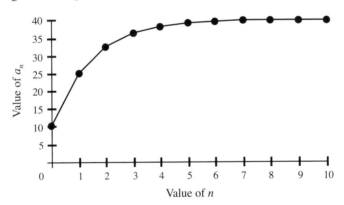

Figure 6.16. How a_n depends on n, where $a_0 = 10$ and $a_{n+1} = 0.5a_n + 20$.

The graph in Figure 6.16 shows the relationship between a_n and n. In this figure we graph a_1 above the number 1, a_2 above the number 2, and so on. This graph is clearly not linear. This same result would be obtained by graphing the functional equation for the sequence.

In Figure 6.17 there is a different kind of graph, showing the relationship between any two successive terms of a sequence defined by the difference equation.[7] Here, we ignore the value of n, and focus on how a_{n+1} depends on a_n. For example, if some a_n is

---

[7] This type of graph was used in Section 6.1. See pages 367 and 368.

6.2. Chaos in Logistic Growth Sequences

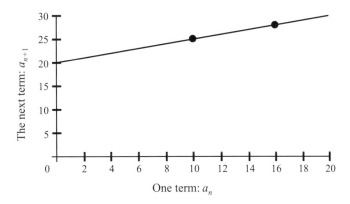

Figure 6.17. How a_{n+1} depends on a_n, where $a_0 = 10$ and $a_{n+1} = 0.5a_n + 20$.

10, then using (6.9), the next term will be $a_{n+1} = 0.5 \cdot 10 + 20 = 25$. So for this difference equation, a value of 10 is always immediately followed by a value of 25. That gives one point on the graph, $(10, 25)$. Similarly, if $a_n = 16$, the next value will be $8 + 20 = 28$, producing the point $(16, 28)$. Repeating this process for many many different choices of a_n, produces the graph in Figure 6.17. For a linear difference equation, it is this latter kind of graph, showing how each term is related to the preceding term, that is linear. As the two graphs illustrate, even if the difference equation is linear, the graph showing how the data actually evolve over time (like Figure 6.16) may well be nonlinear.

Whether a model has a linear or nonlinear difference equation has a special significance for a discussion of chaos. Models with linear difference equations are *never* chaotic. Of all the models we have looked at in the course, only the logistic models have nonlinear difference equations. Of all the models, only logistic models sometimes exhibit chaos. How that can occur, and what it means, will be considered next.

Chaos in the Logistic Model. As a concrete example, we will again consider a fish population model. In Section 6.1, we developed the equation

$$p_{n+1} = 0.000025(90{,}000 - p_n)p_n.$$

Here we will modify the equation slightly, for simplicity's sake, by changing the 90,000 to 100,000. That gives the equation

$$p_{n+1} = 0.000025(100{,}000 - p_n)p_n. \tag{6.10}$$

The numbers in this equation will be easier to work with if we express the population in units of hundred thousands. To be clear how to do this, we can define a new variable a_n to be the population in hundred thousands. For instance, if $p_n = 125{,}000$, then in hundred-thousands, $a_n = 1.25$. Similarly, if $p_n = 200{,}000$, $a_n = 2$. These examples illustrate the more general pattern that $p_n = 100{,}000 a_n$. Of course, it is also true that $p_{n+1} = 100{,}000 a_{n+1}$. In (6.10), if each p_n is changed to $100{,}000 a_n$, and similarly for p_{n+1}, a difference equation is derived for a_n:

$$100{,}000 a_{n+1} = 0.000025(100{,}000 - 100{,}000 a_n)100{,}000 a_n.$$

Dividing both sides by 100,000 produces

$$a_{n+1} = 0.000025(100{,}000 - 100{,}000 a_n)a_n$$

and factoring 100,000 out of the expression in parentheses leads to

$$a_{n+1} = 2.5(1 - a_n)a_n.$$

This is the difference equation for our fish population in units of 100,000s.

In the context of the earlier discussion of logistic growth, we can compare this equation with the general form

$$a_{n+1} = m(L - a_n)a_n,$$

where $L = 1$ and $m = 2.5$. As described in Section 6.1, by looking at $mL = 2.5$, we can see that this model will always lead to a steady population size of $L - 1/m = 0.6$.

Now to explore the way chaos occurs, we will look at variations of this model by changing just the parameter m. From our previous study of logistic models, we already know that mL can give us some information about the way a model operates. In the present example, because $L = 1$, the value of mL is really just m.

When $mL \geq 4$ or $mL < 0$, we know the sequence may produce negative results. Accordingly, we will restrict our attention to $0 \leq m < 4$. For these values of m, if a_0 is in the range $0 \leq a_0 \leq L = 1$ then all of the terms a_n will remain in the same range. Going further, we need not consider cases with $m < 3$. For these values of m the model will eventually level off and chaos will not occur. What remains are values of m between 3 and 4, where we should see future population patterns that remain between 0 and $L = 1$, without leveling off. This is the range of values where chaos is possible. Accordingly, let us consider several examples with m between 3 and 4.

Example: $m = 3.2$. We start with $m = 3.2$. If the initial population is 0.7 (remember, that is in units of 100,000s), what will the future hold for our fish population? The answer is revealed in the top graph of Figure 6.18. It shows that starting from an initial population of 0.7 (actually 70,000) the number of fish oscillates. At first, the oscillations grow, but after about 15 years they steady down to a regular repeated pattern: 0.80 in the even-numbered years and 0.51 in the odd-numbered years.

Although this is different from the pattern we saw earlier, it is still very regular and easily permits future predictions. The situation is similar to the earlier patterns in another way, too. The long-range behavior of the population in the model does not depend on the starting point. In the bottom graph of the figure the same model is used, but this time the starting population size is 0.4 (or 40,000 fish). Once again the same pattern of fluctuation is observed, and after about 6 years, we see the identical behavior as before: year by year the population switches between roughly 0.80 and 0.51.

What the graphs reveal for two starting population sizes can be shown to hold in general, using theoretical methods. We won't go into the details of those methods here, but they are similar to what was used in the last chapter to find the steady population levels. There, for a steady population, we wanted to find $p_{n+1} = p_n$ for any n. This led to an equation that we solved to find the steady population size p_n. Now we are looking for a population that repeats not the next year, but in two years. That is expressed by the condition $p_{n+2} = p_n$. This also leads to an equation for p_n, but it is a little more complicated than the one we found in the previous chapter. Be that as it may, it is possible to find the solutions to the equation (they include 0.7995 and 0.5134) and to show that in the long run the model will fluctuate between these two figures, no matter what the starting population size is.

6.2. Chaos in Logistic Growth Sequences

n	a_n	a_n
0	0.70	0.40
1	0.67	0.77
2	0.71	0.57
3	0.67	0.78
4	0.71	0.54
5	0.66	0.79
6	0.72	0.52
7	0.64	0.80
8	0.74	0.52
9	0.62	0.80
10	0.75	0.51
11	0.59	0.80
12	0.77	0.51
13	0.56	0.80
14	0.79	0.51
15	0.54	0.80
16	0.80	0.51
17	0.52	0.80
18	0.80	0.51
19	0.51	0.80
20	0.80	0.51

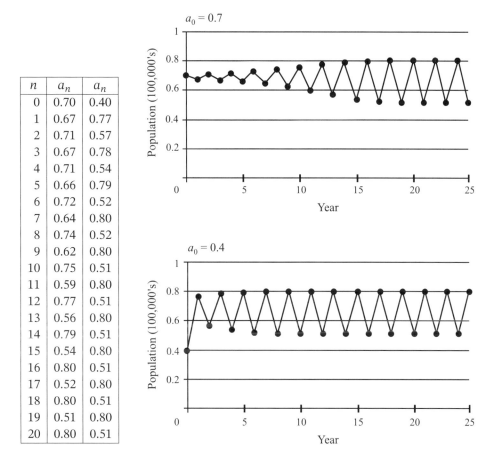

Figure 6.18. Population models with $m = 3.2$. The first a_n column corresponds to the top graph, and the second a_n column corresponds to the bottom graph.

Example: $m = 3.4$. We have examined one example, the case for $m = 3.2$. We already knew that, for $m < 3$, the population model would settle down to a steady size. Now we have seen that for one $m > 3$ the population fluctuates between two steady sizes. Will this same situation occur for other values of m? In Figure 6.19 the situation for $m = 3.4$ is shown. As in the previous figure, there are two examples, one with an initial population size of 0.7, and one at 0.4. The long-term results are very similar to what was seen earlier. Once again the population will fluctuate between two steady values, no matter what the starting population is. This time, though, the fluctuations are a little greater in size, roughly from 0.45 to 0.84.

As these examples show, even when there is not a single steady population, meaningful predictions can be made about the way the population grows in the model.[8] For both examples, a wildlife manager could expect that a year of high fish population would be followed by a year with a low population, but that the population would

---

[8]This sort of oscillation has been observed in controlled experiments with insect populations. See [54].

n	a_n	a_n
0	0.70	0.40
1	0.71	0.82
2	0.69	0.51
3	0.72	0.85
4	0.68	0.43
5	0.74	0.84
6	0.66	0.47
7	0.76	0.85
8	0.62	0.44
9	0.80	0.84
10	0.53	0.46
11	0.85	0.84
12	0.44	0.45
13	0.84	0.84
14	0.46	0.46
15	0.84	0.84
16	0.45	0.45
17	0.84	0.84
18	0.46	0.45
19	0.84	0.84
20	0.45	0.45

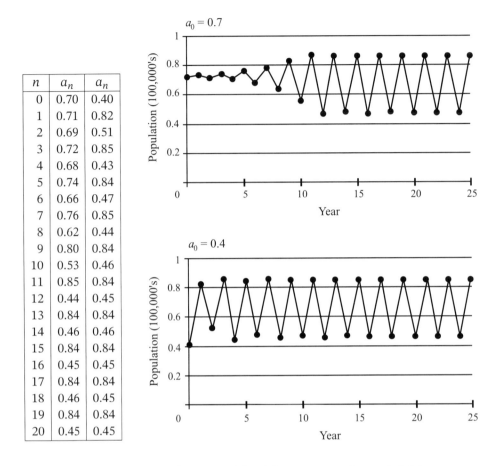

Figure 6.19. Population models with $m = 3.4$, and $a_0 = 0.4, 0.7$.

spring back to the higher level the following year. If the estimate of the initial population is a little off, that has no effect on the long-range behavior. Even if m changes slightly, that doesn't make a significant difference in the way the model behaves. These conditions are not what you would call chaotic. But as we try ever higher values of m, the model begins to change.

Example: $m = 3.5$. For $m = 3.5$ we observe a new behavior (see Figure 6.20). This time the population does not settle down to a pattern of fluctuation between two steady sizes. Instead there are four population sizes that get repeated over and over again. There is still regularity in this pattern, but it is more complicated than the earlier examples, and is not as easy to pick out by eye. It is also still true that the starting population size has little effect on the long-term model behavior. In the top graph of the figure the population starts out at 0.5, and in the bottom it starts at 0.7. But eventually the same cyclic pattern of four population sizes shows up in each case. That is, comparing the two a_n columns of the table, we see an eventual cycle among the same four values: 0.50, 0.87, 0.38, and 0.83.

6.2. Chaos in Logistic Growth Sequences

n	a_n	a_n
0	0.50	0.70
1	0.88	0.74
2	0.38	0.68
3	0.83	0.76
4	0.50	0.64
5	0.87	0.81
6	0.38	0.55
7	0.83	0.87
8	0.50	0.40
9	0.87	0.84
10	0.38	0.47
11	0.83	0.87
12	0.50	0.39
13	0.87	0.83
14	0.38	0.48
15	0.83	0.87
16	0.50	0.39
17	0.87	0.83
18	0.38	0.50
19	0.83	0.87
20	0.50	0.38

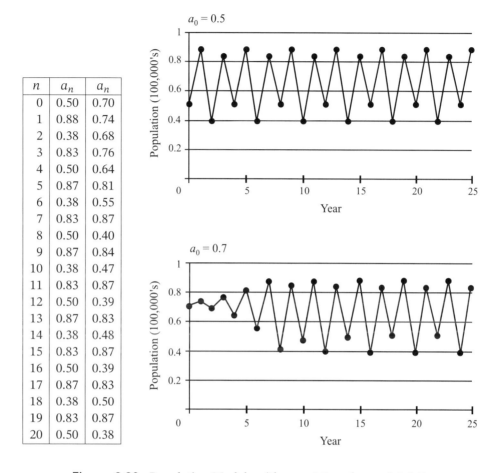

Figure 6.20. Population Models with $m = 3.5$, and $a_0 = 0.4, 0.7$.

This pattern is more complicated than the ones shown earlier, but is still not chaotic. All that has been illustrated up to this point is that as m gets larger and larger, the pattern of the model gets more and more complicated. Interestingly, there is a point at which the behavior is so complicated that all patterns break down. At that point, small differences in m have a tremendous impact on the way the model behaves. To illustrate chaotic behavior, we will consider an example for $m = 3.8295$. The reason for choosing that particular m will be discussed a little later.

An Example of Chaos. In Figure 6.21 there are two graphs for the model with $m = 3.8295$. Each projects the fish population for 50 years. In the first graph, the starting population is 0.9000, in units of hundred thousands, or 90,000 fish. In the second graph, the starting population was taken to be 0.9001, meaning 90,010 fish. The two graphs show very similar patterns of development for about 10 years. But then they separate, and become quite different.

Notice that in the first graph for about six years in a row the population holds very steady at about 0.75. This occurs after about 19 years of fluctuating populations. There is no such pattern in the second graph. There seems to be no order at all in

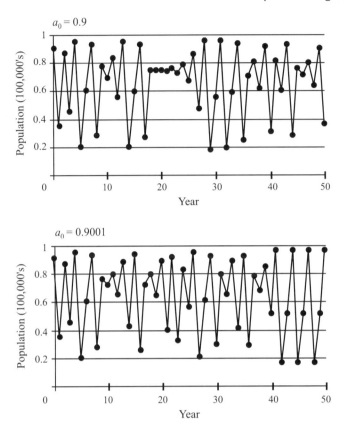

Figure 6.21. Population models with $m = 3.8295$.

the long-term pattern of either graph. For some periods the population goes into a repeated three-year cycle of low, medium, and high populations. But this pattern does not continue long. At another place on the graph there is fluctuation between high and low population amounts, but that too lasts for only a few years. To make it clearer where the models are similar and where they look completely different, the two graphs have been combined in Figure 6.22.

This is a chaotic model. There just isn't any pattern that can be used to predict the future of the model. What is more, even the slightest change in the starting population value leads to completely different long-range patterns of population activity. This is what is referred to as the *butterfly effect*. Remember the example at the start of the chapter about weather prediction? There, a slight error in one temperature leads to a dramatic difference in the weather predicted by a model. With the incorrect temperature the prediction calls for warm sunny weather; with the correct temperature a blizzard is predicted. A variation on this idea involves the beating wings of a butterfly instead of an error in the temperature. One prediction includes the effects on the atmosphere of the butterfly's wings, the other does not. The idea that a single butterfly in Tokyo can alter the predicted weather for New York dramatizes how sensitive

6.2. Chaos in Logistic Growth Sequences

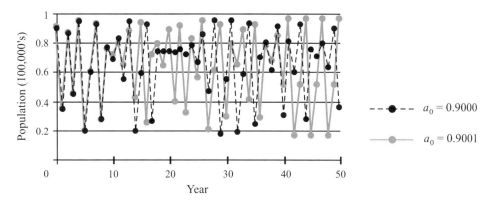

Figure 6.22. Combined graphs for $m = 3.8295$.

the model is to the slightest changes in the parameters. That is what is meant by the butterfly effect.[9]

We can see the butterfly effect at work in the fish population model in the preceding example. In one case the model is based on 10 fish more than the other. After 10 years, the effect of those 10 additional fish is a completely different population projection.

There are two important consequences of the butterfly effect as it appears in the fish population model. First, it shows that longe-range prediction of the real world (as opposed to the model) is essentially impossible. However accurate we can make the model, there will be some slight errors in our measurements. We know now that these errors will lead to major effects in what the model predicts.

The second important point has to do with the complete lack of a predictable pattern in the model for any starting value. If a chaotic model is accurate, looking at real fish population data for several years really provides no help in making future predictions. The actual fish population could remain essentially unchanged for several years (as in the first graph from year 19 to year 25). How tempting to predict that the fish population is in balance, and will remain at the same level for many years to come. But as the graph shows, with no apparent warning or explanation, the population suddenly starts to fluctuate between very high and very low populations. The wildlife manager cannot count on any sensible pattern for future fish population figures. That is chaos.

This model is also very sensitive to small changes in the parameter m. Changing it very slightly can lead to very different looking patterns. In the example earlier, m was set to 3.8295. If, instead, a value of 3.829 or one of 3.830 is selected, the resulting graphs are completely different. They will still follow no particular pattern, and will still be subject to the butterfly effect. But the particular arrangement that appears in the graph will be different from the ones shown in Figure 6.22. The choice of 3.8295 has no special significance for the discussion of chaos. After experiments were made

---

[9]The idea of the butterfly effect is *not* that the beating of the wings of a butterfly in Tokyo *cause* a snow storm in New York. Including the butterfly alters our *prediction* of the weather, not the weather itself. The arrival of a blizzard in New York is the result of billions of tiny effects, not just one. And since we cannot account for all the possible effects, we cannot reliably predict the timing, path, and severity of weather phenomena such as blizzards. The significance of the butterfly is about our inability to capture all of the potential influences of the weather, rather than the impact of just one of these influences.

with many choices of m, 3.8295 was chosen because for that value one graph shows nearly constant populations for one period of several years.

Now we have seen that chaos arises in the logistic growth model for fish populations. It arises in many other kinds of models as well. This is one of the reasons that the study of chaos is so important today. The examples we have seen show that there is essentially no order that can be used to predict future fish populations in the chaotic model. But that is not the same thing as saying that there is no order whatever.

Scientists believe that the chaos we observe in our models can reflect inherent properties of the systems they approximate. And a single system can exhibit both chaotic and nonchaotic behavior under different sets of conditions. When the system is in a nonchaotic condition, our models can accurately reveal predictive patterns. When those same models produce disordered results, lacking discernible patterns, the underlying system can be expected to behave similarly, with wild unpredictable swings from one extreme to another.

Interestingly, it turns out that there is a kind of order in chaos. One aspect of this order is the ability to predict when a system will become chaotic, and in some cases, what to do to keep chaos from occurring. We will take a brief look at this idea as the final topic for this section on chaos.

Order in Chaos: A New Kind of Graph. Take a moment to review what we know about the logistic fish population model. What happens in the long run depends on the parameter m. We assume that m is between 0 and 4, so that the logistic growth sequence will produce only positive values. When m is in that range, there are several possibilities. If m is too small, that is, when $m < 1$, the population will eventually die out completely. For larger values of m, from 1 up to a value of 3, the population will settle down to a steady value and remain the same year after year. Just slightly above 3, we observed a fluctuating pattern between two steady populations. Then for m a bit larger, we found fluctuations among four steady values. Finally, when we made m big enough, chaos appeared. Now we will study the effects of making m larger and larger in a more systematic way. In the process we will see how the logistic population model makes a transition from a non-chaotic to a chaotic system. In this transition to chaos, a new kind of order will emerge.

Chaos theoreticians invented a new kind of graph to display the effects of m and the onset of chaos. The idea is to make a graph showing different values of m on the horizontal axis, and the long-term behavior of the population model on the vertical axis. An example is shown in Figure 6.23.

To understand how this idea works, let us look again at examples we examined earlier. For $m = 3.2$, we saw that eventually the population always reaches a pattern of fluctuation between two steady sizes. At the beginning, the choice of the starting population does have an effect. Different starting values will produce different sequences of population values for the first several years. But in the long run, these first few data points are not really significant. Ultimately, the population will fluctuate between the same two fixed values, no matter what the first few data values are. For this reason, it makes sense to screen out the short-term effects, and focus on what happens in the long run. To do so, we will ignore the predicted population figures for the first 100 years. Then we can graph the next 20 years' worth of data points. That gives the graph shown in Figure 6.23. It indicates that, in the long term, the model produces just two

6.2. Chaos in Logistic Growth Sequences

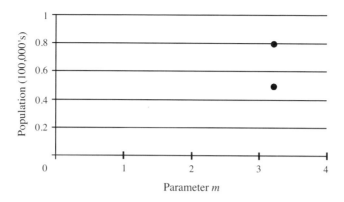

Figure 6.23. Long-term behavior for $m = 3.2$.

data values. For our twenty plotted points, half have one of the data values, and half have the other, giving the appearance of just two points.

This graph captures the long-term behavior of the model with $m = 3.2$. In the long run, there are only two population sizes that occur, and these are shown by the two points on the graph. What is more, we would get exactly the same graph no matter what starting population size we choose. After 100 years, the graphs will all look the same.

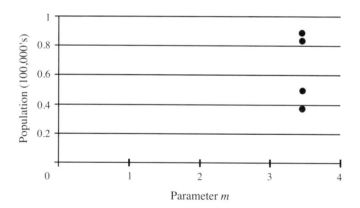

Figure 6.24. Long-term behavior for $m = 3.5$.

In Figure 6.24, a similar graph is shown for $m = 3.5$. As for the preceding graph, we used the model to project 100 years into the future, but did not graph those data points. Then we computed and graphed the data points for the next 20 years. In our earlier discussion for $m = 3.5$, we observed a long-term behavior of fluctuation between four steady values. These four values are clearly revealed in Figure 6.24. The long-term behavior of the model is quite predictable as it cycles among the four steady values.

For a contrasting view, consider $m = 3.8$. That leads to chaos. We project forward 100 years and then plot the next 20, as before, to make Figure 6.25. But this time, there aren't just a few points on the graph. The data points spread across most of the range from 0.2 to 1. The appearance of chaos is revealed in this spread of values. Even after 100 years, the population can be virtually anything.

The last three graphs can all be combined into a single graph. Each has a set of points for a different choice of m, so these points appear on separate vertical lines if we plot them all on one graph. At the same time, we will add data points for $m = 2.9$ and $m = 3.4$. This gives Figure 6.26. There is only one data point for $m = 2.9$. That is because, with $m < 3$, the population eventually settles down to a single steady value. There are two data points for $m = 3.4$, the two steady values that appear in Figure 6.19.

When you look at the combination chart, you start to see some order in the transition to chaos. In Figure 6.27 there is a much more detailed version of the graph. Data were generated for this graph using very closely spaced values of m between 1 and 4, and plotted using very small points.

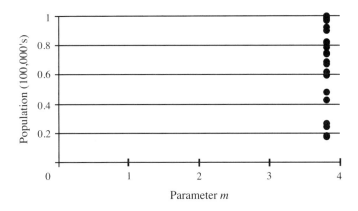

Figure 6.25. Long-term behavior for $m = 3.8$.

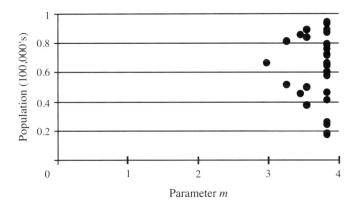

Figure 6.26. Combined graph showing long-term behavior for several values of m.

6.2. Chaos in Logistic Growth Sequences

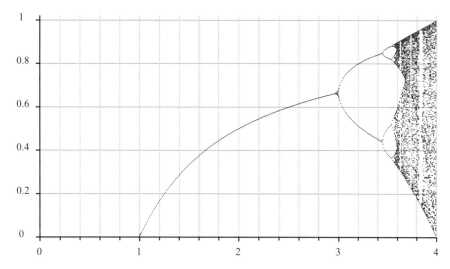

Figure 6.27. Detailed graph showing long-term behavior for $0 < m < 4$.

The graph appears to have a smooth curve between $m = 1$ and $m = 3$. This curve shows how parameter m affects the steady population.[10] But at $m = 3$ something interesting happens. There is a transition from models with a single steady population that holds constant year after year, to one in which the population switches back and forth between two different values. At $m = 3$ this appears as a fork where the curve splits into two parts. A little farther on, each of those forks is itself split. That reveals values of m giving rise to four different population values, with the long-term behavior of the model alternating among the four. We saw that kind of behavior earlier for $m = 3.5$. Although it is hard to see in the graph, farther on each of the four branches splits again, and then again, and eventually the curves degenerate into a kind of cloud of dots. That is where chaos arises. Beyond that point, instead of just a few population values that repeat over and over again, there is just a confused list of future populations with no order or pattern.

Remember that the diagram shows the behavior of the logistic model for many many different choices for m. On a vertical line directly above each m on the horizontal axis, data values are shown for the logistic model for that m. Within the cloud of points, data values on each vertical line are spread all over the place. They do not settle down to one value, or to just a few values. It is the way this chaotic behavior occurs for many different values of m that produces the appearance of a cloud.

Considered from the viewpoint of long-range prediction, the diagram reveals chaos for values of m near 4. For m in that range, the long-term behavior of the logistic model seems to follow no order or pattern. And yet, in this diagram of the onset of chaos, there appear to be patterns of a different type. There are regular gaps and spaces, and some darker areas which appear to reveal curious curves. To see these patterns in even more detail, another graph was created (Figure 6.28). It was produced in exactly the same way as the previous graph, but it concerns only the values of m between 3.5 and 4.0.

---

[10] For instance, with $m = 1.6$, the point on the curve has a y-coordinate of about 0.4. This indicates that, for $m = 1.6$, the model will eventually level off and remain at 0.4, or 40,000 fish.

This has the effect of spreading out the chaotic part of the diagram, and makes the patterns clearer.

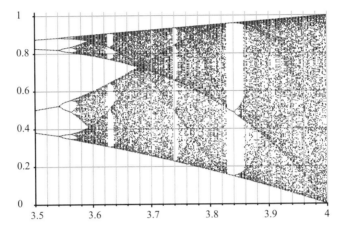

Figure 6.28. Closeup graph for $3.5 < m < 4$.

These patterns have many interesting aspects. In addition to the obvious visual symmetries of dark areas and gaps, there are other less obvious features. Look at the large white gap that is above 3.84 on the m-axis. In the center of the gap there is what appears to be a kind of white circle within a triangular patch of black dots. If that part of the figure is enlarged and explored in greater detail, it turns out to have a shape that is remarkably similar to part of Figure 6.27. That is, there is a miniature chaos diagram within the larger chaos diagram. This is an aspect of chaos that turns up quite frequently, and it is given a special name: self-similarity. When a diagram is self-similar, it is possible to zoom in on one part and find something that looks just like the original figure. Then you can zoom in again on another part and find an even tinier replica of the original figure. In this way, no matter how much the original figure is magnified, it is still possible to recognize similar features to the original graph.

These patterns in the diagrams of chaos for the logistic fish population model can be analyzed in great detail using theoretical methods. Many of the results are typical of a large variety of chaotic models, and are not restricted to logistic growth. Chaos has significance in a large variety of fields, including economics, biology, aerodynamics, medicine, and many others. These applications generally involve models that are more subtle and complicated than the logistic growth models presented here. Nevertheless, even in very complicated models, chaos is accompanied by characteristics that are quite similar to the ones we have seen for the logistic growth model. A thorough understanding of this chapter should give one a pretty good idea of what chaos is.

This completes the discussion of chaos, and it offers an interesting perspective on everything that has come before. Throughout the book we have seen ever more sophisticated models, and how these models can be used to predict future conditions in real problems. For very simple models, it should not be surprising that very simple patterns appear. But as more sophisticated models are considered, the behavior can become more complicated. Logistic growth is very important in applications, and under suitable assumptions it behaves very nicely. But it is a complicated enough model to permit chaos to creep in. Chaos should not be viewed solely as a stumbling block for the modeling process, though. Instead, although it may prevent accurate prediction, it can provide other sorts of insights. For example, the observation of chaos in a model may reflect inherent behaviors of the system it approximates. In some applications, it is valuable to be aware that chaos can occur, and to know how to recognize its onset. In other cases, it may be possible to take steps to prevent the start of chaos. There are even applications where chaos may be desirable, providing a rich range of future possibilities. The details of how that all works out is a subject for another course.

The appearance of chaos in logistic growth models also prompts a different question: does this unpredictable behavior reflect the true nature of the system we are modeling, or is it a result of inaccurate assumptions? In the next and final section, we reconsider the underlying assumptions of logistic growth models, and see a refined version of logistic growth that does not lead to chaos. Although this suggests that chaotic behavior is probably not generally an inherent property of populations in nature, scientists know of many real modeling contexts where chaos can occur. The weather is one such example.

6.2 Exercises

Reading Comprehension.

(1) @Bob is discussing chaos and comments that if initial measurements could be made accurately enough, we would never see chaos in mathematical models. Is this true? Explain.

(2) Which of the following examples shows chaotic behavior as defined in the text? Justify your answer.

- The swarming of a school of fish near the Great Barrier Reef is being predicted based on a model that takes into account the number of fish, the water temperature, the number of predators, and a few other factors. After inputting the data, the researcher sees that the model predicts a 2 minute spiral swarm. She realizes there was a small mistake in the data and corrects it; the new prediction is that the fish will not swarm at all.
- The spread of a flu epidemic is being modeled based on how contagious the strain is, how long a person carries the flu before showing symptoms, how likely a victim is to die from the disease, and a few other factors. How contagious the strain is can be summarized by a number called the case reproduction number, r. One team of experts determines that $r = 2.3$ and the model predicts that the flu will spread through New York state in 2 weeks (14 days). Another team of researchers determines that $r = 2.5$; using this new case reproduction number the model predicts that the flu will spread through New York state in 12 days.

(3) Discuss the role of long-term predictions in analyzing a model. Include an explanation of what is meant by long-term predictions. Also explain what it means to say that long-term predictions for arithmetic growth, geometric growth, and mixed models are *robust*. Contrast this with what can occur in a chaotic model.

(4) @Discuss the implications of a linear difference equation. Include a definition of a linear difference equation, examples of linear difference equations, and examples

of nonlinear difference equations. What is the significance of linear and nonlinear difference equations for the study of chaotic difference equation models?

(5) Explain the concept of self-similarity in a graph, and how it is observed in Figure 6.27 and Figure 6.28.

(6) @One of the models studied earlier in this course is a quadratic growth model. We also know that this model has a linear difference equation. Explain why there is no contradiction in terms in describing these models as both linear and quadratic.

(7) @Explain the butterfly effect. Contrast how this effect pertains to a model with how it pertains to the phenomenon being modeled. Relate the butterfly effect to the examples of the logistic fish population model in the chapter.

(8) Write a one page essay about chaos in logistic models as discussed in this section. Include the following in your answer:
- a discussion of the overall idea of chaos,
- a description of the specific form that chaos takes for logistic models,
- an explanation of the effects of small changes in the initial population size or in the parameters of a chaotic logistic growth model, and
- a comparison of the feasibility of making long term predictions in logistic growth models that are chaotic and in models that are not chaotic.

Add anything else that is significant about chaos in logistic models. Be as complete as possible.

(9) @The graph of Figure 6.27 includes the point (2.0, 0.5). What does this point tell you about the population model? That is, provide a practical interpretation of the point.

(10) @Using Figure 6.27 and Figure 6.28, it is possible to predict the long-term behavior of the fish population model $a_{n+1} = m(1 - a_n)a_n$ for various values of m. Do this for each part below.

   a. @$m = 1.6$, $a_0 = 0.5$
   b. $m = 1.6$, $a_0 = 0.2$
   c. $m = 1.6$, $a_0 = 0.97$
   d. @$m = 3.2$
   e. @$m = 3.52$ (Use both figures)
   f. @$m = 3.56$ (Use Figure 6.28)
   g. @$m = 3.8$ (Use both figures)
   h. @$m = 3.84$ (Use Figure 6.28)

6.3 Refined Logistic Growth

In introducing logistic growth sequences, we observed that a constant growth factor is unrealistic in the long run, and proposed instead that the growth factor should decrease as the population increases. Purely for algebraic simplicity, we first considered models where this decrease is linear. That is, we defined logistic growth in terms of a difference equation $p_{n+1} = r(p_n)p_n$ where $r(p_n)$ is a linear function. We also pointed out some

6.3. Refined Logistic Growth

shortcomings of this type of model. For example, if p_n is large enough, $r(p_n)$ will be negative, leading to a negative value of p_{n+1}. This would clearly be unreasonable in a model of a population.

Nevertheless, we found in Section 6.1 that in some cases logistic growth sequences can effectively model limited population growth, where early rapid growth is followed by eventual approach to an equilibrium. We also saw that they can provide accurate approximations to observed biological growth data. For these reasons, logistic growth sequences can be useful in spite of their shortcomings. And although there are combinations of parameters for which the corresponding logistic sequences do not grow as we expect real populations to grow, the results of Section 6.1 tell us precisely what those combinations are.

In Section 6.2 we studied in greater detail how logistic sequences can behave, observing oscillation among two or more fixed values, and unpredictable variations that are called chaotic. These behaviors have been found in a wide range of applications, and understanding, predicting, and controlling chaotic behavior has become a significant practical concern. Seeing that chaos can arise in our simple logistic growth models affords a glimpse of this important and active area of ongoing research.

But now we turn away from chaos, and return to the problem of accurately modeling limited population growth. We consider a natural question: Do the unexpected behaviors in logistic growth models reflect inherent properties of real populations, or are they a consequence of inaccurate assumptions? In particular, given the previously noted shortcomings of a linear equation for $r(p)$, might a nonlinear $r(p)$ produce models that do not exhibit oscillation or chaos?

Such questions are representative of the way real models are developed. At an initial stage, simplifying assumptions are adopted as a starting point and used to formulate a first version of the model. Results from the model are then compared with observations from the actual system. If the results are inaccurate, the assumptions may be revised to obtain a refined model. It is with a similar spirit that we consider an alternative to logistic growth.

In practice, at this stage of model revision careful observation of the phenomenon or system would play a key role. Here, we are proceeding speculatively. We do not *know* that oscillation or chaos are a consequence of inappropriate modeling assumptions. We simply observe the *possibility* that our assumptions are at fault. Interestingly, oscillatory behavior *has* been observed in real populations. For example, Utida [54] demonstrated such behavior in populations of cowpea weevils under laboratory conditions. Thus, the fact that logistic growth sometimes involves oscillation is not necessarily an error. On the other hand, the fact that some populations do oscillate does not validate the assumptions of logistic growth in all contexts. It remains reasonable to consider a refined version of the logistic growth model.

In our refined version of logistic growth, we will still assume that r decreases as p increases. However, for reasons that will be discussed presently, instead of assuming that $r(p)$ is a linear function, we assume it is the *reciprocal* of a linear function. That is, we assume an equation of the form

$$r(p) = \frac{1}{mp + b} \tag{6.11}$$

with m and b positive constants. This leads to a difference equation for p_n of the form

$$p_{n+1} = \frac{1}{mp_n + b} \, p_n.$$

We can also write the difference equation as

$$p_{n+1} = \frac{p_n}{mp_n + b}.$$

From this form we derive a simple contrast between our original and refined formulations of logistic growth. Both formulations involve a linear function $f(p)$. In the original formulation, each term of the sequence is found by multiplying the preceding term p_n by $f(p_n)$; in the refined formulation each term is found by *dividing* the preceding term p_n by $f(p_n)$.

We will see that this refined version of logistic growth never leads to oscillatory or chaotic behavior. Furthermore, we will derive a general functional equation for these sequences. The following definitions are stated for future reference.

> **Refined Logistic Growth Definition and Difference Equation:** The refined logistic growth assumption for a sequence $p_0, p_1, p_2, \cdots$ is as follows: the growth factor between successive terms p_n and p_{n+1} is given by a function $r(p_n)$ which is the reciprocal of a linear function.
>
> A refined logistic growth difference equation is one expressible in the form
> $$p_{n+1} = \frac{1}{mp_n + b} \, p_n.$$

As noted at the start of the chapter, there is no standard terminology for logistic growth models. Some authors use one definition, and some use another. So, too, for what we are calling the *refined logistic growth* difference equation. It has appeared in other works with various names, including the discrete logistic difference equation, the Verhulst equation, the Beverton-Holt equation, and the Pielou equation. The *refined logistic growth* terminology is used here for convenience, and to emphasize its connection with the version of logistic growth we considered first. As far as we know, this terminology is not used in other works on modeling.

Rationale. The rationale for our revised logistic growth assumption has three aspects, conceptual, graphical, and algebraic. Conceptually, we try to imagine how populations should respond when they grow above a sustainable level. For concreteness, consider again a population of fish in a lake. Suppose the environment has a carrying capacity of 5,000 fish. That means that the production of new members of the population is in balance with elimination of existing members when the population size is 5,000. If somehow the population were increased to 10,000, say by invasion from another lake, we would expect to see a decrease over time. Perhaps one half of the fish would die off, representing a growth factor of $r = 1/2$. But because we are considering a population that is shrinking rather than growing, we can think of $1/2$ as the *survival factor*—if only $1/2$ of the fish survive, then the population size will be multiplied by $1/2$.

6.3. Refined Logistic Growth

The survival factor will depend on the population size. If the hypothetical invasion raised the population even higher, say to 20,000, then we would expect even more of the fish to die off. Then, the survival factor would be a smaller fraction, such as 1/4 or 1/10. These considerations suggest that we might express r as a fraction, $1/f(p)$, where $f(p)$ is a function that gets larger as p increases.

What kind of function might we choose for $f(p)$? Turning to the algebraic aspect, we should begin with a function that is as simple as possible. Following our earlier practice, this suggests taking $f(p)$ to be a linear function. Thus we are led to the refined logistic growth assumption, taking r as the reciprocal of a linear function.

Finally, turning to the graphical aspect, recall the first introduction of logistic growth. There, we observed that the graph of the function $r(p)$ should have roughly the shape of Figure 6.1 (see page 361). The curve should slope down to the right, but should not cross the horizontal axis. (Why?) And that is exactly the sort of curve one finds when $r(p)$ is defined as in (6.11) with positive values of m and b.

The foregoing considerations may help to explain why one might formulate the refined logistic growth assumption. But however the idea arises, once it is decided to make such an assumption, we can explore its consequences in some specific examples.

Examples of Refined Logistic Growth. As a first example, let us revisit the fish population model depicted in Figure 6.3 (page 363). To develop it, we assumed $r = 2$ when $p = 10,000$ and $r = 1$ when $p = 50,000$. Using these same assumptions we now construct a refined logistic growth model by defining r as the reciprocal of a linear function of p. In visual terms, that means we must find m and b so that the graph of

$$r = \frac{1}{mp + b}$$

passes through the points $(10,000, 2)$ and $(50,000, 1)$. But this task becomes simpler if we consider the reciprocal of r. Then our equation becomes

$$\frac{1}{r} = mp + b$$

which defines the reciprocal of r as a linear function of p, and our conditions can be restated as $1/r = 1/2 = 0.5$ when $p = 10,000$ and $1/r = 1$ when $p = 50,000$.

The equations and conditions will be more familiar if we use variables x and y, with $x = p$ and $y = 1/r$. Then the equation says

$$y = mx + b$$

and the conditions are $y = 0.5$ when $x = 10,000$ and $y = 1$ when $x = 50,000$. Thus, we want to find the equation of a line through two given points. We compute the slope

$$m = \frac{y_2 - y_1}{x_2 - x_1} = \frac{1 - 0.5}{50,000 - 10,000} = 0.0000125,$$

and use the point-slope equation $y - y_1 = m(x - x_1)$ with $(x_1, y_1) = (10,000, 0.5)$. That gives

$$y - 0.5 = 0.0000125(x - 10,000).$$

After simplification, we find

$$y = 0.0000125x + 0.375.$$

Restoring the original variables, substitute $1/r$ for y and p for x. Then we have

$$\frac{1}{r} = 0.0000125p + 0.375,$$

and therefore

$$r = \frac{1}{0.0000125p + 0.375}.$$

In this way we have found the desired equation for r.

Now we can formulate our refined logistic growth sequence. The difference equation is

$$p_{n+1} = \frac{1}{0.0000125p_n + 0.375} p_n$$

or more simply,

$$p_{n+1} = \frac{p_n}{0.0000125p_n + 0.375}.$$

If we begin with an initial population $p_0 = 1{,}000$, the difference equation leads to the results shown in Figure 6.29.

It is interesting to compare these results with those we derived earlier using our original logistic growth model. Graphs for the two models appear together in Figure 6.30. The graph and table show that the two models are very similar. This should not surprise us. For both models the initial population p_0 is the same. And because of the way the models are formulated, they have the same fixed population 50,000. Thus the models have to agree at the initial point and after the populations level off. This implies that the graphs will be fairly similar.

We can also compare the growth factor equations for the two models. In each case we have r expressed as a function of p, with r given by a linear function in our original

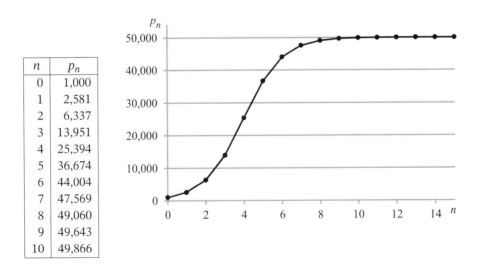

Figure 6.29. Refined logistic growth sequence for the first fish population model of Section 6.1. The variable n represents years, and p_n is the population for year n. The p_n values were calculated with the difference equation $p_{n+1} = \frac{p_n}{0.0000125p_n + 0.375}$ starting from an initial term $p_0 = 1{,}000$.

6.3. Refined Logistic Growth

n	p_n (R)	p_n (O)
0	1,000	1,000
1	2,581	2,225
2	6,337	4,882
3	13,951	10,390
4	25,394	20,678
5	36,674	35,836
6	44,004	48,526
7	47,569	50,314
8	49,060	49,919
9	49,643	50,020
10	49,866	49,995

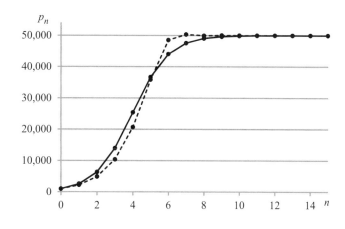

Figure 6.30. Refined (R) and original (O) logistic growth sequences for the first fish population example of Section 6.1. The graph for the refined model is shown with solid lines connecting the points, the graph for the original model is shown with dotted lines. Population figures in the table have been rounded to the nearest whole number.

model and by the reciprocal of a linear function in the refined model. The graphs for these functions appear in Figure 6.31. The line and the curve agree for $p = 10,000$ and for $p = 50,000$ because for each of these populations both models assume the same value of r. Notice that the greatest discrepancies between the two r equations occur for populations greater than 60,000, and these populations never arise in the examples we have considered. This helps to explain why the two models give such similar results.

One part of the preceding analysis illustrates a general method that is worth recognizing. When we want to find r as the reciprocal of a linear function, that is equivalent to finding $1/r$ as a linear function. Therefore, we can proceed as follows:

(1) Begin with two data points of the form (p, r) from the graph of r.

(2) For each point (p, r) compute the corresponding point $(p, 1/r)$.

(3) Find the equation of a straight line through the $(p, 1/r)$ points.

(4) Express the equation in the form $1/r = mp + b$.

(5) Take the reciprocal of each side to obtain an equation of the form $r = \dfrac{1}{mp + b}$.

If we have more than two (p, r) data points, the $(p, 1/r)$ points may not fall exactly on a straight line. In this case we find a straight line that is close to the points as possible.

The Second Fish Population Example. Next we consider the second fish population example from Section 6.1 (page 371). Recall that we made just one change, assuming $r = 2.8$ rather than 2 for a population of size 10,000, and otherwise proceeding as in the first model. This led to results that were surprising at the time—a population model that oscillates between two fixed values rather than approaching an

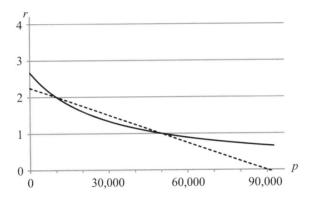

Figure 6.31. Graphs for the growth factor r as a function of population p for the original and refined versions of a fish population model. The solid curve shows r as the reciprocal of a linear function, as specified in the refined model. The dotted line shows r as a linear function, as in the original model. The curve and the line cross for $p = 10{,}000$ and for $p = 50{,}000$ because the models assume identical values of r for these populations.

equilibrium. Let us repeat the same variation again, but this time using a refined logistic growth model.

Proceeding as before, we look at the reciprocals of the specified growth factors: $1/r = 1/2.8$ when $p = 10{,}000$ and $1/r = 1$ when $p = 50{,}000$. Again replacing $1/r$ by y and p by x, we want to find a line passing through the points $(x, y) = (10{,}000, 1/2.8)$ and $(50{,}000, 1)$. The slope of this line is given by

$$m = \frac{y_2 - y_1}{x_2 - x_1} = \frac{1 - 1/2.8}{50{,}000 - 10{,}000}.$$

This can be simplified to

$$m = \frac{9}{560{,}000}.$$

The value of m can be approximated in decimal form as 0.000016, but to assure the greatest possible accuracy, we will continue to work with the fractional form, which is exactly correct.[11]

---

[11] This contrasts with the prior example, where we used decimals rather than fractions at every step. But in that case, the decimal values were exactly correct. The same is not true here. For example m is only approximately equal to 0.000016. Thus, using decimals throughout this example would only produce an *approximate* equation for r. Although using decimals at each step is algebraically simpler, be aware that it may also reduce the accuracy of the results obtained from the model.

6.3. Refined Logistic Growth

Next, we apply the point-slope equation with this value of m and the given point $(50{,}000, 1)$ to find

$$y - 1 = \frac{9}{560{,}000}(x - 50{,}000)$$

$$= \frac{9}{560{,}000}x - \frac{9}{560{,}000} 50{,}000$$

$$= \frac{9}{560{,}000}x - \frac{450{,}000}{560{,}000}$$

$$= \frac{9}{560{,}000}x - \frac{45}{56}.$$

Adding one to both sides of this equation then gives

$$y = \frac{9}{560{,}000}x - \frac{45}{56} + 1$$

$$= \frac{9}{560{,}000}x + \frac{11}{56}.$$

Restoring the original variables, substitute $1/r$ for y and p for x. Then we have

$$\frac{1}{r} = \frac{9}{560{,}000}p + \frac{11}{56},$$

and therefore

$$r = \frac{1}{\frac{9}{560{,}000}p + \frac{11}{56}}.$$

This expresses r as the reciprocal of a linear function of p.

As one further simplification, we can multiply on the right by $560/560$, leading to

$$r = \frac{560}{560} \cdot \frac{1}{\frac{9}{560{,}000}p + \frac{11}{56}} = \frac{560}{\frac{560 \cdot 9}{560 \cdot 1000}p + \frac{560 \cdot 11}{56}} = \frac{560}{0.009p + 110}.$$

In the final step, we are using the fact that the fraction $9/1000$ is exactly equal to 0.009. This provides an exact equation for r because we have not used any decimal approximations.

Proceeding, we now form the refined logistic growth difference equation

$$p_{n+1} = r(p_n)p_n = \frac{560}{0.009p_n + 110} p_n = \frac{560 p_n}{0.009 p_n + 110}.$$

Once again starting with $p_0 = 1{,}000$, we can apply the difference equation repeatedly to generate p_1, p_2, and so on. The results are displayed in Figure 6.32.

The refined logistic model for this example is not much different from the refined model in the first example. The population increases a little more quickly at first, but it still progresses in an orderly manner toward a fixed population of 50,000. There is no sign of the oscillation observed earlier.

Let us again compare the original logistic model and the refined model for this example. Graphs for both models are presented in Figure 6.33. The graphs for the corresponding r equations appear in Figure 6.34.

These first two examples are encouraging. We introduced the refined logistic growth model with the hope that it might lead to results more consistent with our concept of limited population growth. We imagine that a population will grow gradually to its sustainable level, rather than oscillating as in the second fish model example. And indeed, that is what we find when we apply the refined model in both fish population examples. But do all refined logistic growth models behave in a similar manner? Or might they too exhibit oscillatory behavior, or even chaos, for certain combinations of parameters?

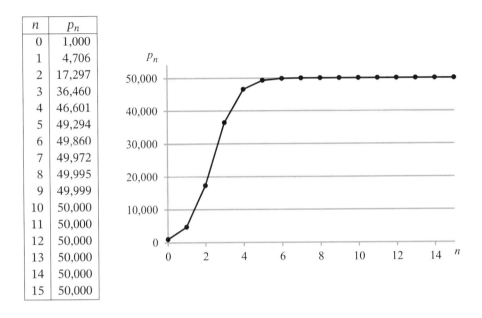

n	p_n
0	1,000
1	4,706
2	17,297
3	36,460
4	46,601
5	49,294
6	49,860
7	49,972
8	49,995
9	49,999
10	50,000
11	50,000
12	50,000
13	50,000
14	50,000
15	50,000

Figure 6.32. Refined logistic growth sequence for the second fish population model. Starting from an initial term $p_0 = 1,000$, the p_n values were calculated using $p_{n+1} = \dfrac{560 p_n}{0.009 p_n + 110}$. The values in the table have been rounded to the nearest whole number.

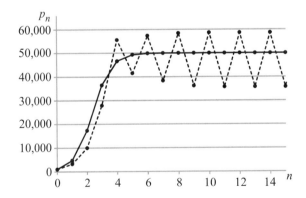

Figure 6.33. Graphs for two models of the second fish population example. The points for the refined logistic growth model are joined by solid lines; the points for the original logistic growth model are joined by dotted lines. Although both models have the same initial value, they diverge almost immediately, and provide qualitatively different predictions of how the fish population varies over time.

6.3. Refined Logistic Growth

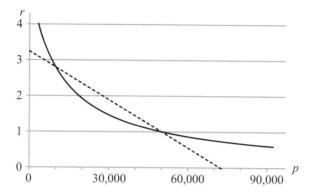

Figure 6.34. Graphs for growth factor r as a function of population p for the original and refined versions of the second fish population example. The curve showing r as the reciprocal of a linear function, and the dotted line showing r as a linear function, are not as close together as they were in the first example. Note too that in the original logistic growth model the population sizes approach 60,000 in even years and 35,000 in odd years, and for each of these figures, the corresponding r values are noticeably different for the curve and the line.

Here is the answer. Refined logistic growth always progresses in an orderly fashion toward an equilibrium level, or carrying capacity, which must be a fixed population. If the initial population size is less than the carrying capacity, then the population will grow consistently larger as it approaches the equilibrium. If the initial population is greater than the carrying capacity then the population will decrease consistently toward the equilibrium.

Another noteworthy characteristic of refined logistic growth is the existence of a general functional equation, in sharp contrast to what was observed with our original version of logistic growth. As we shall see, the description above of refined logistic growth models, either increasing steadily or decreasing steadily, can be justified by properties of the functional equation. Accordingly, we now proceed to investigate that topic.

Refined Logistic Growth Functional Equations. For refined logistic growth models we assume that growth factors are defined by the reciprocal of a linear function. Such a model has a difference equation of the form

$$p_{n+1} = \frac{p_n}{mp_n + b}. \tag{6.12}$$

As we saw earlier, deriving the corresponding equation for r was simplified by considering first the reciprocal $1/r$. Now we will again consider reciprocals, this time on both sides of the difference equation. To illustrate how this leads to a functional equation, we return to the first fish population example.

Consider again the difference equation

$$p_{n+1} = \frac{p_n}{0.0000125 p_n + 0.375}.$$

Taking the reciprocal of each side of this equation gives

$$\frac{1}{p_{n+1}} = \frac{0.0000125 p_n + 0.375}{p_n}$$

$$= \frac{0.0000125 p_n}{p_n} + \frac{0.375}{p_n}$$

$$= 0.0000125 + 0.375 \frac{1}{p_n}.$$

Now two important observations arise. First, the preceding equation is a difference equation for the reciprocal sequence $1/p_0, 1/p_1, 1/p_2, \cdots$. To emphasize this idea, we introduce new notation $q_0, q_1, q_2, \cdots$ for the reciprocal sequence. Thus, for any position number n, the notations q_n and $1/p_n$ are interchangeable. When we substitute q's in the preceding difference equation we find

$$q_{n+1} = 0.0000125 + 0.375 q_n.$$

This brings us to the second observation: what we have found is a mixed growth difference equation. Rewriting it as

$$q_{n+1} = 0.375 q_n + 0.0000125$$

we see it is in the form

$$q_{n+1} = R q_n + D, \qquad (6.13)$$

where we are using upper case letters for the parameters to avoid having two different meanings for the letter r. Notice that the value of R in this example, 0.375, is equal to the parameter b in (6.12). This is true in general. That is, whenever (6.12) holds for a sequence p_n, (6.13) must hold for the reciprocal sequence $q_n = 1/p_n$, and the parameters R and b are equal. However, we will continue the use of R in the next part of the discussion, to emphasize that (6.13) is a mixed model difference equation.

Now we can apply what we learned in Chapter 5 (see page 312). There is an equilibrium value $E = D/(1 - R) = 0.0000125/(1 - 0.375) = 0.00002$, and the functional equation is

$$q_n = E + (q_0 - E) R^n.$$

For the sequence we considered earlier, we had $p_0 = 1{,}000$. Thus $q_0 = 1/p_0 = 1/1{,}000 = 0.001$, and the functional equation becomes

$$q_n = 0.00002 + (0.001 - 0.00002)\, 0.375^n.$$

$$= 0.00002 + 0.00098 \cdot 0.375^n.$$

But we know that $p_n = 1/q_n$. Therefore, we have

$$p_n = \frac{1}{0.00002 + 0.00098 \cdot 0.375^n},$$

which is a functional equation for the refined logistic growth sequence p_n. We can simplify the appearance slightly by multiplying the numerator and denominator of the fraction by 100,000. That leads to

$$p_n = \frac{100{,}000}{2 + 98 \cdot 0.375^n}. \qquad (6.14)$$

These equations can be confirmed by using them to compute particular values

6.3. Refined Logistic Growth

of p_n. For example, (6.14) gives

$$p_5 = \frac{100{,}000}{2 + 98 \cdot 0.375^5} = 36{,}674,$$

to the nearest integer. That agrees with what is in the table in Figure 6.29.

The equilibrium value for p_n can be identified in two ways. First, we know that q_n has an equilibrium value of 0.00002. But $p_n = 1/q_n$, so if q_n is constant at 0.00002, then p_n will be constant at $1/0.00002 = 50{,}000$. This is consistent with the assumptions we used to derive the refined logistic growth difference equation for this example. Alternatively, looking at the functional equation, notice that 0.375^n is an exponential function with base between 0 and 1. We know that this approaches 0 as n increases to very large values. So, for large n, the functional equation will give approximately

$$p_n = \frac{100{,}000}{2 + 98 \cdot 0} = \frac{100{,}000}{2} = 50{,}000.$$

This shows that p_n is approaching an equilibrium of 50,000 as n increases. In addition, because 0.375^n grows consistently smaller as n increases, it can be shown algebraically that the functional equation for p_n is consistently growing larger as n increases. In particular, p_n cannot oscillate. Similar logic applies for any refined logistic growth equation.

The functional equation we have found is one instance of an entire family of functions, referred to as *logistic functions*; their graphs are called *logistic curves*.[12] These can be expressed in a standard form

$$f(n) = \frac{A}{1 + BR^n}, \tag{6.15}$$

where A, B, and R are constants, and $0 < R < 1$. We can put (6.14) into the standard form by multiplying both the numerator and the denominator by $1/2$. That produces

$$p_n = \frac{50{,}000}{1 + 49 \cdot 0.375^n},$$

which is in the standard form with $A = 50{,}000$, $B = 49$, and $R = 0.375$.

Notice that in (6.15), A is always the equilibrium value. Indeed, as n increases, R^n decreases to 0, and $f(n)$ approaches $\frac{A}{1 + B \cdot 0} = A$. This interpretation for the parameter A is valid in any logistic function with $R < 1$. That is one reason that the standard form of the equation is used.

The General Functional Equation. For any sequence with a refined logistic growth difference equation

$$p_{n+1} = \frac{p_n}{mp_n + b},$$

we can find a functional equation just as in the preceding example. Indeed, repeating an approach we have used several times before, we can carry out the same steps as in the example, leaving the parameters m and b as unspecified positive constants. With the additional assumption that $b \neq 1$, this leads to a general functional equation for refined logistic growth.[13]

---

[12]This terminology is quite standard in the literature. Although there are many competing definitions of *logistic difference equation*, there is general agreement on the definitions of logistic functions and curves.
[13]In the exceptional case $b = 1$, the method breaks down because the reciprocal sequence q_n exhibits arithmetic growth rather than mixed growth. In this case, a functional equation can still be derived, but it will not be a logistic function. This is further explored in the exercises.

Proceeding with this plan, we again define $q_n = 1/p_n$. As in the example, we are led to a mixed growth difference equation

$$q_{n+1} = bq_n + m.$$

This has an equilibrium value $E = m/(1-b)$, and the functional equation is given by

$$q_n = E + (q_0 - E)b^n.$$

Replacing q_n by $1/p_n$ and q_0 by $1/p_0$ then leads to

$$p_n = \frac{1}{E + (\frac{1}{p_0} - E)b^n}.$$

Finally, express this in the standard form by multiplying the numerator and denominator by $1/E$. Thus we find

$$p_n = \frac{\frac{1}{E}}{\frac{1}{E}} \cdot \frac{1}{E + (\frac{1}{p_0} - E)b^n}$$

$$= \frac{\frac{1}{E}}{1 + (\frac{1}{E}\frac{1}{p_0} - 1)b^n}$$

$$= \frac{A}{1 + Bb^n},$$

where $A = 1/E = (1-b)/m$ and

$$B = \frac{1}{E}\frac{1}{p_0} - 1 = A \cdot \frac{1}{p_0} - 1 = \frac{A}{p_0} - 1.$$

This shows that a functional equation can always be found in the standard form of a logistic function, and tells us what the parameters are. In summary, we have the following.

> **Refined Logistic Growth Functional Equation:** Suppose a refined logistic growth sequence has initial value p_0 and satisfies the difference equation
>
> $$p_{n+1} = \frac{p_n}{mp_n + b}$$
>
> with positive constants m and b, and with $b \neq 1$. Then there is an equilibrium value $A = (1-b)/m$ and a functional equation for the sequence is given by
>
> $$p_n = \frac{A}{1 + Bb^n},$$
>
> where $B = A/p_0 - 1$.

To illustrate the use of this formulation, recall that our refined logistic growth sequence for the first fish population example had initial value $p_0 = 1{,}000$, $m = 0.0000125$ and $b = 0.375$. We compute the equilibrium value $A = (1-b)/m = 0.625/0.0000125 = 50{,}000$, and $B = A/p_0 - 1 = 50{,}000/1{,}000 - 1 = 49$. Thus, the functional equation is

$$p_n = \frac{A}{1 + Bb^n} = \frac{50{,}000}{1 + 49 \cdot 0.375^n}.$$

And that is what we derived earlier.

6.3. Refined Logistic Growth

For another instance, consider the refined logistic growth model for the second fish population example, where the parameters are $m = 9/560{,}000$ and $b = 11/56$. We know the equilibrium is 50,000, as in the first example, but can also use the equation $A = \frac{1-b}{m}$ to find

$$A = \frac{1 - \frac{11}{56}}{\frac{9}{560{,}000}} = 50{,}000.$$

Also observe $B = A/p_0 - 1$ is again 49 because both A and p_0 are the same in both models. Substituting the parameter values into the general functional equation shows that

$$p_n = \frac{50{,}000}{1 + 49 \cdot (11/56)^n}.$$

This is a functional equation for the refined logistic growth model of the second fish population example.

An Example with Real Data: iPod Sales. The two fish population examples are based on hypothetical values for r and p. Now, as an example based on real data, let us revisit the iPod example (page 380). The data and graph appear again in Figure 6.35. Recall that the *cumulative sales* variable, s, represents the total sales for all the years from 0 (when the iPod was introduced) through year n.

Year (n)	Cumulative Sales (s)
0	0.2
1	1.1
2	5.9
3	28.2
4	68.0
5	119.4
6	174.2
7	228.3
8	279.0
9	322.0
10	357.0
11	383.4

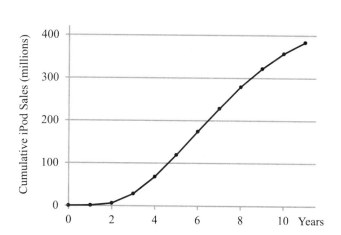

Figure 6.35. Cumulative sales of iPods, in units of millions, versus year. The year, represented by n, is 0 for 2004, 1 for 2005, and so on. The data points on the graph are connected with straight lines for visual emphasis.

Although not a population, the cumulative sales might be expected to behave like one. Often, when a successful new product is introduced, it sells rapidly at first, and then sales taper off as the market becomes saturated. The cumulative sales figures should thus level off approaching an equilibrium. And the table and graph are consistent with this expectation, for growth in cumulative sales is observed to taper off in years 8 through 11. However, when we attempted in Section 6.1 to construct a model for the iPod data using our original version of logistic growth, the results were not very satisfactory. Now we will see that refined logistic growth produces a model that is in much better agreement with the data.

As in the earlier examples, to formulate our model, we will first seek to describe the reciprocals of the growth factors with a linear function of the population variable, s in this setting. In Figure 6.36 we show the computed values of the growth factors, represented by r, and their reciprocals, as well as a graph of $1/r$ versus s. The values in the table have been rounded to a convenient number of decimal places for reading. These calculations, and those below, were performed using a spreadsheet program, carried out to many more decimal places than shown in the table. Unless indicated otherwise, the decimal values throughout this example have been rounded to the number of decimal places shown.

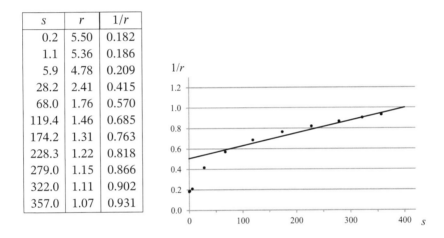

Figure 6.36. Cumulative sales s, growth factors r, and their reciprocals $1/r$ based on the data in Figure 6.35. The straight line in the graph represents an approximate linear equation for $1/r$ as a function of s.

The data points in the graph do not lie on a straight line, so for this application we cannot express $1/r$ exactly as a linear function of s. Instead we determine a straight line that comes close to the data points. The one in the figure represents the equation $1/r = 0.0012s + 0.51$, with $m = 0.0012$ and $b = 0.51$.

This linear equation leads to a refined logistic growth difference equation

$$s_{n+1} = \frac{s_n}{0.0012s_n + 0.51}.$$

With $s_0 = 4.45$, the difference equation produces a refined logistic growth sequence that is pretty close to the original data. This is shown in Figure 6.37.

The authors used a trial-and-error approach to select the parameters m, b, and s_0 in this example. For each trial, we produced a graph like the one in Figure 6.37, and observed how close the curve was to the data points. Although the final parameters we selected for this example produce a model that is pretty close to the original data, we do not claim that they are optimal. In all likelihood, further revision of the parameters

6.3. Refined Logistic Growth

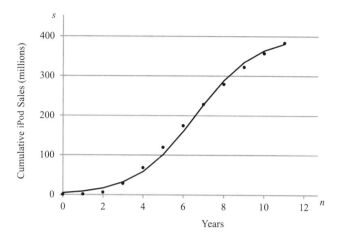

Figure 6.37. Graph of the refined logistic growth sequence for the iPod data. The dots in the graph represent the original data points, from Figure 6.35. The curve was created by plotting the points defined by the difference equation $s_{n+1} = \frac{s_n}{0.0012 s_n + 0.51}$, and connecting them with straight lines. The accuracy of the model is indicated by how close the curve is to the data points.

could produce a curve that does a better job of matching the data. Our goal here has simply been to illustrate the production of a refined logistic growth model to fit a given set of data. Moreover, for this example, a revised logistic growth model approximates the data much better than a model using our original approach to logistic growth.

To complete the analysis of this model, let us find the functional equation. We first find the equilibrium value

$$A = \frac{1-b}{m} = \frac{1-0.51}{0.0012} = 408.33.$$

Next we find

$$B = \frac{A}{s_0} - 1 = \frac{408.33}{4.45} - 1 = 90.76.$$

Thus we have

$$s(n) = \frac{408.33}{1 + 90.76 \cdot 0.51^n}.$$

We use the notation $s(n)$ in place of s_n to emphasize that the functional equation makes sense for fractional values of n. To emphasize this point even more, we could introduce a continuous time variable t (in years), and observe that the graph of the function $s(t) = \frac{408.33}{1+90.76 \cdot 0.51^t}$ is a smooth curve passing through all of the model points (n, s_n).

Properties of Logistic Functions. A logistic function is one that can be expressed in the standard equation

$$f(x) = \frac{A}{1 + B \cdot R^x}, \qquad (6.16)$$

where A, B, and R are constants, and $0 < R < 1$. This is the form we saw in (6.15). In deriving a general functional equation for refined logistic growth sequences, we saw that R is equal to the parameter b. In this section we are interested in logistic functions in general, whether or not they arise in connection with revised logistic growth sequences. Accordingly, we will continue to use R in place of b.

Every logistic function $f(x)$ is defined for all values of x, positive, negative, or zero. When A and B are positive, logistic function graphs all have the same general shape, as illustrated in Figure 6.38. Imagine tracing the graph from left to right. Starting far to the left, the curve appears to lie on the x-axis, though it actually lies just above the axis, sloping slightly uphill at first, and then becoming steadily steeper. Then there is a transition point, called an *inflection point*, after which the curve becomes steadily less steep, even though it is still going uphill. The inflection point in the graph is shown to the right of the y-axis, but it can also occur on or to the left of the axis.

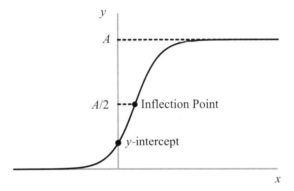

Figure 6.38. Typical graph for a logistic function. Traced from left to right the curve is at first indistinguishable from the x-axis and slopes upward more and more steeply until it reaches an inflection point. Thereafter, the curve becomes less and less steep, eventually approaching an equilibrium at $y = A$. The inflection point is on the line $y = A/2$, and divides the curve into symmetric halves.

The inflection point, which is where the curve is steepest, divides the curve into symmetric halves. If the entire curve is rotated 180° about the inflection point, the left half will perfectly reproduce the original shape of the right half, and vice versa.

We will now see how several of these graphical features can be related to the values of A, B, and R in (6.16).

Equilibrium Level. We have already seen that a refined logistic growth sequence must eventually level off and approach an equilibrium. This corresponds to the horizontal line that the curve approaches as traced far to the right, given by the equation $y = A$. This can be verified using (6.16) and properties of exponential functions. Since we assume $R < 1$, we know that BR^x will get very close to 0 for large positive values of x. In fact, as x increases, BR^x decreases toward 0 and $1 + BR^x$ decreases toward 1. This shows that $f(x)$ *increases* toward A.

Intercept at $y = \frac{A}{1+B}$. The y-intercept of the logistic curve occurs where $x = 0$. Substituting that into (6.16) results in $y = \frac{A}{1+B}$. When the y variable represents a population size, the y-intercept is the initial population. This is consistent with the functional equation for a refined logistic growth sequence, where p_0 is identified as $A/(1+B)$.

6.3. Refined Logistic Growth

Inflection Point. Because of the symmetry of the logistic curve, the inflection point must lie exactly halfway between the x-axis and the horizontal line $y = A$. Thus, at the inflection point, we must have $y = A/2$.

To find the x-coordinate of the inflection point, we set $f(x) = A/2$ in (6.16). This gives

$$\frac{A}{2} = \frac{A}{1 + BR^x}.$$

This equation can be solved for x using known values of A, B, and R.

For example, suppose $f(x) = \frac{30}{1+3 \cdot 0.75^x}$. Then the x-coordinate of the inflection point must satisfy

$$15 = \frac{30}{1 + 3 \cdot 0.75^x}.$$

This leads to

$$15 + 45 \cdot 0.75^x = 30,$$

and after a few more steps to

$$0.75^x = 15/45.$$

Using the methods of Chapter 4, we find

$$x = \frac{\ln(15/45)}{\ln(0.75)} = 3.8188,$$

to four decimal places.

Significance of R. We assume R is between 0 and 1. The closer R is to 0, the steeper the curve will be at the inflection point; the closer R is to 1, the less steep the curve will be at the inflection point.

What if $R > 1$?. Although we generally assume that a logistic function has $R < 1$, it is instructive to ask what happens for $R > 1$. In this case, we know that BR^x will be an increasing exponential function, and so $1 + BR^x$ increases without bound. On the right side of (6.16), therefore, the numerator is a constant and the denominator is growing without bound, so the value of the fraction is approaching zero. This implies that the graph of $f(x)$, when traced far to the right, approaches the x-axis. In fact, the graph in this situation will appear as the mirror image of the one in Figure 6.38, as illustrated in Figure 6.39.

These observations are consistent with our knowledge of refined logistic growth sequences and their functional equations. In that context, the constant R is the same as the parameter b in the refined logistic growth difference equation. Recall that the difference equation takes the general form $p_{n+1} = rp_n$, where r is the reciprocal of the linear function $L(p) = mp + b$. Representative graphs for L and r are shown in Figure 6.40, assuming that m and b are positive. As the graphs show, the intercept on the vertical axis is the low point for the graph of L and the high point for the graph of r. In particular, for any positive population size p the growth factor r is less $1/b$.

Now consider what happens when $b > 1$. The reciprocal, $1/b$, will then be less than 1. But this is the maximum value of the growth factor r. Therefore, as the difference equation is applied, each successive population value is multiplied by a factor less than 1. That implies that the population will steadily decrease and eventually die out completely.

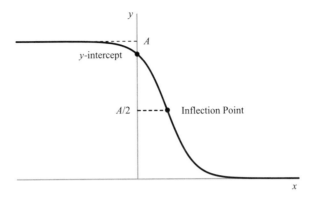

Figure 6.39. Typical graph for a logistic function with $R > 1$. This is the mirror image of the graph for a logistic function having $R < 1$.

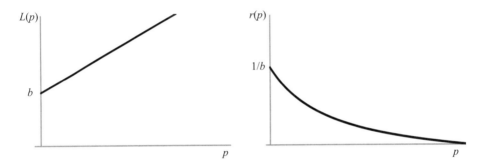

Figure 6.40. Representative graphs for $L(p) = mp + b$ and its reciprocal $r(p) = 1/(mp + b)$, where we assume that m and b are positive. Although the scales for the two figures are not the same, the shapes are correct. In particular, the intercept on the vertical axis is the lowest point on the graph of L and the highest point on the graph of r.

We draw two conclusions from the preceding considerations. First, the logistic function parameter R must be less than 1 if we are modeling a population that is growing toward a positive equilibrium. On the other hand, if we derive a refined logistic growth difference equation from contextual assumptions, and find that our equation has a value $b = R > 1$, we can conclude that this model predicts steady decline of the population, with eventual extinction for very large values of n.

Example Use of Graphical Properties. Suppose we are examining a data set, and notice it has the characteristic shape of a logistic curve. If we can identify the y-intercept and the inflection point, the properties above can be used to find the equation of the curve. As an example, suppose we see a y-intercept at $(0, 100)$ and an inflection point at $(25, 800)$. Since the y-coordinate of the inflection point is $A/2$, we conclude that $A = 1{,}600$. Next, since we know the y-intercept is at 100, we must have

$$\frac{A}{1+B} = \frac{1{,}600}{1+B} = 100.$$

6.3. Refined Logistic Growth

This can be solved to find B as follows:
$$\frac{1{,}600}{1+B} = 100$$
$$1{,}600 = 100(1+B)$$
$$16 = 1+B$$
$$15 = B.$$

Now we know that the equation of the curve must be
$$y = \frac{1{,}600}{1+15R^x},$$
leaving only the parameter R to find. For this purpose, we use the inflection point $(25, 800)$, which must lie on the curve. Therefore
$$800 = \frac{1{,}600}{1+15R^{25}}.$$
We see that this implies that the denominator of the fraction is 2. That leads to several algebraic steps,
$$1 + 15R^{25} = 2$$
$$15R^{25} = 1$$
$$R^{25} = 1/15,$$
and so $R = (1/15)^{1/25}$ which is 0.89734, to five decimal places. The equation of our curve must therefore be approximately
$$y = \frac{1{,}600}{1+15 \cdot 0.89734^x}$$
or
$$y = \frac{1{,}600}{1+15 \cdot \left(\frac{1}{15}\right)^{x/25}}$$
in exact form.

Another Example with Real Data: Yeast Population Growth. In the iPod example, we created a refined logistic growth sequence to approximate observed data. In the process we computed annual growth factors for our data and developed a linear model for the reciprocals of the growth factors. But when our data are not collected at regular time intervals, using growth factors in this way is complicated. In such cases, an alternative is to develop a model directly from the standard logistic function (6.16). Knowing the general form of the functional equation allows us to bypass the formation of a sequence model. The following example illustrates this approach.

The example concerns the growth of a population of yeast, studied by Gause [**28**, pp. 392-396]. In Figure 6.41, tables are shown for two experiments with similar yeast cultures. The first column of each table shows the time t of each measurement, in hours, with $t = 0$ at the time when the yeast were introduced into a growth medium. The second column shows the yeast culture's volume V at each of the corresponding times.[14]

---

[14]The volume values are not expressed in any recognized unit such as cubic centimeters. Rather, they are relative to a standard volume peculiar to the equipment used in the experiments. The size of the standard volume was not provided in the original study, most likely because the absolute volumes were not considered significant in understanding the relative rates of growth of different organisms in different conditions.

Experiment 1

t	V
6	0.37
16	8.87
24	10.66
29	12.50
40	13.27
48	12.87
53	12.70

Experiment 2

t	V
7.5	1.63
15.0	6.20
24.0	10.97
31.5	12.60
33.0	12.90
44.0	12.77
51.5	12.90

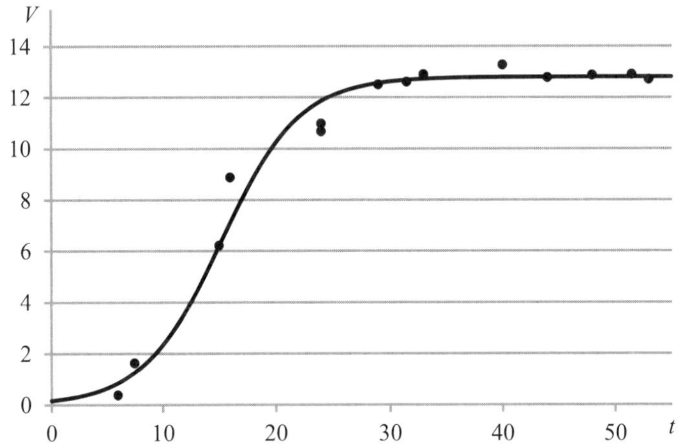

Figure 6.41. Growth data for two experiments with yeast cultures. In each experiment, the size of the culture, measured as a volume, is recorded for several different times, in units of hours. The graph shows data from both experiments, together with an approximating logistic function $V = \dfrac{12.8}{1 + 79.8 \cdot 0.75^t}$.

Gause combined both sets of data into a single graph, as shown in the figure, and then found a logistic function that appears to follow the trend of the data points. The authors replicated this step, using a trial-and-error approach to select the parameters $A = 12.8$, $B = 79.8$, and $R = 0.75$. The corresponding curve, given by

$$V = \frac{12.8}{1 + 79.8 \cdot 0.75^t},$$

appears in the figure. It comes fairly close to the data points. On the average, the dots are about 0.41 units away from the curve.

As always, to start the trial-and-error method it is necessary to choose initial values of the parameters A, B, and R. Applying the ideas from the *Example Use of Graphical Properties* (page 426), these can be selected based on the graph of our data, as follows.

First, observe in Figure 6.41 that the V values appear to be approaching an equilibrium of about 13. This suggests initially taking $A = 13$. Second, based on the graph, we can estimate an intercept on the V-axis at or below 0.37, the lowest V value in the table. So it is reasonable to assume a V-intercept of $V_0 = 0.3$. Combined with $A = 13$, this allows us to compute a value for B, namely $B = A/V_0 - 1 = 13/0.3 - 1 = 42.33$.

Finally, we can estimate from the graph an inflection point at about $t = 12$, and we know that at the inflection point V is supposed to be $A/2 = 6.5$. Substituting these

6.3. Refined Logistic Growth

values for t and V in the equation

$$V = \frac{A}{1 + BR^t} = \frac{13}{1 + 42.33R^t}$$

produces

$$6.5 = \frac{13}{1 + 42.33R^{12}},$$

so we can determine that

$$1 + 42.33R^{12} = 2$$
$$42.33R^{12} = 1$$
$$R^{12} = 1/42.33.$$

Thus we obtain $R = (1/42.33)^{1/12}$ which is about 0.73. Accordingly, we can begin the trial-and-error process with the equation

$$V = \frac{13}{1 + 42.33 \cdot 0.73^t}.$$

The graph of this equation is fairly close to the data points, but is consistently too high up to about $t = 30$. From this starting point, we can proceed by trial and error to seek parameters producing better agreement with the data. This is how the equation in Figure 6.41 was obtained.

This example illustrates that a logistic function can be formulated for a set of data without considering a difference equation or an equation for r as a function of the population size. For this example, the data were collected at irregularly spaced times, so that a consistent set of growth factors cannot easily be determined. However, since we know a general form for the functional equation of refined logistic growth models, we can apply that knowledge without first formulating a difference equation and a sequence. Moreover, we are able to obtain reasonably good first guesses for $A, B,$ and R by estimating geometric features of the graph: the intercept on the V-axis, the point at which the curve levels off, and the inflection point.

Profile for Refined Logistic Growth. As we have seen in this section, refined logistic growth sequences provide an attractive alternative to the original logistic growth sequences discussed in Section 6.1. In addition, our development of the refined model illustrates an important aspect of the design and analysis of models. It shows how studying the properties of one type of model can lead us to reconsider the model's assumptions, and to propose alternative assumptions. In the case of logistic growth and refined logistic growth sequences, that reconsideration leads to an expanded repertoire of model types, and to a new family of functions.

Comparing logistic growth and refined logistic growth sequences, the latter have two significant advantages. They do not oscillate or behave chaotically as sometimes occurs for original logistic growth. And they have nicely behaved functional equations. However, that does not necessarily mean they are a better modeling choice in all contexts. For example, a refined logistic growth sequence will never oscillate, but as mentioned in Section 6.1, there are species of insects for which population sizes can oscillate under laboratory conditions. For such a population, a model based on our first approach to logistic growth might well be appropriate. In many situations, both versions of the logistic growth model will produce similar results. In other cases, one approach may be superior to the other.

Table 6.4. Profile for refined logistic growth sequences.

Verbal Description:	Each term is found by *dividing* the preceding term p_n by a factor that both depends linearly on p_n and is *increasing* as p_n increases. Equivalently, we model the growth factors r by the reciprocal of a linear function of p.
Parameters:	Initial term p_0. Positive coefficients m and b in an equation of the form $r(p) = 1/(mp + b)$.
Difference Equation:	$p_{n+1} = p_n/(mp_n + b)$.
Functional Equation:	$p_n = A/(1 + BR^n)$ where A, B, and R are given by $A = (1 - b)/m$, $B = (A/p_0) - 1$, and $R = b$.
Properties:	Assume the parameters of the difference equation obey the conditions, $m > 0$ and $0 < b < 1$. Then, for any positive p_0, the terms of the sequence will approach an equilibrium value equal to A. If $p_0 < A$ then the terms of the sequence steadily increase toward A; if $p_0 > A$ then the terms steadily decrease toward A.
Graph:	The logistic function $y = A/(1 + BR^x)$ can be graphed for all x, positive, negative, or zero. When $A > 0$, $B > 0$, and $0 < R < 1$, the graph has a shape similar to the curve shown at left with the following properties. • As the curve is traced far to the right, y approaches A. • At $p_0 = A/(B + 1)$ is the y-intercept. • At $y = A/2$ is an inflection point. • The inflection point may be either to the left of, to the right of, or on the y-axis.

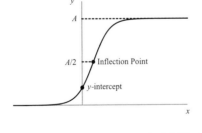

We close this section by summarizing important aspects of refined logistic growth sequences in the following profile. In the interest of brevity, the graph is only shown for the case of a population that is increasing toward a positive equilibrium value. This occurs when $b < 1$ and $p_0 < A = (1-b)/m$, as in all of the examples we have considered.

6.3 Exercises

Reading Comprehension.

(1) In your own words, give a verbal description of logistic growth sequences, and a contrasting verbal description of refined logistic growth sequences.

(2) Explain the rationale for formulating refined logistic growth as we have done. Your answer should include one or more motivations for refining the previous logistic growth model, what new assumptions are made for the refined logistic growth model, and why the new assumptions are reasonable.

(3) @What are the benefits of studying both versions of logistic growth? What insights about modeling would be lost if only one version of logistic growth is studied?

(4) @In both versions of logistic growth, a sequence is defined by a difference equation of the form $p_{n+1} = r(p_n)p_n$ where $r(p)$ is a function of population size.

   a. @What kind of function is $r(p)$ in the original version of logistic growth? What are the advantages and disadvantages of using this kind of function?

   b. @What kind of function is $r(p)$ in the refined version of logistic growth? What are the advantages and disadvantages of using this kind of function?

(5) @The EMM Forensics Society intends to debate the following proposition: *For modeling biological populations regular logistic growth sequences are always inferior to refined logistic growth sequences, because the latter can never oscillate or behave chaotically.* Decide whether you agree or disagree with this proposition, and then write as persuasive an argument as possible in support of your position.

(6) For the refined logistic growth model we found a functional equation and noted that it is usual to refer to this as a logistic function. The assumptions discussed in the chapter and summarized in the profile apply for this problem.

   a. What is the standard equation for a logistic function? What (if any) is the practical interpretation of each parameter?

   b. Describe the general shape of a logistic curve, and explain what is meant by the inflection point.

(7) For the models we have considered, the parameter R of the logistic function is always less than 1. If $R > 1$ then the population model would predict extinction. Explain in your own words how this conclusion can be derived.

Mathematical Skills. In any problem that asks for a graph, the use of graphing software is recommended. Similarly, to compute several terms of a sequence, the use of a spreadsheet or comparable technology is recommended.

(8) @Simplify each of the following functions of the form $r(p) = \dfrac{1}{mp+b}$ as was done on page 415. That is, multiply the numerator and denominator of the main fraction by a suitable number.

a. $@r(p) = \dfrac{1}{(1/2700)p + 5/27}$. [Hint: multiply numerator and denominator by 2,700.]

b. $r(p) = \dfrac{1}{(3/700)p + 2/7}$

c. $r(p) = \dfrac{1}{(1/970)p + 12/97}$

d. $@r(p) = \dfrac{1}{(4/1500)p + 2/5}$

(9) @For each part, find a linear function for $1/r$ consistent with the given information.

   a. $@r = 2.5$ when $p = 50$ and $r = 1.0$ when $p = 200$.
   b. $r = 1.25$ when $p = 10{,}000$ and $r = 0.8$ when $p = 40{,}000$.
   c. $@r(0) = 4/3$ and $p = 380$ is an equilibrium population.

(10) @ For each part of problem 9 find the corresponding refined logistic growth difference equation, simplified as in problem 8. Then calculate the values of p_1 through p_8, and produce a graph of the results. Use these initial values: $p_0 = 160$ in part a, $p_0 = 15{,}000$ in part b, and $p_0 = 50$ in part c.

(11) @For each part of problem 10 find

   i. the parameters A, B, and R
   ii. the functional equation
   iii. the inflection point (both coordinates).

(12) @For each part, find a function of the form $r(p) = \dfrac{1}{mp + b}$ consistent with the given information.

   a. $@r = 1.5$ when $p = 500$ and $r = 1.0$ when $p = 1{,}200$.
   b. The graph of r as a function of p passes through the points $(200, 4)$ and $(500, 2)$.
   c. $@r(0) = 2$ and $p = 250$ is an equilibrium population.
   d. $r = 1.25$ when $p = 100$ and $p = 175$ is an equilibrium population.

(13) @ For each part of problem 12 find the corresponding refined logistic growth difference equation, simplified as in problem 8. Then, with $p_0 = 100$, calculate the values of p_1 through p_8, and produce a graph of the results.

(14) @For each part of problem 13 find

   i. the parameters A, B, and R
   ii. the functional equation
   iii. the values of p_1 through p_4 using the functional equation (these should match the values found using the difference equation)

(15) @For a refined logistic growth sequence it is given that $p_0 = 100$, $p_1 = 250$, and $p_2 = 500$. For this sequence find the following.

   a. @The difference equation
   b. @The functional equation
   c. @The equilibrium population

6.3. Exercises

(16) For a refined logistic growth sequence it is given that $p_0 = 600$, $p_1 = 750$, and $p_2 = 900$. For this sequence find the following.

   a. The difference equation
   b. The functional equation
   c. The equilibrium population

(17) @In each part the equation of a logistic function is given. Using just the equation, determine the equilibrium value, the inflection point, and the y-intercept, and then sketch the graph. Confirm your sketch by using a computer or calculator to produce a graph for the function.

   a. @$f(x) = \dfrac{2{,}500}{1 + 5 \cdot 0.8^x}$

   b. $f(x) = \dfrac{2{,}500}{1 + 0.2 \cdot 0.8^x}$

   c. @$f(x) = \dfrac{800}{1 + 0.04 \cdot 0.4^x}$

   d. $f(x) = \dfrac{800}{1 + 25 \cdot 0.4^x}$

(18) @A logistic curve has an inflection point at $(30, 5{,}000)$ and a y-intercept of 2,000. Find the equation of the curve.

(19) A logistic curve has an inflection point at $(-10, 480)$ and a y-intercept of 400. Find the equation of the curve.

(20) @For a certain logistic function $f(x)$, as the graph is traced from left to right $f(x)$ approaches a value of 300. The inflection point occurs at $x = 8$. The y-intercept is at $y = 50$. Find the equation of the curve.

(21) @Find the equation of the logistic curve shown in the figure below. The y-intercept and the inflection point are marked in the figure.

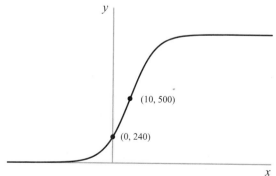

(22) Find the equation of the logistic curve shown in the figure below. The y-intercept and the inflection point are marked in the figure.

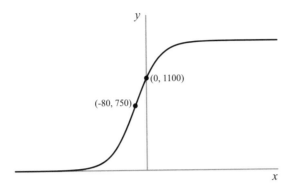

Problems in Context. In any problem that asks for a graph, the use of graphing software is recommended. Similarly, to compute several terms of a sequence, the use of a spreadsheet or comparable technology is recommended.

(23) @Developing a Refined Model for Bacteria. Here we re-visit problem 26 in Section 6.1 (page 387).

Bacteria are grown in a sealed environment in a laboratory. The size of the population of bacteria is measured daily. Early in the experiment, the size grows very rapidly. For example, one day the population was 500, and the next it had doubled to 1,000. (This means the growth factor for that day was 2.) Later, when the population was much larger, the rate of growth slowed down. When the population reached a size of 10,000 it was only growing by about 10 percent each day. (That is a growth factor of 1.1.)

  a. @Find a linear equation for the reciprocal growth factor $1/r$ as a function of population p.
  b. @Use your linear equation to formulate a refined logistic difference equation for the bacteria population.
  c. @Assume a starting population of 500 and find the functional equation for your difference equation. What does it predict about the future growth of the bacteria population?
  d. Explore the future growth of the population according to the difference and/or functional equation. Use numerical and graphical techniques, and assume a starting population of 500. Do your results confirm your predictions from the preceding part of this problem?
  e. @Repeat the preceding item using a variety of different starting populations. Are your results consistent with your predictions in part c based on the functional equation?

(24) Determining How Long the Dissemination of Information Will Take. Here we re-visit problem 27 in Section 6.1 (page 387).

There are about 4,000 students at Anonymous University. Early one morning the student body president tells the vice president that a famous actor (whose identity is supposed to be kept secret) has agreed to be the commencement speaker. In this exercise you will develop a refined logistic model for the way that information is spread to the entire student body.

6.3. Exercises

Let p_n stand for the total number of students who have heard the news after n hours. We will assume that each student tells another student every hour. Initially, that would cause p_n to double every hour. However, as the number of students who have heard the news increases, the situation changes. When a student tells the news to another student, he or she might have already heard the news. In fact, once p_n reaches 4,000, it cannot possibly rise any higher, since there are only 4,000 students at the university. So assuming that p_n doubles initially (when $p_0 = 2$ r also equals 2) and that p_n remains constant when it reaches 4,000, develop a linear equation for $1/r$ and p, as in the previous problem. Then formulate a refined logistic growth difference equation or functional equation for this model, and use it to study how long it takes the news to spread over the whole university.

(25) @Smart Phone Use. Students are studying the growth in the use of smart phones in the US. A data set published by the ITU (a specialized United Nations agency for information and communication technologies) reports the total number of smart phone subscriptions each year from 2000 through 2014. The students develop a model using t for the year, starting from $t = 0$ in 2000, and $y(t)$ for the number of cell phone users, in units of millions, in year t. Looking at a graph of the data, they estimate that there is an inflection point at $(2.9, 165)$ and that the y-intercept is 105. Based on these figures, find an equation expressing $y(t)$ as a logistic function of t. According to your model, what is maximum number of users that can be expected to occur? When will the smart phone market reach 99% saturation? (That is, when will the number of users be 99% of the maximum?)

(26) Developing a Refined Legume Vine Length Model. Here we re-visit problem 28 in Section 6.1 (page 388).

Lupinus albus is a type of legume (bean) found widely in the Mediterranean region, and cultivated in some places as a food crop. The table below shows daily measurements of the length of one particular lupinus vine, as reported in [50, p. 160].

Table 6.5. Measured length of a *lupinus albus* vine, recorded daily. Lengths are in units of millimeters.

day	length	day	length	day	length
4	10.5	10	77.9	16	149.7
5	16.3	11	93.7	17	155.6
6	23.3	12	107.4	18	158.1
7	32.5	13	120.1	19	160.6
8	42.2	14	132.3	20	161.4
9	58.7	15	140.6	21	161.6

Using the same methods as in the iPod example (pages 421–423) and the bacteria exercise above, develop a refined logistic growth sequence to approximate the data in the table. In your answer, let p_n be the model value of the length after n days, and include the values of the parameters m, b, and p_4. Also include a graph like the one in Figure 6.37.

(27) @Beetle Population. A type of beetle, *Rhizopertha Dominica*, was studied under laboratory conditions by Crombie. The beetles were confined to a fixed amount of space, and were given a regularly replenished food supply. The table below shows observed population counts for the beetles over a period of time [**12**, p. 370].

Table 6.6. Population counts for beetles in a laboratory experiment.

week	count	week	count	week	count
0	2	11	130	25	333
2	2	13	175	27	350
4	2	15	205	29	332
5	3	17	261	33	333
6	17	19	302	35	335
7	65	21	330	37	330
9	119	23	315		

Note that the time intervals are an irregular set of one- or two-week periods, so consistent growth factors are not so easily determined. Using the same methods as in the yeast example, formulate a logistic growth function $p(t) = \frac{A}{1 + B \cdot R^t}$ to approximate the data in the table. In your answer, identify the values of the parameters A, B, and R. Also include a graph like the one in Figure 6.41.

Digging Deeper.

(28) @Often, when we derive a logistic growth difference equation, we use decimal approximations for the coefficients. This problem considers the impact of such approximations, by comparing the results from two versions of a model. In the first version the difference equation is expressed in an exact form, in the second in an approximate decimal form. To help us compare the models, we will use a *hat* notation for all of the variables and parameters of the second version of the model, except for n. For example, m and b will represent the exact coefficients in the first model, whereas $\hat{m}$ and $\hat{b}$ will represent the decimal approximations used in the second model. Similarly, we use r for the growth factor in the first version, and $\hat{r}$ for the decimal approximation in the second.

To define the models, we begin with the reciprocal linear model for r. Assuming that $r = 2.4$ when $p = 300$ and $r = 2$ when $p = 500$, we can find a function of the form $r(p) = \frac{1}{mp + b}$. The exact form is

$$r(p) = \frac{1}{(1/2400)p + 7/24}.$$

In approximate decimal form, this becomes

$$\hat{r}(p) = \frac{1}{0.00042p + 0.29}.$$

Thus, we have $m = 1/2400$, $b = 7/24$, $\hat{m} = 0.00042$, and $\hat{b} = 0.29$.

6.3. Exercises

a. Show that the equation for $r(p)$ can be simplified to the alternate exact form
$$r(p) = \frac{2400}{p + 700}.$$

b. Verify that the equation for $r(p)$ in part a is correct by computing $r(300)$ and $r(500)$. Do these values agree with the assumptions of the model?

c. @Using the equation for $r(p)$ in part a, formulate a refined logistic growth difference equation for p_n. Then, with $p_0 = 100$, calculate the values of p_1 through p_8. Use a calculator or computer for these calculations, and record your results rounded to the nearest integer.

d. @Using the equation given above for $\hat{r}$, find the corresponding refined logistic growth difference equation for $\hat{p}$. Then, with $\hat{p}_0 = 100$, calculate the values of $\hat{p}_1$ through $\hat{p}_8$. Again, use a calculator or computer and round your results to the nearest integer. Compare the results with those of part c.

e. @We know that a logistic growth sequence approaches an equilibrium $A = (1-b)/m$. Using the exact values of m and b, compute an exact value of A. Using our decimal approximations $\hat{m}$ and $\hat{b}$, compute $\hat{A}$ on a calculator or computer, rounding your result to the nearest integer. How do your results compare with what you found in parts c and d?

f. @Repeat the analysis above using two additional decimal places in $\hat{m}$ and $\hat{b}$. How does that effect the results found earlier?

(29) Let f and g be functions, and suppose that the graph of f can be obtained by reflecting the graph of g across the y-axis, and vice versa. This happens exactly when the equation $f(x) = g(-x)$ holds for every x, and in that case the graph of f appears as the mirror image of the graph of g.

a. Let $f(x) = \dfrac{100}{1 + 50 \cdot 1.25^x}$ and $g(x) = \dfrac{100}{1 + 50 \cdot .8^x}$. Use the fact stated above to show that the graph of f is the reflection of the graph of g across the y-axis.

b. More generally, show that for any $R > 1$, the graph of $f(x) = \dfrac{A}{1+B \cdot R^x}$ is the mirror image of the graph of the function $g(x) = \dfrac{A}{1+B(1/R)^x}$.

c. Using part b, explain why a logistic function with $R > 1$ must have a graph of the form shown in Figure 6.39.

(30) @An alternate form of the equation for a logistic function is
$$y = \frac{A}{1 + Be^{kx}}.$$
Find an equation relating the values of k and R.

(31) For the logistic function $y = \dfrac{A}{1 + BR^x}$ we know that at the inflection point $y = A/2$.

a. Use algebra to show that the x-coordinate of the inflection point is given by $-\ln(B)/\ln(R)$. There is a hint below[15].

b. Use part a to show that the inflection point is to the right of the y-axis when $B > 1$ and to the left of the y-axis when $0 < B < 1$

---

[15]See *Exponential Equation Solution* in chapter 4 (page 261). For this problem we must also use the properties of exponents on page 257

(32) @In several of the math skills problems, you were given two values of p and the corresponding values of r, and asked to develop a refined logistic growth sequence. It is possible for the value of b to be zero in such a problem. We avoided this situation in our models and focused on cases where R, which is equal to b, satisfies $0 < R < 1$. Now let's consider the case $b = 0$ by answering the following questions.

   a. @If the value of $b = 0$, what is the difference equation? What is the functional equation? What will the graph look like? For a general answer to this problem, the parameter m should be left unspecified. However, if you would like to try an example as a warm up exercise, assume $m = 1.5$ or assume $m = 0.4$.
   b. Assuming $b = 0$ and given any value of p_0, what can you say about the terms p_1, p_2, etc?
   c. Construct an example in which two (p, r) points are given, and where b is found to be 0.
   d. @Find a general rule according to which two (p, r) points lead to $b = 0$.

(33) In previous work we focused on cases where R, which is equal to b, satisfies $0 < R < 1$. Now let's consider the case $b = 1$ by answering the following questions.

   a. An example of such a case is $p_{n+1} = \dfrac{p_n}{2p_n + 1}$; $p_0 = 1{,}000$. Make a list of the first 5 terms. Graph n versus p_n for $n = 1$ through $n = 5$. What do you notice about the graph?
   b. Continuing with the example $p_{n+1} = \dfrac{p_n}{2p_n + 1}$; $p_0 = 1{,}000$, find the functional equation parameter A.
   c. Our derivation of a refined logistic growth functional equation is not valid in the case $b = 1$. And if we attempt to use the parameter definitions in the refined logistic growth profile, that leads to the constant functional equation $p_n = 0$, which is not correct. When $b = 1$ we have to find a functional equation by another method.

   Do so for our example, as follows. We know that $p_{n+1} = \dfrac{p_n}{2p_n + 1}$ and $p_0 = 1{,}000$. Let $q_n = 1/p_n$. Then $p_n = 1/q_n$. Substitute this into our difference equation, and simplify the resulting difference equation for q_{n+1}. Based on this difference equation, determine what type of model q_n is and find its functional equation. Finally, convert this result to a functional equation for p_n by substituting $q_n = 1/p_n$ and simplifying. These steps are similar to the procedure in Refined Logistic Growth Functional Equations. Verify that this functional equation gives the correct value for the first 5 terms of the sequence.
   d. Having found a functional equation for an example with $b = 1$, find a general functional equation for all cases with $b = 1$. To do this, repeat the steps in the preceding part for $p_{n+1} = \dfrac{p_n}{mp_n + 1}$ and with p_0 an unspecified parameter.

(34) @Find a logistic function $y = \dfrac{A}{1 + BR^x}$ that passes through the points $(2, 10)$, $(4, 130)$, and $(6, 250)$. [Hint: this problem becomes much simpler if you assume that the middle point is actually the inflection point of the logistic curve. Note that by symmetry, this can only occur when the middle point is exactly half way between the other two points, as it is in this problem.]

6.3. Exercises

(35) Find a logistic function $y = \dfrac{A}{1 + BR^x}$ that passes through the points $(5, 90)$, $(6, 96)$, and $(7, 98)$. Note that the middle point in this case is not halfway between the other two, so there is no reason to assume it is the inflection point.

Selected Answers to Exercises

Answers are provided here for items marked with @ in the exercise sections of the text.

1.1 Exercises

Reading Comprehension.

(3) A number sequence is a list of numbers in a specific order. Often a data set can be considered to be a number sequence. When the terms of a sequence exhibit a consistent pattern, recognizing and describing that pattern can be used to predict additional terms of the sequence.

(5) Yes. We saw one example in the reading. Here is another: the sequence 1, 4, 9, 16, 25, ... can be described as the squares of the whole numbers 1, 2, 3, 4, ... Alternatively, we can describe the sequence as staring with 1, and with successive terms obtained by adding consecutive odd numbers starting with 3. Thus, starting with one we add 3 to get 4, then add 5 to get 9, then add 7 to get 16, and so on.

Math Skills.

(7) a. One approach is to notice that each term is 3 more than the previous term. The first term is 3.
Another way to say this is that the terms are consecutive multiples of 3. Based on either description, the next three terms are: 18, 21, 24. The 100th term is 300.

b. One approach is to notice that each term is twice the previous term. The first term is 1. Another way to say this is that the terms are powers of 2: 2^0, 2^1, 2^2, 2^3, $\cdots$. Based on either description, the next three terms are: 32, 64, 128. The 100th term is 2^{99}.

c. This is challenging to figure out using the methods learned so far. One approach is to notice that this sequence can be formed by adding consecutive even numbers to the terms. To do this we have to keep track of the even numbers being added as well as the sequence itself. A diagram is helpful:

$$5 \xrightarrow{+2} 7 \xrightarrow{+4} 11 \xrightarrow{+6} 17 \xrightarrow{+8} 25 \ldots.$$

Based on this description, the next three terms are: 35, 47, 61. The 100th term is tedious to find using this description; it is 9905.

e. One approach is to notice that each term is a fraction with denominator 5. The numerators are consecutive whole numbers. The first term is $\frac{1}{5}$. Based on this description, the next three terms are: $\frac{6}{5}, \frac{7}{5}, \frac{8}{5}$. The 100th term is $\frac{100}{5}$.

h. This is a little tough to figure out using the methods discussed so far. One approach is to notice that each term is 3 more than the previous term. The first term is 5. Also, comparing this sequence with the one in part *a* we can observe a close connection: each term of 5, 8, 11, ⋯ is 2 more than the corresponding term of 3, 6, 9, ⋯ This leads to the following verbal description for the new sequence: the terms are each 2 more than the multiples of 3. For example, the 100th multiple of 3 is 300, so the corresponding term of the new sequence is 302. Based on either description, the next three terms are: 20, 23, 26. The 100th term is 302.

j. One approach is to notice each term is an integer all of whose digits are 5's (in every place: ones place, tens place, hundreds place, and so on). Each term is one order of magnitude larger than the previous term. Another way to say this is that each term has one more digit than the previous term. The first term is 5, the second 55. A more algebraic description is that each term can be obtained by multiplying the preceding term by 10 and then adding 5 to the result, with a starting term of 5. In a diagram

$$5 \xrightarrow{\times 10,\ +5} 55 \xrightarrow{\times 10,\ +5} 555 \xrightarrow{\times 10,\ +5} 5555 \xrightarrow{\times 10,\ +5} 55555 \cdots .$$

Based on these descriptions, the next three terms of the sequence are: 555555, 5555555, 55555555. The 100th term is a number represented by one hundred 5s.

(9) One pattern is found by writing out the counting numbers as words: *one, two, three, four, five,* and so on. Now count the letters in each word. That produces the given terms 3, 3, 5, 4, 4, 3, and so on. Using this pattern, the next three terms are 6, 6, 8.

Another pattern is to consider repeating the first nine terms over and over. In this case, the final 3 given in the problem is the start of the second block of terms. Using this pattern, the terms can be listed as 3, 3, 5, 4, 4, 3, 5, 5, 4, 3, 3, 5, 4, 4, 3, 5, 5, 4, 3, 3, 5, 4, 4, 3, 5, 5, 4, ⋯.

Yet another possible pattern is found by shuffling together two sequences: 3, 3, 4, 4, 5, 5, 3, 3, 4, 4, 5, 5, ⋯ and 5, 3, 4, 5, 3, 4, 5, 3, 4, ⋯. The shuffling pattern is to take two terms of the first sequence and one of the second, repeatedly. This gives the same result as the preceding possibility. But it also leads to many other possible variations. For example, if we take the second sequence to be 5, 3, 4, 2, 3, 1, 2, 0, ⋯, that would produce the sequence 3, 3, 5, 4, 4, 3, 5, 5, 4, 3, 3, 2, 4, 4, 3, 5, 5, 1, 3, 3, 2, 4, 4, 0, ⋯.

(11) Using the hint we separate the sequence into 2, __, 4, __, 6, __, 8, ⋯ and 1, __, 4, __, 9, __, 16, ⋯. The first sequence could be even numbers and the second sequence perfect squares. If so, the next three terms of the combined sequence would be 10 (the next even number), 25 (the next square), and 12.

Problems in Context.

(13) a. 15, 21, 28. Notice that this is the same pattern as in problem 7g.

b. One way to look at this is to notice that the rows of coins making up the triangle are consecutive whole numbers. The total number of coins in a triangle of side 4 is the sum of the number of coins in each row (the whole

1.1. Exercises

numbers up to 4): $1 + 2 + 3 + 4 = 10$. A triangle of side 10 is made up of $1+2+3+4+5+6+7+8+9+10 = 55$ coins.

c. The method used above is rather tedious and effectively counts all the previous triangles. However, there is a famous shortcut for finding the sum of natural numbers. Here's an example: $1 + 2 + \cdots + 10 = \frac{10 \times 11}{2} = 55$. Here's another example: $1 + 2 + \cdots + 15 = \frac{15 \times 16}{2} = 120$. This can be applied to 100 to find the number of coins needed is $\frac{100 \cdot 101}{2} = 5050$.

(15) a. Three points needs 3 lines. Five points needs 10 lines. Two points needs 1 line.

b.

Anchor Points	Lines of Web
1	0
2	1
3	3
4	6
5	10

c. The description of how the web is made states that lines go from one anchor point to *another*. If there is no other point, there will be no line. It is not possible to have a web with one anchor point.

d. Each time a new anchor point is added it must be connected to all the other points. The next term will be the sum of the previous term and the number of old anchor points. In this case the previous term is 10 and the number of old anchor points is 5 so the new term is $10 + 5 = 15$. Another approach is to notice that we have seen this sequence in two other problems in this section, 7g and 13, except that there is an extra term at the start of the sequence we are considering now. If this pattern continues to be valid, we would predict that 15 lines are needed for 6 points.

(17) a. No. The relative sizes of the regions can be changed, but the number of pieces will not change.

b. There are 4 regions created. Moving the points does not change the number of regions.

c.

Anchor Points	Maximum Number of Pie Pieces
1	1
2	2
3	4
4	8
5	16

d. One reasonable conjecture is that the next term will be 32.

e. 30 regions

f. 31 regions

g. The largest number of regions possible in this scenario is 31. The pattern of the first five lines in the table is so striking that most people assume it extends to the next line as well. It is possible to predict the correct answer 31 from a careful consideration of the geometry of the problem, but that would require remarkable insight. The problem of regions of a circle is a rich and fascinating one. It has been explored by many and there are several good articles on the topic available on the internet. It is included here to highlight the fact that the first terms of a sequence might follow a pattern that does not extend to the entire sequence. In other words, sometimes patterns peter out!

1.2 Exercises

Reading Comprehension.

(1) A position number for a term of a sequence is an integer that indicates where the term occurs within the sequence. Often the first term of the sequence is considered to have position number 1, and then the second term has position number 2, the third has position number 3, and so on. But sometimes the initial position number is defined to be something other than 1. In many instances we consider the first term to be in position 0. In that case the second term has position number 1, the third has position number 2, and so on. In general, position numbers are consecutive integers assigned to the terms of the sequence beginning with a specified position number for the first term. For example, for the sequence $5, 10, 15, 20, 25, \cdots$, we may specify that the initial term 5 has position number 2. Then the position number for 10 is 3, the position number for 15 is 4, and so on.

Position numbers are used to refer to terms according to their positions in a sequence. For example, in the sequence $1, 4, 9, 16, 25, \cdots$, if we assign a position number of 1 to the first term, we can describe a pattern as follows: each term is equal to the square of its position number. Thus, the term following 16 is in position 5, so the value of the term is $5^2 = 25$. Similarly, the term with position number 15 is $15^2 = 225$.

(3) Some patterns are easily observed visually in the graph of a number sequence. For example, if the dots in the graph form a straight line, that is a striking pattern. If the dots form a curve that consistently slopes downward and levels off along the horizontal axis, that is another kind of pattern. Such patterns may be easier to recognize visually than by considering the numerical values of the terms.

Math Skills.

(6) a. This answer also serves as an example for the other parts of this question.

position	term
1	8
2	80
3	800
4	8,000

1.2. Exercises

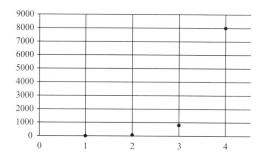

(7) a. Data Table:

position	1	2	3	4	5	...
term	1	2	4	8	16	...

Graph:

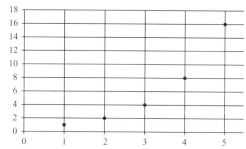

Description (including recursive or direct): Each term is two times the previous term with the first term being 1. This description is recursive. We could also give a direct description as follows. Each term is a power of two. The power is one less than the position number: $2^0, 2^1, \cdots$.

(8) a.

position	1	2	3	4	5	6	7	8	9	10
term	5	10	15	20	25	30	35	40	45	50

(10) Answer for first table: $a_5 = 22$, $a_6 = 25$, and $a_7 = 28$

(12) Answers for the first list: Equations: $a_4 = a_3 - 4$, $a_5 = a_4 - 4$, and $a_{10} = a_9 - 4$
Table: shown at right
Type: This pattern uses the previous term to create a new term so it is recursive.

n	a_n
0	⑰
1	⑰ − 4 = ⑬
2	⑬ − 4 = ⑨
3	⑨ − 4 = 5

(13) Answer for one equation in list c: The equation $c_1 = c_0(c_0 - 1)$ is written with parenthesis notation like this: $c(1) = c(0)(c(0) - 1)$.

Problems in Context.

(15) The table does not show a pattern at first glance. The graph, however, suggests considering a shuffled pattern. (See below.) Note that the points for the terms of the sequence have been joined by straight lines for visual emphasis.

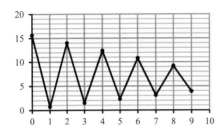

Returning to the table, observe that for even n the terms 15.6, __ , 14, __ , 12.4, __ , 10.8, __ , 9.2 consistently decrease by 1.6. Similarly there is a consistent pattern in the terms for odd n, __ , 0.7, __ , 1.5, __ , 2.3, __ , 3.1, __ , 3.9, increasing by 0.8. Assuming these patterns continue, the next three terms would be 7.6, 4.7, and 6.0. Had we used only the graph to predict the next terms, our predictions would be estimates. Reading data from a graph is inherently less accurate than reading from a table.

(17) a. This is the average temperature in Central Park on April 30.

b. 65

c. 63

d. $T_{25} = 68$

e. May has only 31 days so we know this is a day in the next month: June. On June 2nd the average temperature in Central Park was 70 degrees Fahrenheit.

(19) a. Two descriptions were discussed in the solutions previously. Here they are again using the new terminology. Recursive: Assuming that 1 is in position 1, each term is the sum of its position number and the preceding term. Direct: Each term is the sum of all natural numbers up to and including the position number. This sum can be found quickly using the formula:

$$\frac{(\text{position number})(1 + \text{position number})}{2}.$$

b. Again we will use the new terminology to describe the patterns explained in the previous solutions. Recursive: Each term can be found by adding the next triangular number to the previous term, with an initial term of 1. The requires keeping track of the triangular numbers. In the next section we will explore notation that is helpful in expressing this kind of relationship algebraically. Direct: Each term is the product of three numbers: the position number, one more than the position number, and two more than the position number. The product must then be divided by 6. For example, the 100th term is $\frac{100 \times 101 \times 102}{6}$.

c. We consider the position number to be equal to the number of tosses, so that the initial 2 is in position 1. Recursive: each term is twice the preceding term, with an initial term of 2. Direct: each term can be computed using the formula $2^{\text{position number}}$.

1.3 Exercises

Reading Comprehension.

(2) With difference equations you have to know the value of one term to find the next term. To start the process you have to know at least one term. So you need an initial term to get started. For functional equations, as soon as you know the position number, you can calculate the corresponding term directly. In some cases, an initial value is part of a functional equation, but this does not always occur.

(3) A difference equation will not satisfy the random stranger test unless it is accompanied by an initial value. If it is, then it should always satisfy the random stranger test if the stranger is algebraically literate.

(4) If the stranger is algebraically literate, a functional equation should satisfy the random stranger test every time.

Math Skills.

(6) a. $a_{10} = (-1)^{10} = 1$.
 c. $c_7 = 5 \cdot 7 - 8 = 27$.
 e. $u_2 = 2u_1 - u_0 = 50$. $u_3 = 2u_2 - u_1 = 25$. $u_4 = 2u_3 - u_2 = 0$.

(7) c. $c(n) = n \cdot (n+1)$
 d. $d_{n+1} = d_n + n$.
 f. $v_{n+1} = v_{n-2} + v_n$ is one possibility.

(9) a. recursive, $a_{n+1} = 5 + a_n$
 b. direct, $b_n = n^3$

(10) a. Assuming that the initial term is $a_1 = 81$, a functional equation is $a_n = 3^{5-n}$, and a difference equation is $a_{n+1} = a_n/3$. Not every student is expected to find the functional equation.
 c. Assuming that the initial term is $a_1 = 1$, a functional equation is $a_n = n^2$, and a difference equation is $a_{n+1} = a_n + 2n + 1$. Not every student is expected to find the difference equation.
 e. Assuming that the initial term is $a_1 = \frac{1}{2}$ a functional equation is $a_n = \frac{1}{2n}$. There is no particularly natural difference equation.

(11) a. Each term is 6 more than the preceding term. The first term is 4.
 b. $a_{n+1} = a_n + 6$.
 c. Each term is 2 less than 6 times the position number.
 d. $a_n = 6n - 2$
 e. The functional equation produces $a_8 = 46$. Or, using the difference equation, with $a_4 = 22$, we add 6 four times to reach $a_8 = 22 + 4 \cdot 6 = 46$. Directly counting the lines in a diagram that is eight squares wide and eight squares tall leads to the same result.

(13) a. $a_{n+1} = \frac{1}{3} a_n$, $a_0 = 144$.
 b. Using the difference equation, to go from $a_0 = 144$ to a_7, say, we have to divide by 3 seven times. This shows that $a_7 = 144/3^7$. Using the same logic, $a_n = 144/3^n$ for any n.

Problems in Context.

(15) a. Let's begin with t_0, the number of viable trees on the farm at the start of the model. We know there are 25,000 trees, but only 90% of those trees are viable. Thus $t_0 = 0.90 \times 25,000 = 22,500$. Moving on to t_1 we must subtract the number of harvested trees and add the number of newly planted trees which are viable:

$$t_1 = t_0 - 2,600 + 0.70 \times 5,000 = t_0 + 900 = 23,400.$$

Because the planting and harvesting are repeated each year, we can follow the same algebraic pattern to find $t_2 = 24,300$, $t_3 = 25,200$ and $t_4 = 26,100$.

b. $t_{n+1} = t_n + 900$, $t_0 = 22,500$.

c. $t_n = 22,500 + 900n$.

d. Some possible answers: the number of trees harvested will not be constant because after 8 years some of the new trees will start to be harvested. Similarly, the number of new trees planted might not be constant because as the number harvested increases, the number planted might also be expected to increase.

e. The existing trees are of various ages, and so have already survived several years. Assuming that the trees that fail do so at all different ages, we would expect the newest trees to include a higher percentage of future failures than the existing trees. In addition, the mortality rate for trees is probably highest when the trees are youngest.

(17) a. $v_3 = 10$, $v_4 = 15$.

b. One possible recursive description is that each term is the sum of the previous term and the next natural number. This is especially clear if we think of the first term, $v_1 = 3$, as the sum of the first two natural numbers, $1 + 2$. The next term is $v_2 = v_1 + 3$. One possible direct pattern is that each term is the sum of all the natural numbers up to and including one more than the position number. Using this description we see that $v_4 = 1 + 2 + 3 + 4 + 5 = 15$.

c. $v_5 = 21$, $v_6 = 28$.

d. $v_{n+1} = v_n + n + 2$ or $v_n = \frac{(n+1)(n+2)}{2}$. The functional equation is a famous shortcut for adding the first $n + 1$ natural numbers. You may have seen it in connection with problem 7g or problem 13 of Section 1.1.

(20) a. $b_{n+1} = b_n + 0.02 \cdot b_n = 1.02 \cdot b_n$ or $b_n = 1,000(1.02)^n$.

b. We are assuming there are no deposits or withdrawals from the account and that the interest rate does not change.

c. $c_{n+1} = c_n + 0.02 \cdot c_n + 200 = 1.02c_n + 200$.

(21) a. As these are the triangular numbers, we'll denote them t_n then a difference equation is $t_{n+1} = t_n + n + 1$, $t_1 = 1$ and a functional equation is $t_n = \frac{n(n+1)}{2}$. Compare with problems 17 and 18 of this section.

b. A functional equation is $b_n = \frac{n(n+1)(n+2)}{6}$ and we can use the triangular numbers from part a to build a difference equation: $b_{n+1} = b_n + t_{n+1}$, $b_1 = 1$.

c. We consider that the number of tosses is the position number, so that the terms can be written $c_1 = 2, c_2 = 4, c_3 = 8$, etc. A difference equation is $c_{n+1} = 2c_n$. A functional equation is $c_n = 2^n$.

2.1 Exercises

Reading Comprehension.

(4) A parameter is a letter representing a quantity that is constant within a problem (or part of a problem) but has different values in different problems (or different parts of problems). For example, in the straight line equation $y = mx + b$, the letters m and b are parameters. In discussing any particular line, these letters will be replaced by specific values. The equations $y = 3x+4$, $y = -x+1$, and $y = x$ are all instances of $y = mx + b$ and represent three different lines. Often, a parameter has a particular interpretation or meaning. For the equation $y = mx + b$, the parameters m and b represent the slope and the y-intercept of the line.

Mathematical Skills.

(5) a. This sequence is arithmetic with a common difference of $d = 2$ and an initial term $a_0 = 3$.

c. This sequence is not arithmetic. The difference for the first two terms is $4-1 = 3$, while the difference between the second and third terms is $16-4 = 12$. This shows that the given sequence does not have constant differences, and so is not arithmetic.

e. This sequence is arithmetic with a common difference of $d = -5$ and an initial term $e_0 = 150$.

f. This sequence is not arithmetic. We can see the difference between term 1 and term 2 is less than 5 while the difference between term 4 and term 5 is nearly 10.

(6) a. The initial term is $a_0 = 15 - 3 \cdot 0 = 15$ and the difference equation is $a_{n+1} = a_n - 3$.

e. The initial term is $e_0 = 10(0) - 3 = -3$ and the difference equation is $e_{n+1} = e_n + 10$.

(7) a. The functional equation is $a_n = 1 + 2n$.

e. The functional equation is $e_n = 70 - \frac{1}{2}n$.

(9) b. This sequence is easier to write equations for when viewed as $\frac{3}{12}, \frac{4}{12}, \frac{5}{12}, \frac{6}{12}, \frac{7}{12}, \cdots$. (Verify that this is the same sequence before going on.) The functional equation is $b_n = \frac{3}{12} + \frac{1}{12}n$ and the difference equation is $b_{n+1} = b_n + \frac{1}{12}$.

f. $f_n = 15 - 5n$, $f_{n+1} = f_n - 5$

(10) a. Based on the information provided, the functional equation is $a_n = 5 + 3n$. Using that, we can easily find $a_6 = 23$ and $a_{400} = 1205$.

c. The functional equation is $c_n = 325 - 12n$; $c_7 = 241$; $c_{100} = -875$.

e. Functional equation: $e_n = -12 + 2.5n$; $e_9 = 10.5$; $e_{250} = 613$.

Problems In Context.

(12) a. The terms p_0 through p_5 are 1000, 1500, 2000, 2500, 3000, 3500. The graph should show the points $(0, 1000), (1, 1500)$, etc.

b. After plotting the points of the sequence, we can create a line to graphically extend the pattern. The line reaches 9300 people just before month 17. Thus we can answer, it will take 17 months before at least 9300 people have been infected.

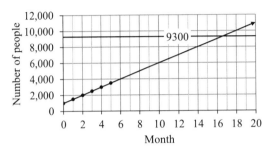

c. For a numerical approach we will make a table.

n	p_n
0	1000
1	1500
2	2000
...	...
15	8500
16	9000
17	9500

We can see that the first time the number of people reaches or exceeds 9300 is at the start of month 17.

(14) a. The terms d_1 through d_4 are 0, 10, 20, 30.

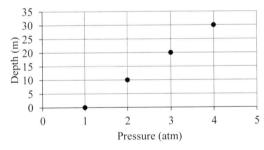

b.

c. Because the terms have a common difference of 10, this sequence is an instance of arithmetic growth. We are modeling ocean depth so it only makes sense to talk about pressure greater than the pressure at the surface of the ocean. The pressure at the surface is 1 atm and is the smallest useful value for n in this model. There will also be some practical upper bound to the depth. The greatest known depth in the ocean is about 11,000 meters in the Mariana Trench, as reported on the internet. It is reasonable to consider whether the arithmetic growth assumption of 10 meters for every 1 atm continues to hold to so great a depth. The internet reports a pressure of 1,070 atm at the bottom of the Mariana Trench. Our model says that a pressure of 1,101 atm

2.2. Exercises

should occur at a depth of 11,000 meters. Therefore the arithmetic growth assumption seems to be fairly accurate to the deepest point in the ocean.

d. $d_{n+1} = d_n + 10$ and $d_n = 10n - 10$ which can also be written $d_n = 10(n-1)$

e. The camera can safely go to a depth of 240 meters. The same may not be true of the diver! Search the internet to find the open sea diving depth record.

f. This requires us to solve $27 = 10n - 10$ and find $n = 3.7$ atms.

Digging Deeper.

(16) a. We can work backwards to find $a_0 = 12 - 4 = 8$. Thus $a_n = 8 + 4n$.

d. To find d we can think of adding d to e_2 repeatedly until we reach e_6. We would have to add d 4 times: $e_2 + 4d = e_6$. Knowing the values for e_2 and e_6 we find:
$8 + 4d = 52$,
$d = \frac{52-8}{4} = 11$.

Working backwards from $e_2 = 8$ we find $e_0 = 8 - 11 - 11 = -14$. Thus $e_n = -14 + 11n$.

An alternate approach: we know that the functional equation will have the form $e_n = e_0 + nd$. This must hold true when $n = 2$, so $e_2 = 8 = e_0 + 2d$. Similarly, with $n = 6$ we get $52 = e_0 + 6d$. These two equations can be solved simultaneously to find e_0 and d.

(17) a. The original and the resulting equation are equivalent.

c. The original and the resulting equation are not equivalent. For the original equation $x = -1$ is a solution. However, $x = -1$ is not a solution for $2x = x - 3$.

2.2 Exercises

Reading Comprehension.

(3) The assumption that initially the body temperature is 98.6°F might be incorrect. If the decedent was feverish at the time of death, that would alter the initial temperature. This in turn would change the predicted time of death. If the initial temperature is actually several degrees *higher* than 98.6, then our predicted time of death will be too late by how ever long it takes the body to cool down to 98.6.

(4) Function evaluation. To find a_3 all that is required is substitution and calculation. That is, we have to compute $2 \cdot 3 + 3$. This is exactly what function evaluation means.

(5) b. It would make sense for n to be continuous because we can certainly consider the fish population after a fractional number of years. On the other hand, the population size must be a whole number. Since p_n is in units of thousands, a whole number of fish corresponds to a p_n value with at most three decimal places. Thus, it is possible for p_n to be 3.146 or 3.147, but nothing between these two values. Therefore, p_n is a discrete variable.

Selected Answers to Exercises

Math Skills.

(6) a. Not at all arithmetic; these are perfect squares. The differences grow consistently larger with each successive term.

c. Approximately arithmetic with a differences between 2 and 3.

e. Approximately arithmetic.

(7) a. Using the numerical approach we compute terms as shown in the table below. We can see that for $n = 4$ we have $a_n = 34$.

n	0	1	2	3	4
a_n	10	16	22	28	34

b. This is a functional equation, so using the theoretical approach we substitute 7 for b_n and solve for n.

$$\begin{aligned} b_n &= 52 - 3n \\ 7 &= 52 - 3n \\ 7 - 52 &= -3n \\ \frac{-45}{-3} &= n \\ 15 &= n \end{aligned}$$

Problems in Context.

(8) a. $q_1 = 27.6$, $q_2 = 25.2$, $q_3 = 22.8$

b. $q_{n+1} = q_n - 2.4$

c. $q_n = 30 - 2.4n$

d. $q_4 = 20.4$. That is, the charge will be 20.4 units 4 hours after launch.

e. First solve $4 = 30 - 2.4n$ for n to find $n = 10.83$. The radio transmitter will continue to work for 10 hours. During the 11th hour the battery charge will fall too low to be of use.

(10) a. We find $h_{n+1} = h_n + 2,000$, $h_0 = 135,000$, and the number of years since 2004, n, must be less than or equal to some value such as 50.

b. $h_n = 135,000 + 2,000n$.

c. The model is represented by the line in the figure, with dots for every fifth term of the sequence.

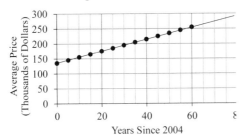

d. The year 2025 is 21 years after 2004 so we use $n = 21$ and find $h_{21} = 177,000$. That is, the average price of a new home in this county in 2025 is predicted to be $177,000.

e. We can solve $250,000 = 135,000 + 2,000n$ for n and find $n = 57.5$. According to the model the average price of a new home in this county in the middle of 2061 (57.5 years after 2004) will be 250,000.

f. Answers will vary. Here are some thoughts. The model predictions are plausible as long as the housing market is stable. Generally housing prices rise every year. We expect irregular changes in the housing prices for a variety of reasons. For example, a change in the community such as a new business hiring hundreds of people or a change in economic conditions such as a recession could produce irregular changes in home prices. Model predictions could be compared with actual data to find changes in the trend. The common difference might be changed based on new data or the type of model changed from arithmetic to something more sophisticated.

(12) a. After presentation of the model, we find the answer to this question: 167.6 degrees.

b. After about 18 and a half hours past 9 AM Standard Galactic Time on Sunday. To be safe, we'll keep it to only 18 hours, that is 3 AM Standard Galactic Time on Monday.

Digging Deeper.

(14) This can be argued conceptually, graphically, or algebraically. The greatest satisfaction comes from finding your own creative explanation. As a reference, here is one algebraic explanation. We have agreed to define d to be the average of all the differences between consecutive data values. We know the sequence values are a_n. Let's call the data points p_n and the largest subscript N. We can break down the differences between data points as follows: $d_1 = p_1 - p_0$, $d_2 = p_2 - p_1$ and so on with the last being $d_N = p_N - p_{N-1}$. We write the average of these values as $\frac{d_1 + d_2 + \cdots + d_N}{N} = d$. With some multiplication and substitution we see: $Nd = (p_1 - p_0) + (p_2 - p_1) + (p_3 - p_2) + \cdots + (p_{N-1} - p_{N-2}) + (p_N - p_{N-1})$. Most of the terms on the right cancel out and we are left with $Nd = p_N - p_0$. Notice that this is not necessarily true for every n, but it is true for the largest subscript N. We agreed that $a_0 = p_0$ exactly, so by adding and substituting we see $a_0 + Nd = p_N$ and recognize from the functional equation that $a_0 + Nd$ is also a_N. So $a_N = p_N$.

(16) Before answering the question, let's be sure we understand it. We are thinking about values for n that are not whole numbers. For example, $n = 2.75$ is three quarters of the way through the second day. The function $f(2.75) = 10 \cdot 2.75 = 27.5$ predicts a fine of 27.5 cents. In fact, the fine would be 20 cents. The function results in an error of $27.5 - 20 = 7.5$ cents. We can see this on a graph as the gap between the graph of the function and the graph of the library fine at a particular n value. We can see that the gap never exceeds 10 cents. Depending on how we choose to round when interpreting the results of the function, we may say the maximum error is 9 cents.

2.3 Exercises

Reading Comprehension.

(1) See page 82.

(4) An equation for h in terms of T can be used to find height given temperature. To find this equation, begin with the given equation for T, and use algebraic steps to

isolate or *solve for h*. To be specific, subtract 72 from both sides of the equation, and then divide both sides by 0.01 to find $h = 100T - 7200$.

(5) a. The slope is a number indicating how steeply a line rises or falls. It is computed using any two points of the line, as the difference in *y*-coordinates divided by the difference in *x*-coordinates. For example, if the points are (3, 1) and (7, 4), the slope is given by $\frac{4-1}{7-3} = \frac{3}{4}$.

Moving between the points from left to right, the slope can be described as the rise over the run, where the rise is how many units higher the second point is compared to the first, and the run is how many units the second point is to the right of the first. The rise is a negative number when the second point is lower than the first.

c. Which ever point is considered to be first, the same result for slope is found. If the computation results in $\frac{2}{3}$ with one order, then it will produce $\frac{-2}{-3}$ in the other order, and we know $\frac{-2}{-3} = \frac{2}{3}$.

(6) a. An intercept is a point where the graph of an equation intersects with a coordinate axis. An *x*-intercept is a point where the graph intersects with the *x*-axis. It is a point of the graph where $y = 0$. Similarly for a *y*-intercept.

b. There are two cases. If the slope of the linear function is not zero, then the graph is a straight line that is neither horizontal nor vertical. In this case there is exactly one *x*-intercept and one *y*-intercept. If the line passes through the origin then the *x*- and *y*-intercepts are the same point, (0, 0). For the second case, when the slope is zero, the linear function is actually a constant function, and the graph is a horizontal line. This line may be the *x*-axis, and then every point is an *x*-intercept and the origin (0, 0) is a *y*-intercept. Otherwise, a horizontal line has a single *y*-intercept and no *x*-intercept.

(8) We have seen that a linear equation with one variable can be solved by using algebraic operations to isolate the variable. It is possible, though, that these operations will eliminate the variable. For example, the equation $3(x + 2) + x = 4x - 11$ reduces to $6 = -11$. In this case, because the resulting equation is false no matter what value we consider for *x*, we see that the original equation has no solution. On the other hand, it may happen that after eliminating the *x* values the resulting equation is true. An example is $3(x + 2) + x = 4x + 6$, which reduces to $6 = 6$. When this happens, the original equation is true no matter what value we consider for *x*, so that the original equation is true for every *x*. That is how an equation may have an infinite number of solutions.

Alternate Answer: A linear equation in one variable, say *x*, can be viewed as taking the form $f(x) = g(x)$ where *f* and *g* are linear functions. Now consider the graphs of these functions, which are straight lines. At any point of intersection of the lines, we must have $f(x) = g(x)$, so the corresponding *x*-coordinate is a solution to the original linear equation. If the two lines are actually the same line, then every point is on both lines and there are infinitely many solutions to the original equation. If the lines are not the same, but are parallel, then there are no points of intersection so the original equation has no solutions.

2.3. Exercises

Math Skills.

(10) b. Taking the square root of an expression involving a variable is not allowable in linear equations.

(11) a. $x = 9$.

c. $x = -1.5$.

e. Non-linear due to square root of x.

g. $x = 20.625$.

i. This is a linear equation but has no solutions. It reduces to the equation $-40 = -36$, which is false no matter what value is assigned to x.

(12) a. $f(x) = 1.2x + 3.4$

c. Non-linear due to square root of x

e. This equation appears to be non-linear due to the x^2 term. However, after algebraic simplification it becomes $y = -5.7x + 12.4$, and so is linear.

g. $y = \frac{2}{3}x$

l. $y = x + 1.6$

(13) a. No solution

c. Every number is a solution

e. $x = 13$

(14) a. $y = 10$

c. $a_5 = 7$

(15) a. Start at the y-intercept, mark 1 on the y-axis. Then move up 3 and right 1. Mark the second point (1, 4). Continue moving up 3 and right 1 then marking points.

(16) a. $y = 3x - 3$

(17) a. slope-intercept form. To graph begin at the y-intercept, in this case -4, proceed as dictated by the slope, in this case up 3 and right 1.

b. point-slope form. To graph begin at the point, in this case (3, 2), proceed as dictated by the slope, in this case (converting 0.5 to 1/2) up 1 and right 2.

e. two-intercept form. To graph, mark both intercepts, in this case (0, 4) and (2, 0), then connect these points.

(18) a. $\frac{x}{5} + \frac{y}{3} = 1$

c. $y = 4x - 6$

e. $y - 7 = -\frac{1}{3}(x - 4)$

Digging Deeper.

(20) In this course we consider only real number solutions, that is, numbers that correspond to points on a number line. For even n, if $y^n = x^n$ then it is also true that $(-y)^n = x^n$. In these cases it is clear that $y^n = x^n$ when $y = \pm x$. This shows the equations $y^n = x^n$ and $y = x$ cannot be equivalent for even n. For odd n the situation is different. Observe that $y^n = x^n$ when $x = y = 0$, or for nonzero x, when $(y/x)^n = 1$. With odd n, this is only possible if $y/x = 1$. We can see that $(y/x)^n$ cannot equal 1 if y/x is negative, or if $0 < y/x < 1$, or if $1 < y/x$. In this way we conclude that when n is odd, the two equations $y^n = x^n$ and $y = x$ are equivalent.

The foregoing remarks hinge on the fact that $z^n = 1$ only when $z = 1$. However, this is not the whole story! The equation $z^n = 1$ actually has n solutions if we only know where to look. To find them, we must move beyond the real numbers to a larger number system, traditionally referred to as the *complex numbers*. While outside the purview of this course, there is plenty of interesting reading available on this topic.

(22)   a. $y = \frac{-3}{2}x + 3$

    c. Solving the two-intercept equation for y we find $y = \frac{-b}{a}x + b$. This equation has a y-intercept of b, as we would expect. Separately we use the two intercepts to compute the slope $m = \frac{0-b}{a-0} = \frac{-b}{a}$, which again agrees with the slope we found by converting the two intercept form into the slope-intercept form.

    d. The equations for lines with a 0 intercept for either x or y cannot be written in two-intercept form. The equations for vertical lines, such as $x = 1$, cannot be written in slope-intercept form. For all other lines, the process developed in the preceding parts of this problem will work. There is a little more to be said here. Writing the equations for vertical and horizontal lines in two-intercept form is not covered in the reading for this section. But there is a nice extension to these exceptional lines. The equations can be written if we simply omit the variable which has no intercept. To be specific, vertical lines have an equation of the form $\frac{x}{a} = 1$ (unless the x-intercept a is zero), and horizontal lines have one of the form $\frac{y}{b} = 1$ (unless b is zero).

(24)   a. The inverted equation, $\frac{c-381}{2} = n$, gives the number of years since 2006, n, necessary to reach a level, c, of CO_2.

    b. We can easily find $n = f/10$ giving the number of days, n, a book is late based on the amount, f, of the fine. However, we must be a bit careful. Library fines are only imposed in multiples of ten cents. It is possible to have a fine of 30 cents or of 40 cents, but not of 35 cents. Therefore, the equation $n = f/10$ is only valid when f is a multiple of 10, in which case the computed value of n will be a whole number.

As discussed on page 76, we can certainly consider fractional values of n. That is, we understand what it means to return a book 3.5 days after the due date. But when we do, we have to recognize that the same fine will be imposed for $n = 3.5$ or $n = 3.6$ or any other value of n between 3 and 4. In this context, the equation $f(n) = 10n$ is not valid for fractional values of n. Moreover, it

2.4 Exercises

doesn't make sense to try to express n as a function of f, because there will be infinitely many valid n's for each valid f. For example, if you are told someone paid a fine of 40 cents, you can conclude that the book was returned sometime between the closing times on the third and fourth day after the due date, but you can *not* determine the specific value of n.

2.4 Exercises

Reading Comprehension.

(3) a. Assuming the validity of proportional reasoning, for a length of 350 mm, exacly half way between the given lengths of 300 and 400 mm, we should expect a weight of 460 grams, because 460 is exactly half way between 290 and 630.

b. Assuming that proportional reasoning is not valid, a graph of all the length and weight data will not appear to fall on or near to a straight line. Similarly, there is no linear equation relating the length and weight of the trout in the study, either exactly or approximately.

Problems In Context.

(4) a. There are several possible forms of the equation. One is $60.58x + 46.02y = 5,000$. The company should purchase 29.4 short tons from the first supplier.

c. First we'll define variables: let c be the number of acres of corn and b be the number of acres of soybeans. If she cultivates 100 acres of corn, she has enough land to cultivate up to $350 - 100 = 250$ acres of soybeans. Can she afford to do this?

We approach this by formulating a linear equation relating b and c, subject to the fact that the farmer has only \$80,000 available. The given information tells us the two intercepts, so we can immediately find the equation $\frac{c}{210} + \frac{b}{408} = 1$. Any choice of b and c that makes the equation true corresponds to using up all the available funds for planting.

Using the equation with $c = 100$ leads to $\frac{100}{210} + \frac{b}{408} = 1$. Solving for b, we find $b = 213.7$, which we see is within the 250 acres available for soybeans. With this plan, the farmer would leave about 36 acres of land unused. From the given information, we can also find that planting these additional acres in soybeans would cost an extra \$7,058.

If the farmer cultivates 50 acres of corn, she has enough land to cultivate up to 300 acres of soybeans. We again use our equation, this time with $c = 50$, obtaining $\frac{50}{210} + \frac{b}{408} = 1$. Solving for b reveals that the farmer can afford to plant $b = 311$ acres of soybeans. But that exceeds the available land. So if she plants 50 acres of corn, she should plant 300 acres of soybeans, which will not completely use the \$80,000 available.

Although the problem does not ask this question, it is reasonable to ask how many acres should be planted of each crop so that all of the available land and money are used. If the farmer uses all the land, then $c + b = 350$, so for any c, we have $b = 350 - c$. Substituting this in our equation leads to

$$\frac{c}{210} + \frac{350-c}{408} = 1.$$

This is a linear equation in a single variable, and can be solved as follows. Multiply on both sides by $210 \cdot 408$ to find

$$408c + 210(350 - c) = 210 \cdot 408.$$

This simplifies to

$$408c - 210c = 210 \cdot 408 - 210 \cdot 350 = 12{,}180.$$

Therefore $c = 12{,}180/198 = 61.5$ acres, and $b = 350 - 61.5 = 288.5$ acres.

(6) We introduce the variables h and t representing the height from which she throws the paper plane and the time the plane takes to reach the ground, with h in units of feet and t in units of seconds. We are given that $t = 8$ when $h = 4$, and by logic, $t = 0$ when $h = 0$. This gives us two data points, and assuming a linear model, that is enough to find an equation relating t and h. The question we want to answer gives information about h and asks us to find t, so in our model we will express t as a function of h. That means our data points will be expressed in the order (h, t) and are given by $(4, 8)$ and $(0, 0)$. The slope is therefore $m = (8 - 0)/(4 - 0) = 2$. The intercept is $b = 0$ because we know that $(0, 0)$ is one point on our line. These observations lead to the linear equation $t = 2h$. Now normally Abigail's shoulder is four feet above the ground. But if she stands on an 18 inch stool her shoulder will be raised by 18 inches, which is the same as 1.5 feet. This shows that $h = 5.5$, and our equation produces $t = 2 \cdot 5.5 = 11$. According to the model, when Abigail stands on the stool and throws the paper plane, it will take 11 seconds to reach the ground.

Reflections: One objection to the way this problem is stated is that the length of time it takes a paper plane to reach the ground depends heavily on how the plane is thrown. It will stay aloft longer if thrown slightly uphill than if it is thrown slightly downhill. Similarly, it will descended faster if simply dropped, rather than thrown. Unless Abigail is extremely careful to throw the plane exactly the same way each time, it is unlikely that the time to reach the ground will be completely determined by how high the plane is when thrown. In fact, we expect that different times will be observed if the plane is thrown repeatedly from the same height. Another significant limitation is that there is no justification given for assuming a linear model. With several data points, we might create a graph and see whether a linear model appears to be valid. But here we have only two data points, and that doesn't give us any basis for accepting or rejecting a linearity assumption.

(7) a. Let T represent temperature, in degrees Fahrenheit, and P represent pressure, in units of psi. We are given two data points, $(T, P) = (70, 15)$ and $(T, P) = (247, 20)$. We compute the slope as $m = (20 - 15)/(247 - 70) = 5/177$. The point-slope formula now gives

$$P - 15 = \frac{5}{177}(T - 70).$$

This is a linear equation relating P and T.

b. Setting $T = 150$ our equation becomes $P - 15 = \frac{5}{177}(150 - 70)$, which simplifies to $P = 15 + \frac{5}{177} \cdot 80 = 17.26$. That is, we predict the pressure is 17.26 psi when the temperature is 150 degrees.

c. In the equation above, neither variable is a function of the other, because neither variable has been isolated. On the other hand, if we rewrite the equation as
$$P = 15 + \frac{5}{177}(T - 70).$$
then P is isolated, so it is expressed as a function of T.

(9) a. The model assumes the number of customers will continue to grow at the about the same rate as it grew from 1998 to 2014.

b. The assumption of a constant growth rate is probably reasonable in the short term, as long as conditions do not change suddenly. The given data points are fairly consistent with a linear model. There are many possibilities that could cause a sudden change of conditions. For example, in some areas, consumers can choose between two or more providers of electricity. If a new electric company starts up in the town in question, that would probably change the rate at which customers enroll with the company of the study. Also, the number of customers depends significantly on the population size for the town. If the population experiences a growth spurt or a sudden decline, that would also be expected to change the growth rate for new customers.

c. The graph shows the company will reach their limit when n is a little less than 29. This is confirmed using the equation, $30{,}000 = 22{,}850 + 250n$ which can be solved to find $n = 28.6$. This corresponds to year $1996 + 28.6 = 2024.6$ or about half way through 2024.

d. Compare new data with predictions made by the model. If they do not match, update the model.

(11) a. $p = \frac{5}{11}d + 15$
b. 60.45 psi
c. 2,167 feet

(13) One form of the model is $p = \frac{-5}{2}(n - 10) + 125$ where p is the price in dollars and n is the number of bags sold.

Digging Deeper.

(14) There were 1277 adults and 3123 children.

(16) a. 6 seconds
b. $L = 10m$, $L = \frac{1}{6}s$, and $L = 600h$
c. It would take her 1.84 hours.

3.1 Exercises

Reading Comprehension.

(2) The parameters d and e each represents a quantity that is constant within a problem (or part of a problem) but has different values in different problems (or different parts of problems). For example, when we use the quadratic growth difference equation $a_{n+1} = a_n + d + en$ in a particular problem, the d and e will be replaced

with numerical values that are constant for that part of the analysis. In contrast, n and a_n are variables and will take on many different values within the same problem or problem part.

(4) The functional equation has the form $a_n = a_0 + dn + en(n-1)/2$, which can be algebraically rewritten as $a_n = a_0 + dn + en^2/2 - en/2$. The quadratic term $en^2/2$ cannot be algebraically eliminated, and shows that the functional equation is not linear. The graph of a quadratic growth sequence reveals nonlinearity because it is in the form of a parabola rather than a straight line.

Mathematical Skills.

(6) a. The next three terms are $-6, -15, -26$
   b. The next three terms are 344, 513, 730. This is not quadratic growth, but there is a simple pattern in the second differences column that can be used to extend the sequence.

(9) a. The sequence begins 6, 6.5, 7.2, 8.1, 9.2.
   b.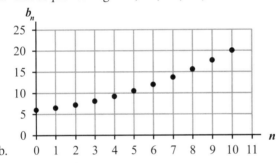
   c. $d = 0.5$ and $e = 0.2$
   d. $b_n = 6 + 0.5n + 0.2(n-1)n/2$, $b_4 = 9.2$, $b_{10} = 20$

(13) a. $a_n = 105 - n + 4n(n-1)/2$;
     $a_{20} = 105 - 20 + 2 \cdot 20 \cdot 19 = 845$
   b. $n = 29$

(14) a. We can use the initial first difference $d = 0.5$ and the common second difference $e = 2.4$ in various ways to find $a_{16} = 306$.
   b. $n = 20$

(17) a. Quadratic with $a_{n+1} = a_n + 2 + 1n$ and $a_n = 3 + 2n + 1(n-1)n/2$
   b. Neither

(18) b. This cannot be quadratic growth because the graph is a straight line.

(19) $50 \cdot 51/2 = 1{,}275$

Problems in Context.

(22) a. Computing the first and second differences for the terms e_1 through e_4 we find constant second differences. Based on that pattern, we can extend the table to

3.1. Exercises

e_5 and e_6, as shown in the table below.

n	e_n	1st Difference	2nd Difference
1	3		
		6	
2	9		3
		9	
3	18		3
		12	
4	30		3
		15	
5	45		3
		18	
6	63		

b. $e_5 = 45, e_6 = 63$.

d. Had the table above started with $n = 0$, we could have used it to construct a functional equation. The table does not include that term because it does not make sense in the context of triangles. However, for purely algebraic purposes, we can extend the pattern of the table to $n = 0$, keeping the constant second difference e equal to 3. Then d, the first of the first differences, must also be 3, and e_0 has to be 0. These parameter values lead to the functional equation $e_n = 0 + 3n + 3(n-1)n/2$ and we find $e_{30} = 1395$. (Note it is incorrect to interpret the top entries of the columns of our original table as e_0, d, and e. Doing so leads to the erroneous functional equation $e_n = 3 + 6n + 3(n-1)n/2$. By substituting numerical values for n we can see this equation produces incorrect values for e_n. More specifically, substituting any value for n produces not e_n but rather e_{n+1}. In particular, this incorrect formula gives e_{30} as 1,488, which is actually e_{31}.)

(23) It is possible to analyze this problem using a quadratic growth sequence. However, a more direct method is to use the formula for the sum of the first 64 natural numbers: $64 \cdot 65/2 = 2,080$ pieces of gold.

Digging Deeper.

(26) a. 120, 165, and 220

b. The first differences are $3, 6, 10, 15, 21, 28, 36, \cdots$. The second differences of that sequence (which are the third differences of the tetrahedral numbers) are constant. Alternatively, these first differences can be recognized as the sums $1 + 2, 1 + 2 + 3, 1 + 2 + 3 + 4$, and so on. We know that these sums are given by a quadratic function, and so comprise a quadratic growth sequence.

c. The first differences are listed above. Let's call them d_n. As they are a quadratic sequence, we can find a functional equation for them. However, it should be noted that $d_1 = 3$ and to construct the functional equation we need to know d_0. Once that is sorted out, we can find the difference equation for the tetrahedral numbers, $t_{n+1} = t_n + n^2/2 + 3n/2 + 1$.

(28) a. $a_2 = 36\frac{2}{3}$ and $a_3 = 56$

b. Given three terms from a quadratic growth sequence, it is always possible to find the functional equation. We know that will have the form $a_n = a_0 + dn + en(n-1)/2$. Each given term of the sequence has values of n and a_n that can be substituted into this equation, leaving a_0, d, and e as unknowns. For example, if we are told that $a_5 = 11$, that gives us the equation $11 =$

$a_0 + 5d + 10e$. Notice that it is a linear equation. If we have three given terms, we obtain three linear equations in the three unknown parameters. The equations can be solved simultaneously to determine the parameters, and hence the functional equation.

(30) $68 \cdot 69/2 - 22 \cdot 23/2 = 2{,}093$

3.2 Exercises

Reading Comprehension.

(1) If the second differences are constant or nearly constant; if the graph shows points that follow a parabolic curve, or are distributed around such a curve.

(3) Proportional reasoning is not valid for quadratic growth sequences. In general, proportional reasoning about a sequence is only valid when the graph of the sequence is a straight line, or closely approximated by a straight line. Quadratic growth sequences have graphs that are parabolic, not linear, so proportional reasoning is not valid for them. An example is provided by the computer network problem (see page 152), where we found that the number of lines necessary to connect n computers is given by $t_n = (n-1)n/2$. For example, $t_{10} = 9 \cdot 10/2 = 45$, and $t_{20} = 19 \cdot 20/2 = 190$. Here we see that adding 10 computers results in an increase of 145 in the necessary number of connecting lines. Using proportional reasoning, we would expect that every additional increase of 10 computers would likewise require an additional 145 lines. Thus, we expect to find $t_{30} = t_{20}+145$, $t_{40} = t_{30}+145$, and $t_{50} = t_{40} + 145$. This leads to the prediction that $t_{50} = 190 + 3 \cdot 145 = 625$. But the actual value is $t_{50} = 49 \cdot 50/2 = 1{,}225$, almost twice as much as the estimate. This shows dramatically that using proportional reasoning in a quadratic growth model can be extremely inaccurate.

Math Skills.

(7) $n = 78$

(9) a. Remembering to start with $a_0 = s_0 = 5$ and noticing $d = 8$ and $e = 3$ we have $s_n = s_0 + dn + e(n-1)n/2 = 5 + 8n + 3(n-1)n/2$.

b. Observe that the original sequence $5, 8, 11, \cdots$ is an instance of arithmetic growth, so the functional equation for the sum sequence can be obtained directly from (3.13). We first formulate the functional equation $a_n = 5 + 3n$, corresponding to (3.12), showing that $e = 3$. We also know that $a_0 = 5$ and $a_1 = 8$. Therefore, (3.13) becomes $s_n = 5+8n+3n(n-1)/2$, the same equation we found in part a.

c. First, we identify 155 as a_{50}, either by a numerical method, or by solving the functional equation $a_n = 155 = 5 + 3n$ for n. Then we recognize that the desired sum is s_{50}, which we compute as 4,080 using the functional equation for the sum sequence.

Problems in Context.

(11) a. Let a_n be the total number of feet fallen n seconds since Johnny began recording data. We should also extend the table to include $n = 0$. It makes sense

3.2. Exercises

to say after 0 seconds the ship would have fallen 0 meters; $a_0 = 0$. There are nearly constant second differences, so we will proceed with a quadratic model. The initial first difference (for a table starting with $n = 0$) is $d = 0.81$. Because the second difference is not exactly constant we have a choice as to the parameter e. Using the average of the second differences is a reasonable approach: $e = 1.615$. This leads to $a_{n+1} = a_n + 0.81 + 1.615n$ and $a_n = 0 + 0.81n + 1.615\frac{n(n-1)}{2}$.

c. Answers will vary. For the model above, a little over 35 seconds.

(13) a. The diagram is essentially the same as Figure 3.11 (page 154), with K and Q representing the server computers, and the other three dots representing user computers. However, one additional line joining K and Q must be added to the figure. 19 wires are needed

b. $T_0 = 4, T_1 = 8$, and $T_2 = 13$

c. $T_{n+1} = T_n + 4 + n$

d. $T_n = 4 + 4n + n(n-1)/2$

e. 5,354 connection wires

f. 220 users

(15) Using a recursive analysis, suppose that n teams have been scheduled, and one additional team arrives. That new team must play against each of the original n teams, requiring n new games. This shows that the number of games required for $n + 1$ teams is n more than the number for n teams. So, if t_n represents the number of games for a tournament with n teams, we have $t_{n+1} = t_n + n$, which is a quadratic growth difference equation with $d = 0$ and $e = 1$. For an initial condition, we note that $t_2 = 1$. Then the difference equation leads to $t_3 = t_2 + 2 = 3$, $t_4 = t_3 + 3 = 6, t_5 = t_4 + 4 = 10$, and so on. This information can be entered into a table, with first and second differences, starting with $n = 2$. Then, continuing the table's pattern upward to $n = 1$ and $n = 0$, we find that $t_1 = t_0 = 0$. Thus $t_n = 0 + 0n + 1(n-1)n/2 = (n-1)n/2$, and we can calculate that $t_{10} = 45$ and $t_{20} = 190$. This is another good example of the fact that proportional reasoning does not apply to non-arithmetic sequences. The number of teams doubled, but the number of games more than quadrupled.

(17) **Answer:** We can follow the same analysis as used in the House Construction example. Let n be the number of years since the company started logging (so the first year is $n = 0$), A_n be the number of acres logged year in year n, and s_n be the total number of acres the company has logged through year n. By assumption, the sequence A_n is arithmetic: $A_n = 1,200 + 200n$. A difference equation for the sum is $s_{n+1} = s_n + A_{n+1} = s_n + 1,200 + 200(n + 1) = s_n + 1,400 + 200n$ and the total number of acres logged in the first year is $s_0 = 1,200$. Now we can construct a functional equation: $s_n = 1,200 + 1,400n + 200(n-1)n/2$. Since the first year of operation corresponds to $n = 0$, for the tenth year $n = 9$, and we find $s_9 = 21,000$. Finally, $s_n = 50,000$ for n approximately equal to 16.5. This means the company will have logged the entire tract of land after about seventeen and one half years. An alternative approach is to let $n = 1$ for the first year of operation. The analysis will then look a little different, but it leads to the same answers.

Digging Deeper.

(20) a. Because $1 + 3 + 5 + 7 = 16$ it will take 4 quarters of a second, or 1 second to fall 16 feet.

b. 7 feet

d. It will take 8 quarter seconds, which is 2 seconds. The diver falls only twice as long. The diver is falling just over twice as fast (60 feet per second instead of 28 feet per second) in the last quarter second.

3.3 Exercises

Reading Comprehension.

(2) The expressions define the same function if, for every value of the variable(s), both expressions produce the same result. This situation can be recognized if one expression can be algebraically manipulated into the other expression. Alternatively, if the two expressions produce identical graphs, they define the same function (at least for the domain values represented in the graph).

(4) If they have made no errors, the two functions are actually the same function. It is a special property of quadratic functions that agreement for three or more values implies the functions are the same.

(6) The list of $f(x)$ values has constant second differences. This always occurs in a table of values for a quadratic function provided the x values are regularly spaced. In this example, the x values are consistently 0.4 units apart.

(8) Hasan and Hongwei are using different variants of proportional reasoning. These are not appropriate with quadratic growth. Esperanza's logic is also not valid for quadratic growth. This can be verified by applying her method for several repetitions. Expressing the results in $100,000$s, this produces the sequence 1, 7, 49, 343, 2,401, $\cdots$. Checking second differences shows that this sequence is not at all like quadratic growth. Natasha's prediction is the most consistent with quadratic growth. As explained in the reading, for a quadratic function $f(x)$, if you double the size of x, you should expect to roughly *quadruple* the size of $f(x)$.

Mathematical Skills.

(9) Only graphs a and e show the symmetric curved shape with a single maximum or minimum of a quadratic graph.

(10) a. This is not a quadratic function because it includes a term equal to $x^2 \cdot x^2 = x^4$ that cannot be eliminated by algebraic simplification. Therefore the given expression cannot be written in the form $ax^2 + bx + c$.

c. This is not a quadratic function because x^2 is in the denominator.

(11) a. $-3x^2 - 3x + 7$

c. $2x^2 + x - 15$

(12) a. $x = \pm 1.83$

c. $x = 1.26$ or $x = -0.26$

3.4. Exercises

(13) a. Factored form. $x = 2$ is the solution for each of the factors and is the only solution for the equation.

c. Quadratic formula. $\dfrac{-3 \pm \sqrt{229}}{22}$

(14) a. Rewrite as $3x^2 - 7x - 5 = 0$. Quadratic formula: $x = \dfrac{7 \pm \sqrt{109}}{6}$. Since 109 is not a perfect square, the square root is irrational.

c. no real solutions

(15) a. Using the quadratic formula (or factoring) we find the x-intercepts are $x = 3$ and $x = 1$. The constant coefficient, in this case $c = 15$, is the y-intercept. The coefficient of x^2, in this case $a = 5$, is positive so we know the orientation is $\smile$. The center line is a vertical line at $x = -b/2a$ in this case $x = 2$. The vertex is the point on the parabola where the parabola crosses the center line. We already know $x = 2$, substituting that into the equation we find $y = -5$. Thus the vertex is $(2, -5)$. All this is verified by the graph below.

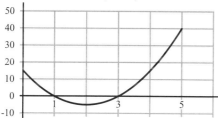

(16) a. $(x + 1)(x + 2)$

c. $(x + 0.5)(x + 8)$

e. $5(x + 1/5)(x - 7) = (5x + 1)(x - 7)$

(17) a. $10(x - 17)(x - 23) = 0$

Digging Deeper.

(18) Let's begin by considering when we can use the quadratic formula. It requires an equation with 0 on one side. This can be obtained by subtracting y from both sides of the given equation: $0 = x^2 - 4x - y$. Now we have a quadratic in standard form with coefficients $a = 1$, $b = -4$, and $c = -y$. The quadratic formula gives us $x = \dfrac{-b \pm \sqrt{b^2 - 4ac}}{2a}$. In this case $x = \dfrac{4 \pm \sqrt{16 + 4y}}{2} = 2 \pm \sqrt{4 + y}$. We can restate this information as follows: the inverse of the function $y = x^2 - 4x$ is $x = 2 \pm \sqrt{4 + y}$. It is also worth noting that for this inverse x is not a *function* of y because it has two values.

3.4 Exercises

Reading Comprehension.

(1) Revenue is the income from an activity or business. Profit is what remains when expenses are subtracted from the revenue. This can be a negative quantity. A break even point is one at which the revenue and expenses are equal. At a break even point there is neither a profit nor a loss.

(3) One significant limitation of a linear demand model (with a negative slope) is this: such a model will produce negative values for demand for sufficiently high prices. This is evidently invalid, because demand cannot be a negative quantity. A linear model is also questionable for very low values of price. For moderate prices, the demand may well vary in an approximately linear fashion. But if the price is made too low, it is reasonable to expect the demand to increase disproportionately. For example, at a public event, demand may be limited for a fairly priced food item, such as a hotdog. But if the price is set far below the item's value, say charging a nickel for a hotdog, then almost everyone will want one.

Problems in Context.

(5) a. To maximize profits, they should charge $1.34 for a cup of yogurt to attain $9,090.00 in profit (or 90.9 hundred dollars).

b. The range of profitable prices is $0.05 to $2.63.

(7) The new models are $C = 0.25(-186.7p + 264) + 25 = -46.675p + 91$ and $P = (-186.7p^2 + 264p) - (-46.675p + 91) = -186.7p^2 + 310.675p - 91$. The price that should be charged per drink is $p = 0.832$ and is predicted to result in $P = \$38.24$ average profit per day. The price is the same as in the original model and the predicted profit is $25 lower. So in effect, the extra flat fee comes straight out of the profits. As in the original model, we would probably choose to charge $0.80 or $0.85 for ease of making change.

(9) a. $s = -2p + 900$

b. $R = ps = p(-2p + 900) = -2p^2 + 900p$

c. The greatest revenue will result from charging a price of $225. The maximum revenue is $101,250 which is $31,250 more than the current revenue.

d. We can use this information to create a profit model: $P = R - C = ps - 175s = -2p^2 + 900p - 175(-2p + 900) = -2p^2 + 1250p - 157,500$. The maximum profit (of $37,812.50) results from a price of $312.50.

Digging Deeper.

(11) a. $R = ps = p(92p^2 - 355.12p + 337.69) = 92p^3 - 355.12p^2 + 337.69p$
$C = 0.25s = 0.25(92p^2 - 355.12p + 337.69) = 23p^2 - 88.78p + 84.4225$
$P = R - C = 92p^3 - 355.12p^2 + 337.69p - (23p^2 - 88.78p + 84.4225) = 92p^3 - 378.12p^2 + 426.47p - 84.4225$

b. At a price of approximately $0.63 there is a maximal revenue of $94.80. This is nearly identical to the maximal revenue found in the original model, though at a slightly lower price. In both cases, it is likely that the students would charge a more convenient price than the one that is theoretically optimal.

c. At a price of approximately $0.79 there is a maximal profit of $61.86. Again, a more convenient price would likely be charged. The maximal profit results of using a quadratic model for demand are practially the same as the results of using a linear model for demand. The graph below shows the profit curves

4.1. Exercises

based on both the original linear demand model and the new quadratic demand model introduced in this problem. Up to a price of 1.15 they are very close to each other.

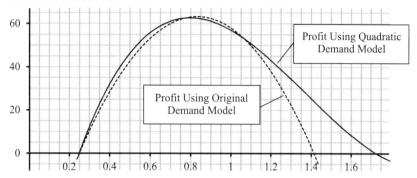

d. Answers will vary. One key point is that in both cases the two models agree then diverge. The further away from the data we move, the larger the difference in predictions. We considered a large range of times for the ice model and so found divergent results. We considered a small range of prices for the soda model and so the two model results remained more similar. Moreover, the goals for the two models are different. For the sea ice problem the goal is to predict what may happen years or decades in the future. In contrast, for the sales model, we wish to find the best selling price, and that occurs in the center of our data range.

4.1 Exercises

Reading Comprehension.

(1) a. In a geometric growth sequence, each term is obtained by multiplying the preceding term by a positive constant. If the constant is less than 1, the terms of the sequence decrease. While this can still be considered to be geometric growth, it is also sometimes referred to as geometric decay.

b. A geometric decay sequence is a geometric growth sequence in which the terms actually decrease in size, or in the case of a sequence with negative terms, decrease in absolute value. That is, geometric decay is geometric growth with a constant multiplier that is between 0 and 1.

(2) a. A geometric growth sequence is one in which each term is a contant multiple of the preceding term.

b. The general form is $a_{n+1} = ra_n$ where r represents a positive constant. For example, if $r = 1.3$ then the difference equation becomes $a_{n+1} = 1.3a_n$. This says that each term is a constant multiple of the preceding term, and so is a direct representation of the definition.

(4) a. With an initial term of 150, the curve would be shifted upward and become steeper. The new y-intercept would be 150, rather than 10.

c. This change has the effect of flipping or reflecting the curve across a vertical axis. On the original curve, each unit moved to the right doubles the height of the curve above the horizontal axis. In the new curve, each unit moved to the *left* doubles the height.

Math Skills.

(8) a. This sequence has a constant ratio of 1.3, meaning 2.6/2, 3.38/2.6, 4.394/3.38, etc., all equal 1.3. Therefore the sequence is geometric with a difference equation of $a_{n+1} = 1.3a_n$ and a functional equation of $a_n = 2(1.3)^n$.

d. This is neither arithmetic, quadratic, nor geometric. Neither the first nor second differences are constant, ruling out arithmetic and quadratic growth. (There is a pattern in the second differences, but they are not constant.) The ratios are 4/10, 1/4, 3, 4, and 30/12, which are not equal, nor even approximately equal. Thus the sequence is not geometric.

(9) 16

(10) a. The growth factor is $4.8/3.2 = 1.5$ and the percentage change is $1.5 - 1 = 0.5 = 50\%$.

(11) a. Yes. The ratios of successive terms are constant and equal 10/11.

c. No. The ratios of successive terms are not constant: $(1/4)/(1/3) = 3/4$, $(1/5)/(1/4) = 4/5$, $(1/6)/(1/5) = 5/6$, and so on.

(12) a. The growth factor is $1 + 0.72 = 1.72$.

d. The percentage change is $0.78 - 1 = -.22 = -22\%$. This is a decrease of 22%.

(13) a. This is a geometric growth sequence with difference equation $a_{n+1} = 1.3a_n$.

c. Arithmetic growth; $c_{n+1} = c_n - 1.3$.

(14) a. Geometric growth; $a_n = 14(0.3)^n$.

d. Geometric growth; $a_n = 287 \cdot 0.97^n$.

(15) a. The common ratio is $150/100 = 1.5$. The next three terms are $a_2 = 225$, $a_3 = 337.5$, and $a_4 = 506.25$. The difference equation is $a_{n+1} = 1.5a_n$. The functional equation is $a_n = 100(1.5)^n$.

c. The common ratio is $c_1/c_0 = 1.1$. The next three terms are $c_2 = -9.68$, $c_3 = -10.648$, and $c_4 = -11.7128$. The difference equation is $c_{n+1} = 1.1c_n$. The functional equation is $c_n = -8(1.1)^n$.

(16) a. Parameter set iii.

b. Parameter set i.

(17) a. $2 \cdot 1.05^{10} = 3.2578$, rounded to four decimal places

b. Starting with 3.2, multiply by 1/10 three times: $b_3 = 0.0032$

c. $c_{12} = 2{,}600(1.05)^{12} = 4{,}669.226$, rounded to three decimal places

(19) c. No term of the sequence equals -0.001. We find $a_{10} = -0.0012$ has not yet reached -0.001 and $a_{11} = -0.0002$ has gone beyond -0.001 (in the sense of getting closer to 0). All following terms will continue to get closer to 0.

(21) $a_6 = 196.16$, rounded to two places

4.2. Exercises

Digging Deeper.

(23) a. From $r^2 = 0.25$ we find $r = 0.5$. Therefore, the difference equation is $a_{n+1} = 0.5a_n$ and the functional equation is $a_n = 100 \cdot 0.5^n$.

c. From $r^3 = 16$ we find $r = 16^{1/3}$ which is approximately 2.520. The difference equation is $c_{n+1} = 16^{1/3} \cdot c_n$ exactly, or $c_{n+1} = 2.520 \cdot c_n$ approximately. Similarly, the functional equation is $c_n = 100 \cdot (16^{1/3})^n$ exactly, or $c_n = 100 \cdot 2.520^n$ approximately.

(25) a.

n	0	1	2	3	4	5	6	7	8	9	10
b_n	70	38	22	14	10	8	7	6.5	6.25	6.125	6.0625

b. $b_{n+1} = (1/2)b_n + 3$

c.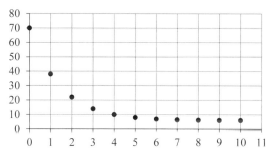

d. $b_n = 70(0.5)^n + 3(0.5^{n-1} + 0.5^{n-2} + \cdots + 0.5 + 1)$
$= 70(0.5)^n + 3(1 - 0.5^n)/(1 - 0.5) = 70(0.5)^n + 6(1 - 0.5^n)$

e. The functional equation for c_n can be found in several ways. One approach is to make a table of values for c_n and observe that it is a geometric growth sequence. That leads to the functional equation $c_n = 64(0.5)^n$ so $b_n = 64(0.5)^n + 6$. This allows us to derive the functional equation for b_n without finding an expression for $0.5^{n-1} + 0.5^{n-2} + \cdots + 0.5 + 1$. Compare this new functional equation with the one found in part d, and verify that the two equations are equivalent.

4.2 Exercises

Reading Comprehension.

(3) a. In equal periods of time, the variable is multiplied by equal growth factors. Or, for equal changes in the variable t, the function $f(t)$ is multiplied by equal growth factors.

c. We know the growth factor for a year is 9, and the growth factor for each six months is equal. Call this amount c. A year is equal to two six-month periods, so the growth factor for those two periods would be c^2. This shows that $c^2 = 9$, so $c = 3$. In general, the growth factor for half an interval is the square root of the growth factor for the full interval, under the strong geometric growth assumption.

(4) b. In many cases, populations have large numbers of members, and we use units of thousands or millions or other large amounts. In such a case, a decimal

value of the variable can represent a whole number of population members. For example, if p is the population in thousands, then the statement $p = 8.23$ means the population is 8,230. Also, we can consider a geometric growth assumption to hold approximately. In this case, the exact fractional value of the variable in the model can be rounded off to a whole number if necessary to predict a population size. For example, we can interpret a geometric growth model for which the terms grow by a factor of 1.2 as modeling a population that grows by approximately 20% at each observation. Generally, the numerical difference between a fractional value predicted by the model and the nearest whole number will not be significant. For example, if the model predicts a population of 12,348.77 and we interpret this as a population of 12,349, the slight error involved is not significant relative to the size of the population.

(5) a. If interest is compounded over a shorter period than a year, say monthly or daily, the banks use proportional reasoning to determine the amount of interest for each compounding period. Thus, for monthly interest payments, each payment is determined using one twelfth of the annual percentage rate. Due to these compounding payments, the actual percentage of interest for a full year will be greater than the interest rate cited in advertisements. That is, if the advertisement cites an annual interest rate of 4%, compounded monthly, the accumulated effect of monthly interest amounts of 4/12% over a year will be more than 4%. The actual amount of interest accumulated for a year is referred to as the *effective* rate, and for savings accounts, this is also referred to as the *yield*.

b. The calculations the banks use to compute interest over periods shorter than a year are *not* an example of the strong geometric growth assumption. Under the strong assumption, if the annual rate is 4%, corresponding to a growth factor of 1.04, then for one month the growth factor would be $1.04^{1/12} = 1.00327$, corresponding to 0.327% interest per month. In the answer to part a, the monthly interest rate would be one twelfth of 4%, that is, 0.333%. The slight difference between these two percentages reflects the fact that the banks do *not* use the strong geometric growth assumption to compute interest payments for fractional parts of a year.

Math Skills.

(8) a. The decimal form of 150% is 1.50. $r = 1 + 1.5 = 2.5$.
c. $r = 1 - 0.95 = 0.05$
e. $r = 1.053$

(9) a. The ratio varies but if rounded to the first decimal place is $r = 2.3$. The nearly constant ratio mean a geometric growth model would be a good approximation for the data.

(11) a. $T = 40 + H$ or $H = T - 40$
b. $t = n/3$ or $n = 3t$. (Note there are three 20-minute intervals in an hour.)
c. $T = 40 + 60(2/3)^{3t}$

(12) a. It should be 4. This way when applied twice the combined factor is $4^2 = 16$.

4.2. Exercises

b. It should be 2 so that when applied 4 times the combined factor is $2^4 = 16$.

e. Answers will vary but should include that fact that the answers from parts a and b were derived using the strong geometric growth assumption, and that these are the same answers obtained with the calculator for $n = 1/2$ and $n = 1/4$. So the calculator programming for fractional exponents gives the same results as reasoning from the strong geometric growth assumption for the two examples considered.

Problems in Context.

(13) b. To better see the various values, two graphs are shown below. Notice the different scales on the vertical axis.

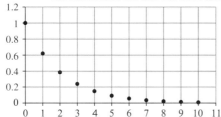 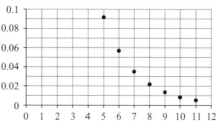

We can see that the sequence reaches 0.1 between $n = 4$ and 5. It is especially clear on the second graph that for $n = 5$ the sequence is less than 0.1. We can also see that the sequence reaches 0.01 between $n = 9$ and $n = 10$.

c. $p_n = 0.62^n$

e. There are various valid answers, including: We assumed the new water entering the lake is pollution-free. That is a simplifying assumption. We also assumed that 38% of lake water is replaced each year and that the percentage will remain constant for the next several years. It might be helpful to check the level of pollutant in the lake then compare our model predictions with future measurements. It might be helpful to focus on one type of pollutant.

(15) a. $R_{n+1} = (25/3)R_n$ and $R_n = 30{,}000(25/3)^n$ with $R_2 = 2{,}083{,}333$, $R_3 = 17{,}361{,}111$, and $R_4 = 144{,}675{,}926$.

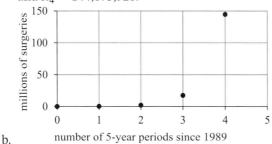

b. number of 5-year periods since 1989

c. 30,000 each year

d. $R_5 = 1{,}205{,}632{,}716$; over a billion surgeries.

e. The model offers one forecast for the future market for RK surgeries. So, an investor might use that to estimate how profitable a chain of clinics might be. On the other hand, basing a geometric growth model on a single pair of data points is questionable. Rapid growth at the beginning of a new technology or service must eventually slow down, invaldating the assumption of a constant growth factor. This is seen in the projection for 2014, which exceeded the 2014 population of the US by nearly a factor of four.

(17) a. $\$15{,}000(1 + 0.023/12)^{60} = \$16{,}826.25$
b. $\$15{,}000(1 + 0.023/4)^{20} = \$16{,}822.56$
c. The effective rate of the first investment is 2.3199% per year. For the second investment the effective rate is 2.5% because there is no compounding during the year. The second investment is a better deal.
d. It will take 32 quarters or 8 years.

(19) a. See below
b. For the model answers will vary. Below is an example using the parameters $r = 1.17596$ and starting value 160.

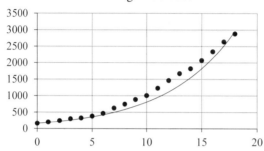

c. The functional equation for the model in the prior answer is $a_n = 160 \cdot 1.17596^n$. For the year 2020, we have $n = 40$ for the first quarter, 41 for the second quarter, and so on. The values of a_{40}, a_{41}, a_{42}, and a_{43} range from 104,670 to 170,216. This projection is probably not very reliable since the model does not fit the data particularly well. In fact, at the end of original data, the model seems to be growing steeper whereas the data seem to be flattening out. If several more years of data seem to confirm the model, that would increase our confidence in the prediction for 2020.

(21) a. The modified table is

n	1	2	3	4	5	6
H_n	59.39	39.66	25.26	18.15	13.56	10.62

Following the hint in the footnote we start with $r = 0.6$ and $H_0 = 99$, and obtain the functional equation $H_n = 99 \cdot 0.6^n$. This is represented by a curve in the graph below, where the dots represent the given data from the temperature probe.

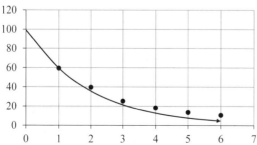

Visually we see that the curve is below the data points, and getting further away as the curve progresses to the right. After trial and error, a better fit to the data is found with $H_0 = 86$ and $r = 0.68$. This has functional equation $H_n = 86 \cdot 0.68^n$.

b. See below

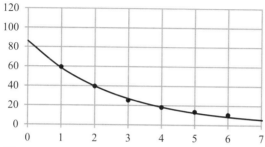

c. The parameter $H_0 = 86$ is the value of H at time zero, when the probe was first placed in the water. That means the probe was 86°C above the temperature of the water, 10°C. Therefore the probe was initially at a temperature of 96°C.

d. The modified table for R_n is

n	1	2	3	4	5	6
R_n	54.39	34.66	20.26	13.15	8.56	5.62

Trial and error revealed a very good fit with parameters $R_0 = 83$ and $r = 0.64$. A graph is shown below.

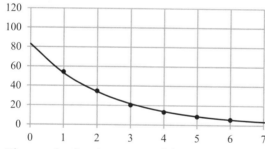

e. The graphs for the two models are very similar, though close examination shows that the R graph is a slightly better fit to the data than the H graph. This can be verified numerically. On the average, the distance between the data points and the model values is 1.03 for the H model, and 0.77 for the R model. Using each model to compute an initial temperature for the probe, we obtain 96°C with the H model as compared with $R_0 + 15 = 83 + 15 = 98$°C with the R model.

f. Answers will vary due to the different models chosen. But for all models, t and n are related by the equations $t = 5n$ and $n = t/5$. Substituting in the functional equation for H_n, this produces $H = 83(0.64)^n = 83(0.64)^{t/5}$.

g. Again, answers will vary due to the different models chosen. But in all models, H and T are related by $H = T - 10$ or $T = H + 10$. Substituting $T - 10$ for H in $H = 83(0.64)^{t/5}$ leads to $T - 10 = 83(0.64)^{t/5}$, hence $T = 10 + 83(0.64)^{t/5}$.

(23) a. Defining n to be the number of 4-hour intervals since the drug was taken we find the amount of drug in the blood is $a_n = 160(3/4)^n$. Using trial and error,

we find $a_{17} \approx 1.20$ and $a_{18} \approx 0.90$. This means the drug will be undetectable by this test between 68 and 72 hours after taking the drug.

b. Because $a_{34} \approx 0.0090$ the test will be effective for just under 136 hours, or about 5 days and 16 hours.

Digging Deeper.

(25) a. A very large number indeed: 9.256×10^{20} or about 925,600,000,000,000,000,000 or more than 900 quintillion.

b. Using 1 square inch = 2.49×10^{-10} square miles, we find 5.76×10^{10} square miles are needed. This is nearly the land area of 300 Earths!

c. Nothing says that all of the ancestors in the queen's family tree are *distinct*. In fact, they cannot be. As the total number of ancestors in a single generation increases to impractical proportions, it must be the case that certain ancestors appear in many different positions in the geneology.

4.3 Exercises

Reading Comprehension.

(2) An exponential function can be expressed in the form $y = Ab^x$ where A and b are constants, and $b > 0$. The y-intercept is equal to A, and unless $A = 0$, there is no x-intercept.

(5) Evaluating a function means computing its value for a numerical value of the variable. For example, if $f(x) = 100e^{1.06x}$, then evaluating the function for $x = 2.8$ means computing $f(2.8)$. Inverting the function has two meanings. If we find the value of x that produces a particular value of $f(x)$, that is an instance of inverting the function. For instance, using the earlier example, we might find the value of x for which $f(x) = 5$, by solving the equation $100e^{1.06x} = 5$ for x.

The second meaning of inverting the function is to find an equation that expresses x as a function of $f(x)$. Or, if we denote the function value by y, then inverting the function means to solve the equation $y = f(x)$ for x as a function of y. Turning again to our example, that would mean solving the equation $y = 100e^{1.06x}$ for x. The correct answer is $x = \frac{1}{1.06} \ln\left(\frac{y}{100}\right)$.

(8) The logarithm of a number c means the x value for which 10^x equals c. As the graph illustrates, there is no x for which $10^x \leq 0$, because the entire curve $y = 10^x$ is *above* the x-axis. Accordingly, there can be no logarithm of c if $c \leq 0$.

Math Skills.

(9) a. Exponential because it can be written $y = 10(1.3^{1/2})^x$ which is in the standard form.

c. None of these. The 2^x term shows that the function is neither linear nor quadratic, and the added 100 shows that the function is not exponential.

d. Linear. The equation can be written $y = \frac{2}{7}x - \frac{3}{7}$, which is the standard form for a linear function with $m = 2/7$ and $b = -3/7$.

4.3. Exercises

(10) a. $y = 2^x$

(11) a. $p(t) = 1{,}000 \cdot 16^{t/4} = 1{,}000 \cdot (4^2)^{t/4} = 1{,}000 \cdot 4^{2t/4} = 1{,}000 \cdot 4^{t/2}$
b. $p(t) = 1{,}000 \cdot 16^{t/4} \approx 1{,}000 \cdot (10^{1.204})^{t/4} = 1{,}000 \cdot 10^{1.204t/4} = 1{,}000 \cdot 10^{0.301t}$
c. $p(t) = 1{,}000 \cdot 16^{t/4} \approx 1{,}000 \cdot (e^{2.773})^{t/4} = 1{,}000 \cdot e^{2.773t/4} = 1{,}000 \cdot e^{0.69325t}$
d. $p(t) = 1{,}000 \cdot 16^{t/4} = 1{,}000 \cdot (16^{1/4})^t = 1{,}000 \cdot 2^t$

(13) a. $f(t) = 25e^{3\ln(1.04)\cdot t} \approx 25e^{0.117662t}$
c. $f(t) = 48.5e^{1.57\ln(2)\cdot t} \approx 48.5e^{1.088241t}$

(14) a. $f(t) = 52(e^2)^t \approx 52 \cdot 7.389056^t$
c. $f(t) = 85.4(1.012^{1/5})^t \approx 85.4 \cdot 1.002389^t$

(15) b. The first and last are equivalent: $30(1.44)^{0.5(t+1)} = 30(1.2)^{(t+1)} = 30(1.2)(1.2)^t = 36(1.2)^t$.

(16) a. Because $1.5^{2.709511} = 2.9999996$ is closer to 3 than $1.5^{2.709512} = 3.0000009$, the solution is 2.709511 to 6 places.

(17) a. By systematic trial and error we find that $10^{1.11} < 13$ and $10^{1.12} > 13$, so the exact solution is between 1.11 and 1.12. We then calculate $10^{1.115} = 13.03$, so the correct answer must be less than 1.115, and therefore closer to 1.11 than to 1.12. This shows that, rounded to two decimal places, the answer will be 1.11.
b. By definition, the solution equals log(13).
c. The calculator gives $\log(13) = 1.1139\cdots$, which does equal 1.11 when rounded to two decimal places.

(19) a. By systematic trial and error we find that $7^{0.82} < 5$ and $7^{0.83} > 5$, so the exact solution is between 0.82 and 0.83. We then calculate $7^{0.825} = 4.97$, so the correct answer must be greater than 0.825, and therefore closer to 0.83. This shows that, rounded to two decimal places, the answer will be 0.83.
b. $\dfrac{\log 5}{\log 7}$ or $\dfrac{\ln 5}{\ln 7}$
c. The calculator shows 0.8270874753 which rounds to 0.83, and agrees with the answer of part *a*.

(21) If $1.2t = u$, then the equation becomes $3^u = 7$. We know the solution to this is $u = (\log 7)/(\log 3)$. Therefore, we must have $1.2t = (\log 7)/(\log 3)$. Solving for t produces $t = \dfrac{\log 7}{1.2 \log 3}$.

(22) a. $t = \dfrac{\log(12)}{3\log(1.04)} \approx 21.118999$
c. $t = \dfrac{5\log(10)}{2\log(0.8)} = 5/(2\log(0.8)) \approx -25.797128$

(23) a. 85.387377
c. 1.262648

(24) a. We know that the equation will take the form $f(x) = 60b^x$, and need to find b. The given value for $f(4)$ implies that $60b^4 = 303.75$. This shows that $b^4 = 303.75/60 = 5.0625$. Therefore $b = 5.0625^{1/4} = 1.5$. Final answer: $f(x) = 60 \cdot 1.5^x$.

c. Applying a similar method as in a, we begin with $f(x) = Ab^x$ with both A and b unknown. The coordinates of each given point must satisfy the equation. Therefore $Ab^2 = 90$ and $Ab^5 = 2{,}430$. Dividing, $\frac{Ab^5}{Ab^2} = \frac{2{,}430}{90}$, which simplifies to $b^3 = 27$. This shows that $b = 3$. Using either of the original equations, for example $A \cdot 3^2 = 90$, we find $A = 10$. Thus $f(x) = 10 \cdot 3^x$.

e. Note that the given information implies that $f(5) = 1.9 \cdot 0.78$, so the same method used in a applies. Final answer: $f(x) = 0.78(1.9^{1/5})^x$ exactly, or $f(x) = 0.78 \cdot 1.136974^x$ approximately.

(25) a. $y = 2^{0.5x}$

c. Substituting the coordinates of the given points into $y = A \cdot 2^{kx}$ shows that $48 = A \cdot 2^{k \cdot 4}$ and $384 = A \cdot 2^{k \cdot 7}$. Dividing produces $\frac{384}{48} = \frac{A \cdot 2^{7k}}{A \cdot 2^{4k}}$. This simplifies to $8 = 2^{3k}$, so $3k = \frac{\log 8}{\log 2}$ and $k = \frac{\log 8}{3 \log 2}$. In this case the decimal form is also an exact solution $k = 1$. Using either of the original equations we find $A = 3$. Thus $y = 3 \cdot 2^x$.

(26) a. $y = 250e^{(\ln(2)/10)x} \approx 250e^{0.069315x}$

Digging Deeper.

(28) Set $\sqrt{1/5}(0.5 + 0.5\sqrt{5})^n = 54{,}321$ and solve for n. The answer is $24.3\cdots$. Now find F_{24} as the closest integer to $\sqrt{1/5}(0.5+0.5\sqrt{5})^{24}$, and similarly F_{25}. The results show that $F_{24} = 46{,}368$ and $F_{25} = 75{,}025$. Therefore, 54,321 is not a Fibonacci number, and F_{24} is the closest Fibonacci number to 54,321.

(30) Solving the equation $b^x = a$ with the Exponential Equation Solution that uses the log function (page 261), we find $x = \log(a)/\log(b)$. This is equal to $\log_b(a)$ by definition. Similarly, we can solve $b^x = a$ with the Exponential Equation Solution that uses the ln function (page 263), finding $x = \ln(a)/\ln(b)$. Again by definition, this must equal $\log_b(a)$.

4.4 Exercises

Reading Comprehension.

(1) For $p(t) = 1{,}000 \cdot 8^{t/6}$, the function value is 1,000 at $t = 0$, and increases by a factor of 8 every 6 hours. For $p(t) = 1{,}000 \cdot 2^{t/2}$, the function value is again 1,000 at $t = 0$, but this function increases by a factor of 2 every 2 hours. Similarly, $p(t) = 1{,}000 \cdot (1.414)^t$ is 1,000 at $t = 0$ and increases by a factor of 1.414 every hour. To see that the first two equations describe the same function, notice that $8 = 2^3$. Substituting this into the first equation, we find $p(t) = 1{,}000 \cdot 8^{t/6} = 1{,}000 \left(2^3\right)^{t/6} = 1{,}000 \cdot 2^{3t/6} = 1{,}000 \cdot 2^{t/2}$. This shows that the first two equations are equivalent. In a similar way, the third equation can be seen to be (approximately) equivalent to the others, because 1.414^2 is very nearly equal to 2.

4.4. Exercises

(3) Doubling time refers to the length of time over which the value of an exponential function doubles. That would not be meaningful in a radioactive decay model because the exponential function in that context is actually shrinking, not growing, so from a given observation we would never see the amount of radioactive material double. In the trout model, the equation relates length and weight, neither of which is time. In this context also, doubling time does not make sense. We are not modeling how the weight or length of a trout increases over time, so the model would not address the question of how long it takes the weight or length to double. For a bank account that is gathering interest, doubling time make perfect sense. We know that the balance is growing larger over time, and that it is modeled by an exponential function. If the balance is observed to double in 7 years, then we can conclude it will continue to double *every* 7 years. In this case we say that the bank account has a doubling time of 7 years.

(5) Relative error expresses a given error as a fraction or percentage of the correct value. This is useful because the significance of an error is usually best understood in relation to the correct value. If a bookkeeper makes a $100 error in a fund that only totals $200, that is much more serious than if the fund total is one million dollars. In the first case, the relative error is 50% because the error divided by the true value is $100/200 = 0.50$. In the second case the relative error is only $100/1,000,000 = 0.0001 = 0.01\%$. In words, that is an error of one one-hundredth of a percent.

Relative error might very well be useful in a temperature model of the sort considered for the probe data. In that context, we are probably most interested in the magnitude of an error as a percentage. Saying the model is only off by one degree when the modeled temperature is 450° (in an oven, for example) reflects higher accuracy than an equal error when the modeled temperature is 5° (in a freezer, for example).

Math Skills.

(7)   a. $a(t) = 150 \cdot 2.3^{t/5}$

   c. $a(t) = 480 \cdot 1.04^{t/60}$

(8)   a. First equation: $a(t) = 590 \cdot 2^{t/3.5}$
   Second equation: $a(t) = 590 e^{(\ln(2)/3.5)t} \approx 590 e^{0.198042 t}$.

(9)   a. $a(t) = 20 \cdot 1.8^{t/0.4}$, so $a_0 = 20$, $r = 1.8$, and $d = 0.4$. The function is initially 20 and increases by a factor of 1.8 every 0.4 hours.

   c. $a(t) = 450 \cdot 1.5^t / 1.5^2 = (450/1.5^2) \cdot 1.5^{t/1}$, so $a_0 = 200$, $r = 1.5$, and $d = 1$. The function is initially 200 and increases by a factor of 1.5 every hour.

(10)   a. Situation 1 : $P = 50 \cdot 8^{t/3}$. Situation 2 : $P = 50 \cdot 2^t$. These are equivalent because $8 = 2^3$. Therefore $8^{t/3} = \left(2^3\right)^{t/3} = 2^{3t/3} = 2^t$.

(11) a. The solution can be found as follows.
$$450 \cdot 1.5^{t/3} = 2{,}000$$
$$1.5^{t/3} = 2{,}000/450 = 40/9$$
$$t/3 = \frac{\ln(40/9)}{\ln(1.5)}$$
$$t = 3\frac{\ln(40/9)}{\ln(1.5)}$$

c. $t = \dfrac{\ln 5}{2.5 \ln 1.8}$

Problems in Context.

(12) a. We can read from the equation $A = 4 > 0$ and $b = 0.1 < 1$ so this will match the upper right graph.

c. Population size is positive so $A > 0$ and the population is growing so $b > 1$. We expect a graph above the horizontal axis that increases from left to right. This matches the upper left graph.

e. The height of a deer is a positive quantity. We expect the data to follow this rule: the larger the distance between footprints, the taller the deer. These two observations imply a curve that is above the horizontal axis and sloping upward from left to right, matching the graph in the upper left.

(14) $t = 8.020031$ days, $y = 10(1/2)^{t/8.020031}$

(16) $5\ln(1/2)/\ln(0.85) \approx 21.325121$ years

(18) Setting $c_n = 5{,}000$ leads to a value of n between 25.24 and 25.25. This shows that $c_{25} < 5{,}000$ and $c_{26} > 5{,}000$. Therefore, the number of infections first exceeds 5 million for year 26, which corresponds to 2019.

(20) The initial temperature is 70 degrees and the food temperature is decreasing by 22% every 30 minutes.

(22) a. $p(m) = 1.45(1.028)^{m/6}$
b. $p(t) = 1.45(1.028)^{2t}$
c. $p(t) = 1.45 e^{(2\ln 1.028)t}$

(24) $S(137) = 20(1 - 0.028)^{137} = 0.408609$ parts per million rounded to six decimal places

(26) a. $C(t) = 0.003(1/2)^{t/5{,}730}$
b. Solve $0.003(1/2)^{t/5{,}730} = 0.0007$ for t: 12,030 years
c. 12,030 years ago

(28) $B(t) = 156(140/156)^{t/5}$ and 127 hours (or 5 days and 7 hours)

(29) a. The table is

t	2	5	10	15	20	25	30
H	70.47	59.39	39.66	25.26	18.15	13.56	10.62

The figures below both show a plot of the data points. These figures also include curves that are referred to in the answer for part b.

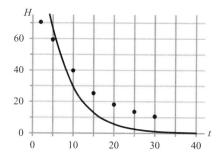 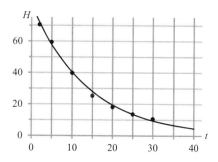

b. The figure on the left above shows a curve representing the function $H(t) = 150 \cdot 0.85^t$. The curve is not a close approximation to the data, but it provides a starting point for finding a better approximation by trial and error. A fairly accurate approximation is provided by the function $H = 83 \cdot 0.93^t$. The figure above on the right shows the graph of this equation and the data points. The table below shows the values of t, the values of H from the data, the corresponding values of H from the model equation, and the error (model − data). Although the computations were carried out to many more decimal places of accuracy, the figures in the table have been rounded for readability. For this reason, there are a few rows of the table where subtracting the reported model and data values does not exactly equal the reported error value.

t	data	model	error
2	70.5	71.8	1.3
5	59.4	57.7	−1.6
10	39.7	40.2	0.5
15	25.3	27.9	2.7
20	18.2	19.4	1.3
25	13.6	13.5	0.0
30	10.6	9.4	−1.2

c. The largest absolute error is 2.7; the average absolute error is about 1.2.
d. This is $H(0) + 10 = 93$.
e. $T = H + 10 = 83 \cdot 0.93^t + 10$. We solve $83 \cdot 0.93^t + 10 = 20$ to find $t = \frac{\ln(10/83)}{\ln(0.93)} \approx 29.16$.

4.5 Exercises

Reading Comprehension.

(1) It means that the decimal expression of e goes on forever, without a consistent repeating pattern in the digits. It also indicates that e cannot be expressed exactly as a ratio of whole numbers.

(3) The $y = e^x$ curve crosses the y-axis at a 45° angle. For any $b \neq e$, the curve $y = b^x$ crosses the y-axis at a different angle.

(5) To approximate e^x, Euler's method is to compute
$$1 + \frac{x}{1} + \frac{x^2}{2 \cdot 1} + \frac{x^3}{3 \cdot 2 \cdot 1} + \frac{x^4}{4 \cdot 3 \cdot 2 \cdot 1} \cdots.$$
For any $b \neq e$, to approximate b^x we must first multiply x by a special number that is itself difficult to compute. This preliminary step can be avoided only when the base is e.

(6) A calculator shows that $2.5^{3.1} > 3.1^{2.5}$ but $1.5^{2.2} < 2.2^{1.5}$. Without a calculator these comparisons would be very difficult to determine. However, e has the special property that e^x is always greater than or equal to x^e. Therefore, we know that $e^{2.9} \geq 2.9^e$ without making any calculations. In addition, $e^x = x^e$ only when $x = e$. Therefore, we can say with certainty that $e^{2.9} > 2.9^e$.

Math Skills.

(7) First, using the constant 0.693 given on page 291, we compute $y = 0.126 \cdot 0.693 = 0.087318$. Next we compute our approximation as
$$2^{0.126} \approx 1 + \frac{y}{1} + \frac{y^2}{2 \cdot 1} + \frac{y^3}{3 \cdot 2 \cdot 1} + \cdots.$$
This can be carried out for as many terms as desired—the more terms, the better the approximation. For example, if we use 5 terms (up to $y^4/24$), we find $2^{0.126} \approx 1.091244$. Direct calculation using the exponent key on a calculator produces 1.091264, so the approximation found with Euler's approach is correct to four decimal places.

(9) First, we compute $y = 2.3 \cdot \ln(1.06) = 0.134018 \cdots$. Store the result in the calculator's memory so the calculations of step 2 are completed to the calculator's full accuracy. Next we compute our approximation as
$$1.06^{2.3} \approx 1 + \frac{y}{1} + \frac{y^2}{2 \cdot 1} + \frac{y^3}{3 \cdot 2 \cdot 1} + \cdots.$$
The number of terms to use can be determined as we make the calculations: if we want 4 decimal place accuracy we continue until the first four digits of the results stop changing. After adding the fifth term ($y^4/24$) the result is 1.143413591. Adding one more term produces 1.143413951. These two results agree for their first six decimal places, so the estimate $1.06^{2.3} \approx 1.143414$ is probably correct to 5 or 6 decimal places. Direct calculation of $1.06^{2.3}$ using an exponent key produces 1.14341396. This shows that our estimate using six terms was actually correct to seven decimal places.

Digging Deeper.

(10) a. The decimal form of the interest is 0.04, but we are compounding quarterly so the rate each period is $0.04/4 = 0.01$. The balance will grow by a factor of 1.01 every 1/4 of a year, so the balance after t years will be $B(t) = 100 \cdot 1.01^{t/(1/4)}$ or more simply, $B(t) = 100 \cdot 1.01^{4t}$.

b. Compounding monthly means each period is 1/12 of a year, so the rate each period is $0.04/12 = 0.003333$. The balance will grow by a factor of 1.003333 every 1/12 of a year, so the balance after t years will be $B(t) = 100 \cdot 1.003333^{t/(1/12)}$ or more simply, $B(t) = 100 \cdot 1.003333^{12t}$.

5.1. Exercises

c. Compounding daily means each period is 1/365 of a year, so the rate each period is 0.04/365. The balance will grow by a factor of $1+0.04/365$ every 1/365 of a year, so the balance after t years will be $B(t) = 100(1 + 0.04/365)^{t/(1/365)}$ or more simply, $B(t) = 100(1 + 0.04/365)^{365t}$.

d. $B(t) = 100(1 + 0.04/m)^{mt}$

e. $B(t) = 100(1 + 0.04/8760)^{8760t}$

f. The $B(t)$ functions found above become successively better approximations to $B(t) = 100e^{0.04t}$ as the number of compounding periods increases (although to see any difference at all between these different functions you have to adjust the graph window carefully—try $67 \leq t \leq 69$ and $1{,}500 \leq y \leq 1{,}510$). In other words, the shorter the compounding period, the closer $B(t)$ comes to the function $B(t) = 100e^{0.04t}$. If this trend continues for larger and larger values of m, it is plausible to consider $B(t) = 100e^{0.04t}$ as a formula for the accrual of interest that is compounded instantaneously.

5.1 Exercises

Reading Comprehension.

(2) The shape of the graph for a mixed model is a vertically shifted exponential graph. This is reflected in the functional equation for a mixed model by the presence of an added constant. That is, whereas the functional equation for a geometric growth sequence has the form $a_n = A \cdot r^n$ for some constants A and r, the functional equation for a mixed growth sequence has the form $a_n = A \cdot r^n + B$ where B is also a constant. The graph of this latter equation can be obtained by shifting the graph of the geometric growth functional equation vertically upward by B units (when $B > 0$) or downward by $|B|$ units (when $B < 0$).

(4) Here is an outline of key points. (1) A fixed point is a value of a model variable (such as population) for which the difference equation produces no net change. That is, when p_n is a fixed point, and when we use the difference equation to find p_{n+1}, we find $p_{n+1} = p_n$. (2) An equilibrium is always a fixed point. That is, if there is a constant sequence $c, c, c, \cdots$ that arises from a difference equation, then the value c must be a fixed point of the difference equation. (3) A fixed point can be found using the difference equation by replacing both p_n and p_{n+1} by x, and solving for x. For example, for the difference equation $p_{n+1} = 1.3p_n - 100$, a fixed point can be found by solving the equation $x = 1.3x - 100$ for x. (4) When a difference equation has a fixed point, that indicates that there will be a constant sequence that satisfies the difference equation. We can say more for a mixed growth sequence with a fixed point of E when the parameter r is positive and less than one: if the initial value is $p_0 = E$ the sequence will be $E, E, E, \cdots$, and otherwise, the terms of the sequence will get closer and closer to E as n increases. In particular, if $p_0 > E$, the terms will steadily decrease toward E and if $p_0 < E$, the terms will steadily increase toward E.

Math Skills.

(5) a. $r = 1.3$, $d = 2$, $a_{n+1} = 1.3a_n + 2$, $a_n = 1(1.3)^n + 2\frac{1.3^n - 1}{0.3}$ or $a_n = -6.67 + 7.67(1.3)^n$. Graph shown below.

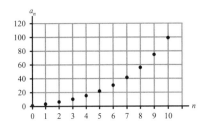

(6) a. Checking first and second differences and ratios we find the ratio of the first differences is constant. This is a mixed growth model with $r = 0.5$. Using this and and two consecutive terms, we find $d = 12$. Thus $a_{n+1} = 0.5a_n + 12$ and $a_n = 8(0.5)^n + 12\frac{1-0.5^n}{0.5}$ or $a_n = 24 - 16(0.5)^n$.

c. Arithmetic; $a_{n+1} = a_n + 2.4$; $a_n = 12.45 + 2.4n$

e. Quadratic; $a_{n+1} = a_n - 20 + 4n$; $a_n = 100 - 20n + 2n(n-1)$

(7) a. The points curve and level off. We can rule out arithmetic growth as the points do not form a straight line. We can rule out quadratic growth as the points level off rather than curving around to the opposite direction. This leaves geometric or mixed growth. For geometric growth the curve will level off near the x-axis. This curve does not do that. Rather it levels off around $y = 10$. Thus it most resembles mixed growth.

(9) a. 90, 36, 18, 12, and 10

b. Although the description was given in a new way, this is indeed a mixed growth sequence. We can see this algebraically in the difference equation. As described, the difference equation is $a_{n+1} = (a_n + 18)/3$. This is equivalent to $a_{n+1} = \frac{a_n}{3} + \frac{18}{3} = \frac{1}{3}a_n + 6$. Now we have the difference equation in standard form. The functional equation is $a_n = 90(1/3)^n + 6\frac{1-(1/3)^n}{2/3} = 9 + 81(1/3)^n$.

(10) a. The fixed point can be found two ways: (1) using $E = d/(1-r) = 10/0.25 = 40$ or (2) solving $x = 0.75x + 10$ to find $x = 40$. Because $r = 0.75$ is less than 1 we expect the sequence to level off at 40. This can be seen in the graph below with $a_0 = 20$.

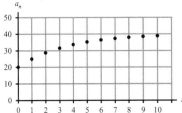

(11) a. Using the parameters in the difference equation, $r = 1.6$ and $d = -95$ we find $a_n = 500(1.6)^n - 95\frac{1.6^n - 1}{0.6}$ or, with $E = d/(1-r)$, we find an equivalent form $a_n = \frac{475}{3} + \frac{1,025}{3}(1.6)^n$.

b. $a_{20} = 4{,}130{,}655$ (rounded to nearest whole number)

c. We will solve $10{,}000 = \frac{475}{3} + \frac{1{,}025}{3}(1.6)^n$ for n. Following the steps in the text we find $n = \ln((10{,}000 - \frac{475}{3})/\frac{1{,}025}{3})/\ln(1.6)$ or, to six decimal places $n = 7.150184$.

5.1. Exercises

Of course we want a whole number for n. We can find, to one decimal place, $a_7 = 9{,}329.9$ and $a_8 = 14{,}832.8$ to see that the sequence first exceeds 10,000 when $n = 8$.

d. Because $r > 1$ the graph does not level off. Because $a_0 > E$ the graph curves steeply up to the right.

(13) a. $a_n = 125 \cdot 0.4^n + 200 \frac{1-0.4^n}{1-0.4} = 125 \cdot 0.4^n + 200(1 - 0.4^n)/0.6$. This simplifies algebraically to $a_n = \frac{1{,}000 - 625 \cdot 0.4^n}{3}$. Or, rounding off the fractions to three decimal places, $a_n = 333.333 - 208.333 \cdot 0.4^n$.

b. In exact form, the value is $(1{,}000 - 625 \cdot 0.4^{20})/3$. That is 333.333 to three decimal places.

c. Using the difference equation, we find that $a_1 = 0.4 a_0 + 200 = 250$, and $a_2 = 0.4 a_1 + 200 = 300$. So the sequence first reaches 300 for $n = 2$.

d. Yes. We can see this numerically by computing many terms with the difference equation, or we can see in the simplified form of the functional equation that the exponential part, 0.4^n, becomes insignificant for large n. The easiest way to answer the question is to observe that the parameter r is less than 1, and that implies the sequence will level off.

(15) a. $a_n = 100 \cdot 1.05^n - 50 \left(\frac{1.05^n - 1}{0.05} \right) = 100 \cdot 1.05^n - 1{,}000(1.05^n - 1)$
$= 100 \cdot 1.05^n - 1{,}000 \cdot 1.05^n + 1{,}000 = 1{,}000 - 900 \cdot 1.05^n$

(17) a. $a_n = 12 + 8(0.75)^n$

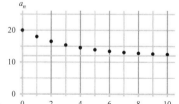

b.

c. As $r = 0.75$, $r < 1$. Yes, the graph is leveling off.

d. As $a_0 = 20$ and $E = 12$ we have $a_0 > E$. The graph is above the equilibrium line.

(19) a. Giving the parameters to 2 decimal places we have $a_n = 52.63 - 12.63(1.95)^n$.

b. Either by solving $0 = 52.63 - 12.63(1.95)^n$ and rounding or by computing the first few terms we find the first value of the sequence below 0 is $a_3 = -41.03$. This is exactly correct, as you can verify by using the difference equation without rounding.

c. The exact value for E is $50/0.95$ which, to six decimal places, is 52.631579.

d. Because $r = 1.95$ is greater than 1, the sequence does not level off.

e. The most straightforward way to answer this is to compute a few terms of the sequence and note that they are decreasing. We can use the parameters to determine this behavior by comparing $a_0 = 40$ with $E = 50/0.95$. Because $a_0 < E$ and $r > 1$ we know the sequence will decrease.

f. The graph is shown below. It does confirm the answers to parts c (by observing a horizontal asymptote far to the left in the graph, although the vertical scale does not permit verification of the actual numerical value of the equilibrium value), d, and e.

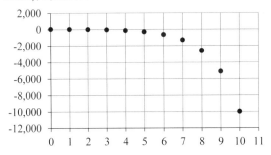

(21) a. The graph appears to be leveling off at 3, so $E = 3$. The graph crosses the y-axis at 7 so $a_0 = 7$.

b. The graph is leveling off (going from left to right) so $r < 1$.

c. This indicates that $a_0 > E$ and indeed $7 > 3$.

Digging Deeper.

(25) a. $a_n = 5(10)^n + 5\frac{10^n - 1}{9}$

b. $E = 5/(1 - 10) = -5/9$ so $a_n = -5/9 + (5 + 5/9)(10)^n$

c. Keep $r = 10$, change to $d = 3$ and $a_0 = 3$: $a_n = 3(10)^n + 3\frac{10^n - 1}{9}$

(27) a. There are several correct approaches to verify the functional equation. One is to notice that the form above is $b_n = E + (b_0 - E)r^n$. We know $E = d/(1-r) = 150/0.25 = 600$ and $b_0 - E = 200 - 600 = -400$ which verifies the functional equation.

b. (One of many possible solutions). We can begin by grouping the terms that make up s_{n+1}: $s_{n+1} = (b_0 + b_1 + \cdots b_n) + b_{n+1}$. Then notice that the terms in the parenthesis are exactly s_n so now we have $s_{n+1} = s_n + b_{n+1}$. Finally, use the functional equation above to replace b_{n+1}: $s_{n+1} = s_n + 600 - 400(0.75)^{n+1}$.

c. $s_n = 200 + 600n - 300(1 - 0.75^n)/0.25$

5.2 Exercises

Reading Comprehension.

(1) A structural basis is a rationale for using a mixed model that reflects our knowledge of the modeling context. For repeated loan payments, it is certain that a mixed model is valid because that is how interest charges are defined by lenders. That is an extreme example of a structural basis. In the repeated drug dosage model, our understanding of how each dose is eliminated from the body leads us to a mixed model. In contrast, if we depend only on the shape of a graph or an approximate mixed growth pattern in a set of data, that is an empirical basis. Usually, we have more confidence in a model that has a structural basis because that provides an explanation of *why* a mixed model is appropriate.

5.2. Exercises

(2) The functional equation does *not* give accurate predictions for fractional values of n because the mixed model does not take into account the way the amount of drug in the body decreases between doses. See **Fractional Values of *n*** on page 322.

(4) a. Matt's balance will remain constant because one month's interest on $1,500 will be $0.03 \cdot 1,500 = 45$. After he makes a $45 payment, his balance will remain what it was orginally.

(5) We know that the terms of a mixed growth sequence have this property: the first differences form a geometric growth sequence. We can use this by examining the first differences of the sum sequence s_n defined in the problem statement. Each such difference is of the form $s_{n+1} - s_n = a_{n+1}$, so the sequence of first differences for s_n is the same as the sequence of terms of a_n. That is assumed to be a geometric growth sequence. This shows that the sum sequence s_n is a mixed growth sequence.

Math Skills.

(8) a. $666.67 - 566.67 \cdot 0.85^n$
   c. $500 + 100 \cdot 0.7^n$
   e. $80 - 30 \cdot 1.015^n$

(9) a. $E = 2.5$, decrease
   c. $E = -925$, increase

(10) a. $d = 220$

(11) a. $a_n = 10{,}000 - 9{,}650 \cdot 1.005^n$. This is zero for $n = 7.14323$, approximately, and 8 is the smallest n for which $a_n \leq 0$.
   c. $c_n = 560{,}000 - 346{,}050 \cdot 1.00375^n$. This is zero for $n = 128.60146$, approximately, and 129 is the smallest n for which $c_n \leq 0$.

(12) a. $c_8 = 0.4(1.6)^8 + 25 = 42.18$ and $c_n = 0.4(1.6)^n + 25$
   d. We expect $0.4(1.6^0 + \cdots + 1.6^n) + 25(n+1)$. Simplifying using the geometric series formula (page 306) results in $0.4\frac{1.6^{n+1}-1}{0.6} + 25(n+1)$.

Problems in Context.

(13) a. $a_1 = 1{,}050$, $a_2 = 1{,}090$, and $a_3 = 1{,}122$
   b. $a_{n+1} = 0.8a_n + 250$
   c. $E = d/(1-r) = 250/0.2 = 1{,}250$
   d. Using the numerical approach is one way to find that for $n \geq 15$ the sequence stays above 1,240 units.
   e. No, the level of drug will dip between doses. In this case the level will dip to approximately $1{,}250(0.8) = 1{,}000$ units just before the next dose. More generally, we know that drug elimination mixed models are not valid for fractional values of n.

(15) a. $E = 100/(1 - 0.75) = 400$ units

b. $E = 30/(1 - 0.75) = 120$ units

c. $280 = d/0.25 \rightarrow d = 70$ units. Note that this is consistent with the calculations above as it is between 30 and 100 units.

d. The equilibrium level is 4 times the repeated dosage. This is valid because we know $E = d/(1 - r) = d/0.25 = 4d$. Equivalently, the repeated dosage is one quarter the equilibrium level. So, if the equilibrium level is to be 280, the repeated dosage must be one quarter that amount, or 70.

(17) a. Here is a partial answer with one possible justification for adopting the model. We are assuming that the PCBs are being broken down or washed out of the lake at a constant percentage rate and that a constant amount of new PCBs are being added each month. If the inflow and outflow from the lake amount to a fixed percentage, as in the tank model, and if the amount of polution flowing into the lake is roughly constant each month, these assumptions are reasonable.

b. $p_n = \frac{1}{36} - \frac{7}{900}(0.82)^n$

c. If the inflow and outflow replace a constant percentage of the lake's volume each month, and if the amount of PCBs added to the lake each month also remains constant, the amount of PCBs in the lake will level off at 1/36 parts per million.

(19) a. $b_n = 16{,}000 - 11{,}000(1.0125)^n$

b. $b_{24} = 1{,}179.14$

c. Using a numerical approach is one way to find the balance first goes below 0 for $n = 31$. This means that it will take 31 months or 2 years and 7 months to pay off the loan.

d. Immediately after the 30th payment, the balance will be $b_{30} = \$32.25$. With an additional month's interest, that becomes $\$32.25 \cdot 1.0125 = \$32.66$. That is the amount of the final payment.

e. Mariel has paid $\$200 \cdot 30 + \$32.66 = 6{,}032.66$. After deducting the original \$5,000 we find a total interest payment of \$1,032.66. A nice gift indeed!

(21) $s_{n+1} = 1.06 s_n + 500$. This is a mixed growth model because the difference equation is expressed in the standard form for a mixed model.

(23) He will pay off his loan in 11 years. Note that doubling the payment results in retiring the debt in less than half the time.

(25) We are told that a_n is the amount in the bank after n deposits. On the day the account is opened, there is an initial deposit of \$100, after which the balance is \$100. That is the value of a_1, because it is the amount after one deposit. For every additional deposit, we will have $a_{n+1} = 1.015 a_n + 100$. This is a mixed growth difference equation with $r = 1.015$ and $d = 100$.

One way to find a functional equation is to define an appropriate value of a_0. This is the amount after making no deposits, and so is zero. Consider that to be the balance one month before the account is opened. Then one month later the balance will be $a_1 = 1.015 a_0 + 100 = 100$. Thus we see that $a_0 = 0$ leads to the correct value of

5.2. Exercises

a_1, and we can express the functional equation as $a_n = a_0 r^n + d(r^n - 1)/(r - 1) = 100(1.015^n - 1)/0.015 = \frac{20{,}000}{3}(1.015^n - 1)$.

Note that the second deposit is made one month after opening the account. Therefore a_2 is the balance after one month. Similarly, a_3 is the balance after two months, a_4 the balance after three months, and so on. In particular, a_{61} is the balance after sixty months, or five years. With the functional equation, we find $a_{61} = \$9{,}865.79$. Similarly, the balance after 10 years will be $a_{121} = \$33{,}725.75$, and after 15 years will be $a_{181} = \$92{,}020.89$.

The first balance in excess of \$100,000 will be $a_{187} = \$101{,}242.58$. This corresponds to 186 months after opening the account, or 15 years and 6 months.

(27) a. $\hat{p}_{n+1} = 1.25\hat{p}_n$

b. $\hat{p}_0 = 250$ and the sequence continues 312.5, 390.625, 488.281, 610.352, 762.939, 953.674. The running totals are $\hat{s}_0 = \hat{p}_0 = 250$, followed by 562.5, 953.125, 1,441.406, 2,051.758, 2,814.697, 3,768.372. (Figures rounded to three places.)

c. The first differences have a common ratio of 1.25.

d. $\hat{s}_{n+1} = 1.25\hat{s}_n + 250$ and $\hat{s}_n = -1{,}000 + 1{,}250(1.25)^n$

e. $\hat{s}_{10} = 10{,}641.532$ thousand procedures

f. Answers will vary. All should include that the model using a 52.8% annual growth assumption results in a prediction of $s_{10} = 49{,}721.48$ thousand people.

(29) One form of the model is $T_{n+1} - 65 = (2/3)(T_n - 65)$. This simplifies to $T_{n+1} = (2/3)T_n + 65/3$. The functional equation is $T_n = 285(2/3)^n + 65$.

(31) With T_n representing the temperature after n five-second intervals, one reasonably close model is given by $T_n = 86 \cdot 0.63^n + 15$. This implies the probe had an initial temperature of $T_0 = 101$ and an ambient temperature of 15 in the cold water bath.

Digging Deeper.

(33) a. After six hours the growth factor would be $(0.64)^2 = 0.4096$ so 40.96% would remain.

b. Suppose the amount remaining after one hour is p times the original amount. Then after three hours what remains will be p^3 times the original amount. Using $p^3 = 0.64$ we find, to five decimal places, $p = 0.86177$, so 86.177% of the drug remains after one hour. After only one hour not much of the drug would have been eliminated. It makes sense that a high percentage of the drug remains in the system.

c. Letting p be the percentage of drug remaining after one hour as above, the percentage remaining after n hours is p^n.

d. $a_{n+1} = 0.4096 a_n + 200$, $a_0 = 500$

e. The exact value for r is $0.64^{4/3}$. To 6 decimal places, this is 0.551535. The difference equation for the sequence is $a_{n+1} = 0.551535 a_n + 200$, $a_0 = 500$.

(35) a. One approach is to find a difference equation for s_n. Using the given definitions and a bit of algebra we find

$$\begin{aligned}
s_{n+1} &= p_0 + p_1 + p_2 + \cdots + p_{n+1} \\
&= p_0 + p_0 r^1 + p_0 r^2 + \cdots + p_0 r^{n+1} \\
&= p_0 + (p_0 + p_0 r^1 + \cdots + p_0 r^n) r \\
&= p_0 + (p_0 + p_1 + \cdots + p_n) r \\
&= p_0 + s_n r \\
&= s_0 + s_n r.
\end{aligned}$$

This is a difference equation for s_n in the form $a_{n+1} = d + a_n r$ making the parameters easily identifiable. In particular the constant growth factor is r, as stated, and the added constant is $d = s_0 = p_0$.

b. This is shown in the answer to the preceding part.

c. We are familiar with the functional equation form $a_n = a_0 r^n + d \frac{1-r^n}{1-r}$. In this case that becomes

$$s_n = p_0 r^n + p_0 \frac{1-r^n}{1-r} = p_0 \left(\frac{r^n(1-r)}{1-r} + \frac{1-r^n}{1-r} \right) = p_0 \left(\frac{r^n - r^{n+1} + 1 - r^n}{1-r} \right)$$

$$= p_0 \left(\frac{1 - r^{n+1}}{1-r} \right)$$

which is equivalent to what is shown in the question.

6.1 Exercises

Reading Comprehension.

(3) A logistic growth difference equation only produces positive results when applied to p_n values within a restricted range. If p_n is too large, then p_{n+1} will be negative. The limiting case means the largest possible value for p_n that does not lead to a negative p_{n+1}. That largest value is L.

(7) This example is a slightly modified version of an earlier fish population example. In the earlier example the population leveled off, approaching an equilibrium. In this modified version, the growth factor for a population size of 10,000 was raised from 2 to 2.8. This changes the behavior of the model significantly: it now settles into an oscillating pattern and never approaches a constant equilibrium. The significance of the example is two fold. First, it shows that logistic growth sequences do not always approach an equilibrium. Second, it shows that a seemingly minor change in one data value can produce a significant qualitative change in the behavior of the model.

(8) a. This is a graph with points of the form (p_n, p_{n+1}). The horizontal axis and vertical axis both represent population values, and a point of the graph shows the effect of a single application of the difference equation. For example, if the point $(3, 5)$ appears on the graph, that indicates that a population of 3 will lead to a population of 5 after one iteration. This is useful in studying logistic growth because it reveals global properties of the model. For example, the

6.1. Exercises

high point of the curve shows the largest possible population size that can result from the difference equation. Also, the x-intercepts show the endpoints within which the difference equation can only produce positive next terms. Note that this graph can be viewed as the graph of the difference equation, or of the recursive process. That is, if we consider the difference equation to be of the form $p_{n+1} = f(p_n)$, this is the graph of the function f.

b. This is a graph with points of the form (n, p_n). The horizontal axis represents the position number, which is usually some measure of time, and the vertical axis represents population values. A point of this graph shows the term of the sequence for a given value of n. For example, if the point $(3, 5)$ appears on the graph, that indicates $p_3 = 5$. This graph shows how the model evolves over time (or as n increases), so it can be used to answer questions about what will happen in the future. For example, if the curve levels off it indicates that the population approaches an equilibrium; if the curve zigzags up and down, that indicates the population will oscillate. Note that this graph can be viewed as the graph of the functional equation.

c. This is a graph with points of the form (p, r). The horizontal axis shows possible population sizes and the vertical axis shows possible growth factors. A point of the graph shows what growth factor would apply for a given population size. For example, if the point $(3, 5)$ appears on the graph, that indicates that a population of 3 will experience a growth factor of 5 when the difference equation is applied, leading to a population of 15 at the next iteration. This is useful in studying logistic growth because it helps us understand how the growth factor changes as the population changes. We can use the graph to help find an equation for the function $r(p)$, and that in turn is used in the difference equation $p_{n+1} = r(p_n)p_n$.

Mathematical Skills.

(9) b. logistic
 d. mixed growth
 f. logistic
 g. quadratic

(10) Make a graph showing population (p) on the x-axis and growth factor (r) on the y-axis. We have two data points: $(1,000, 3)$ and $(25,000, 1)$. The slope of the line joining these two points is $m = \frac{1-3}{25,000-1,000} = -\frac{2}{24,000} = -\frac{1}{12,000}$. Now we can use the point-slope form for the equation of the line: $(y - 3) = -\frac{1}{12,000}(x - 1,000)$ so $y = -\frac{1}{12,000}x + \frac{1}{12} + 3 = -\frac{1}{12,000}x + \frac{37}{12}$. But really, the variable x should be p, and y should be r, so the equation is $r = -\frac{1}{12,000}p + \frac{37}{12}$ or in approximate decimal form, $r = -.00008333p + 3.083333$.

(12) a. For population in thousands $r = -0.003p + 0.03$
 c. For population in thousands $r = -0.01p + 0.09$

(13) a. $a_1 = 22.5$, $a_2 = 43.59375$, $a_3 = 61.474$, $a_4 = 59.209$. The first two values are exactly correct, the other two values have been rounded.

b. Because $mL = 2.5 < 3$, as long as the initial population is less than $L = 100$, we expect the population to level off at the fixed point $L - 1/m = 60$.

c. This can be found in various ways. One way is to call the fixed point x and solve $x = 0.025(100-x)x$ for x. As discussed in the text, one solution is $x = 0$. For the other solution, because we know $x \neq 0$, our equation is equivalent to assuming $r = 1$ or solving $1 = 0.025(100 - x)$. In the end we find $x = 100 - 1/0.025 = 60$.

(15) We see $m = 0.0035$, $L = 220$, and $r_0 = mL = 0.0035(220) = 0.77$. Notice that $0 < mL < 1$. For $0 < p_0 < 220$ the population will die out.

(17) Because $mL = 2.8$ is between 0 and 3, the model is expected to level off at $L - 1/m = 11{,}250$ for $0 < p_0 < L$. The other fixed point is $p = 0$.

(19) a. Fixed points at 0 and $L - 1/m = -160$. Because $mL = 0.2 < 1$, this model will decrease and level off at 0, for any p_0 with $0 < p_0 < 40$.

c. Fixed points at 0 and $L - 1/m = 500/7 \approx 71.43$. Because $mL = 3.5$ is between 3 and 4, from any initial term p_0 between 0 and 100, all the following p_n values will remain between 0 and 100, but may not ever level off.

Problems in Context.

(22) Below is a graph including all 5 sequences. The points of each sequence have been connected. We can see that regardless of starting population, all models level off at $p = 600$. Sequences with an initial value close to 600 level off sooner than those with an initial value further from 600.

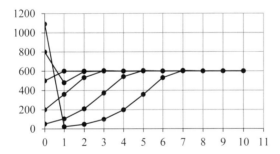

(24) a. Because $4 < mL = 5.2$ this difference equation will lead to negative values of a_n for many values of a_0. Our methods do not provide a long range prediction in this case.

b. Because $1 < mL = 2.6 < 3$ we expect the model to level off at $L - 1/m = 800$. This does make sense: it is possible that the weeds will grow until they are in balance with their environment.

c. Because $mL = 0.91 < 1$ we expect the model to tend toward zero. This does make sense: it is possible that all the weeds will die off. This could occur, for example, if the conditions at one time promoted the growth of weeds, allowing some amount to grow, but that later changes made the lake inhospital to the weeds. A logistic model under these new conditions could reflect an initial weed population that diminishes over time eventually dying out completely.

6.1. Exercises

(26) a. Exact form: $r = -\frac{9}{95000}p + \frac{389}{190}$. Approximate decimal form: $r = -0.0000947p + 2.0474$.

b. Approximate decimal form: $p_{n+1} = 0.0000947(21{,}611.11 - p_n)p_n$.

c. The table and graph below show what the model produces with an initial term of $p_0 = 500$.

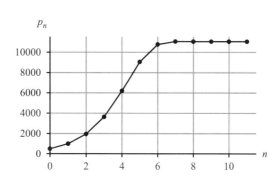

n	p_n
0	500
1	1,000
2	1,953
3	3,637
4	6,193
5	9,045
6	10,768
7	11,061
8	11,055
9	11,056

d. The table and graph suggest the population is leveling off at about 11,000. With $mL = 2.05$, we know this logistic growth model will level off at $L - 1/m = 99{,}500/9$ which rounds to 11,056. This is consistent with the values in the table.

e. With different starting values, the model always appears to level off at about 11,056. From the properties of logistic growth, we know that any initial term between 0 and $L = 21{,}611.11$ produces a sequence that levels off at $L - 1/m$ because $1 < mL < 3$.

(28) The data are approximated pretty well by a logistic growth model with $p_4 = 10$, and with $m = 0.003$, $L = 493.3333$, so that the difference equation is $p_{n+1} = 0.003(493.3333 - p_n)p_n$. The parameters are based on the linear equation $r = -0.003p + 1.48$. Graphs for this linear equation and for the sequence of p_n values appear below.

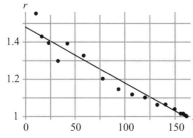 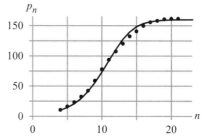

Below are two tables. The one on the left shows the growth factors derived from the data values, as well as the r values produced by the linear equation. The table on the right shows the observed length data, as well as the p_n values produced by the difference equation. In both tables the model and error columns were computed using more decimal places than are shown in the table. Subtracting the displayed entries for observed and model values does not always produce the corresponding displayed error entry due to the effects of rounding.

Length	r obs	r model	error
10.5	1.55	1.45	−0.104
16.3	1.43	1.43	0.002
23.3	1.39	1.41	0.015
32.5	1.30	1.38	0.084
42.2	1.39	1.35	−0.038
58.7	1.33	1.30	−0.023
77.9	1.20	1.25	0.043
93.7	1.15	1.20	0.053
107.4	1.12	1.16	0.040
120.1	1.10	1.12	0.018
132.3	1.06	1.08	0.020
140.6	1.06	1.06	−0.007
149.7	1.04	1.03	−0.009
155.6	1.02	1.01	−0.003
158.1	1.02	1.01	−0.010
160.6	1.00	1.00	−0.007
161.4	1.00	1.00	−0.005
161.6		1.00	

day	p obs	p_n model	error
4	10.5	10.0	−0.5
5	16.3	14.5	−1.8
6	23.3	20.8	−2.5
7	32.5	29.5	−3.0
8	42.2	41.1	−1.1
9	58.7	55.7	−3.0
10	77.9	73.2	−4.7
11	93.7	92.2	−1.5
12	107.4	111.0	3.6
13	120.1	127.3	7.2
14	132.3	139.8	7.5
15	140.6	148.3	7.7
16	149.7	153.5	3.8
17	155.6	156.5	0.9
18	158.1	158.1	0.0
19	160.6	159.0	−1.6
20	161.4	159.5	−1.9
21	161.6	159.7	−1.9

Digging Deeper.

(30) The difference equation will be given by

$$s_{n+1} = \frac{s_n}{0.00122 s_n + 0.51}.$$

Thus, from an initial term of 4.4, the difference equation leads to the table and graph below. Both visually and numerically, this model is a much more accurate approximation to the data than any of the models considered in problem 29.

n	Data	s_n
0	0	4
1	1	9
2	6	17
3	28	32
4	68	58
5	119	101
6	174	160
7	228	227
8	279	289
9	322	335
10	357	364
11	383	381

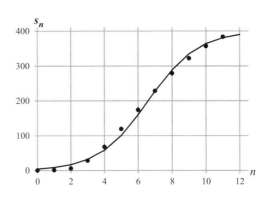

(31) a. $m = 1/1200$ and $b = 5/12$

b. $p_{n+1} = \dfrac{1200}{p_n + 500} p_n.$

c. This new model does not match the original model very well. In fact, the new model seems to level off around 700, where the original model leveled off at 600.

d. We wish to solve the equation $x = \dfrac{1200x}{x + 500}$. One possible answer is $x = 0$. For $x \neq 0$ we can divide both sides of the equation to obtain $1 = \dfrac{1200}{x + 500}$.

This leads to $x = 700$, verifying what we found graphically and numerically in part c.

6.2 Exercises

Reading Comprehension.

(1) This is not true, for two reasons. First, as a matter of nomenclature, when we speak of a mathematical model, we are referring to a complex of interrelated mathematical properties, relationships, functions, and so on. These exist as part of an idealized abstract mathematical universe, and do not directly depend on measurements. When we study a model, we may assign values to parameters based on measurements, but the measurements are not made within the model. Chaos in a mathematical model is a mathematical construct. It is a property that is defined in terms of characteristics of the model, and it's presence or absence is not dependent on measurement. Rather, the fact that some mathematical models are chaotic can be logically deduced from the defining properties of the model. By way of analogy, consider the concept of an equilateral triangle in geometry. Bob could argue that if the sides of triangles could be measured accurately enough, we would never observe an equilateral triangle, because there would be inescapable miniscule differences in the lengths of the legs, and with accurate enough measurement these could be detected. And Bob is right. But that does not mean equilateral triangles do not exist in geometry.

But even if we amend Bob's statement to say that sufficiently accurate measurements would permit accurate predictions even in chaotic models, he would still be wrong. Part of the idea of chaos is that measurements of genuine data are generally inexact. (An exception is an integer variable that can be directly counted.) For example, measurements of weights, lengths, periods of time, temperatures, voltage, air pressure, humidity, and the like can only be measured imperfectly. We know this because repeated measurements of the same quantity always show some variability. For a chaotic system, the small inescapable errors that occur in measurements can lead to qualitative differences in what a model predicts. And since we know that our measurements will have errors, we have no reason to believe that the model's predictions are even approximately correct. That is the consequence of chaos.

(4) In general, the difference equations we have studied all have the form $a_{n+1} = f(a_n)$ for some function f. Conceptually, this simply means that we can compute the next term of sequence by performing some operations on the preceding term. In the case of quadratic growth, we actually perform operations on both the preceding term and on the position number, so that the difference equation

takes the form $a_{n+1} = f(a_n, n)$. If the function f is linear, meaning that it involves only addition, subtraction, and multiplying or dividing by fixed constants, then we say the associated difference equation is linear. In our studies, arithmetic growth, geometric growth, and mixed models all have linear difference equations. For example, in the mixed growth difference equation $a_{n+1} = 3a_n + 2$, the function f is given by $f(x) = 3x + 2$, which is a linear function. In contrast, logistic growth difference equations are not linear. For example, the difference equation $a_{n+1} = 1.2(100 - a_n)a_n$ can be expressed in the form $a_{n+1} = f(a_n)$ by defining the function as $f(x) = 1.2(100 - x)x = -1.2x^2 + 120x$. That is a quadratic function, and is nonlinear.

The significance of linearity or nonlinearity in the context of chaos is this: chaos can only arise in models with nonlinear difference equations.

(6) We have studied both difference equations and functional equations for our sequence models. The difference equation can be expressed in the form $a_{n+1} = f(a_n)$ (or possibly in the form $a_{n+1} = f(a_n, n)$). It tells how each term is related to the preceding term (and possibly to the position number). The functional equation for the same sequence would have the form $a_n = g(n)$. This tells how each term is related to its position number. For quadratic growth sequences, the *difference* equation is a linear equation, because f is a linear function. But the functional equation is nonlinear because g is a quadratic function.

We can put this distinction in visual terms by considering two different graphs for a sequence. For this idea, rather than looking at quadratic growth, let us consider a geometric growth sequence with difference equation $a_{n+1} = 1.5a_n$, and initial term $a_0 = 0.10$. The most familiar graph for this sequence has points of the form (n, a_n). This is the graph of the *functional* equation, $a_n = 0.10 \cdot 1.5^n$. We know that this is an exponential function and the graph is not a straight line. For the second graph we depict points of the form (a_n, a_{n+1}). This is the graph of the *difference* equation $a_{n+1} = 1.5a_n$. This is a linear equation and the graph is a straight line. These are two different ways of visualizing a sequence, and they look different. One is linear, the other is not. There is no contradiction here because we are considering two different aspects of the sequence.

(7) The butterfly effect refers to the idea that a tiny change in the parameters or initial conditions for a model can lead to major changes in what the model predicts. The name refers to an example where a model is supposed to predict the weather in the future based on measurements of the current conditions and equations that govern how the weather changes over time. In the example, the model is run with two sets of parameters and initial conditions, differing only in the inclusion of the beating wings of a butterfly in one of the models, and its exclusion from the other model. But that tiny difference leads the two versons of the model to make vastly different predictions: fair weather in one case and a blizzard in the other.

While we can attribute variations in the predictions of a model to a single insignificant change in the value of a parameter, that is not the same thing as saying that a corresponding change in the phenomenon being modeled would cause an equally dramatic change to that phenomenon in the future. It is a misconception to say that beating of the wings of a butterfly can *cause* a blizzard. That is not what the butterfly effect implies. Rather, the blizzard is the cumulative effect of thousands or millions of factors, any one of which appears to be as insignificant as the butterfly's wings. And although we can run the *model* twice with initial values identical except for that butterfly, in actual fact there will be undetermined tiny inaccuracies in all of our initial inputs. The butterfly effect is a dramatic warning that the model is sensitive to every one of those tiny inaccuracies. That tells us that the predictions we obtain from running the model can be completely incorrect, in spite of all of our efforts to make the equations and initial measurements more accurate.

We saw the butterfly effect in the logistic fish population model, with an intial population of 90,000 fish in one case, and 90,010 in the other. The two models produced very similar results for about 5 years, but then diverged and produced completely different predictions. In particular, one model predicted a nearly constant population of around 70,000 for years 8 through 13, while the other version predicted wild oscillations between 30,000 and 90,000 for the same years.

(9) Because the point $(2.0, 0.5)$ is the only point in the diagram for $m = 2$, it tells us that for this value of m, the model always approaches an equilibrium value of 0.5. This is subject to the assumption that $0 < a_0 < 1$.

(10) a. In all parts of this problem, you have to find the value of m on the horizontal axis of the graph, then imagine a vertical line going up to the graph from that m. If the vertical line just hits one point of the graph, that must be a steady population value for that m. If there are two points, the population will go back and forth between them. If there are four points, there will be a cycle repeated among all four. And if there is just a cloud of points, the future behavior will be chaotic.

For $m = 1.6$ we find a single point in the diagram, with the population size about 0.38. We can therefore predict that from a starting term of 0.5, the sequence will approach an equilibrium of approximately 0.38 in the long run.

d. With $m = 3.2$ there are two points on the graph, and the population will eventually alternate between them: 0.5 and 0.8, approximately.

e. There are 4 values for this m. The population will cycle among them, repeating every four years: 0.36, 0.5, 0.82, and 0.88, approximately.

f. For $m = 3.56$ there appear to be 8 values.

g. For $m = 3.8$, chaos!

h. It looks as though there might be about 3 points here. So the long-term behavior would likely be to jump back and forth among three different population sizes.

6.3 Exercises

Reading Comprehension.

(3) By studying both models we see how reflecting on the properties of a model might lead us to modify our assumptions and thus be led to a new sort of model. This is an important aspect of the development of models.

(4) a. Originally we used a linear model for $r(p)$. This has the advantage that linear equations are familiar and algebraically simple. We have had a lot of experience working with linear equations, contributing to our ability to find a linear equation for $r(p)$. A disadvantage of the linear model is that we really have no structural justification for assuming that the growth factor depends linearly on the population size. In greater detail, a linear equation will lead to negative values of r for some population sizes, and that in turn means the model can result in negative population sizes. That is unrealistic for populations, as well as many other kinds of models where a logistic model might apply.

b. The refined logistic growth model uses an equation for r that is the reciprocal of a linear function. This has the advantage that it never leads to negative values of r. A disadvantage is that such functions are more complicated algebraically, and less familiar. However, because this type of equation is closely related to linear equations, we were able to adapt some of what we know about linear functions to this new setting. Although not anticipated when we adopted the refined model for r, it also has the advantage that it leads to useful functional equations, and simpler patterns of terms in refined logistic growth sequences.

(5) We disagree with the proposition. We saw in the reading that oscillation of population sizes has been observed experimentally in a study of a certain kind of weevil. Such a population could never be successfully modeled with refined logistic growth, which never exhibits oscillatory behavior. On the other hand, the original logistic growth model can, with appropriate parameters, produce sequences that appear very similar to the observations of the weevil experiment. For this population, therefore, the original logistic growth model would be superior to the revised logistic growth model. Both versions of logistic growth involve an assumption about how the growth factor r varies with the population size. For any given modeling context, we should use the model whose assumptions most nearly reflect the actual behavior of the population under study.

Mathematical Skills.

(8) a. $r(p) = 2700/(p + 500)$

d. $r(p) = 1500/(4p + 600)$

(9) a. $1/r = (1/250)p + 1/5$

c. Equilibrium population implies that $r = 1$, so we have $r(380) = 1$. Equation: $1/r = (1/1{,}520)p + 3/4$.

(10) a. Difference equation: $p_{n+1} = 250p_n/(p_n + 50)$. Values of p_n rounded to one decimal place: $p_1 = 190.5$, $p_2 = 198.0$, $p_3 = 199.6$, $p_4 = 199.9$, $p_5 = 200.0$, $p_6 = 200.0$, $p_7 = 200.0$, $p_8 = 200.0$. Graph: see below at left.

c. Difference equation: $p_{n+1} = 1520p_n/(p_n + 1{,}140)$. Values of p_n rounded to one decimal place: $p_1 = 63.9$, $p_2 = 80.6$, $p_3 = 100.4$, $p_4 = 123.0$, $p_5 = 148.1$, $p_6 = 174.7$, $p_7 = 202.0$, $p_8 = 228.8$. Graph: see below at right.

6.3. Exercises

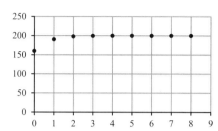 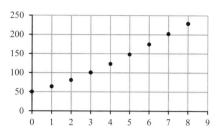

(11) a. $A = 200$, $B = 1/4$, and $R = 1/5$; $p_n = 200/(1 + 0.25(0.2)^n)$. The inflection point is at $p = A/2 = 100$; solving $100 = 200/(1+0.25(0.2)^n)$ we find $n = \ln(4)/\ln(0.2) \approx -0.861353$. Rounding to two decimal places, the inflection point is $(-0.86, 100)$.

c. $A = 380$, $B = 6.6$, and $R = 3/4$; $p_n = 380/(1+6.6(3/4)^n)$. At the inflection point, $n = \ln(1/6.6)/\ln(3/4) \approx 6.559566$. Rounding to two decimal places, the inflection point is $(6.56, 190)$.

(12) a. $r(p) = \dfrac{1}{(1/2100)p + 9/21}$

c. $r(p) = \dfrac{1}{(1/500)p + 1/2}$

(13) For both parts below, sequence terms that are not exact integers have been rounded to one decimal place.

a. Difference equation: $p_{n+1} = \dfrac{2100}{p_n + 900} p_n$. Terms: $p_1 = 210$, $p_2 = 397.3$, $p_3 = 643.1$, $p_4 = 875.2$, $p_5 = 1035.3$, $p_6 = 1123.4$, $p_7 = 1165.9$, $p_8 = 1185.2$. Graph: see below left.

c. Difference equation: $p_{n+1} = \dfrac{500}{p_n + 250} p_n$. Terms: $p_1 = 142.9$, $p_2 = 181.8$, $p_3 = 210.5$, $p_4 = 228.6$, $p_5 = 238.8$, $p_6 = 244.3$, $p_7 = 247.1$, $p_8 = 248.5$. Graph: see below right.

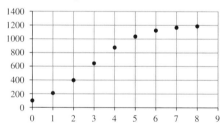 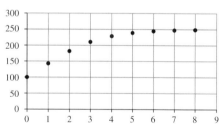

(14) a. $A = (1-b)/m = (1-9/21)/(1/2100) = 1200$, $B = A/p_0 - 1 = 1200/100 - 1 = 11$, $R = b = 9/21$; $p_n = 1200/(1 + 11(9/21)^n)$. The values of p_1 through p_4 do indeed match those given above.

c. $A = 250$, $B = 1.5$, $R = 1/2$, $p_n = 250/(1 + 1.5(1/2)^n)$.

(15) a. $p_{n+1} = 3{,}000 p_n/(2p_n + 1{,}000)$. Note that values of r were found as follows: for $p = 100$, $r = p_1/p_0 = 250/100 = 2.5$ and for $p = 250$, $r = p_2/p_1 = 500/250 = 2$.

b. $p_n = 1000/(1 + 9(1/3)^n)$

c. $A = 1{,}000$

(17) a. Notice $A = 2{,}500$, $B = 5$, and $R = 0.8$. The equilibrium value is $A = 2{,}500$. The inflection point is found using $y = A/2$ and solving for x; it is $(7.21, 1{,}250)$. The y-intercept is $y = A/(1+B) = 416.67$. The graph is below.

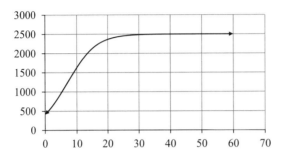

c. $A = 800$, $B = 0.04$, and $R = 0.4$. The equilibrium value is $A = 800$. The inflection point is $\left(\frac{\ln(25)}{\ln(0.4)}, 400\right) \approx (-3.513, 400)$. The y-intercept is $y = 800/1.04 \approx 769.23$. (The graph is not provided here.)

(18) The inflection point is at $A/2 = 5{,}000$ so $A = 10{,}000$. The y-intercept is $2{,}000 = A/(1+B)$ so $B = 4$. The point $(30, 5{,}000)$ is on the curve so $5{,}000 = A/(1+BR^{30}) = 10{,}000/(1+4R^{30})$. This leads to $R^{30} = 1/4$ hence $R = (1/4)^{1/30} = 0.954842$ to six decimal places. The equation of the curve is $y = \dfrac{10{,}000}{1+4(0.25)^{x/30}} \approx \dfrac{10{,}000}{1+4(0.954842)^x}$.

(20) $y = 300/(1 + 5(0.2)^{x/8}) \approx 300/(1 + 5(0.817765)^x)$.

(21) $y = 1000/[1 + (19/6)^{1-x/10}] = 1000/[1 + (19/6) \cdot (6/19)^{(1/10)x}]$

Problems in Context.

(23) a. Using the points $(500, 1/2)$ and $(10{,}000, 1/1.1)$ we find $1/r = (9/209{,}000)p + 1{,}000/2{,}090$

b. $p_{n+1} = p_n/((9/209{,}000)p_n + 100/209)$

c. $p_n = \dfrac{109{,}000/9}{1+(209/9)(100/209)^n}$ shows the population will level off at $109{,}000/9 = 12{,}111$.

e. Regardless of the starting population, the function will level off at 12,111 as predicted. Starting populations below 12,111 will result in graphs that rise from left to right. Starting populations above 12,111 will result in graphs that fall from left to right.

(25) Based on the graphical information we find $A = 330$, $B = 15/7$, and $b = R = (7/15)^{1/2.9}$. Saturation level is $A = 330$. The model gives $y(t) = 0.99 \cdot 330$ when $t = 2.9\left(1 + \dfrac{\ln(1/99)}{\ln(7/15)}\right) = 20.384775$ to six decimal places.

(27) The original data appear as dots in the graph below. There are also three curves, shown with a solid line, a line with long dashes, and a line with short dashes. We will refer to these as the solid curve, long-dash curve, and short-dash curve.

6.3. Exercises

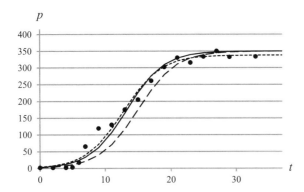

From the graph of the originial data we can estimate the following characteristics: p-intercept equal to 2, equilibrium level of 350, and an inflection point for $t = 15$. This leads to an equation with parameters $A = 350$, $B = 174$, and $R = (1/174)^{1/15}$, which is a good starting point for trial-and-error refinement. It corresponds to the equation $p(t) = 350/(1+174 \cdot (1/174)^{t/15})$, represented by the long-dash curve in the figure. For the authors' best trial-and-error refinement the parameters are $A = 350$, $B = 347/3$, and $R = 0.7$, which produce the solid curve. Compared to the original data, this curve has an average error of about 15.5, disregarding signs. The original study [12] included an equation with parameters $A = 338$, $B = 87$, and $R = e^{-0.35}$, that is, $p(t) = 338/(1 + 87e^{-0.35t})$, shown by the short-dash curve. This equation has an average error of about 11.9, and so is a slightly better approximation to the data than the solid curve.

Digging Deeper.

(28) c. The difference equation is $p_{n+1} = 2400p_n/(p_n + 700)$. This produces $p_1 = 300$, $p_2 = 720$, $p_3 = 1217$, $p_4 = 1524$, $p_5 = 1644$, $p_6 = 1683$, $p_7 = 1695$, $p_8 = 1699$.

d. The difference equation is $\hat{p}_{n+1} = \hat{p}_n/(0.00042\hat{p}_n + 0.29)$. This produces $\hat{p}_1 = 301$, $\hat{p}_2 = 723$, $\hat{p}_3 = 1218$, $\hat{p}_4 = 1520$, $\hat{p}_5 = 1637$, $\hat{p}_6 = 1675$, $\hat{p}_7 = 1686$, $\hat{p}_8 = 1689$. The values of p and $\hat{p}$ diverge. As usual with rounding, the resulting discrepancies may or may not be acceptable depending upon the intended use for the model.

e. The exact value is $A = 1{,}700$. The approximation is $\hat{A} = 1{,}690$. This is consistent with the numerical results, where the p_n results seem to level off at 1,700 while the $\hat{p}_n$ results level off at 1,690.

f. With the additional accuracy in $\hat{m}$ and $\hat{b}$ the equilibrium value of $\hat{p}_n$ becomes 1,699.784, which rounds to the true value of 1,700. For both versions of the $\hat{p}_n$ model, the values very quickly approach the equilibrium value, so that the difference $p_n - \hat{p}_n$ is generally in the same range as $A - \hat{A}$. For the more accurate version, $A - \hat{A}$ is less than 0.25, as compared to about 9.5 for the less accurate version.

(30) We want $e^{kx} = R^x$, so $e^k = R$. Therefore, $k = \ln R$.

(32) a. Difference equation: $p_{n+1} = \frac{p_n}{mp_n} = \frac{1}{m}$. Functional equation: $p_n = 1/m$ (for $m \geq 1$). Graph: p_0 can be anything, but all the rest of the points fall on the horizontal line with a constant p value of $1/m$.

d. $r_1 p_1 = r_0 p_0$

(34) Assuming the inflection point is at $y = 130$ we find $A = 260$. Substituting two pairs of values for x and y into $y = \frac{260}{1+BR^x}$ the result is a pair of equations in the unknowns B and R. Solving these we find $R = 0.2$, and $B = 5^4 = 625$. Thus $y = \frac{260}{1+625(0.2)^x}$

Bibliography

[1] Kevin Allocca. *PSY Passes Bieber; 'Gangnam Style' New Most-Viewed Video of All Time*. YouTube Trends, 11/24/2012. Accessed online at http://youtube-trends.blogspot.com/2012/11/psy-passes-bieber-gangnam-style.html, 6/15/2015.

[2] A. P. Anderson. Minnesota Botanical Studies, *Geological and Natural History Survey of Minnesota*, bulletin 9, part V (1895), p 238.

[3] Raymond Ayoub. What is a Naperian Logarithm?, *American Mathematical Monthly*, vol. 100, no. 4 (1993), pp 351-364.

[4] Sandra G. Boodman. 10-Year Study of Surgery For Myopia Has Mixed News; Better Distance Vision but Earlier Farsightedness, *Washington Post*, October 18, 1994, p z-08.

[5] *BP Statistical Review of World Energy June 2014*, Oil Section (web page). Accessed online at http://www.bp.com/content/dam/bp/pdf/Energy-economics/statistical-review-2014/BP-statistical-review-of-world-energy-2014-oil-section.pdf, 9/5/2014.

[6] K. M. Bliss, K. R. Fowler, and B. J. Galluzzo. *Math Modeling: Getting Started and Getting Solutions*. Society for Industrial and Applied Mathematics, Philadelphia, 2014. Accessed online at http://m3challenge.siam.org/about/mm/pdf/siam-guidebook-final-download.pdf, 5/6/2014.

[7] *CalcTool: Pressure at Depth* (web page). Accessed at http://www.calctool.org/CALC/other/games/depth_press, 5/29/2014.

[8] Patrice Christmann. Mineral resources: should we fear a shortage?, *Paris Tech Review*, February 12, 2013. Accessed online at http://www.paristechreview.com/2013/02/12/mineral-resources-shortage/, 6/12/2015.

[9] Patterson Clark. Containing Ebola: What it would take, *Washington Post*, October 10, 2014, p A1. Accessed online at http://apps.washingtonpost.com/g/page/national/containing-ebola-what-it-would-take/1366/, 1/10/2015.

[10] *CO2Now.org* (web page). Accessed at http://co2now.org/current-co2/co2-now/annual-co2.html, 5/6/2014.

[11] *CRC Handbook of Chemistry and Physics*, 95th Edition. CRC Press, Boca Raton, FL, 2014.

[12] A. C. Crombie. On Competition between Different Species of Graminivorous Insects, *Proceedings of the Royal Society of London*, Series B, Biological Sciences 132, n0. 869 (1945), pp. 362-95. Accessed online at http://rspb.royalsocietypublishing.org/content/132/869/362, 6/27/2016.

[13] Deepak. *500 millionth iPhone sold?* The Frustrum Blog, 3/25/2014. Accessed online at http://thefrustum.com/blog/2014/3/25/500-millionth-iphone-sold, 6/10/2016.

[14] John Derbyshire. *Unknown Quantity: A Real and Imaginary History of Algebra*. Joseph Henry Press, Washington, DC, 2006.

[15] William Dunham. *Euler, The Master of Us All*, Mathematical Association of America, Washington, DC, 1999.

[16] The Economist. *International schools, the new local* (web page). From the print edition 12/20/2014. Accessed at http://www.economist.com/news/international/21636757-english-language-schools-once-aimed-expatriates-now-cater-domestic-elites-new, 3/14/2015.

[17] Albert Einstein, "Geometry and Experience", in Stuart Brown, John Fauvel, and Ruth Finnegan, eds. *Conceptions of Inquiry*. Routledge, London, 1989.

[18] Anne Elixhauser, Ph.D. and Claudia Steiner, M.D., M.P.H. Infections with Methicillin-Resistant Staphylococcus Aureus (MRSA) in U.S. Hospitals, 1993 – 2005. Statistical brief number 35, Agency for Healthcare Research and Quality, Center for Delivery, Organization, and Markets, Healthcare Cost and Utilization Project, July 2007. Accessed 12/11/2014 at http://www.hcup-us.ahrq.gov/reports/statbriefs/sb35.pdf.

[19] Jordan Ellenberg. *How Not to Be Wrong: The Power of Mathematical Thinking*. Penguin Press, New York, 2014.
[20] Ericsson Mobility Report, November 2014. Accessed 12/11/2014 at http://www.ericsson.com/res/docs/2014/ericsson-mobility-report-november-2014.pdf.
[21] Elementary Mathematical Models 2nd Edition Excel Spreadsheet Collection, December 2018. Accessed 12/1/2018 at http://emm2e.info.
[22] Elementary Mathematical Models 2nd Edition Resources, December 2018. Accessed 12/1/2018 at http://emm2e.info.
[23] Elementary Mathematical Models 2nd Edition Technology Guide, December 2018. Accessed 12/1/2018 at http://emm2e.info.
[24] File: Copper - world production trend.svg, *Wikimedia Commons*, August 21, 2013. Accessed online at https://commons.wikimedia.org/wiki/File:Copper_-_world_production_trend.svg, 6/12/2015.
[25] Ross L. Finney, Franklin D. Demana, Bert K.Waits, and Daniel Kennedy. *Calculus: Graphical, Numerical, Algebraic*. Prentice Hall, Boston, 2012.
[26] Galileo Galilei. *Dialogues Concerning Two New Sciences,* translated by Henry Crew and Alfonso de Salvio, New York, Dover Pub., 1954.
[27] Martin Gardner. *The Night Is Large. Collected Essays, 1938 1995*. St. Martin's Press, New York, 1997.
[28] G. F. Gause. Experimental Studies on the Struggle for Existence, *Journal of Experimental Biology*, vol. 9, no. 4 (1932), pp. 389-402.
[29] James Gleick. *Chaos: Making a New Science*. Viking, New York, 1987.
[30] Ronald L. Graham, Donald E. Knuth, and Oren Patashnik. *Concrete Mathematics*. Addison Wesley, Reading, Mass., 1989.
[31] Denny Gulick. *Encounters with Chaos*. McGraw, New York, 1992.
[32] Sir Thomas Little Heath. *A History of Greek Mathematics*, Vol. 2. Clarendon, an imprint of Oxford University Press, London, 1921.
[33] Dan Kalman and Daniel J. Teague. Galileo, Gauss, and the Green Monster, *Mathematics Teacher*, vol. 106, no. 8 (2013), pp. 580-585.
[34] M. Kameche, H. Benzeniar, H. Henna, F. Derghal, and N. Bouanani, ALSAT-2A Mission: Experience of three years of station keeping orbit maintenance, paper presented at the 2nd IAA Conference on Dynamics and Control of Space Systems, Rome, March 2014. Accessed online at http://www.dycoss.com/program/final/IAA-AAS-DyCoSS2-14-14-10.pdf, 5/6/2014.
[35] Steven G. Krantz. *Mathematical Apocrypha: Stories and Anecdotes of Mathematicians and the Mathematical*. Mathematical Association of America, Washington, DC, 2002.
[36] D. P. Lyle MD. *Timely Death* (web page). Accessed at http://www.dplylemd.com/Articles/timelydeath.html, 6/27/2014.
[37] Eli Maor *e: The Story of a Number*. Princeton University Press, Princeton, NJ, 1994.
[38] J. P. McKelvey. Kinematics of Tape Recording by J. P. McKelvey, *American Journal of Physics*, vol. 49, no. 1 (1981), pp. 81-83.
[39] Meeting with Ford Engineer. Clean MPG - an Online Community. Web page: http://www.cleanmpg.com/forums/showthread.php?t=46466. Page last accessed January 2015.
[40] NOAA ESRL Data: Mauna Loa CO_2 annual mean data. National Oceanic and Atmospheric Administration, Earth System Research Laboratory, Global Monitoring Division. Available at ftp://aftp.cmdl.noaa.gov/products/trends/co2/co2_annmean_mlo.txt. Downloaded 6/16/15.
[41] National Snow and Ice Data Center (NSIDC), Sea Ice Index Archives. Available at http://nsidc.org/data/seaice_index/archives.html. Downloaded 8/14/14.
[42] National Weather Service Climate, National Weather Service Forecast Office, New York, NY. Web page: http://www.nws.noaa.gov/climate/index.php?wfo=okx. Page last accessed June 2014.
[43] Tony Norfield. *T-Shirt Economics: Labour in the Imperialist World Economy* (web page). Accessed at http://column.global-labour-university.org/2012/08/t-shirt-economics-labour-in-imperialist.html, 7/3/2014.
[44] Quantitative Environmental Learning Project (QELP), Data Set 023. Available at http://www.seattlecentral.edu/qelp/sets/023/023.html. Downloaded 11/30/14.
[45] T. Brailsford Robertson. *The Chemical Basis of Growth and Senescence*. J. P. Lippencott, Philadelphia and London, 1923, pp 72-74.
[46] *Uniform Crime Reports*. San Antonio Police Department. Accessed at online at http://www.sanantonio.gov/SAPD/UniformCrimeReports.aspx#lt-302629-2011, 1/11/2014.
[47] James T. Sandfur. *Discrete Dynamical Modeling*. Oxford University Press, New York, 1993.

Bibliography

[48] South Australian Science Teachers Association, Water Quality Tests Summary. Available at http://www.sasta.asn.au/v2/adc/datalogging/DataSinglePagePDFs/ADCBookDatalog13-23.pdf. Downloaded 11/25/14.

[49] Simon Singh. *The Simpsons and their Mathematical Secrets.* Bloomsbury USA, NY, 2013.

[50] D'Arcy W. Thompson. *Growth and Form.* Cambridge University Press, Cambridge, 1942.

[51] John W. Tuckey. *Exploratory Data Analysis.* Addison–Wesley, Reading, MA, 1977, p 138.

[52] U. S. Department of the Interior. News Release: Five-Year Survey Shows Wetlands Losses are Slowing, Marking Conservation Gains and Need for Continued Investment in Habitat, October 6, 2011. Accessed online at http://www.fws.gov/wetlands/Documents/Status-and-Trends-of-Wetlands-in-the-Conterminous-United-States-2004-to-2009-News-Release.pdf, 1/11/2014.

[53] U. S. Energy Information Administration, Short Term Energy Outlook. Web page: http://www.eia.doe.gov/steo. Page last accessed June 2014.

[54] Syunro Utida. Damped Oscillation of Population Density at Equilibrium, *Researches on Population Ecology*, vol. 9, no. 1 (1967), pp. 1–9.

[55] Historical Tuition Data | William & Mary. Accessed 2/2/2019 at https://www.wm.edu/offices/financialoperations/sa/tuition/historical-tuition-data/

[56] Gary Witt. Using Data from Climate Science to Teach Introductory Statistics, *Journal of Statistics Education*, vol. 21, no. 1 (2013), pp. 1-24. Accessed 8/27/2014 at http://www.amstat.org/publications/jse/v21n1/witt.pdf.

Index

added multiples of three sequence, 16–22, 28
algebraically equivalent, 85, 170
algorithm, 58
APR, 326
arithmetic growth
   conceptual statement, 39
arithmetic growth, 39
   compared to other models, 393
   compared to quadratic growth, 124
   difference equation, 42
   functional equation, 49
   sum, 155, 158
asymptote, 251
axis of symmetry, 173, 175

base of an exponential function, 248
bee ancestor example, 227
bee ancestor model, 227
break-even points, 193
Burgi, 260
butterfly effect, 400

center line for parabola, 173, 175
chaos, 234, 373, 390
   butterfly effect, 400
   future prediction, 400
   graph of onset, 402
   nonlinear difference equations, 394, 395
   patterns in, 405
   self-similarity, 406
   slight changes in parameters, 392
coefficients, 169
   quadratics, 174
common logarithm, 258
comparison of different kinds of models, 393
complex numbers, 177
computer network example, 150
continuous
   variable, 271
continuous variable, 63, 75
cost, 191
counting numbers, 4

data table, 15
decay, 42
deductive proof, 131
demand, 107, 190

dependent variable, 74
descending form, 169
difference equation, 29
   and functional equation, 47, 393
   arithmetic growth, 42
   computer model, 151
   different types compared, 393
   family, 42
   geometric growth, 202, 203
   initial value, 30
   linear and nonlinear, 393, 394
   loan payments, 327
   logistic growth, 363, 365, 366
   mixed growth, 298
   quadratic growth, 124, 127
      standard form, 126
   refined logistic growth, 410
   water tank, 223
differences, 122
   second, 123
direct rule, 20
discrete variable, 63, 75
drug dosage model, 319

e, 262, 290
effective rate, 232
equation
   difference, 29
   functional, 28
equations, functions, and expressions, 82
equilibrium value, 307, 308
equivalent, 85, 170
even number reciprocal sequence, 15, 18, 20, 21, 28
examples
   added multiples of three sequence, 16–22, 28, 121, 126, 134
   bee ancestors, 227
   carbon dioxide, 2, 39, 43, 345
   cell phone company, 109
   college tuition, 64
   compound interest, 231
   computer network, 150, 174
   copper demand, 339
   even number reciprocal sequence, 15, 18, 20, 21, 28

fish population, 360, 411
flu epidemic, 41
free fall, 142
grain mixture, 96, 105
handshakes, 153
heat transfer, 234, 275, 340
house construction, 155
internet video, 342
iPod sales, 380, 421
Laffer curve, 45
library fines, 41, 49
mold growth, 374
mouse population, 220
oil consumption, 182
oil reserves, 158
polar ice cap, 147
population growth, 233
pumpkin growth, 377
radial keratotomy, 336
radioactive decay, 233, 270, 271
repeated drug doses, 298, 319
repeated loan payments, 325
satellite fuel, 41, 44, 49, 102
second fish population, 371, 413
soft drinks, 107, 112, 190
T-shirts, 69
tape recorder, 143
time of death, 59
trout length and weight, 279
water quality, 272
water tank, 222, 232
yeast population, 427
exponent rules, 254
    in box, 257
exponential equation, 248
exponential function, 247
    asymptote, 251
    base, 248
    base e, 262
    definition, 248
    graphs, 250
    no x-intercept, 251
    shifted, 277, 281, 307
    solving equations, 257
    two-step computation method, 293
    y-intercept, 251
expressions, equations, and functions, 82

factored form, 184
falling body example, 142
Fibonacci numbers, 230
find-a_n question, 73
find-n question, 73
fixed point, 310, 311
    in logistic growth, 369
fixed population, 369
    formula, 370
function evaluation, 74
function inversion, 74

functional equation, 28
    and difference equation, 47, 393
    arithmetic growth, 49
    different types compared, 393
    geometric growth, 209, 210
    inversion, 258
    mixed growth, 302, 306
    quadratic growth, 129, 134
    refined logistic growth, 418, 420
        derivation, 419
functions, equations, and expressions, 82

Galileo, 142
Gauss, Karl F., 131
geometric decay, 233
geometric growth, 201
    applications, 231
        compound interest, 231
        population growth, 233
        radioactive decay, 233
        tank models, 232
    compared to other models, 393
    constant growth factor definition, 204
    constant percentage increase or decrease, 212
    continuous and discrete variables, 224
    continuous model, 271
    data based, 275
    difference equation, 202, 203
    empirical examples, 227
    examples
        bee ancestors, 227
        heat transfer, 234
        mouse population, 220
        temperature probe data, 275
        trout length and weight data, 279
        water quality, 272
        water tank, 222
    functional equation, 209, 210
    graphs, 205
    growth factor, 202
    recognizing instances of, 203
    sequence definition, 201
    strong statement, 225
    structural examples, 220
geometric series, 304, 306
geometric sum, 335, 339, 340
grain mixture problem, 96, 105
graph, 17
graphical method, 46
    for quadratic equation, 135
graphs
    exponential functions, 250
    geometric growth, 205
    linear equations, 89
    logistic functions, 424
    logistic growth, 376
    mixed growth, 307
    quadratic functions and equations, 172
    quadratic growth, 128

Index

relating two successive data values, 394
shifted exponential function, 307
transition to chaos, 402
growth factor, 202
   straight-line model, 361
   variable, 361

half-life, 233
heat transfer, 234, 340
house construction example, 155

independent variable, 74
inflection point, 424
initial term, 30
initial value, 30
integers, 4
intercept
   x-intercept
      exponential function graph, 251
   x-intercept, 89
      quadratic graph, 175
   y-intercept, 89
      exponential function graph, 251
      quadratic graph, 174
interest
   compound, 231
inverting a function, 183
inverting a functional equation, 258
irrational number
   base e, 262
   from a square root, 177
isotope, 233

limited population growth, 359, 364, 409, 415
linear
   difference equation, 393, 394
   equation, 82
      graphs, 89
      many solutions, 87
      no solution, 87
      point-slope form, 94
      slope-intercept form, 84, 92
      standard form, 84
      two-intercept form, 95
   function, 82
   operational definition, 83
linearity assumption, 102
loan payments, 325
   difference equation, 327
logarithm
   base 2, 262
   base e, 262
   common, 258
   natural, 263
logarithmic functions, 258
logistic
   curve, 419
   function, 419, 423
logistic growth, 359

chaos, 373, 390
compared to other models, 393
difference equation, 363, 365, 366
examples
   fish population, 360
   iPod sales, 380
   mold growth, 374
   nonexample, 380
   pumpkin growth, 377
   real data, 377, 380
   second fish population, 371, 413
fixed population, 369
fixed population formula, 370
graphs, 376
leveling off, 364, 369, 371, 373
no functional equation, 364, 374
not leveling off, 371, 372, 396
oscillation, 372
parameters, 365, 366
statement in box, 364

mathematical model, 1, 57
   computer network, 150
   continuous, 271
   drug dosage, 319
   falling bodies, 142
   families compared, 393
   family, 121
   fish population, 360, 411, 413
   heat transfer, 234, 340
   house contruction, 155
   iPod sales, 380, 421
   loan payments, 325
   logistic, 359
   mixed, 297
   quadratic growth, 121
   refined logistic, 408
   yeast population, 427
mixed growth, 297
   difference equation, 298
   difference equation definition, 299
   equilibrium value, 307, 308
   examples
      drug dosage, 319
   functional equation, 302, 306
   graphs, 307
   recognizing instances of, 299
   sequence verbal definition, 297
   structural examples, 319
   suggested by data, 342
mixed model
   compared to other models, 393
   heat transfer, 340
   leveling off, 320
   loan payments, 325
modeling methodology., 390
mouse population example, 220

Napier, 260

natural log, 263
negative growth, 42
Newton's Law of Cooling, 234, 236, 277
nominal rate, 232, 326
nonlinear, 46
nonlinear difference equation, 393, 394
nonlinearity
   quadratic functions, 174
numerical method, 43, 49
   for quadratic equation, 135

open upward or downward, 173
oscillation, 373

parabola, 128, 173
parameter, 42
   effect on predictions, 391
   in monthly payment formula, 334
   logistic growth, 365, 366
   quadratic growth, 127
   utility in family of models, 335
parenthesis notation, 19
point-slope equation, 94
polynomial approximation
   for exponential functions, 290
population growth, 233
   limited, 359, 364, 409, 415
position number, 15
proof, 131
proportional reasoning, 112
   invalid for quadratic functions, 174
   invalid for quadratic growth models, 153

quadratic
   equation
      standard form, 169
quadratic equation, 136
   no solution, 178, 181
quadratic formula, 181
quadratic function
   structural, 190
quadratic graphs
   axis of symmetry, 173, 175
   center line, 173, 175
   high and low points, 175
   summary in box, 184
   vertex, 175
   x-intercepts, 175
   y-intercept, 174
quadratic growth, 121
   alternate verbal description, 126
   and proportional reasoning, 153
   and sums of arithmetic growth models, 158
   applications, 142
   compared to arithmetic growth, 124
   compared to other models, 393
   constant 2nd difference definition, 123
   difference equation, 124, 127
      standard form, 126

falling bodies, 142
functional equation, 129, 134
graphs, 128
house construction example, 155
network problems, 150
parameters, 127
structural, 155
verbal definition, 122

radioactive decay, 233
random stranger test, 8
recursion, 20
recursive analysis, 153
   quadratic growth, 150
recursive rule, 20
refined logistic growth, 408
   difference equation, 410
   examples
      fish population, 411
      iPod sales, 421
      real data, 421, 427
      second fish population, 413
      yeast population, 427
   functional equation, 418, 420
      derivation, 419
refined logistic growth assumption, 410
relative error, 280
revenue, 191
robustness, 391
root, 176
rule
   direct, 20
   for a pattern, 4
   recursive, 20
rules of exponents, 254
   in box, 257
running total, 155, 156, 336

second differences, 123
self-similarity, 406
sequence, 4
   terms, 4
shifted exponential function, 277, 281, 307
shortcut
   geometric series, 304
   sums of powers, 304, 306
   sums of whole numbers, 130, 132
slope, 90
slope formula, 91
slope-intercept equation, 92
slope-intercept form, 84
solving equations
   exponential, 257, 261, 264
   for one variable in terms of another, 88
   graphical and numerical methods, 257
   in factored form, 185
   linear, 86
   quadratic, 176, 181, 184
square root, 177

Index

strong geometric growth assumption, 225
structural linearity, 106
structural quadratic function, 190
structural quadratic growth, 155
subscript, 18
subscript notation, 19
sum
   for arithmetic growth model, 155
   for geometric growth sequence, 335, 339, 340
   shorthand notation, 131
   three-dot notation, 131
sums, 156, 336
symmetry
   in quadratic graph, 173

tank model, 232
temperature probe data example, 275
terms of a sequence, 4
tetrahedral numbers, 11, 123, 168
theoretical method, 50
three dot notation, 131
trend line, 147
triangular numbers, 11
trout weight and length example, 279
two-intercept equation, 95

variable
   continuous, 63, 75
   dependent, 74
   discrete, 63, 75
   independent, 74
vertex of a parabola, 175

water quality example, 272
water tank model, 222
whole numbers, 4

yield, 232

zero of a function, 176